W0055415

Springer
Tokyo
Berlin
Heidelberg
New York
Barcelona
Hong Kong
London
Milan
Paris

H. Asama, T. Arai,
T. Fukuda, T. Hasegawa (Eds.)

Distributed Autonomous Robotic Systems 5

With 100 Figures

Springer

Hajime Asama
Head of Instrumentation Project Promotion Division
Advanced Engineering Center
RIKEN (The Institute of Physical and Chemical Research)
2-1 Hirosawa, Wako-shi, Saitama 351-0198, Japan

Tamio Arai
Professor
Department of Precision Machinery Engineering
The University of Tokyo
7-3-1 Hongo, Bunkyo-ku, Tokyo 113-8656, Japan

Toshio Fukuda
Professor
Department of Micro System Engineering
Nagoya University
Furo-cho, Chikusa-ku, Nagoya 464-8603, Japan

Tsutomu Hasegawa
Professor
Department of Intelligent Systems
Graduate School of Information and Electrical Engineering
Kyushu University
6-10-1 Hakozaki, Higashi-ku, Fukuoka 812-8581, Japan

ISBN 978-4-431-65943-3 ISBN 978-4-431-65941-9 (eBook)
DOI 10.1007/978-4-431-65941-9

Library of Congress Cataloging-in-Publication Data applied for.

Printed on acid-free paper

Softcover reprint of the hardcover 1st edition 2002

Typesetting: Camera-ready by the editors and authors

SPIN: 10869189

Preface

A decade has already passed since the first International Symposium on Distributed Autonomous Robotic Systems (DARS) was held in Wako-shi, Saitama, Japan. The DARS symposia have been held every two years since then. DARS 92, DARS 94, and DARS 96 were held in Japan; DARS 98 in Karlsruhe, Germany; and DARS 2000 in Knoxville, Tennessee, USA. In 2002, DARS came back to Japan and was held June 25–27, 2002, in Fukuoka, Japan. DARS 2002 was planned in conjunction with the robot soccer competition, RoboCup-2002, which was held also in Fukuoka and in Busan, Korea, because the scope of multirobot cooperation encompasses both events. This book includes all the papers accepted after review and presented at DARS 2002.

In DARS symposia, various aspects of distributed autonomous robotic systems have been discussed, such as architecture, communication, control, sensing, planning, and learning. While the size of the conference (the number of presentations and participants) has remained almost constant, interests seem to have shifted gradually in 10 years due to technological progress and concurrent change in the social environment. The following tendencies have been observed:

(1) Practical papers have increased in comparison with theoretical ones.
(2) Learning issues have become more important.
(3) Application of DARS has become a matter of greater concern.

In DARS 2002, 46 technical papers were presented in sessions of Modular Robotic Systems, Communication for Cooperation, Human-Machine Cooperative Interaction, Multi-Robot Coordination, Robot Soccer, Distributed Sensing and Mapping, Distributed Control, Multi-Agent and Group Systems, Multi-Robot Motion Planning, Emergence in Mobility, and Learning in Distributed Robotic Systems.

We express great thanks to Prof. Kazuhiro Kosuge, Prof. Satoshi Murata, program vice co-chairs, and program committee members who reviewed papers submitted to DARS 2002 and contributed to enhancing the quality of the program. We would like to thank Prof. Masafumi Yamashita, vice general chair, and the organizing committee and local arrangement committee members, as well as members of the Institute of Systems & Information Technologies, for successful organization of DARS 2002. We appreciate the cooperation of Prof. Minoru Asada, Chair of the RoboCup-2002 symposium, in setting up the joint program of the RoboCup-2002 symposium and DARS 2002.

We express our gratitude to the co-sponsoring organizations, IEEE RAS (Robotics & Automation Society), RSJ (Robotics Society of Japan), JSME (Japan Society of Mechanical Engineers), SICE (The Society of Instrument & Control

Engineers of Japan), RIKEN (The Institute of Physical and Chemical Research), as well as FANUC FA and Robot Foundation and the Tateisi Science and Technology Foundation, which contributed to DARS 2002 in the way of financial support. We also thank the IEEE Fukuoka Section for their cooperation.

We thank Prof. Rolf Pfeifer (University of Zurich, Switzerland), and Dr. Lynne E. Parker (Oak Ridge National Laboratory, USA) for their plenary talks, which were programmed as part of the joint program of DARS 2002 and the RoboCup Symposium.

Finally, we are grateful to Dr. Kuniaki Kawabata and to Ms. Chieko Takahashi for their help in secretariat administration.

Hajime Asama
Tamio Arai
Toshio Fukuda
Tsutomu Hasegawa

Contents

Chapter 6: Robot Soccer

Chapter 7: Distributed Control

Chapter 8: Distributed Sensing and Mapping

Chapter 9: Multi-Agent and Group Systems

Chapter 10: Multi-Robot Motion Planning

Chapter 11: Emergence in Mobility

Chapter 12: Learning in Distributed Robotic Systems

DARS 2002 Organization

General Chair:

Tsutomu Hasegawa (Kyushu University, Japan)

Vice General Chair:

Masafumi Yamashita (Kyushu University, Japan)

Advisory Committee:

Toshio Fukuda (Nagoya University, Japan)
Tamio Arai (The University of Tokyo, Japan)

Organizing Committee

Members:

Shun-ichi Amari (RIKEN, Brain Science Institute, Japan)
Yuichiro Anzai (Keio University, Japan)
Ronald C. Arkin (Georgia Institute of Technology, U.S.A.)
Minoru Asada (Osaka University, Japan)
George A. Bekey (University of Southern California, U.S.A.)
Gerardo Beni (University of California, Riverside, U.S.A.)
Raja Chatila (LAAS, France)
Rüdiger Dillmann (University of Karlsruhe, Germany)
Koji Ito (Tokyo Institute of Technology, Japan)
Shuzou Katayama (KEPCO, Japan)
Oussama Khatib (Stanford University, U.S.A.)
Pradeep Khosla (CMU, U.S.A.)
Akira Kitahara (Kyushu Matsushita Electric Co., Ltd., Japan)
Shigeru Kokaji (AIST, Japan)
Paul Levi (University of Stuttgart, Germany)
Takenori Morimitsu (ISIT, Japan)
Tadashi Nagata (Kyushu University, Japan)
Lynne E. Parker (Oak Ridge National Laboratory, U.S.A.)
Arthur C. Sanderson (Rensselaer Polytechnic Institute, U.S.A.)
Kazuo Tanie (AIST, Japan)
Masafumi Yano (Tohoku University, Japan)
Kazuhiko Yokoyama (Yaskawa Electric Corporation, Japan)

Local Organization:

Yoshihiko Kimuro (ISIT, Japan)

Program Committee

Chair: Hajime Asama (RIKEN, Japan)

Vice Co-Chairs:

Kazuhiro Kosuge (Tohoku University, Japan)
Satoshi Murata (Tokyo Institute of Technology, Japan)

Members:

Norihiro Abe (Kyushu Institute of Technology, Japan)
Arvin Agah (University of Kansas, U.S.A.)
Rachid Alami (LAAS, France)
Yoshikazu Arai (Iwate Prefectural University, Japan)
Martin Buss (Technical University of Berlin, Germany)
Gregory S. Chrikjian (Johns Hopkins University, U.S.A.)
Dario Floreano (EPFL, Switzerland)
Aarne Halme (Helsinki University of Technology, Finland)
Yasuhisa Hasegawa (Nagoya University, Japan)
Hideki Hashimoto (The University of Tokyo, Japan)
Hiroshi Hashimoto (Tokyo University of Technology, Japan)
Masafumi Hashimoto (Hiroshima University, Japan)
Toshimitsu Higashi (Murata Machinery, Ltd., Japan)
Hirohisa Hirukawa (AIST, Japan)
Sumiaki Ichikawa (Tokyo Science University, Japan)
Akio Ishiguro (Nagoya University, Japan)
Hiroshi Ishiguro (Wakayama University, Japan)
Hidenori Ishihara (Kagawa University, Japan)
Masaru Ishii (Fukuoka Institute of Technology, Japan)
Kuniaki Kawabata (RIKEN, Japan)
Shin'ya Kotosaka (Saitama University, Japan)
Tetsuo Kotoku (AIST, Japan)
Yasuharu Kunii (Chuo University, Japan)
Daisuke Kurabayashi (Tokyo Institute of Technology, Japan)
Yoji Kuroda (Meiji University, Japan)
Christian Laugier (INRIA, France)
Akihiro Matsumoto (Toyo University, Japan)
Sadayoshi Mikami (Future University of Hakodate, Japan)
Natsuki Miyata (AIST, Japan)
Hiroshi Mizoguchi (Tokyo University of Science, Japan)
Makoto Mizukawa (Shibaura Institute of Technology, Japan)

Yasushi Nakauchi (National Defense Academy, Japan)
Martin Nilsson (Swedish Institute of Computer Sciences, Sweden)
Itsuki Noda (AIST, Japan)
Hisashi Osumi (Chuo University, Japan)
Jun Ota (The University of Tokyo, Japan)
Koichi Ozaki (Utsunomiya University, Japan)
Chris J. Paredis (CMU, U.S.A.)
Igor Paromtchik (RIKEN, Japan)
Kosuke Sekiyama (Fukui University, Japan)
Wei-Min Shen (USC, U.S.A.)
Takanori Shibata (AIST, Japan)
Yoshiki Shimomura (The University of Tokyo, Japan)
Peter Stone (AT&T Labs-Research, U.S.A.)
Ichiro Suzuki (University of Wisconsin Milwaukee, U.S.A.)
Tsuyoshi Suzuki (Tokyo Denki University, Japan)
Satoshi Tadokoro (Kobe University, Japan)
Tsuyoshi Ueyama (DENSO Wave Inc., Japan)
Kazunori Umeda (Chuo University, Japan)
Manuela Veloso (CMU, U.S.A.)
Ljubo Vlacic (Griffith University, Australia)
Zhi-Dong Wang (Tohoku University, Japan)
Heinz Wörn (University of Karlsruhe, Germany)
Ning Xi (Michigan State University, U.S.A.)
Motoji Yamamoto (Kyushu University, Japan)
Mark Yim (PARC Inc., U.S.A.)
Kazutaka Yokota (Utsunomiya University, Japan)
Eiichi Yoshida (AIST, Japan)
Kazuo Yoshida (Keio University, Japan)
Hideo Yuasa (The University of Tokyo, Japan)
Alexander Zelinsky (Australian National University, Australia)

Secretariat:

Kuniaki Kawabata (RIKEN, Japan)
Chieko Takahashi (RIKEN, Japan)

Chapter 1
Introduction

Perspective of Distributed Autonomous Robotic Systems

Hajime Asama

Advanced Engineering Center, RIKEN (The Institute of Physical and Chemical Research), Hirosawa 2-1, Wako-shi, Saitama 351-0198, Japan
asama@cel.riken.go.jp

The first International Symposium on Distributed Autonomous Robotic Systems (DARS) was held in Wako-shi, Japan, in 1992. During the symposium, my first child was born. I have been looking at the growth of my daughter and DARS concurrently. The intelligence of a robot certainly could not be so realized as sufficiently in these ten years as that of a human child. However, the problems in DARS have been well investigated and clarified, and steady technical progress has been achieved owing to the progress of IT, especially the availability of powerful processors and convenient OS and tools for system development, and to the popularization of network technology.

From Theory to Practice

When the first DARS symposium was held ten years ago, DARS was expected to be a new paradigm to realize flexible and robust robot systems. In comparison to a single robot, or centralized system, characteristic problems for DARS to solve were recognized. In the early years, theoretical issues were discussed, such as design schemes for functional distribution, distributed problem solving including motion planning, and deadlock resolution in asynchronous multiple robots, decentralized self-organization, and so on. Swarm intelligence was one of the key concepts of DARS. Methodologies to organize cooperative multiple mobile robots were actively discussed. Though various cooperative motions were realized, they are only specific motions in limited conditions.

In parallel to the progress in information technology, the concept of DARS has become rather popular, and the interests of DARS researchers have moved from basic studies to more practical issues. On the premise of organizing actual multiple autonomous robotic agents, architecture of the system, sensors for multirobot environment, and control hardware for cooperative robots have been discussed as well as motion planning and adaptive behaviors taking care of embodiment. Robotic agents are not the only logical ones in a computer simulation any more, but are recognized as rather physical ones in a real world.

Expansion of Scope

The society for DARS is very interdisciplinary and includes mechanical engineering, electrical engineering, control engineering, computer science, biology, covering robotics, AI, and A-life. Through the interaction of researchers from a variety of disciplines, diverse and epochal ideas and scopes have been brought.

Concerning cooperation of robotic agents (robots, modules), a larger number of robots or modules are used in comparison with the early discussion of cooperation of a few agents. The scope of cooperative robots has been expanded from mutual cooperation of multiple robots to consideration of their interaction with the environment or human operator. That means a ubiquitous computing environment should be taken into account for feasible and purposive cooperative robots.

Importance of Learning Function

Adaptive motions are essential for robots which operate in a real environment. Since it is almost impossible to model or predefine the multirobot environment and to code the programs of agents in advance, the learning function of robotic agents is recognized as being more and more important. Emergent systems are seen as a new concept to organize decentralized autonomous systems. Learning or emergence is becoming one of the main topics in DARS now.

More Applications

As seen in humanoid robots in Japan, robot technology (RT) has matured to some extent. It can already be regarded as infrastructure technology for general mechatronic systems. RT is utilized everywhere. Concurrently, much attention is given to application of DARS. For the task of covering certain areas, DARS is quite effective. From such a viewpoint, DARS technology is expected to be applied to rescue operations, deactivation of mines, space, ITS, and so on. RoboCup, the robot soccer competition, is becoming a major event, where development of technology for cooperative robots is demanded.

As seen in the trends and perspective of DARS mentioned above, we may conclude that the research on DARS will be kept active not only from the academic viewpoint to clarify the function of decentralized autonomous systems, but also from the practical viewpoint to apply the DARS technology to actual needs.

Chapter 2
Modular Robotic Systems

Navigating modular robots in the face of heuristic depressions

Kazuo Miyashita* and Shigeru Kokaji

National Institute of Advanced Industrial Science and Technology
1-2-1, Namiki, Tsukuba, Ibaraki 305-8564, Japan

Abstract. A modular robotic system is composed of a number of modules, each of which can change its position relative to its neighboring modules under certain physical constraints. The modular robotic system can move itself flexibly by repeated motions of its component modules. However, a huge size of possible combinations of subsequent module motions and tight physical constraints among them cause difficulty in making an appropriate plan for the modular robotic system to reach a designated goal position, especially when it is located in unfamiliar environments with obstacles. Heuristic search methods fail to find a plan efficiently due to poor cost estimation used in the search (i.e., *heuristic depression* problem). In this paper, we propose a method for navigating the modular robotic system by extending a real-time heuristic search algorithm to overcome the heuristic depression problem. The experimental results show that the proposed method is effective for navigating the modular robots in several problem settings.

Key words: modular robots, real-time heuristic search, heuristic depression

1 Introduction

A modular robotic system consists of independently controlled modules, which can connect, disconnect and walk over/through the neighboring modules. Dynamic mobility of component modules results in self reconfiguration and locomotion capabilities of the modular robotic system. Because of its modularity, redundancy and homogeneity, the modular robotic system is (1) adaptive to dynamically changing environment, (2) robust against malfunction and (3) suitable for mass production.

There have been many research activities in developing modular robotic systems [3,6,2,9,4,12] and several types of the hardware systems have been proposed, and at the same time, computational intractability in planning and controlling motions of the modular robotic systems has been pointed out and tried to be resolved [10,13,8]. Until now the motion planning systems developed for the modular robotic systems make specific assumptions based upon their hardware constraints to reduce search space for planning, hence their algorithms might not be valid for other types of modular robotic systems.

* e_mail: k.miyashita@aist.go.jp

And most of the planning systems make plans off-line before execution and do not assume the existence of known/unknown obstacles. When the modular robots work in unknown environments, planning and execution must be interleaved for gathering information through sensing, thus planning should be done in real-time.

Online algorithms for controlling motions of a robot in unknown terrains have been developed by researchers in the fields of theoretical robotics and computer science [1,11]. But they have strong limitations on types and locations of a robot and obstacles in the environments for eliminating the needs to consider several constraints arising from interactions among the robot and the environment. Therefore, their methods cannot be readily applied to motion planning for the modular robots that have complicated physical constraints on possible movements. In this paper, we propose a phased search architecture, *RPM* (Real-time Planner for Modular Robots), for navigating modular robots in unknown terrains based upon the real-time heuristic search algorithm. In RPM, local path planning for each module and global decision making for the entire system movement are systematically integrated to prevent producing redundant moves of the modules.

In the following of this paper, first, our 2D model of modular robots navigation problem is described and the *Real-Time-*A^* *algorithm* (RTA*) [7] is introduced to be applied to the problem. Then, we explain a ***heuristic depression*** which causes the original RTA* algorithm to produce a superfluously lengthy plan in the problem. We propose a phased search architecture for making a better heuristic estimate of each module with local search and deciding an appropriate movement of the modular robots with global search. And experimental results are shown to validate effectiveness of the proposed architecture for navigating modular robots. And, finally conclusions and future research directions are discussed.

2 Modular Robots Navigation Problem

In this research, we concentrate ourselves into developing computational architecture for real-time navigation of modular robots in unknown environments, where the modular robots are required to reach the goals avoiding the obstacles. In this paper, we assume that the modular robots can move around in the environment but the obstacles and the goals never change their positions. And the modular robots are informed of their initial positions and the location of goals but not about obstacles. The module recognizes existence of obstacles in its neighborhood using its sensing capability and shares the information with the other modules. Hence, the modular robots gradually build an accurate map of the environment as they move around.

To be free from several idiosyncratic hardware constraints and acquire generally applicable findings, we conducted our research using a simple model of the virtual 2D modular robotic system and built our algorithm on the foundation of a standard heuristic search method.

2.1 2D Modular Robots Model

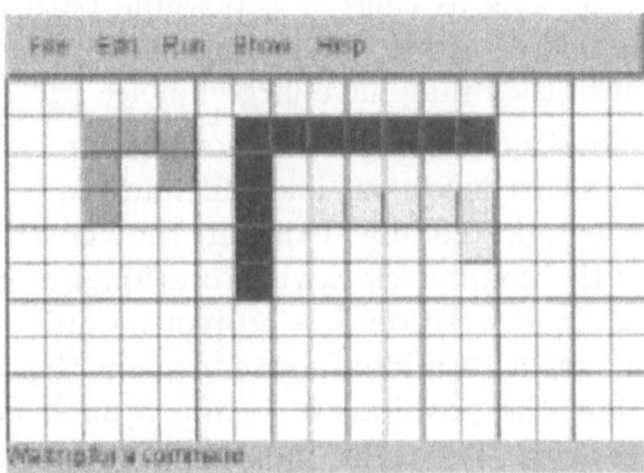

Fig. 1. 2D modular robots simulator

We developed a model of the 2D modular robotic system and implemented a simulator (Fig. 1) in JAVA. In Fig. 1, dark gray blocks in the left represent modular robots, black blocks in the center are obstacles, and light gray blocks below the obstacles show goals.

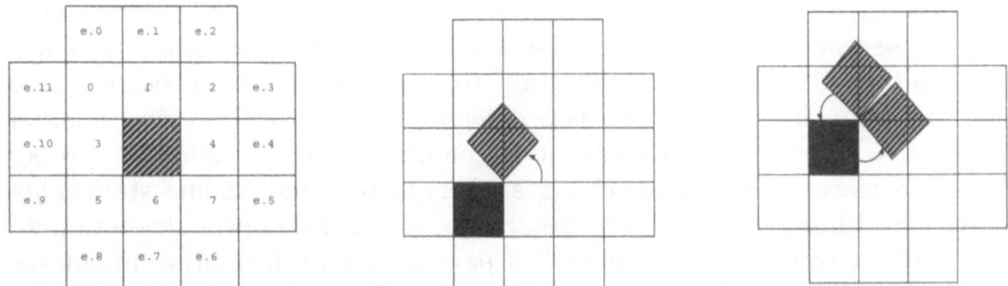

Fig. 2. Module's sensing capability, and two types of module's movement: (1) 90 degree rotation and (2) 180 degree rotation

The movement of the module is constrained by the existence of other modules and obstacles. As essential constraints of general modular robots' movements, we presume the followings:

1. A module moves relative to another module, called as a *pivot* module.
2. A module must be attached to the pivot module after the movement.
3. The whole modules need to be connected with each other after the motion.

4. While moving, a module must not collide with other modules or obstacles.

We do not assume a module that does not move, which is often called as a *fixed base* module in the several research of the modular robots. In this paper, the module robots can move to whatever destination far away from their initial location, as long as there is a possible path to the destination. And a module is capable of rotational/rolling motions, but sliding motions are not allowed because they may have large kinetic friction.

To make autonomous movements in unknown environments, each module is capable of sensing its surroundings. The left figure in Fig. 2 shows the area which can be sensed by the module sitting in the center of the figure. The central figure depicts the counter-clockwise rotational motion of the module which makes the module at position 5 as a pivot module. As is shown in the figure, the module makes a rotational movement on the vertex of the pivot module. In this motion, the positions 0, 1, and 3 must not be occupied by other modules or obstacles. The right figure in Fig. 2 puts another example of a module's movement. To make this motion feasible, the positions 0, 1, 2, 4, *e*.0 and *e*.1 must be empty of other modules and obstacles. As explained in Sec. 3.1, these tight physical constraints make the modular robots' path planning problem intractably hard for a brute-force search algorithm.

2.2 Real-Time Search Algorithm

The problem of motion planning or path planning in robotics can be solved as a state-space search problem. The goal of the state-space search problem is to find a path from an initial state to a goal state. State-space search algorithms can be divided into two categories: one is *off-line* and another is *real-time*. When the environment is completely known before the robot begins its traverse and never changes afterwards, some off-line state-space search algorithms, such as the A* algorithm, can find an optimal solution for the problem. But when the robot has partial or no information about the environment before it begins its traverse, the off-line algorithms usually fail to find a solution, not to mention that they can't find an "optimal" path.

Real-time algorithms, such as RTA* [7], perform sufficient computation to determine a plausible next move making the best of currently available information about the environments, execute that move, then perform further computation to determine the following move, and so on, until the goal state is reached. Hence, the real-time search algorithms allow the robot with sensors to accumulate information of the environments during its movement and exploit the accumulated information to make a plausible plan. These algorithms cannot guarantee to find the optimal solution, but usually find an acceptable solution in a reasonable amount of time. Thus, we apply RTA* for navigating the modular robots in the unknown environments. The outline of the algorithm is as follows[1]:

Step 1: Calculate $f(x') = h(x') + k(x, x')$ for each neighbor x' of the current state x, where $f(x')$ is the estimated distance from x to the goal via x', $h(x')$ is the current *heuristic estimate* of the distance from x' to the goal state, and $k(x, x')$ is the distance between x and x'.
Step 2: Move to a neighbor with the minimum $f(x')$ value. Ties are broken *randomly*.
Step 3: Update the value of $h(x)$ to the second-best of $f(x')$ value.

The algorithm can backtrack to a previously visited state when the estimate of solving the problem from that state plus the cost of returning to the state is less than the estimate cost of going forward from the current cost. By updating $h(x)$ in step 3, the algorithm prevents infinite loops while permitting backtracking when it appears favorable, resulting in a form of single-trial learning through exploration of the problem space. RTA* is guaranteed to be *complete* in the sense that it will eventually reach the goal under the following conditions: (1) the problem space is finite, (2) all the edge costs are positive, (3) there exists a path from every state to the goal state, and (4) the value of heuristic estimates are finite.

3 Phased Search Architecture

We apply RTA* to navigation of the modular robots in the unknown environments. In this paper, we assume that search is done in a centralized fashion by the modular robotic system and only one module can make a movement at each execution. The application of RTA* to navigation of the modular robots is straightforward. Each search state x is a configuration of the modular robot in a specific location. And the heuristic estimate $h(x)$ of the distance between the modular robots and the goal can be a sum of the Diagonal Distance [2] between each module and its closest unoccupied goal location. This is based upon the conjecture that conjunctive goals of modules are completely independent.

3.1 Heuristic Depressions

However, except for the very simplistic problems, this naive formulation results in lengthy plans, which are not desirable in the robotic applications because, in general, execution cost of a plan is much higher than planning cost in robotic applications.

[1] The shown algorithm is for the case where the depth of the look-ahead horizon is 1.

[2] The Diagonal Distance is the maximum of the X-axis distance and the Y-axis distance between the state and the goal.

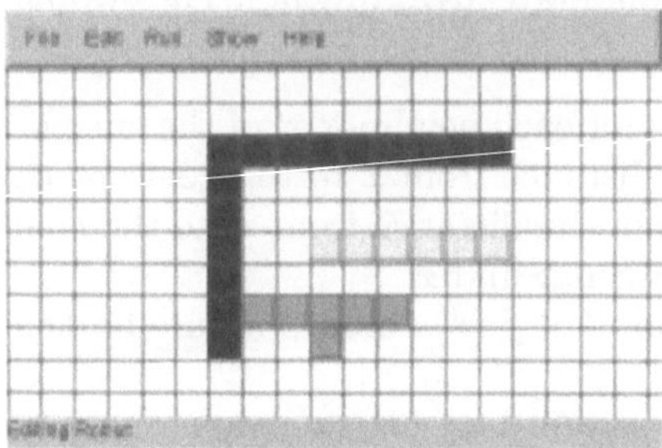

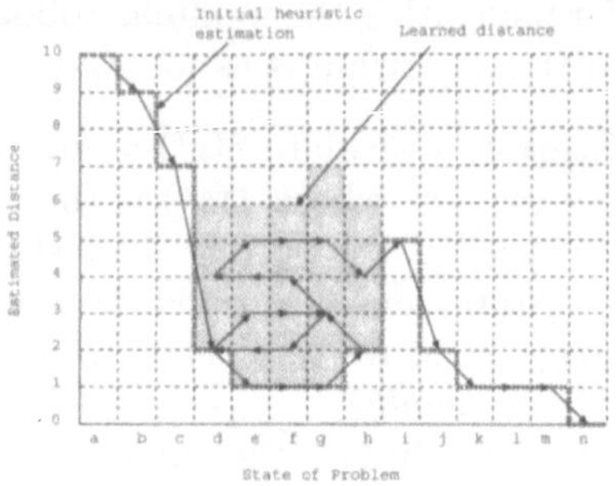

Fig. 3. Modular robots trapped in heuristic depressions and schematic view of search process escaping from heuristic depressions

The constraints on module's motions described in Sec. 2.1 make the modular robots trapped with the obstacles as shown in the left figure of Fig. 3. In the figure, the left-most module attached to the obstacle can leave the obstacle by moving into a direction of either 2, 4, or 7 in Fig. 2. For moving into a direction 2 or 7, a pivot module is required at a location 1 or 6, which are attached to the obstacle. To move into a direction 4, a pivot module is needed at a location 2 or 7. Since every module needs to be connected with each other, the trapped module also needs to be connected with a module either at a location 2 or 7. But since locations behind the module (i.e., locations 0, 3 and 5) are obstacles in Fig 3, for the module to be connected with a module at a location 2 or 7, there must be a module at a location 1 or 6, either of which is attached to the obstacle. This means that modules attached to the obstacles cannot leave them freely. For the modular robots to leave the obstacles, the entire modular robots must go downwards in the figure. Since this is the opposite direction to the goals, the modular robots waste huge steps of motions wondering in the proximity of the obstacles trying to leave them. Therefore, once a motion makes the modular robots attached to the obstacles as shown in Fig. 3, a resultant plan becomes very lengthy. Difficulty of avoiding being trapped in local minima is not unique to this problem, but in general, inaccuracy of heuristic estimates, called a ***heuristic depression*** [5], causes a search based planner to make an inefficient plan for filling gaps between a heuristic estimate and a correct distance of the states during search as schematically shown in the right figure of Fig. 3.

3.2 Refining Heuristic Estimate

In Step 1 of the algorithm described in Sec. 2.2, a heuristic estimate of the distance from the current state to the goal state is calculated as a sum of the Diagonal Distance of each module to its closest goal location. The Diagonal Distance metric is excessively underestimated when there are obstacles blocking module's movements in the environment. With more accurate heuristics, efficiency of the search is expected to improve drastically avoiding being trapped in heuristic depressions. Thus, as a remedy of the original RTA* algorithm, the better estimate for each module in the state is obtained by searching the shortest path to the goal when new obstacles are discovered after a motion of the modular robots. We call this search as *Module Path Planning.*

To estimate a module's distance to the goal, we assume that a single module can move by itself (i.e., it can move without help from a pivot module and it does not have to consider about connectivity and collision with the other modules) from its current position to the designated goal. Without the obstacles, the Diagonal Distance gives an accurate distance estimate for the module, which is the shortest path length from the module to the goal. But when there are obstacles in the environment, it might be impossible for the module to reach the goal in the shortest path. In general, a module needs to detour the obstacles along the way to the goal, thus the actual distance from the module to the goal is larger than the Diagonal Distance.

For calculating a more accurate distance than the Diagonal Distance, in Module Path Planning, A* is applied to search for the shortest path from the current position of a module to the unoccupied closest goal when new obstacles are discovered in the environments. And, if a newly estimated value is larger than the original value, the heuristic estimate value of the state is replaced by the new value. Since A* is the best-first heuristic search, it takes exponential time to run in practice. Then, in Module Path Planning, the depth of search can be limited to a certain threshold value to guarantee real-time execution of the algorithm. To avoid being trapped in heuristic depressions described in Sec. 3.1, the special attentions need to be paid to certain topological relationship among goals, obstacles and modules in searching for a path. As explained before, since the module robots tend to be trapped in the proximity of the obstacles, the following heuristics are developed: (1) when a module is attached to a line of obstacles, it should move to the nearby goal that is attached the obstacles, or without such a goal it should go to the closer edge of the obstacles, and (2) when a goal is attached to a line of obstacles, a module should move to the goal.

To be noted is that in Module Path Planning each module is assumed to move independently of the other modules, which keeps the computational complexity of search linear to the number of the modules. Moreover, since Module Path Planning can be executed independently by each module without communicating each other, modules with sufficient computational capability can perform the search fully in parallel. Hence, we propose a phased search architecture (see Fig. 4) for navigating modular robots. In the architecture, when new information of the environment is obtained after a movement, a heuristic estimate of the state is updated in parallel by each module. And based upon the updated estimate, a next move is determined in a centralized computation after checking connectivity of the modules.

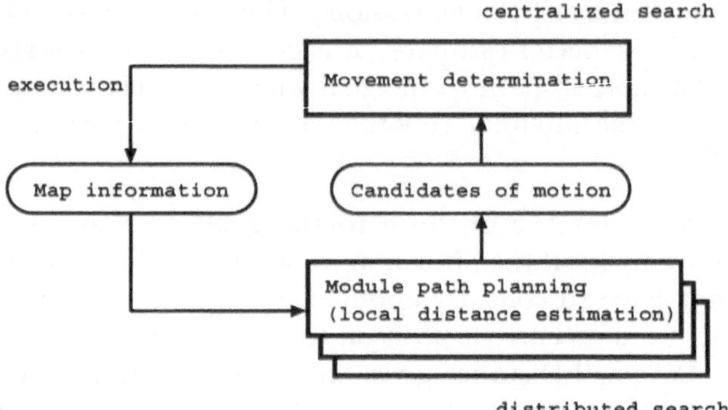

Fig. 4. Phased search architecture for navigating modular robots

4 Experiments

We incorporated the above described ideas in the original RTA* algorithm and developed the real-time path planner for the modular robots, called RPM. In order to evaluate effectiveness of RPM experimentally, we made some preliminary experiments of navigating 2D modular robots in the unknown environments.

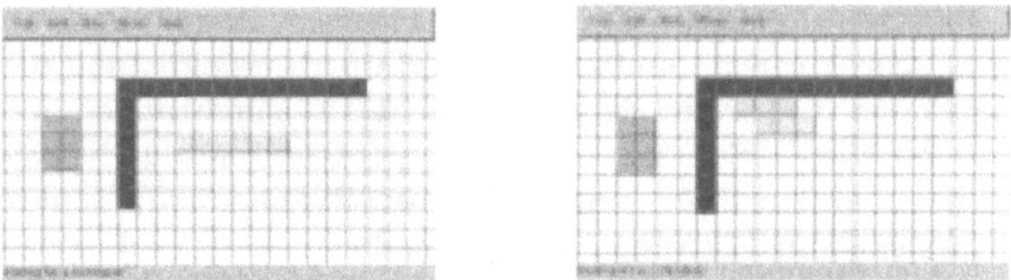

Fig. 5. Example Problems (1) Expl: goals are apart from obstacles and (2) Exp2: goals are attached to obstacles

The problems we used for experiments are shown in Fig. 5. The modular robots in the left of the figures are requested to move to the goals in the right avoiding the L-shaped obstacles in the center. In an initial state, the modular robots have location information about themselves and the goals but they are ignorant about existence of the obstacles, thus they are put in the unknown environment.

4.1 Results and Analysis

We compare the results of the original RTA* algorithm and RPM (with no limitation on the search depth of Modular Path Planning). In the experiments there are some possibilities that some states are in ties and in such cases the

Table 1. Experiment results of Exp1 and Exp2

	Exp1		Exp2
	RTA*	RPM	RPM
Ave. length	3435.2	166.5	352.7
S.D. length	2703.2	39.9	146.3
Min. length	153	112	208
Max. length	6937	452	893

	Exp1		Exp2
	RTA*	RPM	RPM
0 - 99	0	0	0
100 - 199	40	95	0
200 - 299	0	4	51
300 - 399	0	0	18
400 - 499	0	1	15
500 - 599	0	0	10
600 - 699	0	0	3
700 - 799	0	0	2
800 - 899	0	0	1
900 - 999	0	0	0
1000 -	60	0	0
Sum	100	100	100

ties are broken randomly. Therefore we repeated solving the same problem 100 times with the different random seed values in the experiments.

The results of the experiments are summarized in Table 1. In the tables, RTA* means the original Real-Time A* algorithm and RPM indicates the RPM algorithm that combines RTA* with estimate refinement by Module Path Planning. Since, in the experiments of Exp2, the original Real-Time A* algorithm failed to find a solution within limitation of available memory size (1G byte), only the results of RPM are shown in the tables.

The experiments are evaluated using the length of resultant plans. The first row of the left table in Table 1 shows the average of plan length in 100 experiments, The second row presents the standard deviation of plan length. The third and fourth rows are the minimum and maximum plan length obtained in 100 experiments respectively. As shown in the table, RPM outperforms the original RTA* algorithm considerably in Exp1. The right table in Table 1 shows the distribution of plan length in 100 experiments. Each row in the table shows the number of plans whose length is within the designated range. The table presents that in Exp1 RPM succeeded to find good solutions (less than 200 moves) in most of the cases of 100 random trials. On the contrary, RTA* produces redundant plans (over 1000 moves) in 60 cases. This shows that in RTA* the modular robots are caught in heuristic depressions and forced to wander around for getting them over. From the table, it is also clear that Exp2 is more difficult than Exp1 for modular robots. This is because in Exp2 some goals are attached to the obstacles and modules must approach those goals along the obstacles due to physical constraints explained in Sec. 2.1, thus versatility of motion selection for modules are more restricted than in Exp1. In such a difficult problem, RPM still succeeded to find a good solution for many cases.

5 Conclusions

In this paper, it is experimentally shown that RPM is the effective algorithm for navigating the modular robots in the unknown environments. RPM extends RTA* by refining heuristic estimate of the states by executing local search upon discovery of new obstacles, thus overcoming difficulty of heuristic depressions. In the future study, we are planning to develop a distributed version of RPM that also allows simultaneous motion executions by multiple modules. Then, RPM will be embedded in the 3D hardware modular robots.

References

1. Piotr Berman. On-line searching and navigation. In Amos Fiat and Gerhard J. Woeginger, editors, *Online Algorithms: The State of the Art*, pages 232–241. Springer, 1998.
2. G. S. Chirikjian. Metamorphic hyper-redundant manupilators. In *Proc. of JSME International Conference on Advanced Mechatronics*, pages 467–472, 1993.
3. T. Fukuda and S. Nakagawa. Approach to the dynamically reconfigurable robotic system. *Journal of Intelligent and Robot Systems*, 1:55–72, 1988.
4. G. J. Hamlin and A. C. Sanderson. Tetrobot modular robotics: Prototype and experiments. In *Proc. IEEE/RSJ Int. Conf. Intelligent Robots and Systems (IROS'96)*, pages 390–395, 1996.
5. Toru Ishida. *Realtime Search for Learning Automous Agent.* Kluwer Academic Publishers, Boston, 1997.
6. Shigeru Kokaji. A fractal mechanism and a decentralized control method. In *Proc. of USA-Japan Symp. Flexible Automation*, pages 1129–1134, 1988.
7. Richard E. Korf. Real-time heuristic search. *Artificial Intelligence*, 42:189–211, 1990.
8. K. Kotay and D. Rus. Algorithms for self-reconfigurable molecule motion planning. In *Proc. 2000 IEEE/RSJ Int. Conf. on Intelligent Robots and Systems*, 2000.
9. Satoshi Murata, Haruhisa Kurokawa, and Shigeru Kokaji. Self-assembling machine. In *Proc. IEEE Int. Conf. on Robotics and Automation*, pages 441–448, 1994.
10. Amit Pamecha, Imme Ebert-Uphoff, and Gregory S. Chirikjian. Useful metrics for modular robot motion planning. *IEEE Transactions on Robotics and Automation*, 13(4):531–545, 1997.
11. Nageswara S. V. Rao, Srikumar Kareti, Weimin Shi, and S. Sitharama Iyengar. Robot navigation in unknown terrains: Introductory survey of non-heuristic algorithms. Technical Report ORNL/TM-12410, Oak Ridge National Laboratory, 1993.
12. Daniela Rus and Marsette Vona. Crystalline robots: Self-reconfiguration with compressible unit modules. *Autonomous Robots*, 10:107–124, 2001.
13. Mark Yim, Ying Zhang, John Lamping, and Eric Mao. Distributed control for 3D metamorphosis. *Autonomous Robots*, 10:41–56, 2001.

A Self-Reconfigurable Modular Robot (MTRAN) – Hardware and Motion Planning Software –

Akiya KAMIMURA†, Eiichi YOSHIDA†, Satoshi MURATA‡, Haruhisa KUROKAWA†, Kohji TOMITA† and Shigeru KOKAJI†

†National Institute of Advanced Industrial Science and Technology (AIST): Namiki 1-2-1, Tsukuba, Ibaraki, 305-8564 Japan, kamimura.a@aist.go.jp
‡Tokyo Institute of Technology: 4259 Nagatsuta-cho, Midori-ku, Yokohama, 226-8502 Japan, murata@dis.titech.ac.jp

Abstract. In this paper, we present our latest modular robot "MTRAN" (Modular Transformer) and its motion planning algorithm based on a four-module block method. The developed module is composed of two semi-cylindrical parts connected by a link. Each part has three connecting surfaces where another module can be connected by magnetic force. Each module has only two degrees of freedom, but a group of modules can not only configure static 2-D or 3-D structures but also generate robotic motions. We realized several robotic motions and transformations by using several modules. Block motions prepared by a proposed motion planner are also verified by hardware experiments.

Key Words. Modular Robot, Self-Reconfiguration, Robotic Motion Generation, Motion Planning

1 Introduction

In recent years, the feasibility of reconfigurable robotic systems has been examined through hardware and software experiments [1]–[8]. Self-reconfigurable robots, especially homogeneous ones, can adapt themselves to the external environment by changing their configurations or repairing themselves by using spare modules. This type of robot is useful in harsh environments that humans are unable to access. In such areas, adaptability and self-maintainability are significant factors.

Reconfigurable modular robots are classified into two types, a lattice type [2,4,8] and a thread type [6,7]. While the former can transform itself into various static structures like a jungle gym, it has difficulty generating dynamic robotic motions. In contrast, the latter has a snake-like shape that can generate various dynamic motions, though it has difficulty in self-reconfiguration.

We have developed a new type of modular robot system, called *Modular Transformer* or *MTRAN* [9], which is designed to realize the merits of both the lattice type and the thread type. This has been achieved by adopting a simplified module design, and a lightweight but strong connection mechanism.

There have been several studies on distributed [10,11] and centralized [4,8] motion planning methods. However, these methods cannot directly be applied to MTRAN due to its restricted degrees of freedom. We propose a method using global information of modules, which enables a class of MTRAN clusters to move along a desired trajectory. The proposed motion planner can output an efficient sequence of module motion where several modules are driven in parallel.

In this paper, we outline the module hardware. We describe its control system architecture in chapter 2, its motion planning method in chapter 3 and hardware experiments using several modules in chapter 4. The experiments confirmed the feasibility of the hardware and the motion planning method.

2 Hardware

We have developed a module hardware that has the simple structure shown in Fig. 1. The module is composed of a link part and two semi-cylindrical parts, one active the other passive. Each semi-cylindrical part can rotate about its axis by 180 degrees using a servomotor embedded in the link and has three connecting surfaces with four permanent magnets. Since the polarities of magnets between the active and the passive parts are different, each module can connect to other modules' surfaces that have the opposite polarity. Therefore, the active and the passive parts form a 3-D checkerboard-like structure when they are constructed in a lattice structure. Reconfiguration is achieved by repeating basic operations such as detaching, changing a position of a module and reconnecting.

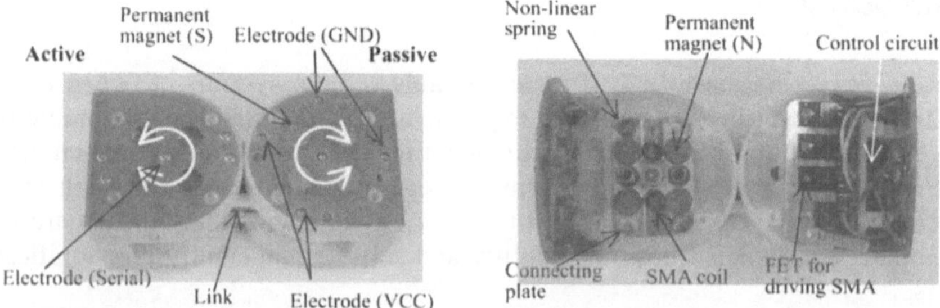

Fig. 1. Photos of appearance and inner structure of the developed module

2.1 Mechanical and Electrical Design

The specifications of the module are summarized in Table 1.

The link part includes two sets of geared motors and servo circuits. Each servo has sufficient maximum torque to lift one module as shown in Fig. 2.

All connecting surfaces have electrodes for power supply and serial communication. The electrodes are electrically connected when each surface is connected. Therefore, it is only necessary to connect wires to a single module to supply power and to communicate with all the modules.

The active part has a connection mechanism composed of non-linear springs, Shape Memory Alloy (SMA) coils, and four permanent magnets (N poles on the surface), which are fixed on the connecting plate. This mechanism is based on the Internally Balanced Magnet Unit (IMBU) technique [12]. The connecting plate moves automatically to connect to another module's surface by attractive force of the magnets. To detach the modules, we heat the SMA coils by electric current. The modules are then easily detached with the help of the springs [9].

The passive part contains control circuits, including an onboard microcomputer (BASIC STAMP II (BS II, Parallax, Inc)). It controls motor positioning and switches electric current to the SMA coils. A host computer issues control commands corresponding to these operations through the serial communication line as shown in Fig. 3.

Table 1. Module specifications

Item	Value
Dimension	66x132x66 mm
Weight	0.44 kg
CPU	BASIC STAMP II
Power supply	DC 12V
Maximum torque of each axis	23 kg·cm
Connecting force	25 N
Elapsed time for detachment	5 seconds
Power consumption for detachment	180 J

2.2 Control System Architecture

The control system consists of the host PC, the relay processor (the same one as the onboard microcomputer), and the control circuits in the modules connected as shown in Fig. 3. All elements communicate through 4800bps asynchronous (RS-232C) serial communication using token passing.

The host PC broadcasts control commands that include module ID numbers. Only the module with the broadcast ID executes the task and returns a validation signal to the host. Modules are currently controlled in a centralized manner.

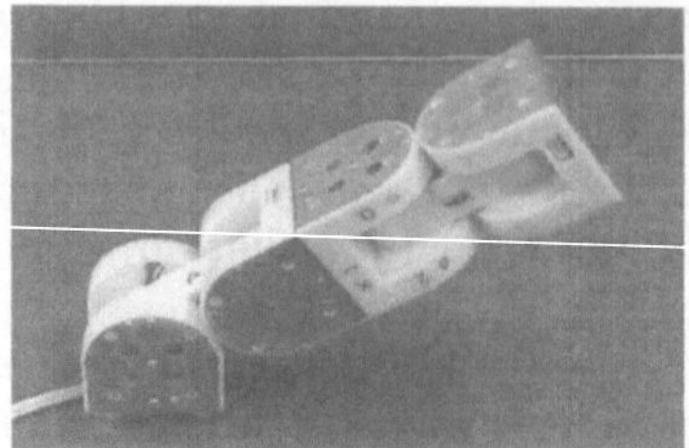

Fig. 2. Lifting motion with two modules

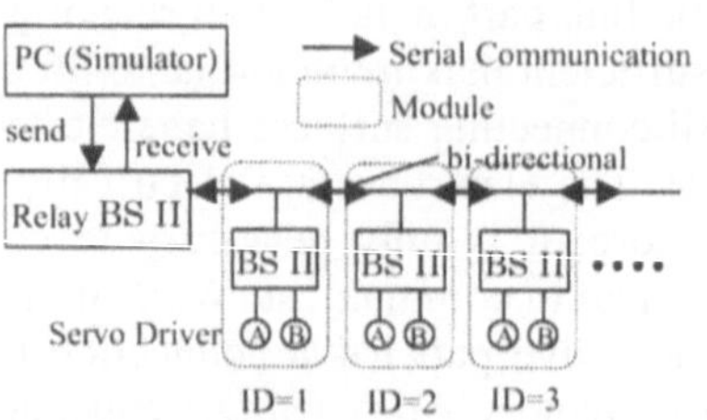

Fig. 3. Schematic of control system architecture

2.3 Motion-Planning Interface

We developed interface software on the host PC to control the modules (Fig. 4). The motion planning process is as follows [13]. First, we create an initial configuration of modules using a configuration editor control panel (not shown). Next, we prepare the module motion plan manually using the motion control panel in Fig. 4. The generated plan is called a *motion sequence* and is a sequence of basic motions. Collisions between modules and connectivity of the whole configuration are checked automatically. In the current software, we implement quasi-static motion by taking gravity and sliding friction into account. Finally, the motion sequence is translated into a series of hardware control commands, which are sent to the modules via the serial line.

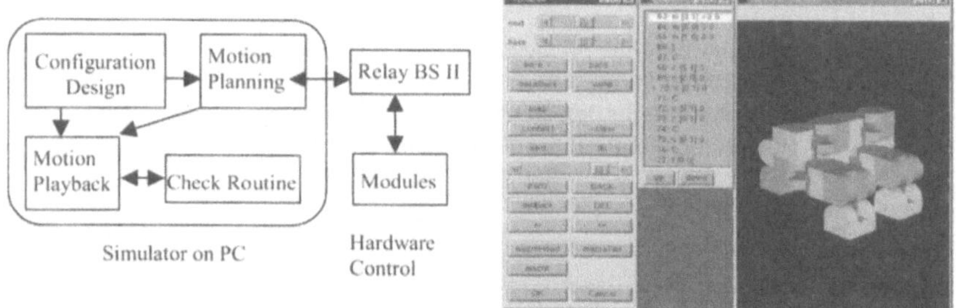

Fig. 4. Diagram of interface software (left) and its screen shot (right)

3 Motion-Planning Method

Besides manual planning with the interface software in 2.3, we have been developing an automatic planner. As it is not easy to find a general plan for going from any given initial configuration to a final configuration, we have developed an automatic planner for a particular class of reconfigurations. Suppose we have a variable-length chain, called a cluster, composed of several four-module *blocks*

(Fig. 5). Each block looks like a large cube as shown in Fig. 5. In addition, a couple of modules called a converter are attached to the cluster. The converter changes the direction of rotation of other modules in the cluster chain.

The *motion planner* outputs a series of reconfigurations, in each of which one module block is transferred to another place along any given three-dimensional trajectory in the lattice grid (Fig. 6).

In our previous paper [14], the motion of a cluster was generated by a two-layered motion planner. The upper layer decomposes the planning problem into subproblems solvable by the lower layer. The lower layer is designed to solve simplified planning problems based on a database of rules for each local reconfiguration.

As the generated motion allows only one module to move in a single step, we introduce a new motion parallelism algorithm by a *motion scheduler*. It processes the planned sequence into a motion plan, including motion steps that can be executed in parallel. This makes use of the concurrent feature on the modular robot's reconfiguration process and increases the efficiency by reducing the execution time required for the plan.

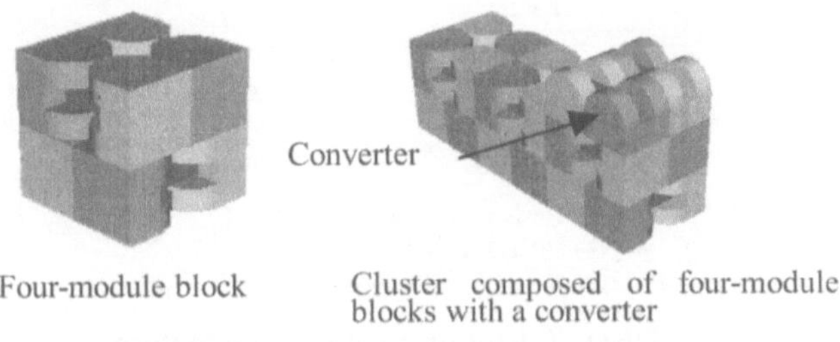

Fig. 5. Four-module block and a cluster

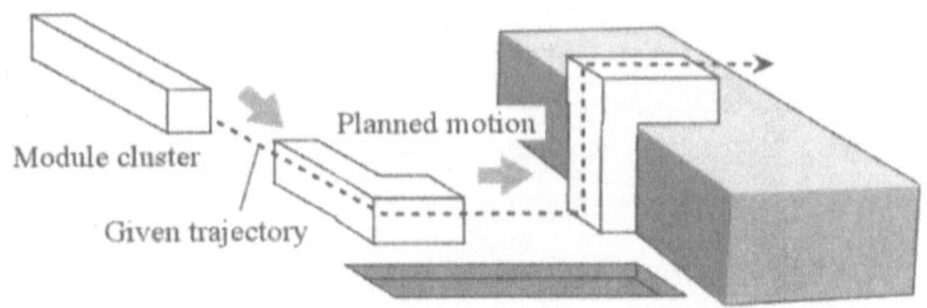

Fig. 6. Planning of cluster motion

3.1 Motion-Planner Architecture

This section briefly outlines the previously developed motion planner [14]. The upper and lower layers of the motion planner are called the *global-flow planner* and the *local-motion scheme selector* respectively. As shown in Fig. 7, the global-flow planner searches for possible module paths and motion orders to provide the global cluster movement, called *flow*, according to the desired trajectory. This is realized as a motion of a block such that the tail block is transferred toward the given heading direction via the side of the cluster. We adopt simple motion schemes sending modules one by one towards the head.

The local-motion scheme selector checks if the paths generated by the global planner are valid for each *member* module of the block. If a given path proves to be valid, the selector chooses the motion plan by adding a set of locally coordinated motion sequences called *motion schemes* from a rule database. If not, it tries another possible path. Note that this is a centralized planning method assuming that all the information of modules in the cluster is available.

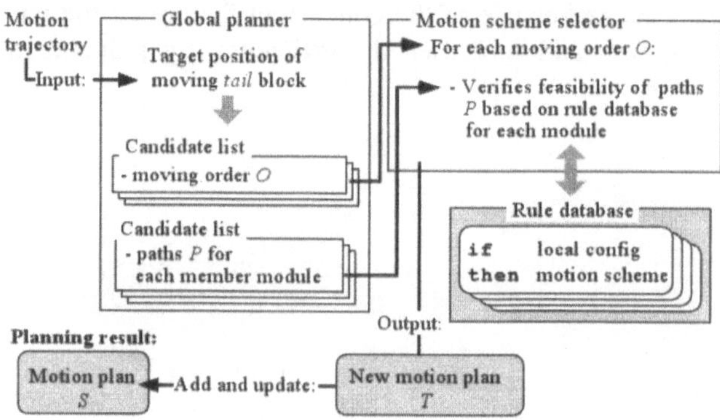

Fig. 7. Motion-planner architecture

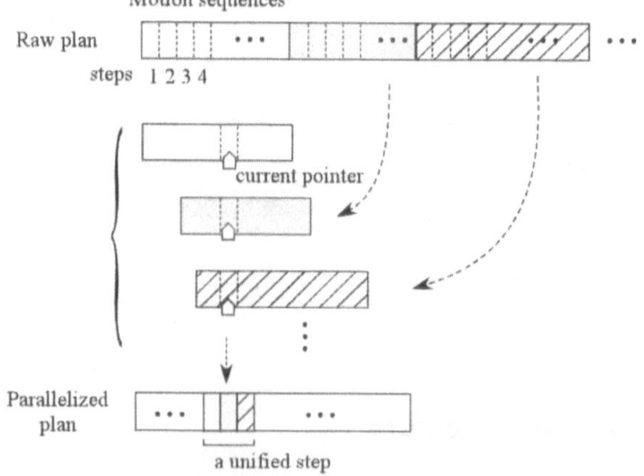

Fig. 8. Parallelization of plan by motion scheduler

3.2 Motion Scheduler

The output of the planner described so far is a sequence of motion schemes to achieve the desired trajectory. However, only one motion scheme is allowed at a time. The motion scheduler is devised to improve the efficiency through parallel execution of multiple motion schemes.

3.2.1 Parallelizing a motion plan

The output plan made by the original motion planner is a series of motion sequences, each of which corresponds to a single path of a block from tail to head. When a block is far enough from the tail, the next block can start its motion without interference. The motion scheduler tries to parallelize the plan by merging a motion sequence with the following ones (Fig. 8). During this process, collisions and total connectivity are checked. The scheduler also checks whether the parallelized motion plan ends up with the same configuration as the original plan. Those motion steps executable in parallel are unified into one motion step. The scheduler repeats these procedures throughout the original plan to derive a parallelized motion sequence.

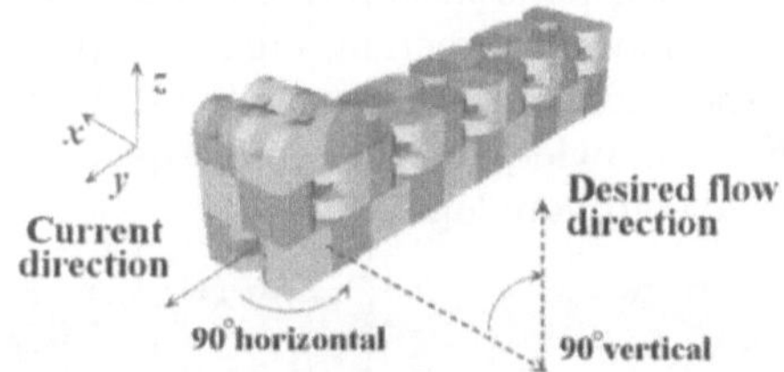

Fig. 9. Desired trajectory of module cluster

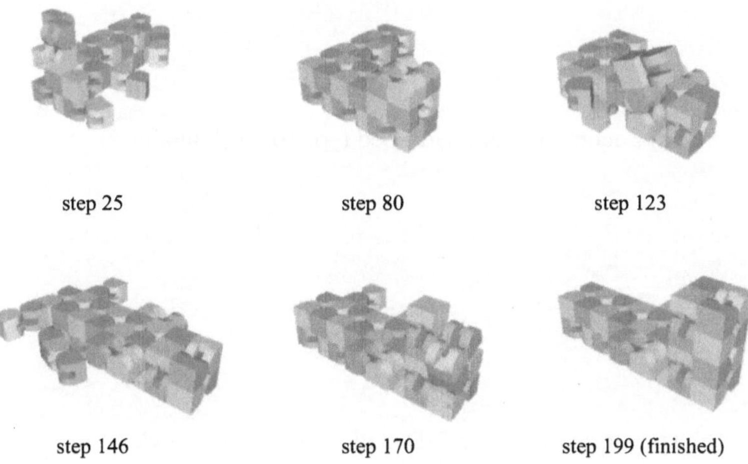

Fig. 10. Generated plan for desired motion in Fig. 9

3.2.2 Example of generated motion

The motion-planning framework described so far is applied to a module cluster composed of 22 modules. The desired trajectory includes horizontal and vertical direction changes of cluster flow as shown in Fig. 9. Figure10 shows some snapshots of the generated motion.

The raw plan generated by the motion planner takes 354 motion steps, where only one motion scheme is allowed at one step. After rescheduling, the length was reduced to 199 steps.

4 Hardware Experiments

4.1 Experiment of Robotic Motion and Transformation

Figure 11 (a) shows the locomotion of a quadruped robot using eight modules. This robot can walk by using two of the leg parts and then turn about its vertical center axis by folding two leg parts in opposite directions. Arrows in the photographs indicate the robot's moving direction. Figure 11 (b) shows the transformation and locomotion experiment using nine modules. This robot is transformed from a 2-D structure into a crawler, and then into a quadruped robot by releasing the closed module rings to configure legs.

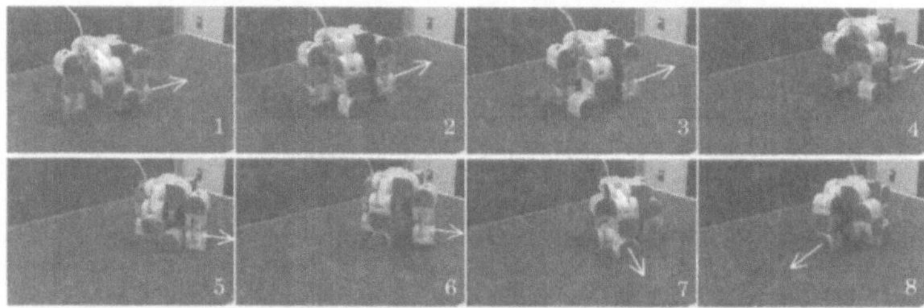

(a) Locomotion of quadruped type robot (8 modules)

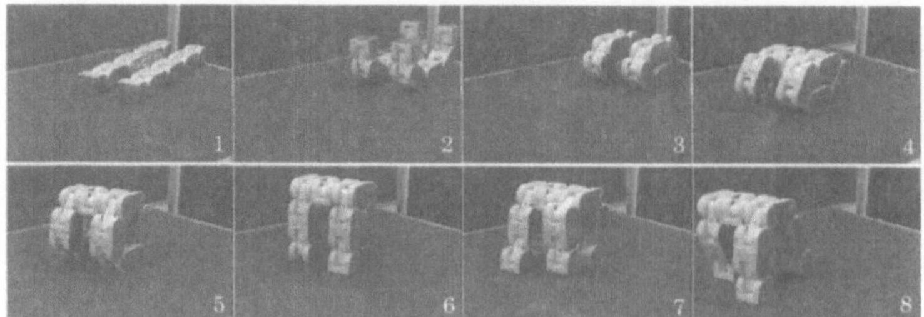

(b) Transformation from 2-D static structure to crawler robot and then to quadruped robot (9 modules)

Fig. 11. Robotic motion and transformation experiments

4.2 Experiment of Block Motion

Figure 12 shows the experiment of an eight-module cluster flow motion. The output plan made by the generator was slightly modified to avoid a current hardware problem.

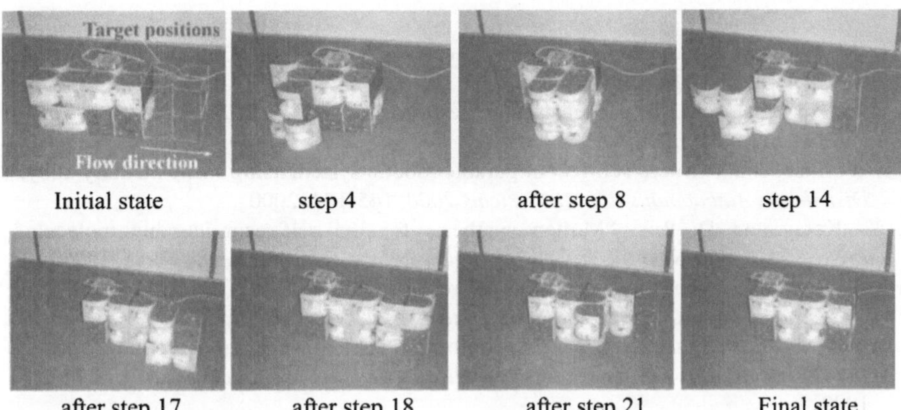

Fig. 12. Experiment of cluster motion of block structure using 8 modules

5 Future Works

External and internal sensors are needed in order to realize a modular robot system that can adapt itself to the external environment and change its configuration or locomotion modes. In addition, the tether connected to the modular robot becomes a problem for practical use when the robot moves or changes its configuration. We will also reduce the power consumption of SMA coils and embed a battery on the next version.

The current modular robot system is controlled in a centralized manner, but to realize fault tolerance we must implement distributed algorithms in the hardware to enable each module to move autonomously in a cooperative manner. To realize this, the modules must have local communication capability.

From the application point of view, it will be important for us to find suitable configurations for specific tasks so that we can show the usefulness of this reconfigurable system.

6 Conclusions

In this paper, we described the module hardware for a modular robot system and its associated motion-planning algorithm. We confirmed by hardware experiments that a variety of configurations and transformations are feasible and that robotic motions are possible. The motion planner with the scheduler improved parallelism for changing configurations of the modular system.

References

1. S. Murata, et al.: "Self-assembling machine, " *Proc. IEEE Int. Conf. on Robotics and Automation*, 441-448, 1994.
2. S. Murata, et al.: "A 3-D self-reconfigurable structure," *Proc. IEEE Int. Conf. on Robotics and Automation*, 432–439, 1998.
3. E. Yoshida, et al.: "Micro self-reconfigurable robotic system using shape memory alloy," *Distributed Autonomous Robotic Systems 2000*, 145-154, 2000.
4. K. Kotay and D. Rus: "Motion synthesis for the self-reconfigurable molecule," *Proc.1998 IEEE/RSJ Int. Conf. on Intelligent Robots and Systems*, 843–851, 1998.
5. P. Will, et al.: "Robot modularity for self-reconfiguration," *Proc. SPIE, Sensor Fusion and Decentralized Control in Robotic Systems II*, 236–245, 1999.
6. A. Casal and M. Yim: "Self-reconfiguration planning for a class of modular robots," *Proc. SPIE, Sensor Fusion and Decentralized Control in Robotic Systems II*, 246–257, 1999.
7. A. Castano, et al.: "Autonomous and self-sufficient CONRO modules for reconfigurable robots," *Distributed Autonomous Robotic Systems 4*, Parker L E, et al. eds., Springer, 155–164, 1998.
8. C. Ünsal, et al.: "A modular self-reconfigurable bipartite robotic system: implementation and motion planning," *Autonomous Robots*, 10-1, 23–40, 2001.
9. A. Kamimura, et al.: "Self-reconfigurable Modular Robot – Experiment on reconfiguration and locomotion," *Proc. 2001 IEEE/RSJ Int. Conf. on Intelligent Robots and Systems (IROS2001)*, 606–612, 2001.
10. E. Yoshida, et al.: "A distributed method for reconfiguration of 3-D homogeneous structure," *Advanced Robotics*, 13-4, 363–380, 1999.
11. K. Tomita, et al.: "Self-assembly and self-repair method for distributed mechanical system," *IEEE Trans. on Robotics and Automation*, 15-6, 1035–1045, 1999.
12. S. Hirose, et al.: "Internally-Balanced Magnet Unit," *Advanced Robotics*, 1-3, 225-242, 1986.
13. H. Kurokawa, et al.: "Motion Simulation of a Modular Robotic System," *Proc. IECON 2000*, 6, 2000 (in CD-ROM).
14. E. Yoshida, et al.: "A Motion Planning Method for a Self-Reconfigurable Modular Robot," *Proc. 2001 IEEE/RSJ Int. Conf. On Intelligent Robots and Systems (IROS2001)*, 590–597, 2001.

Automatic Parameter Identification for Rapid Setting Up of Distributed Modular Robots

Kohsei Matsumoto, HuiYing Chen, Kenichi Shimada, Jun Ota, and Tamio Arai

Department of Precision Engineering, School of Engineering,
The University of Tokyo, 7-3-1 Hongo, Bunkyo-ku, Tokyo, 113-8656 JAPAN

Abstract. This paper describes a concept of a distributively placed modular robot system and a methodology of parameter identification of the system. The robot system is consist of modules which have mono-function(e.g. sensor or actuator) and radio communication. Depending on tasks, the system is constructed by attaching modules to an object or in a working area distributively. Then the modules achieve a task by their cooperation. Generally it is difficult and troublesome for users to set up a distributed system, because it is necessary to measure many parameters(e.g. position and orientation of modules) in practical use.

In our algorithm, the parameters are identified rapidly by executing motions with actuators, and observing the changes of the state with sensors. The motion which we call testing motion influences accuracy of identification because of sensing errors. Accounting for this, especially for the case of a mobile robot and a ceiling camera module, the design of modules' testing motion is made by evaluating the influence of camera's sensing error and dead-reckoning's locating error. The effectiveness of the algorithm is shown in the simulation and a basic experiment.

1 Introduction

There are many researches on cooperative multiple robots from the viewpoint of emphasizing of adaptability, throughput, flexibility, and robustness to various tasks. The followings are the previous works on distributed robot systems which is consist of mono-functional robots. Mono-functional robots are proposed to construct a suitable and compact robot for various tasks, they are often called modular robots.

Hashimoto et al. proposed a position estimation method by dead-reckoning and a control method of localization for a transportation system [1]. The system was consist of modular robots to deal with the changes of payload. Burke et al. showed omni-directional motion with three modules which were connected together [2]. Each of them was equipped with processing unit, a wheel and a steering. In these two researches, it was assumed that each system part was connected closely and also mechanically via wire. This constrains the free placement of modules to objects. Then it is not effective to deal with size changing of objects for transportation. These approaches were to place functions together by connecting sensors or actuators. On the other hand, there is another approach to deal with tasks, attaching sensors or actuators in specific places distributively. Sato et al. demonstrated a robotic room which had an intelligent bed, a manipulator, and so on [3]. The bed was equipped with pressure sensors to monitor a patient's condition. The manipulator moves on a rail fixed on the wall, and helps human. They support human-tasks immediately by placing robotic elements distributively in the environment. The mentioned robot systems were basically static from the viewpoint of systems composition. Fukuda et al. have proposed CEBOT which can reconfigure the robot system. The system is consist of heterogeneous robots called CEBOT, and they can renovate the relation among units for cooperation dynamically [4]. Lee et al proposed resource sharing architecture for mobile robots in a room which are equipped with sensors [5]. Their system can renovate the logical connection among physical components by using network protocol Jini. However the sensor location is static. Therefore flexibility is not high.

In conclusion, an ideal distributed robot system is relocatable and changeable dynamically for tasks. In this paper, we propose modular robots which satisfy relocatability in a working area. To realize the relocatability, it is necessary to get each parameters(e.g. relative positions and orientations) quickly for cooperation of distributively placed sensors and actuators. Regarding this viewpoint, the purpose of this research is to propose an algorithm to identify the parameters as quick and accurate as possible among relocatable sensor and actuator modules. Supposing the modular robots cooperate as if they were a robot or a robotic environment, it is significant and useful for users to identify parameters of modules automatically from the viewpoint of reducing troublesome of measuring parameters.

Section 2 summarizes a concept of our robot system. Section 3 explains an algorithm to realize automatic parameter identification. Section 4 describes a simulation and an experiment. And the conclusion is described in the last section.

2 Concept of Distributedly Placed Modular Robot System

2.1 Module Architecture

It is defined that each module is equipped with a processing unit and a radio communication unit for its function control, as well as a mono-function(e.g moving, operating and sensing). And each of them can control its function.In the concept of our modular robot system, operators can make the objects autonomous as if they were robots by attaching modules to them. That is to say, designing a specific complicated robot is replaced by placing functional modules on an object and environment for the tasks.

As an example for modules, Fig.1 shows a camera module which is for sensing positions. The camera module is consist of an image processing unit, radio communication unit, and a camera. Fig.2 shows a wheel module which provides ability of moving. The wheel module is equipped with a processing

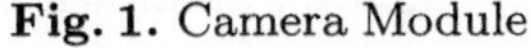

Fig. 1. Camera Module **Fig. 2.** Wheel Module

unit, a radio communication unit, PWS(Power Wheel Steering), and a battery. It will also have an adapter with a degree of freedom for attaching to an object. Additionally, the wheel module can measure its velocity of each wheel by encoders. A concrete usage of modules will be shown in next subsection.

2.2 Procedures for Using Modular Robot System

In this subsection, procedure for using modular robot system is described with an example of table transportation shown in Fig.3. The procedures are consist of 4 steps as follows.

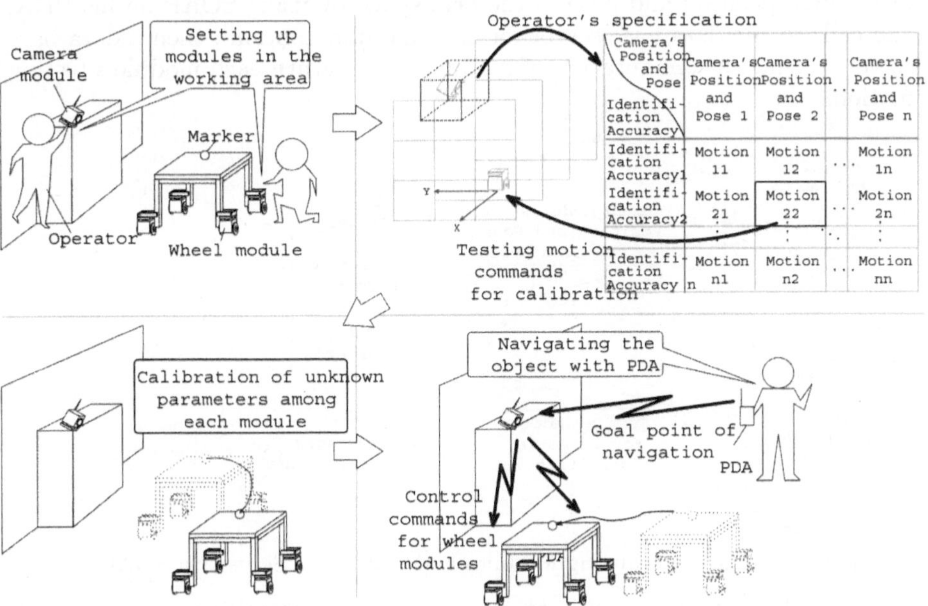

Fig. 3. Procedures of Using Modular Robot System in the Example of Object Transportation by Attaching Camera and Wheel Modules

1st: An operator sets up a system by attaching modules to an object and to the working area. In the example of Fig.3, wheel modules are attached to a table and a camera module is attached in the environment to observe the marker on the table.

2nd: The operator measures the rough position and pose of each module with eyes in the working area divided into grids which has the origin on the center point of a leader wheel module. For illustrating easily, the second step of Fig.3 shows the case which the operator decides the camera module's position and pose. Then the operator decides the parameter identification accuracy for executing task. Here, testing motion is defined. Testing motions are consist of motion commands for each module. By executing testing motion with actuators and observing their changes by sensors, unknown parameters are identified. Testing motions are calculated in advance for saving calculation time and are recorded in data-base. By deciding each modules' position, pose for certain identification accuracy, testing motions are transmitted to each module from data-base via the Internet.

Additionally, from the viewpoint of implementation, SOAP(Simple Object Access Protocol) is used to tapping into a data-base server. SOAP enables modules to execute remote procedure calls among remote hosts. The protocol is defined by XML, and is transmitted by HTTP. Therefore it is flexible to work on various platforms and communicate over fire walls on the Internet. Fig.4 shows the communication among PDA(Personal Digital Assistants), modules and a data-base. At first, the operator sends each modules' rough position and pose in the grid space by using SOAP on his PDA, and calls to the data-base server for each module' motion , then a data-base server recognizes configuration of modules, and return each module's testing motion.

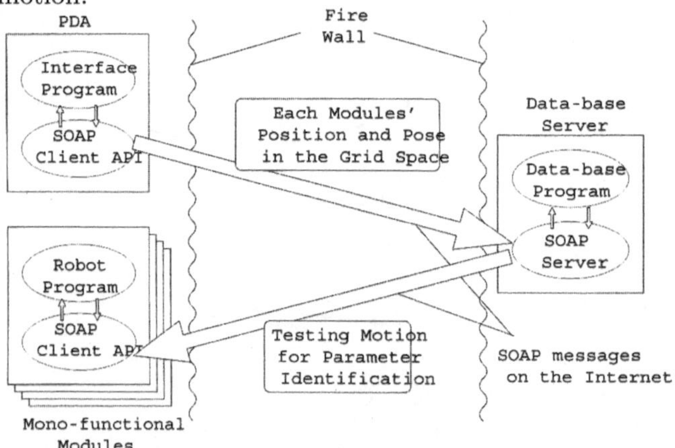

Fig. 4. Communications among Modules and Data-base Server via SOAP

3rd: The system executes testing motion by controlling actuator modules, and observes the changes of marker's position. By using odometry with marker's position data, the unknown parameters(positions and orientations) of the system are identified.

4th: The operator navigates the "robotic" table by pointing the goal position on the camera module's image captured in his PDA.

Testing motions of modules should be designed to reduce the influence of sensing errors for accurate identification. The concrete algorithm of designing the testing motion and parameter identification are described in the following section.

3 Automatic Parameter Identification

3.1 Outline of Algorithm

As it was mentioned in previous section, unknown parameters of our robot system are identified by observing testing motion of wheel modules by camera module. Generally, a great number of inputs is used to estimate parameters accurately in the system identification approach. However this traditional approach takes much time to identify, and it can not be applied to our robot system, because many wheel modules' motions as inputs lead parameter identification error. On the other hand, robot systems are required to be identified with expedition in certain occasions. Though the traditional approach estimates parameters accurately, we consider that to get parameters which satisfy the accuracy to support human tasks rapidly is much important. Therefore the traditional approach is not effective for practical use of a robot system, because it takes much time even if it identifies parameters accurately. From these reasons, we propose a parameter identification method which satisfies accuracy for tasks with the least inputs to save time. In this section, an automatic parameter identification method and motion design for the accurate identification for a modular robot system is described.

As a simple case, the system which is consist of a ceiling camera and a wheel module is shown in the left picture in Fig.5. The right figure shows the relationships between coordinate systems. The wheel module is on the ground, and global coordinates X-Y-Z is on the initial position and orientation of the wheel module, X_c-Y_c-Z_c is camera coordinates.

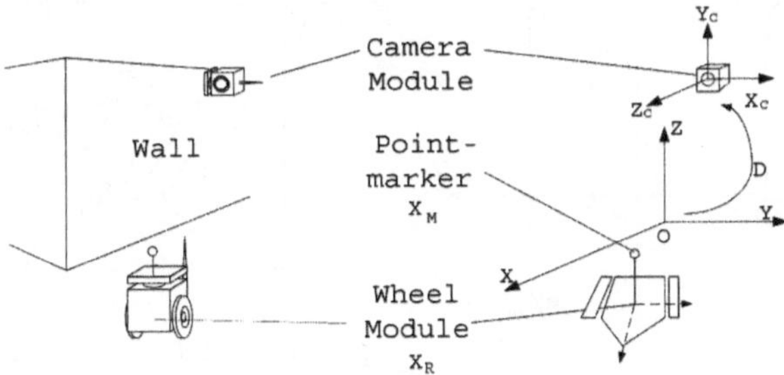

Fig. 5. Coordinate Systems of Whole System

A basic strategy of parameter identification is: at first, the whole system model is constructed with the knowledge of a pin-hole camera model and kinematics of wheel module. Then the equation of the model is linearized by assuming that the point-marker's position is measured at every sampling point when the wheel module moves. After that, standard deviation of parameter which is calculated from the sensing errors of the camera and the moving wheel module's velocity are evaluated. Finally, the nonlinear optimization method is applied.

Table 1 shows the notation and the status of each parameter. f,k_u, k_v,α, u_0, v_0 are camera intrinsic parameters. T is the tread of the wheel module. They have been calibrated in advance through experiments, because these parameters can be considered to be unchangeable when the system works. v_{r_i}, v_{l_i} are velocities of every wheel of the wheel module. They are measured by encoders. The rest are unknown parameters to be calibrated.

Table 1. Notation and Status of Parameters

parameter	*meaning*	*status*
f	Focal distance	known
k_u, k_v	Pixel size	known
α, u_0, v_0	Origin of u,v coordinates	known
T	Tread of wheel module	known
v_{r_i}, v_{l_i}	Velocity of wheels	known
$x_c, y_c, z_c, \phi, \theta, \psi$	Camera module's position and pose	unknown
x_M, y_M, z_M	Marker's relative position	unknown
$x_{R_0}, y_{R_0}, \theta_{R_0}$	Wheel module's initial position	unknown
$x_{R_i}, y_{R_i}, \theta_{R_i}$	Wheel module's position	unknown

3.2 Modeling Whole System

Here, the camera module's modeling and the wheel module's modeling are described. Then each module model is combined as a whole system's model.

F is camera intrinsic matrix which means conversion from U-V pixel coordinates of CCD plane to X_s-Y_s camera coordinates on the focal plane(Fig.6). **D** is rigid transformation matrix where ϕ, θ, ψ are roll, pitch, yaw, and where $C_{[\phi,\theta,\psi]}$ is $cos[\phi, \theta, \psi]$, $S_{[\phi,\theta,\psi]}$ is $sin[\phi, \theta, \psi]$, x_c, y_c, z_c are camera's position in the global coordinates.

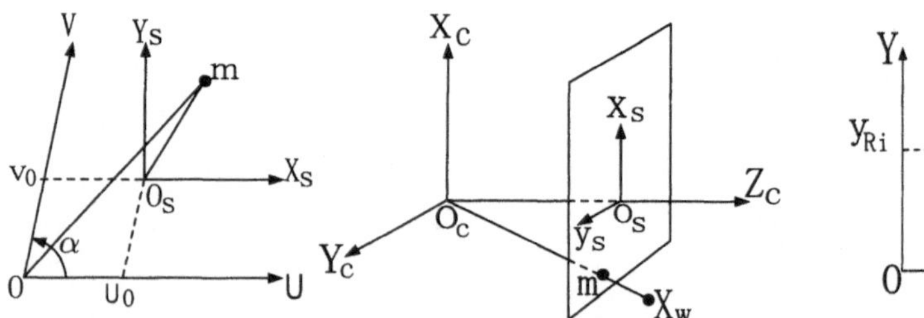

Fig. 6. Parameter Definitions of Camera Module

Fig. 7. Parameter Definitions of Wheel Module

$$\mathbf{F} = \begin{bmatrix} fk_u & -fk_u cot\alpha & u_0 & 0 \\ 0 & fk_v/sin\alpha & v_0 & 0 \\ 0 & 0 & 1 & 0 \end{bmatrix} \tag{1}$$

$$\mathbf{D} = \begin{bmatrix} C_\phi C_\theta & C_\phi S_\theta S_\psi - S_\phi C_\psi & C_\phi S_\theta C_\psi + S_\phi S_\psi & x_c \\ S_\phi C_\theta & S_\phi S_\theta S_\psi + C_\phi C_\psi & S_\phi S_\theta C_\psi - C_\phi S_\psi & y_c \\ -S_\theta & C_\theta S_\psi & C_\theta C_\psi & z_c \\ 0 & 0 & 0 & 1 \end{bmatrix} \tag{2}$$

Here, we define the position of the moving wheel module and a point-marker's position as follows

$$\mathbf{x}_{R_i} = \begin{bmatrix} x_{R_i} \\ y_{R_i} \\ 0 \\ 0 \end{bmatrix} = \begin{bmatrix} x_{R_{i-1}} \\ y_{R_{i-1}} \\ 0 \\ 0 \end{bmatrix} + \begin{bmatrix} \frac{v_{r_{i-1}}+v_{l_{i-1}}}{2} cos\theta_{R_i-1}\Delta t \\ \frac{v_{r_{i-1}}+v_{l_{i-1}}}{2} sin\theta_{R_i-1}\Delta t \\ 0 \\ 0 \end{bmatrix} \tag{3}$$

$$\mathbf{x}_M = \begin{bmatrix} x_M \; y_M \; z_M \end{bmatrix}^T \tag{4}$$

$$\mathbf{x'}_{M_i} = \begin{bmatrix} x_M cos\theta_{R_i} - y_M sin\theta_{R_i} \\ x_M sin\theta_{R_i} + y_M cos\theta_{R_i} \\ z_M \\ 1 \end{bmatrix} \tag{5}$$

where $1 \leq i \leq n$ and the orientation of the moving wheel module is

$$\theta_{R_i} = \theta_{R_{i-1}} + \frac{v_{r_{i-1}} - v_{l_{i-1}}}{T} \tag{6}$$

in the global coordinates(Fig.7). In addition, $\mathbf{x}_M$ is a relative position to the local coordinates which is at the center point of the tread. We denote the initial position of the wheel module and a point-marker as $\mathbf{x}_{R_0}$ where $n = 1$.

$$\mathbf{x}_{R_0} = \begin{bmatrix} x_{R_0} \; y_{R_0} \; 0 \; 0 \end{bmatrix}^T \tag{7}$$

Equation (1) to (6) yield the whole system model as follows

$$s \begin{bmatrix} u_i \\ v_i \\ 1 \end{bmatrix} = \mathbf{FD}(\mathbf{x'}_{Mi} + \mathbf{x}_{Ri}) \tag{8}$$

By dividing the first row and the second row by s which is calculated from the third row of equation(8), u_i,v_i can be calculated as $f_i(\mathbf{x})$,$g_i(\mathbf{x})$.

Here, $\mathbf{y}$ is positions on the camera screen of the point-marker which are measured by the camera module at each point when the wheel module moves.

$$\mathbf{y} = \begin{bmatrix} u_1 \; v_1 \; \dots \; u_n \; v_n \end{bmatrix}^T \in R^{2n} \tag{9}$$

It gives

$$\mathbf{y} = \mathbf{Ax} \tag{10}$$

where

$$\mathbf{x} = \begin{bmatrix} \mathbf{z} \; \mathbf{v} \end{bmatrix}^T \tag{11}$$

$$\mathbf{z} = [\phi \quad \theta \quad \psi \quad x_c \quad y_c \quad z_c \quad x_M \quad y_M \quad z_M \quad x_{R_0} \quad y_{R_0} \quad \theta_{R_0}]^T \tag{12}$$

$$\mathbf{v} = \begin{bmatrix} v_{r_1} \; v_{l_1} \; \dots \; v_{r_{n-1}} \; v_{l_{n-1}} \end{bmatrix}^T \in R^{2(n-1)} \tag{13}$$

$\mathbf{A}$ is Jacobian matrix by which equation(8) is linearized. $\mathbf{x}$ is consist of $\mathbf{z}$ which are the initial parameters and $\mathbf{v}$ which are velocity measured by encoders on each point.

3.3 Design of Wheel Module Motion

Accuracy of the parameters can be changed depending on the wheel module's velocities. To identify $\mathbf{z}$ accurately, it is necessary to design wheel module's motion which reduce influences of sensing and positioning errors on $\mathbf{z}$. Equation (10) leads to

$$\mathbf{y} = \mathbf{A}_1\mathbf{z} + \mathbf{A}_2\mathbf{v} \tag{14}$$

$\mathbf{A}_2$ can be represented as

$$\mathbf{A}_2 = \begin{bmatrix} a_{2,1,1} & \cdots & a_{2,1,2(n-1)} \\ \vdots & \ddots & \vdots \\ a_{2,2n,1} & \cdots & a_{2,2n,2(n-1)} \end{bmatrix} \in R^{2n\times 2(n-1)} \tag{15}$$

Equation (15) leads the error variance matrix of $(\mathbf{y} - \mathbf{A}_2\mathbf{v})$ as

$$\begin{aligned} \mathbf{\Sigma_{yv}} = diag.({\sigma_y}^2 + (\sum_{j=1}^{2(n-1)} {a_{2,1,j}}^2){\sigma_v}^2, \ldots, \\ {\sigma_y}^2 + (\sum_{j=1}^{2(n-1)} {a_{2,2n,j}}^2){\sigma_v}^2)) \in R^{2n\times 2n} \end{aligned} \tag{16}$$

where σ_y is the standard deviation of the camera module's sensing error, σ_v is that of the wheel module's measuring velocity. The error variance matrix of the parameters is estimated as follows

$$\mathbf{\Sigma_z} = ({\mathbf{A}_1}^T {\mathbf{\Sigma_{yv}}}^{-1}\mathbf{A}_1)^{-1} \tag{17}$$

Suppose $\mathbf{z}'$ is the parameters contain error. $\mathbf{z}'$ can be represented as a function h of initial parameters $\mathbf{z}$ and their errors $\mathbf{\Sigma_z}$, that is

$$\mathbf{z}' = h(\mathbf{z}, \mathbf{\Sigma_z}) \tag{18}$$

While the value $f_i(\mathbf{z},\mathbf{v})$,$g_i(\mathbf{z},\mathbf{v})$ is a function of parameters, to make error variance of parameters smaller,

$$\sum_{i=1}^{n}\{(f_i(\mathbf{z},\mathbf{v}) - f_i(\mathbf{z}',\mathbf{v}))^2 + (g_i(\mathbf{z},\mathbf{v}) - g_i(\mathbf{z}',\mathbf{v}))^2\} \rightarrow min \tag{19}$$

is evaluated to obtain much precise sensing.

3.4 Parameter Identification

Testing motion to minimize identification error was described in the previous subsection. After executing the testing motion of the wheel module, parameters are identified. Unknown parameter $\mathbf{z}$ is estimated by applying the steepest descent method which minimizes the following performance index **PI**.

$$\mathbf{PI} = \sum_{i=1}^{n} \{(u_i - f_i(\mathbf{x}))^2 + (v_i - g_i(\mathbf{x}))^2\} \rightarrow min \tag{20}$$

where u_i,v_i is the marker's position captured by the camera module, and $f_i(\mathbf{x})$,$g_i(\mathbf{x})$ is an estimated marker's position by the whole system model.

4 Parameter Identification Experiment

A basic experimental result are described in this section. The experimental configuration was set up as Fig.8. A wheel module with a marker was put on the origin of the world coordinates in x-direction. And a camera module was attached on the ceiling facing to the wheel module. The camera module's position and orientation are approximately measured with eyes. Finally, we used our algorithm to identify these inaccurate parameters.

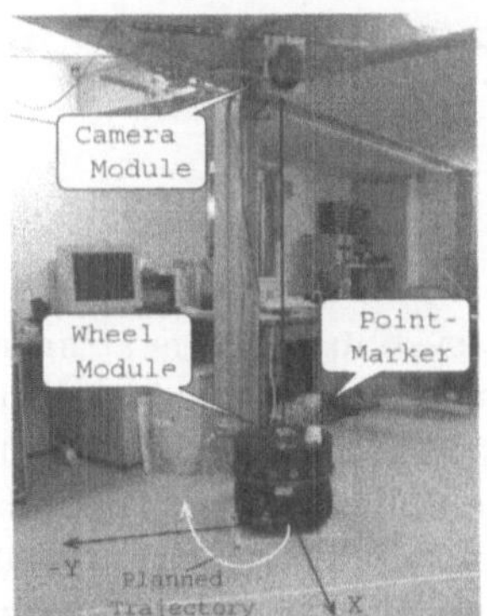

Fig. 8. Experimental Configuration

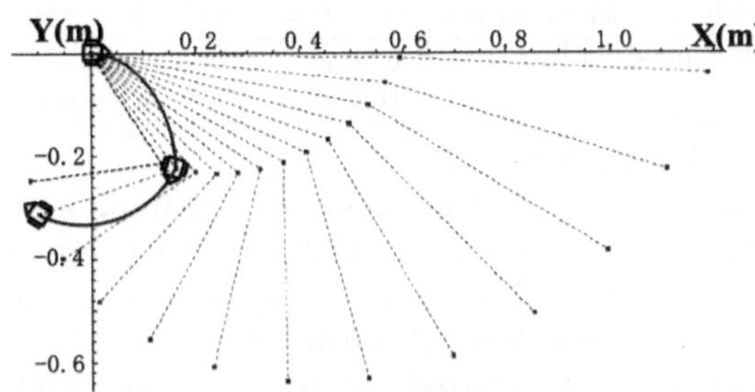

Fig. 9. Testing Motions of Wheel Module

Fig.9 shows a simulation for deciding wheel module's optimized motion to identify parameters accurately. The wheel module starts at the origin, and it moves via 2 points. The motions are assumed as arcs with the left wheel uniform velocity at 0.6(m/s), and with changing the right wheel velocity from 0.07(m/s) to 0.6(m/s). The reason of the assumption is to decrease the influence of position estimation error to accuracy of identification. The thick line is the practical track taken by the wheel module in the optimized motion. Here, let us explain why at that time the wheel module's motion is considered as the best. The moving distance of the wheel module has the influence on the error. The longer the wheel module moves, the larger the locating error becomes. Therefore, in order to make the error smaller, the wheel module should not move too farther. On the other hand, when we make the moving distance shorter, the sensing will be difficult and inaccurate, and it makes the sensing error bigger. Therefore, from these viewpoints, the most important thing is to find out the best combination point to reduce both locating and sensing error. As a result, the optimized motion is noticed as the thick line shows. With this data, parameters can be identified.

An experiment was made with the simulation result. Using the testing motion in Fig.9, the marker was and captured by the camera. Then a gravity point of the marker on each image is estimated, and the parameters are identified with eye measurement value ($x_c = 0.0$, $y_c = 0.0$, $z_c = 1.80(m)$, $\phi = 90.0$, $\theta = 0.0$, $\psi = 180(deg)$) as initial value. The identification result is shown in Table 2. The identification leads the likelihood length to the measured value by using the approximate original data. There are two reasons of the error of θ. One is the locating error of the operator. When the camera module is placed on the ceiling, the base of the module can not be attached accurately. Moreover, the camera module can pan and tilt, the mechanism influences the error. The other is the positioning error by slippage which is occurred when the wheel module execute testing motion. By moving with lower velocity, the error will be smaller.

Table 2. Result of Parameter Identification

Identified Value with Our Method	$x_c = 2.70 \times 10^{-2}$, $y_c = 4.30 \times 10^{-2}$, $z_c = 1.70(m)$, $\phi = 91.3$, $\theta = -20.5$, $\psi = 180(deg)$
Measured Value	$x_c = 0.0$, $y_c = 0.0$, $z_c = 1.75(m)$, $\phi = 90.0$, $\theta = 0.0$, $\psi = 180(deg)$

5 Conclusion

This paper proposes a concept of a modular robot system which is attached distributively to environment. For realizing the concept, a parameter identification method is presented to obtain relationships among modules. In the method, deviations of sensing errors of the camera and the wheel module's velocity are evaluated. To minimize the influence, a set of velocities is obtained. It means the wheel module's motion to acquire accurate parameters. The effectiveness is shown by the experiment.

References

1. Hashimoto, M. et al. (1995) Dynamic Object Transportation Control Method by Multiple Robots. *Journal of the Robotics Society of Japan.* **13-6**. 152–159 (in Japanese)
2. Burke, T. et al. (1993) Kinematics for Modular Wheeled Mobile Robots. *Proc. of Int. Conf. on Intelligent Robot and Systems.* **2**. 1279–1286
3. Sato, T. et al. (1996) Robotic room: Symbiosis with human through behavior media. *Robotics and Autonomous System.* **18**. 185–194
4. Fukuda, T. et al. (1996) The Meaning of Functional Reconfiguration under the Dynamic Environment and System Behavior. *Proc. of Int. Conf. on Intelligent Robot and Systems.* **3**. 1676–1683
5. Lee, B.J. et al. (2001) New Architecture for Mobile Robots in Home Network Environment Using Jini. *Proc. of Int. Conf. on Robotics and Automation.* **1**. 471–476

Distributed replication algorithms for self-reconfiguring modular robots

Zack Butler[1], Satoshi Murata[2], and Daniela Rus[1]

[1] Dartmouth College, Hanover NH USA
[2] Tokyo Institute of Technology, Tokyo Japan

Abstract. Self-reconfiguring modular robots have the ability to reform themselves into a wide variety of different shapes to accomplish their tasks. In addition, a group of self-reconfiguring modules can divide up into several smaller groups to perform operations (such as exploration) in parallel. In either instance, due to the large number of independent modules in the system, distributed algorithms are highly desirable. In this paper, we describe a set of homogeneous distributed algorithms for self-reconfiguring modular robots that allow division and locomotion in two and three dimensional systems as well as recombination in two dimensions. The algorithms are written in a rule-based style inspired by cellular automata, and allow for the development of correctness analyses, which are also presented here.

1 Introduction

Self-reconfigurable robots are modular robots that can morph into different shapes without outside assistance. The ability to change shape allows these robots to adaptively execute a variety of different tasks, such as locomotion through small passages, climbing tall obstacles, and moving while supporting large payloads. Self-reconfiguring robots can also function as parallel distributed machines. That is, in almost all existing systems, each module has a processor on-board, and to take best advantage of the modularity, the algorithmic processing as well as module control should take place in a distributed fashion. This also allows the system to be more robust to failures of individual modules and communications, and supports the partition of the robot. Several self-reconfigurable systems have been designed and built [4,8,9,15,16] and centralized algorithms have been proposed for most [3,6,14], but purely distributed algorithms have been less explored [5,12,13,17].

In this paper we explore a novel feature of self-reconfiguring robots: *self-replication.* We consider self-replication ability in which a large robot can divide itself in several independent smaller robots with the same basic functionality (but not identical size). For example, a system consisting of 100 modules could function as one large robot or 10 smaller robots each consisting of 10 modules, or any number of other configurations. Self-replicating robots are useful in tasks where the overall effectiveness and task completion time is improved by parallelism, such as distributed surveillance or exploration. Although self-replication has been studied extensively in software systems [11], it has been less examined in hardware systems such as robots [10].

In this paper we address the vision of self-replicating robots by developing distributed algorithms for dividing a self-reconfiguring robot into smaller independent robots, and recombining several smaller robots into a larger robot. The algorithms we propose are homogeneous and purely distributed, in that each module runs the same algorithm (with no central controller) and uses only local information to determine its actions. This is important here, as groups of different sizes will be created, and the algorithms must work correctly regardless of the overall group size.

We also describe simulations of these algorithms within a locomotion task, which can lead to application in more general domains. Consider the tasks displayed in Fig. 1. The left image shows a simulation snapshot of a system exploring a maze-like environment, with groups splitting up to more efficiently cover the maze. Here, the number of groups could be increased as the complexity of the maze increases. The right image shows a simulation snapshot from a system in which several groups are covering an open piece of terrain. In this case, a larger number of groups is desirable to speed up exploration, but small groups cannot navigate large obstacles. Therefore, the system can use the ability to dynamically change group size and number to more efficiently conduct exploration.

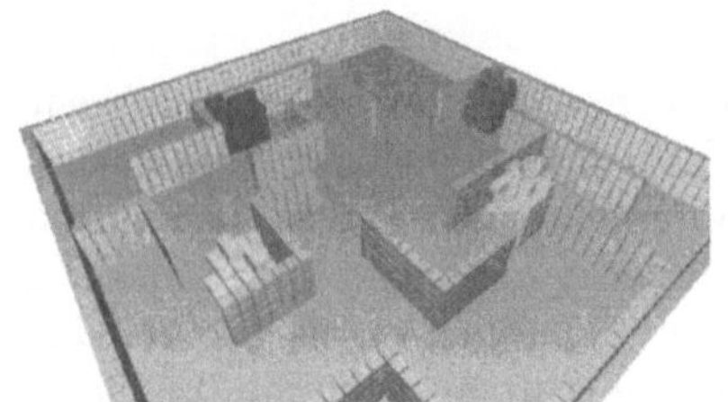

Fig. 1. (Left) Four groups of modules traversing a maze. (Right) Groups of modules exploring rough terrain.

More specifically, we develop distributed generic algorithms for dynamically changing the size of the robot, and the number of robots. By "generic" we mean algorithms that are independent of the specific architecture of the basic robot module. Our algorithms can be instantiated to any self-reconfiguring robot system where an individual module has the ability to move linearly and make convex and concave transitions on a substrate of identical modules[1]. We use the abstract geometric model first described in [1], where each module is represented as a cube. A set of geometric rules that is evaluated by each module determines the movement behavior of each cube, and thus the overall behavior of the robot. This cellular-automata inspired approach leads to completely distributed and homogeneous algorithms that can be proven correct and be instantiated on a variety of hardware systems

[1] Most existing self-reconfiguring systems, e.g. [5,7,9,15], have these abilities.

Algorithm 1 Instantiation of D_1

```
while (1) do
  flag = 0 in all modules
  while (Any flag == 0) do
    Choose unflagged module at random
    Set flag = 1 for current module
    while (Not all rules tried) do
      Evaluate random rule for current module
      If rule applies, use it and break
```

for specific tasks. In our previous work [2] we present rule sets, correctness results, and instantiations for the locomotion task. In this paper we develop algorithms and correctness results for the self-replication task.

2 2D algorithm

In this section we describe a distributed algorithm for division of two-dimensional systems, using the geometric model described in Sec. 1. We develop rule sets that implement several tasks: division of a group into two halves, locomotion in either +x or -x, and recombination of two groups that meet. These rule sets are then composed to accomplish correct combinations of redivision and recombination.

2.1 Locomotion

In [2] we presented rule sets for the locomotion task in which a modular robot can move over obstacles. The robot can be viewed as a collection of cells in a particular type of cellular automata. The evaluation of this distributed system is not performed in the traditional way, but instead models the natural delays in actuating physical systems. The evaluation is done with the D_1 activation model defined in [2]. In this model, the cells are evaluated with some random asynchrony subject to the constraint that a cell can delay activation at most one cycle relative to another cell. This is instantiated in simulation as shown in Algorithm 1. In [2] we also described locomotion rules that can be evaluated asynchronously, and we are working at integrating those and other more general locomotion rules with division rules.

The basic idea of the locomotion rules is for the modules to start in a rectangular group and move like a tank tread – those in the back column move over the top of the group and form the next column in front. The local rules that generate this motion are presented in Fig. 3. These rules are presented as conditions on the neighborhood of the active module, which is denoted by the starred box in each rule. The action of the rule is either to move the module (in this figure) or set a variable within the module (during division). In all cases, the diagrams follow the legend presented in Fig. 2.

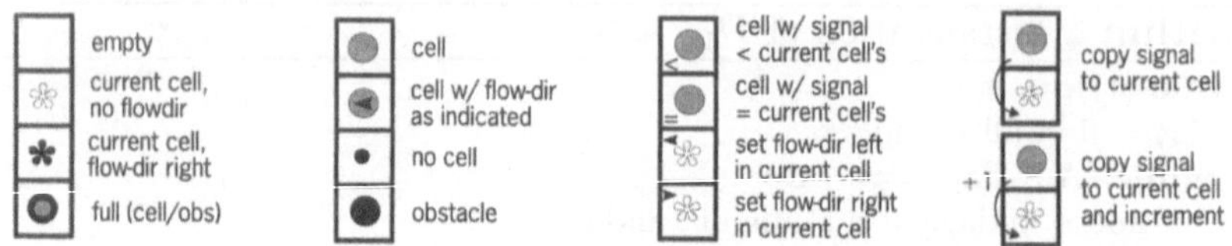

Fig. 2. Legend used for all rule figures, showing the meaning of the central symbols (left two columns) and other predicates and postconditions (right columns).

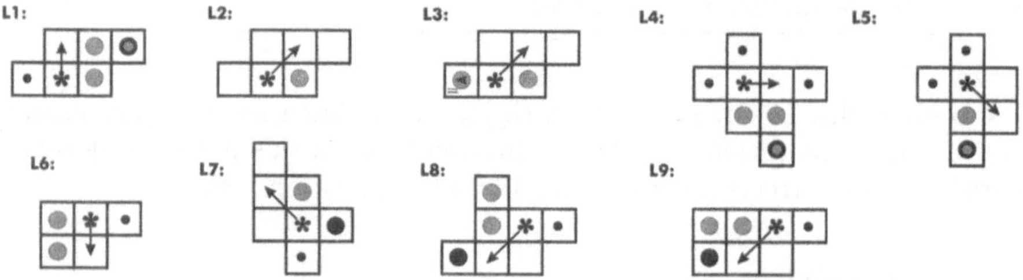

Fig. 3. Rules used for locomotion (group motion will be to the right). Each rule corresponds to a particular local geometry and results in module motion. The arrow in each rule represents the direction of motion of the module upon application.

Each locomotion rule corresponds to a possible local geometry for a module and defines the next move for the module. For example, rule L1 makes a module in the back column of the group move upward if there is not another cell above it. Rules L2-L6 will move that module along the top of the group and down the front side[2].

2.2 Division / self-replication

We would like to empower modular robots with the ability to split up when they need to explore different directions, or when the surveillance task is enhanced by parallelism. The *division* operation splits up a robot into two smaller groups. This kind of robotic self-replication can be controlled in a distributed fashion, using the same approach as for locomotion. The key is to define a set of rules that select which robot modules belong to which smaller group by the local neighborhood only. The end result will be local control that is guaranteed to split an undifferentiated modular robot into differentiated groups.

The basic division control algorithm splits a robot with no holes into two smaller robots of equal size. It requires a state variable (*flow-dir*) corresponding to the moving direction. Fig. 4 shows snapshots from a simulation that

[2] The locomotion rule set in Figure 3 includes rule L3 which is not part of the rule set described in [2]. The addition of L3 does not change the correctness of the locomotion rules but it facilitates the transition between the control for dividing a robot into two smaller robots that can move in opposite directions.

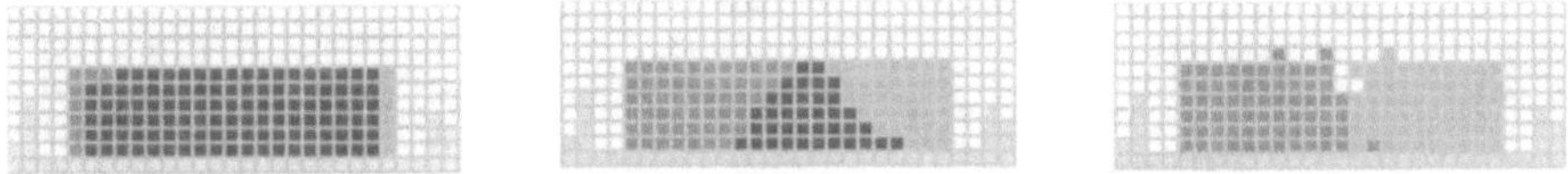

Fig. 4. Simulation of initial differentiation, showing the progress of group selection and the start of locomotion. Light gray modules (in this and following figures) have *flow-dir* = 1, dark gray modules *flow-dir* = -1 and black modules *flow-dir* = $\emptyset$.

implements self-replication. All robot modules begin in an undifferentiated state (*flow-dir* = $\emptyset$) and divide into two groups: left-facing (*flow-dir* =-1) and right-facing (*flow-dir* =1). The module differentiation is done by propagating of an integer variable called a *signal* through the system. As the modules finish dividing (which they can detect by comparing signals in a local neighborhood), they begin to locomote using the rules of Fig. 3.

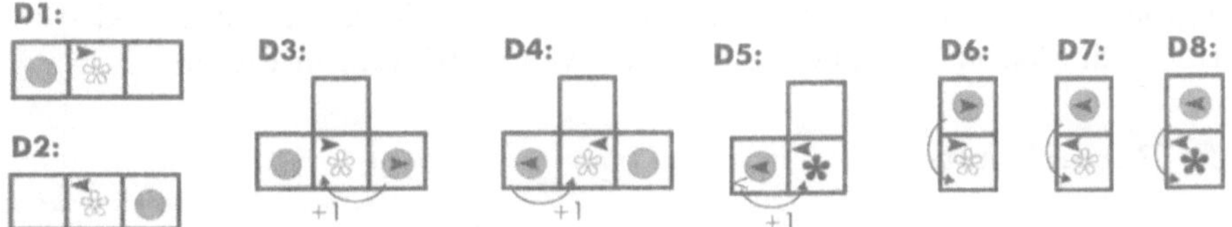

Fig. 5. Rules used for initial differentiation. Rules D1 and D2 set the direction of cells at the sides of the group, rules D3-D5 set direction for cells on the top row, and D6-D8 copy the direction to cells in the interior of the group.

At the outset, all modules are undifferentiated, and use the rules of Fig. 5 to divide. The modules on the edges choose to move toward empty space (using rule D1 or D2 and setting *flow-dir* left or right as appropriate) and set their signal to zero. The modules on the top row copy their neighbor's direction (as in the middle picture of Fig. 4) and signal level, and increment the signal (rules D3 and D4). In addition, after setting *flow-dir*, a module will wait one cycle before moving, to ensure that its *flow-dir* is seen by the module below. Finally, when the two groups meet along the top row, the modules may need to change *flow-dir* until the two groups are about the same size (i.e. the modules at the boundary have about the same signal value). All modules not in the top row will copy their direction from the module above using D6-D8, resulting in a configuration like that in the right-hand picture of Fig. 4.

To split a group in motion, it is sufficient for one module at the front edge of the group to reset its *flow-dir* to $\emptyset$ (based on a sensor signal, higher-level directive or some other trigger). This is a local operation and requires no collective decision from the modules. The initiating module simply resets and waits for a number of cycles before evaluating its rules[3], to ensure that all

[3] The waiting time is determined by the size of the group, which the module estimates based on the length of its last traverse.

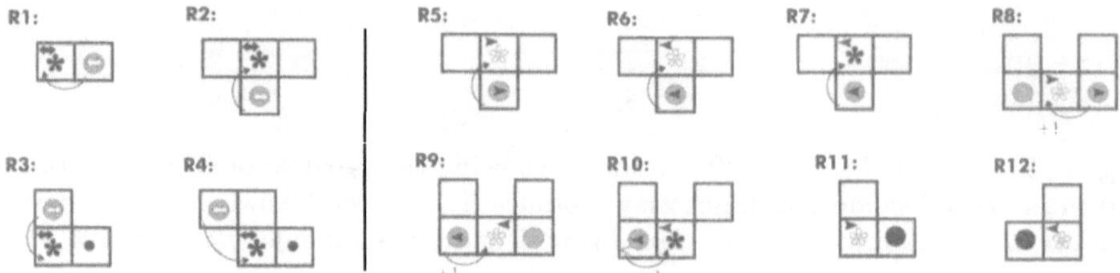

Fig. 6. Rules used for resetting (R1-R4, in which a module unsets its direction in response to an undirected neighbor) and dividing non-rectangular shapes (R5-R12, in which a modules copies its direction from a neighbor).

modules reset. After a module becomes undifferentiated, rules R1-R4 of Fig. 6 ensure that all the modules behind it will be undifferentiated. Additionally, the modules wait for all to reset by copying the waiting time from the neighbor ahead. At this point, the modules may choose a new moving direction. Since the group may be in a non-rectangular shape, additional rules are needed to perform differentiation. These rules are given as R5-R12 in Fig. 6.

2.3 Recombination

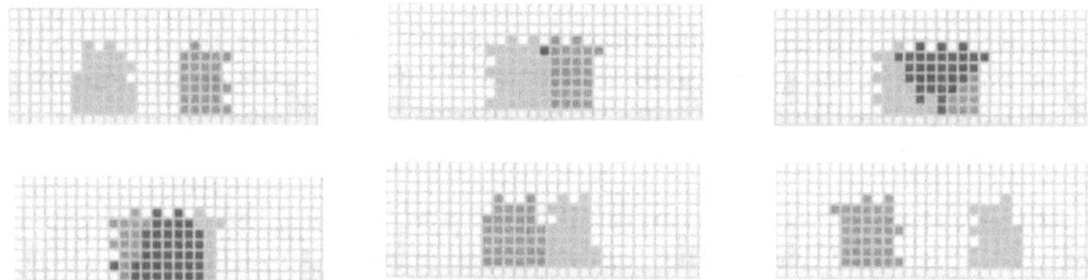

Fig. 7. Simulation of recombination and redivision.

Two robot groups can also be merged to generate a bigger robot. The recombination operation merges two groups that meet each other face-to-face into a single undifferentiated group. This group may also divide again, as shown in Fig. 7. Recombination can be formulated with local rules much like the division operation. Each module requires a counter we call *stuck*. A module increments its *stuck* if it has an opposite-directed module in its and no motion rule can be applied. The *stuck* counter allows groups to wait until they have collected with no holes in the middle[4]. Once *stuck* reaches a small threshold in a module at the top of the group, the module sets its *flow-dir* to $\emptyset$, as shown in Fig. 7(top, middle). This is sufficient to reset both groups as shown in Fig. 7(top, right) since the resetting module is "in front of" both groups. The group can then redivide and resume locomotion, as shown in Fig. 7(bottom).

[4] Recall that the division rules assume a group without holes.

2.4 Simulation

A graphical simulation of this algorithm has been developed, images from which are shown in Fig. 8. In this simulation, the evaluation loop of Algorithm 1 is instantiated, and a large group of modules can be seen to divide, first into two groups, then into four, each of which move through the environment, demonstrating self-replication.

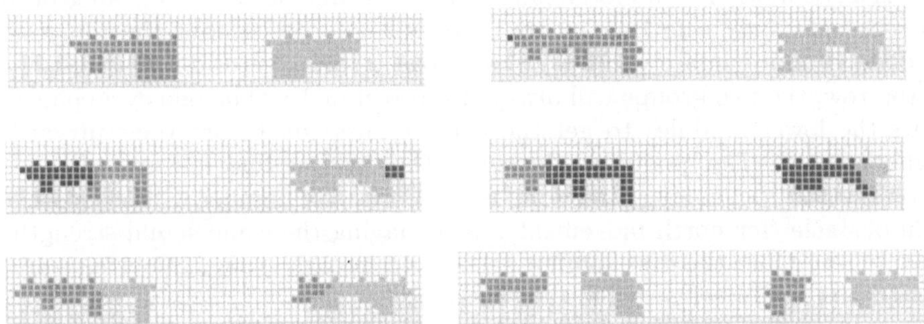

Fig. 8. Simulation of locomotion and recursive split. Two groups locomote away from each other (top), reset (middle) and each divide into two (bottom left), after which all four groups resume locomotion.

2.5 Analysis

One of the advantages of working with the simplified geometry and rule types of the current system is that correctness analysis can be produced in a straightforward fashion. In this instance, based on the rules in Figs. 3, 5 and 6, we can show the following Propositions:

1. Locomotion proceeds over obstacles of (groupheight-2).
2. Initial split produces two locomoting groups.
3. Recursive split creates two locomoting groups with probability ≈ 1.
4. Recombination will cause reset and redivision of all modules.

In this discussion, we describe system geometry using compass directions (in which North is up and East to the right, relative to the rule figures). We also define *regular locomotion* as the case in which a contiguous group of modules all have the same (non-null) *flow-dir*.

Proposition 1 is based on the proof of the original locomotion rules as presented in [2]. That proof shows that the rules, when applied to an initially rectangular group, produce forward motion without disconnecting the group or creating deadlock. The locomotion rules presented in Fig. 3 are the same with one extra rule (L3) and additional conditions to two other rules (L4 and L5). The extra rule applies only in the case where two groups are touching, and so will not fire during regular locomotion. As for the extra conditions in rules L4 and L5, as long as the top two rows have no obstacles, these are redundant in regular locomotion (they are there to assist redivision).

Proposition 2 can be shown as follows: In the first evaluation cycle, all modules on the two sides will set the correct direction. Then, over several more cycles, the top row will divide itself in close to half using rules D1 through D5. The modules at each side cannot move, since they are all at the "front" of their respective groups. Therefore, the only motion that can take place is a NE motion by a module in the top row, and since the overall group is a rectangle, only rule L3 can fire, and will only do so once the top row has divided approximately in half. Since *flow-dir* propagates downward from the top row, the two groups will always form rectangles. The one cycle pause allows the lower modules to get the correct direction before their upward neighbor moves. The modules adjacent to the dividing line will move N and then NE without changing direction, as their opposite number will be treated as an obstacle (for north movement), or as having the same signal strength (for rule L3). Once the first column has moved in each group, the groups are isolated and each begins regular locomotion.

We must also show that none of the other rules can affect this process. In particular, the reset rules have the potential to unset *flow-dir* in modules with undifferentiated neighbors. However, a module will only reset if the module above it is reset, or if it is on top and the module in front of it resets. Since the propagation of the division happens in these same directions, once a module is set, it cannot be reset during initial division.

Proposition 3 is shown in two parts. First we show that the reset of one module does in fact reset all modules. We then show that division of non-rectangular groups proceeds correctly if all modules start at the same time. Finally, we discuss under what conditions the division fails if the modules do not begin the process simultaneously.

The reset itself will always complete, as follows: Since the reset activates after a SE move, there cannot be any modules to the east of the resetter, and there will always be a module directly to its west. The module to the west will reset on its next activation, and the reset will propagate back and down. Modules in the same column as the original resetter may not be adjacent to it, in which case the topmost one will use rule R4 to propagate the reset. Then, because each module waits after resetting, there will always be an undifferentiated module in each row to continue the propagation.

Division of a non-rectangle begins as for a rectangle with the modules on each edge choosing *flow-dir*. Note that this may now include moving modules sticking off the two sides of the group. The modules on the top row will all identify themselves as such by noting an empty space either above them or above and to the side (since two moving modules cannot be adjacent above them). These modules then determine the division line for the group. Note in addition that since the signal strength is propagated down with the direction, the modules adjacent to the division line will have the same strength as each other, even though the various rows may have different widths and center locations. Therefore, the modules along the division line will move N and NE without changing direction or interfering with each other.

Because of the way modules wait after the reset, it is possible that nearby modules will restart rule evaluation at different times. Correctness of regular locomotion relies on the D_1 activation model to prevent disconnection at the NE corner, but the D_1 model is violated when one module restarts several cycles before another. However, this failure requires either a particular order of activation or a two-cycle difference in the waiting times of neighbors. These are both very unlikely occurrences, and in fact we have yet to see such a disconnection in several hundred trials.

Proposition 4 is shown as follows: First of all, for a module to have its opposite number directly ahead of itself, there must be two groups meeting face-to-face. Note that this module then cannot move anywhere except S. It will move S, however, if there is space to do so. If there is any space in its column, note that this space will be filled in at most (height-1) cycles. Therefore, since this module waits (height) cycles before resetting, it will not reset with a hole underneath it. Then, note that it will necessarily have one module of each *flow-dir* on either side of it. Each will see that module as being in front and undifferentiated, so they will each reset, and so on. Once the reset reaches the edge of the top row, the corner modules will wait (so that all modules on the edges reset), then choose the new *flow-dir* which propagates inward and downward.

3 3D algorithm

Based on the insights of the planar division and locomotion algorithms, we have also extended our algorithms to three-dimensional systems. In this work, we model each module of a three-dimensional system as a cube, and develop similar geometry-based rules as in two dimensions. We have implemented division of one large group into four smaller groups (using two signals rather than one) and locomotion in the plane (via straight line motion in cardinal directions and 90° turns) in a single rule set. This rule set consists of 36 rules (20 for division, 10 for straight-line motion, and 6 for turning) and has also been implemented in simulation, pictures from which are shown in Fig. 9.

The division of a 3-D group is performed much the same as in 2-D. The modules on each XY corner choose their direction first, as seen in the top left of Fig. 9 and set both signals (*across* and *back*) to zero. Their directions are copied (a) back along the edge opposite the direction of motion (incrementing the *back* signal) and (b) across the front edge (perpendicular to direction of motion, incrementing the *across* signal). The signals are again used to ensure that the groups each end up about the same size and approximately square (in XY cross-section), as can be seen in the top right of Fig. 9. Interior modules

Fig. 9. 3-D simulation: In the top row, differentiation begins from the corners and filters into the center, at which point four groups locomote away from the center. In the bottom row, a group moving up and to the right stops, changes direction, and moves away heading up and to the left.

on the top (Z) layer obtain their direction when two neighbors have the same direction. Modules with a +Z neighbor copy direction from that neighbor.

The locomotion uses all of the same rules as in 2-D, with a change to the new NE motion rule (rule L3). This is due to the use of two signals rather than one — since adjacent groups have perpendicular directions of travel, one group's *back* should be the same as the other's *across* to ensure that they are the same size. In addition, a third NE rule is introduced that applies to modules that are next to two different groups (i.e. in the corner of their group). Although the locomotion rules for the 3-D system are still planar in nature (each group locomotes as a set of planar layers), additional conditions can be added to the NE motion rules to ensure that the layers stay connected.

Finally, a new set of rules has been developed that allows a group of modules to reform a prismatic shape and turn 90°, provided they are on flat ground. A new binary state, *halt*, is used to signal the group that it is time to stop locomotion. This signal is generated by a module moving to the front of the group, like the reset signal in 2-D, and is propagated through the system. The halt prevents the start of the back-to-front locomotion process (although modules in motion can continue, so that a prism is once again formed). Once a module is halted, it can discern from its neighborhood whether locomotion has completed, and if so, selects a new *flow-dir* and turns off *halt*. Since the group is prismatic, only the modules at the back can restart locomotion, and do so in the new common *flow-dir*.

3.1 Analysis

In 3-D, we can show that a prismatic block of modules (with one even-length X or Y side) will split into four prismatic groups, each successfully locomoting in a different direction. This proof is carried out much the same as in 2-D, in that we show that the groups created by division are prismatic (cf. rectangular in 2-D) and locomotion begins only at the group boundaries, and only once those boundaries have reached equilibrium. This proof also relies on the rules' checking of signal levels in neighboring modules, although here two neighbors and two signal levels must be checked before initiating locomotion.

We have also developed a proof of correctness for the turning rule set, which demonstrates that a group of modules on flat ground will necessarily reform a prism upon a proper halt signal, and will turn to a new direction and begin locomotion without leaving any modules behind.

4 Conclusion

In this paper, we have presented algorithms for a generic type of self-reconfiguring robot that allow division, locomotion and recombination in a single distributed program. These algorithms can form the basis for more general exploration algorithms by such systems, in which the system of modules could organize into different groupings to best suit the environment. In addition, as the locomotion algorithms for generic geometry have been instantiated on to different hardware platforms, it is our intent that these algorithms can be likewise instantiated. In the near future, we are working toward general 3-D locomotion along arbitrary trajectories and over arbitrary terrains, as well as 3-D recombination algorithms, to allow for more functional exploration.

References

1. Z. Butler, K. Kotay, D. Rus, and K. Tomita. Cellular automata for decentralized control of self-reconfigurable robots. In *ICRA 2001 Workshop on Modular Self-Reconfigurable Robots*, 2001.
2. Z. Butler, K. Kotay, D. Rus, and K. Tomita. Generic decentralized control for a class of self-reconfigurable robots. In *Proc of IEEE ICRA*, 2002.
3. C.-H. Chiang and G. Chirikjian. Modular robot motion planning using similarity metrics. *Autonomous Robots*, 10(1):91–106, 2001.
4. T. Fukuda and Y. Kawakuchi. Cellular robotic system (CEBOT) as one of the realization of self-organizing intelligent universal manipulator. In *Proc. of IEEE ICRA*, pages 662–7, 1990.
5. K. Hosokawa, T. Tsujimori, T. Fujii, H. Kaetsu, H. Asama, Y. Koruda, and I. Endo. Self-organizing collective robots with morphogenesis in a vertical plane. In *Proc. of IEEE ICRA*, pages 2858–63, 1998.
6. K. Kotay and D. Rus. Locomotion versatility through self-reconfiguration. *Robotics and Autonomous Systems*, 26:217–32, 1999.
7. S. Murata, H. Kurokawa, E. Yoshida, K. Tomita, and S. Kokaji. A 3-D self-reconfigurable structure. In *Proc. of IEEE ICRA*, pages 432–9, May 1998.
8. S. Murata, E. Yoshida, K. Tomita, H. Kurokawa, A. Kamimura, and S. Kokaji. Hardware design of modular robotic system. In *Proc. of the Int'l Conf. on Intelligent Robots and Systems*, pages 2210–7, 2000.

9. A. Pamecha, C-J. Chiang, D. Stein, and G. Chirikjian. Design and implementation of metamorphic robots. In *Proc. of the 1996 ASME Design Engineering Technical Conference and Computers in Engineering Conference*, 1996.
10. L. Penrose. Self-reproducing machines. *Scientific American*, 200(6):105–14, 1959.
11. M. Sipper. Fifty years of research on self-replication: An overview. *Artificial Life*, 4(3):237–57, 1998.
12. K. Stoy, W.-M. Shen, and P. Will. Global locomotion from local interaction in self-reconfigurable robots. In *Proc. of IAS-7*, 2002.
13. K. Tomita, S. Murata, H. Kurokawa, E. Yoshida, and S. Kokaji. Self-assembly and self-repair method for a distributed mechanical system. *IEEE Trans. on Robotics and Automation*, 15(6):1035–45, Dec. 1999.
14. C. Ünsal, H. Kiliççöte, and P. Khosla. A modular self-reconfigurable bipartite robotic system: Implementation and motion planning. *Autonomous Robots*, 10(1):23–40, 2001.
15. Cem Ünsal and Pradeep Khosla. Mechatronic design of a modular self-reconfiguring robotic system. In *Proc. of IEEE ICRA*, pages 1742–7, 2000.
16. M. Yim, D. Duff, and K. Roufas. PolyBot: a modular reconfigurable robot. In *Proc. of IEEE ICRA*, 2000.
17. M. Yim, Y. Zhang, J. Lamping, and E. Mao. Distributed control for 3D shape metamorphosis. *Autonomous Robots*, 10(1):41–56, 2001.

Chapter 3
Communication and Cooperation

Communication Mechanism in a Distributed System of Mobile Robots

Andrzej Kasiński and Piotr Skrzypczyński

Technical University of Poznań, Department of Control, Robotics, and Computer Science ; ul. Piotrowo 3A, PL-60-965 Poznań, Poland
e-mail {akas,ps}@ar-kari.put.poznan.pl

Abstract. This paper addresses problems of communication and information sharing in the system of heterogenous mobile robots and stationary sensors. We analyze information sharing between robots and sensors treated as agents, and propose a framework based on the Contract Net Protocol. We define the software architecture for the communication system. This architecture, based on an object-oriented library, provides agents with unified communication interfaces.

1 Introduction

Control of co-operative tasks of the team of mobile robots involves complex information processing activity requiring intensive communication. The communication traffic is generated by data sharing, task allocation and co-ordination. While designing a communication system one must account for the properties of the physical layer, and for the logical nature of the activity which is to be supported.

In the article, information sharing mechanisms in a group of autonomous mobile robots and stationary sensors are considered. These mechanisms are implemented in a dedicated communication system with the Contract Net Protocol adopted for negotiation. The architecture of the communication system is described and protocols are specified. Results of tests validating the correctness of the design are reported. Performance comparison is provided.

2 System architecture

The considered collective perception and world modelling task is performed by *robot agents* (**RA**), navigating in the laboratory. The robots move along pre-planned routes. Information gathered from sensors of the robot, from other robots (on request), and from stationary devices, is used to construct or to update the internal world model of the robot. When the obstacle is encountered, the robot is able to detect it and to make a necessary detour by means of reactive navigation. Each robot uses its odometry and the on-board sensors (laser scanner) to determine its own position and orientation. When the localization uncertainty exceeds the acceptable value, the robot asks for the positioning service from the external perception agents.

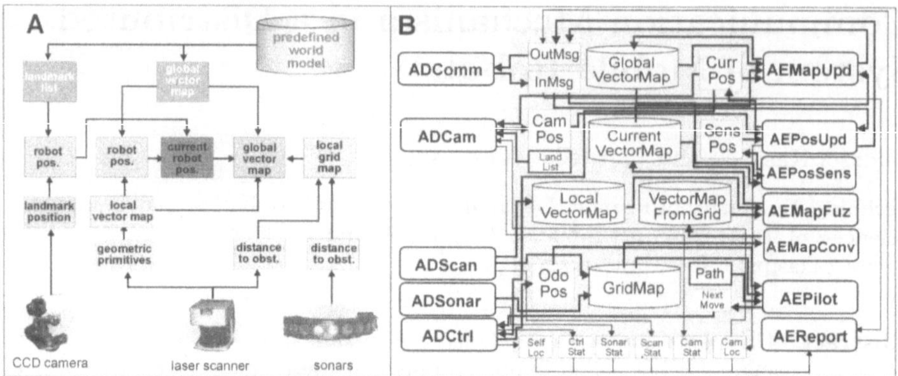

Fig. 1. The perception network (A) and the blackboard architecture of the mobile robot (B)

Perception agents (**PA**) compete for serving navigation data to robots. Perception agents have been defined as information sources, consisting of the data source (a physical sensor) and of some data-processing methods. These methods are used to extract the desired information out of raw sensory data (e.g. a camera image), and to produce the results accompanied with the calculated characteristics of measurement uncertainty.

Operator agent (**OA**) initializes and configures robots and perception agents. In a dedicated message exchange protocol, the operator program sets selected parameters of robots and perception agents enabling them to join the system automatically. Such a solution complies with the postulate of system openness (the number of elements working in the system can change dynamically).

In our previous works we have introduced the *perception network* as a framework for multi-sensor systems in robotics [6,8]. The software running on the mobile robots is based on the perception network structure (Fig. 1A), and comply with the multi-agent blackboard architecture introduced in [3]. The sensor/actuator drivers, data fusion, and data transformation modules work as software agents communicating through the blackboard [5]. The blackboard contains different descriptions of the robot environment and task (Fig. 1B). Agents are coupled to physical devices – sensors and actuators (*ADxxx*) or to processing tasks – experts (*AExxx*). The device agents execute their actions concurrently, observing time constraints of respective sensors and actuators. Around the blackboard, there are device agents representing sonars, the laser scanner, the on-board camera, and the robot controller. The data processing agents are : the map conversion and map fusion agents, the self-localization agent, and the pilot agent providing the reflexive navigation. Monitoring of the whole data-flow and the execution of operator commands is the duty of the report agent (Fig. 1B). Agents detect events in the system by observing the changes of data on the blackboard [5]. The infor-

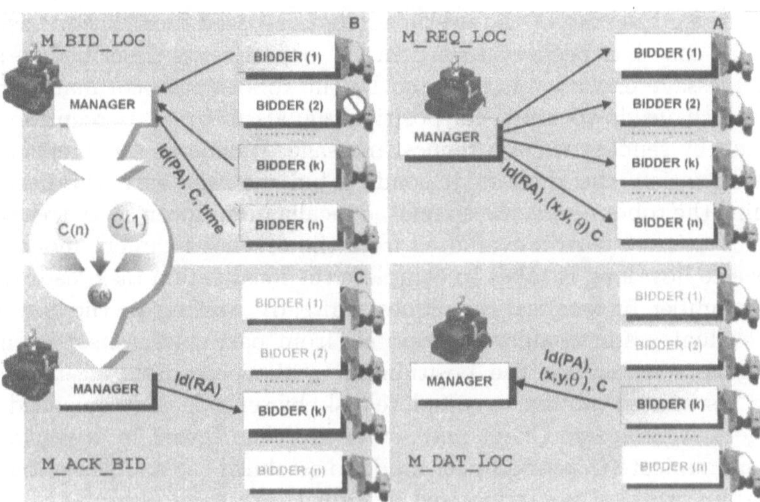

Fig. 2. The Contract Net Protocol for the localization task

mation needed to arrange control protocol is implemented with specialized flags. The blackboard-agents cannot communicate directly each-other or with other robots and perception agents. The communication within the robot's body relies on the blackboard, while the inter-agent communication is implemented by the specialized device agent *ADComm*, which is the manager of the physical communication channel.

3 The logical structure of the inter-agent communication

Particular modules (programs) in the described system communicate via TCP/IP network. A dedicated communication system has been implemented to run over mixed cable/radio LAN. After logging into the system, each agent is automatically recognized by the Operator Agent, which assigns it an identification number *Id(Agent)*, and creates for it a separate communication thread. Such a solution enables concurrent execution of the communication actions by many agents, under the supervision of the **OA**. For each agent a specific communication time-out has been defined in order to solve the access conflicts in the network.

The robots use the Contract Net Protocol [14] to choose the best data from the available external sources (Fig. 2). Such a solution has been adopted in order to avoid problems with global, distributed blackboard management [12]. The blackboard-agents and the data structures specific to the particular robot (with its specific sensing capabilities) are encapsulated within the **RA**. Also the Perception Agents hide their internal structure, providing only the requested navigation data to the robots. Here we describe the negotiation procedure for the case of the overhead camera-based localization task. This particular case has been evaluated in the experiments described in section 5. The message exchange is initiated by the robot (protocol manager). The robot which needs to know its position signals it to all perception agents (bidders), by sending them a request message. Whenever a perception agent is able to satisfy the request, it sends a bid message with the parameters informing the robot about the estimated localization uncertainty (represented by the covariance matrix evaluated from the overhead camera model [2]). It signals also how long it takes to complete the localization task, as it depends on the number of localization requests already waiting in the queue. The robot evaluates and compares proposals from particular sensor agents and awards the contract for the positioning to the perception agent with the highest (estimated) ability to complete task, by sending the acknowledgement message to this agent. Client and server are thus linked by a contract and they communicate in peer-to-peer mode to establish the transfer of data. The negotiation protocol is summarized in Tab. 1.

message	mnemonic	sender	receiver	parameters
request	M_REQ_LOC	robot agent	perception agent	$Id(\mathbf{RA})$ – robot's identifier (x_0, y_0, θ_0) – position and orientation of the robot $C_{(x,y,\theta)}(\mathbf{RA})$ – uncertainty (covariance matrix)
bid	M_BID_LOC	perception agent	robot agent	$Id(\mathbf{PA})$ – perception agent's identifier $C_{\mathbf{RA}}(\mathbf{PA})$ – estimated uncertainty of localization T_{loc} – estimated time of the localization task
acknowledgement	M_ACK_LOC	robot agent	perception agent	$Id(\mathbf{RA})$
data	M_DAT_LOC	perception agent	robot agent	$Id(\mathbf{PA})$ (x_n, y_n, θ_n) – new position and orientation $C_{(x,y,\theta)}(\mathbf{PA})$ – uncertainty of new position and orientation

Table 1. Contract Net Protocol messages with parameters for the localization task

4 Implementation of the communication mechanisms

The main objectives to achieve in the implementation of the communication system were :

- robustness to failures and communication errors;
- encapsulation of TCP/IP addressing by introduction of the proprietary addressing scheme, based on human-readable names (`AR_LABMATE1`, `AP_OVERHEAD2`, etc.);
- introduction of the Operator Agent as the central node of the communication system, enabling users to watch the communication activity and the status of agents;
- precisely defined programming interface to the communication system (by using OOP).

The *User Datagram Protocol* (UDP) is the basic layer of the communication. The main reason for this choice is the necessity to distinguish particular processes (programs running on the same machine) in the address.

Operator Agent's address is known to all agents (stored in their configuration files). An agent entering the system contacts the **OA**, sending it's symbolic na- me, type (**RA** or **PA**), IP address, and port number. The **OA** verifies the data and registers the new agent, giving it an unique identification number *Id(Agent)*. The operator updates it's Agent Data Base (ADB) containing information on all agents, currently active in the system, and sends a copy of this ADB to all the agents. Thus, the knowledge of the current system configuration is distributed among all agents.

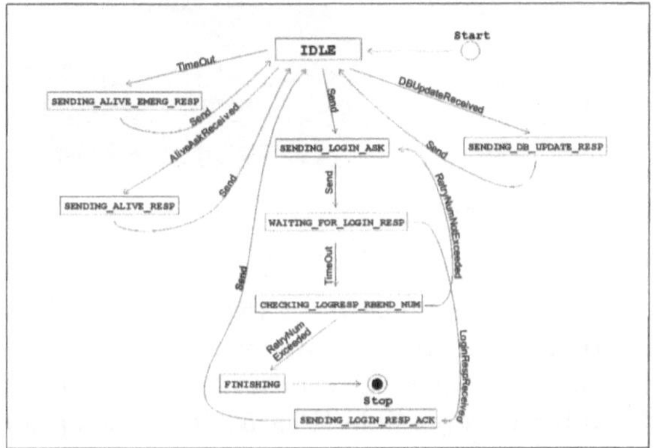

Fig. 3. Operator Agent communication protocol

In the case of a failure of the Operator Agent, human supervisor can run another **OA**, which can query the necessary data from other agents. The task of the **OA** is also to periodically check if the agents are "alive". It is accomplished by sending every 30[s] the special message `M_ALIVE_ASK` to all agents registered in ADB. The agent who does not respond to this message several times, is treated as "dead", and removed from ADB (Fig. 3). The agents communicate by using point-to-point message passing. They use their local copies of the ADB to translate the symbolic names into real IP addresses. It is also possible to send a message through the **OA**, using only the symbolic name of the receiver. In both modes of communication the data integrity is ensured by hand-shaking and timeouts. The communication mechanisms have been implemented as a library of classes. By that approach, the details of TCP/IP programming can be hidden, and such functions of the communication system as the translation of symbolic names are transparent to the application-level programmer.

5 Experimental validation

The multi-agent perception system has been tested using two heterogeneous robots and overhead cameras as stationary sensors (Fig. 4A). Two robot-agents were differently configured mobile platforms of Labmate type (Fig. 4C). Both robots had on-board PC computers, which are nodes of the LAN. Either *Aironet* wireless Ethernet cards or radio-modems (configured as PPP connection) were used to connect the mobile robots to the network. Robots were equipped with laser scanners and proximity sensors. One of the robots was also equipped with sonars, the other one had on-board vision subsystem with one CCD camera (not used in the experiments presented here).

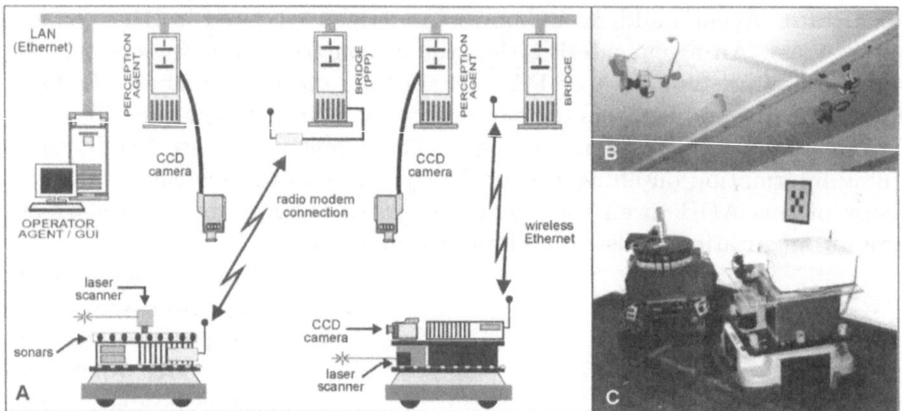

Fig. 4. The multi-agent system of mobile robots

The hardware of the perception agents consists of two overhead cameras monitoring the scene (Fig. 4B). Both cameras are connected to framegrabbers on stationary computers. The cameras are provided with fish-eye lenses to make the field of view as large as possible (Fig. 5). The perception agents task is to localize robots with respect to the room reference frame. The robots are equipped with LED markers to simplify the localization [2].

The programs on robot-agents run under Linux, while the camera-based perception agents have been implemented as Win32 applications.

We validated our approach to inter-agent communication on the task of localization and world-model building. Below we present results of experiments, in which two Labmate mobile agents have been supported by the stationary camera-based perception agents.

Figure 6 shows the process of the occupancy grid update and the parametric interpretation of the grid model, leading to the symbolic representation. The occupancy grid built without the self-localization from external sensors (upper row) allows to distinguish particular objects, but is too "blurred" to be useful for parametric interpretation. Also, the uncertainty of the robot's position (visualized as the ellipse) is getting high, after traveling relatively

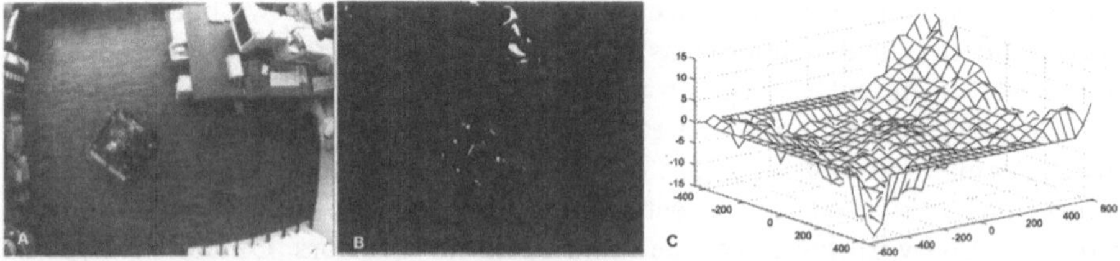

Fig. 5. Images from the overhead camera agent (A,B) and the uncertainty model for the camera-based localization (C) – see [2] for details

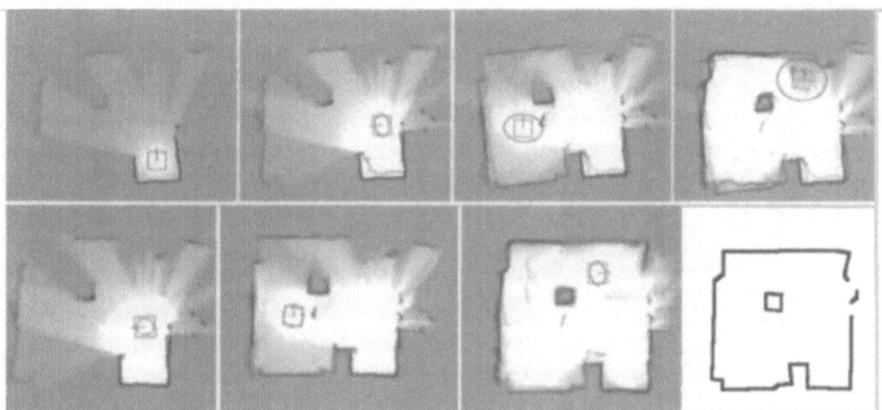

Fig. 6. Occupancy grid update without (above) and with localization (below)

short path. By recoursing to the information from the camera-based perception agent, it was possible to keep the position uncertainty within reasonable limits, and to build much more precise occupancy grid (lower row). This grid map is post-processed by using Hough transform [8], to obtain the vector-based map of the environment (Fig. 6, last picture).

Figure 7 compares the results of the overhead camera-based localization for the case, when the negotiation procedure described above is used, and the case when simple choice of the "first available" perception agent is made. Using the Contract Net Protocol, the robot made 43 requests for localization, 24 of which were successful, i.e. one of the two camera-based agents localized the robot within the given uncertainty area (Fig. 7B). When the robot did not use the bid evaluation mechanism, but simply asked the first available camera-based agent to localize it, the number of requests was 80, among them only 21 were successful (Fig. 7A). The robot performed much worse without the bid evaluation mechanism, because in many cases it initialized the localization in places, which were very hard for the co-operating camera-agent, e.g. where only three of four robot's LEDs were visible (Fig. 7C). In contrary, the bid evaluation ensures that the robot uses the best offered localization service, and initialize localization, when its position with respect to the field of view of the co-operating camera-based agent is good (Fig. 7D).

We have performed also some experiments with the concurrent localization of both mobile robots by the perception agents (Fig. 8). In these experiments one of the robots used both laser scanner-based self-localization and the information from external agents, while the other one relied only on the odometry and information from the overhead cameras. The ratio of successful localizations was similar for both robots, and on similar level as in the single robot experiment. This experiment shows that sharing the pool of perception agents by several robots does not significantly degrade the system performance. However, it remains an open question whether the communica-

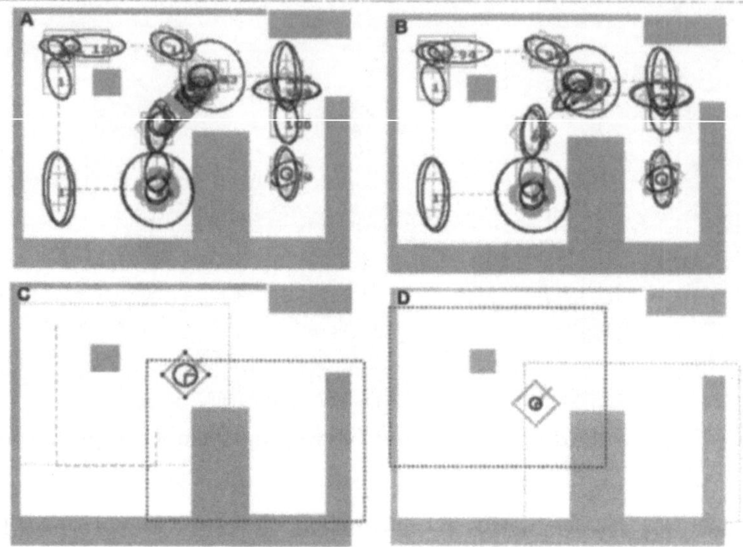

Fig. 7. Results of the camera-based localization experiment

tion architecture scales to larger numbers of robots and stationary sensors. One of the possible problems is the time the Contract Net Protocol manager (the robot) has to wait to complete all the bids. This time grows with the number of agents involved in the negotiation. However, some solutions to this problem are already known (e.g. [10]).

Fig. 8. A sequence of images from the localization experiment with two robots

6 Conclusions and related work

While there is a considerable work done on such aspects of multi-robot systems like task scheduling and coordination, formation control, and cooperative localization, the issues of communication and sharing high-level information (e.g. world model) received as so far much less attention.

Several researchers have studied the problems of communication in groups of behavior-based robots. In [1] the importance of communication in robotic societies has been investigated. The conclusion was, that the benefit from the *explicit* communication is considerable in such systems and tasks, where no *implicit* communication is available through observation of other agents. An example of the behavioral framework for the co-operation of multi-robot teams is Parker's ALLIANCE [11]. In this framework, the role of active (explicit) communication is minimized. However, robots broadcast the information about the task performed at the given moment[1], the communication is not necessary in order to complete a task by the robot. The architecture proposed in [11] gives robots high robustness to the communication failures, but at the expense of the deliberative capabilities and the taskability. Also the system proposed in [15] is behavior-based and uses specific, hand-coded communication channels to implement dependencies between particular behaviours of different robots making a team. The cited approaches contrast with that of our system, where robot-agents are provided with hybrid deliberative/reactive architecture [3], and co-operate with stationary sensors. The stationary sensors are also treated as agents, but they cannot perform any *physical* action, except communication over network. This is why we cannot rely on implicit communication. The architecture and communication mechanism we propose are not designed for any specific type of task (e.g. foraging), we rather look for a more general procedure of "seeking advice" in such cases, when the robot cannot complete its task because of the information deficit (e.g. lack of a map or localization data).

The information sharing issues have been investigated in [4], where a concept of distributed sensing and map sharing among a group of mobile robots has been presented. As an example, the simulated process of occupancy grid building has been implemented. The case reported in [4] can be classified as the case of simple, homogeneous robots making use of the common representation of the environment. In our approach the information sharing among robots and sensors takes place at the definitively higher level of knowledge representation than in [4].

The notion of *bids*, somewhat similar to the one presented here, has been introduced in [13], where it was used to coordinate multi-robot exploration. However, this solution differs from our approach, because in [13] a centralized unit (the mapper module) has been used, which evaluated the bids and coordinated robots.

The concept of the communication system presented in [9] is in principle similar to our approach, but in our case the communication is focused on sharing the data which directly support the navigation, rather than on the task selection. We also use the unified low-level mechanism for all kinds of messages in that system, while in [9] two different types of communication co-exist.

[1] what in this case is rather equivalent to implicit communication – the broadcasting was used, because simple robots used by Parker had no means to observe each other's state

The work reported here extends our earlier research [3,6–8], providing the validation of the communication mechanism, being the backbone of the multi-agent system. In particular, the results of experiments have shown that the negotiation mechanism contributes to the quality of the overhead camera-based localization. The obtained communication architecture is transparent, open, easily expandable, and implemented in a heterogenous system.

Acknowledgement

This work was supported by the State Committee for Scientific Research under grant 8T11A00319

References

1. Balch T., Arkin R. C. (1994) Communication in Reactive Multiagent Robotic Systems. Autonomous Robots, **1**, 1–25
2. Bączyk R., Skrzypczyński P. (2001) Mobile Robot Localization by Means of an Overhead Camera. Proc. Automation 2001, Warsaw, 220–229
3. Brzykcy G., Martinek J. et al. (2001) Multi-Agent Blackboard Architecture for a Mobile Robot. Proc. IROS 2001, Maui, 2369–2374
4. Cai A., Fukuda T., Arai F. (1997) Information Sharing among Multiple Mobile Robots for Cooperation in Cellular Robotic System. Proc. IROS 97, Grenoble, 1768–1773
5. Engelmore R., Morgan T. (Eds.) (1988) Blackboard Systems. Addison-Wesley.
6. Kasiński A., Skrzypczyński P. (1998) Cooperative Perception and World-Model Maintenance in Mobile Navigation Tasks. In: DARS 3, T. Lueth *et al.*, (Eds.), Springer-Verlag, Berlin, 173–182
7. Kasiński A., Skrzypczyński P. (2000) Experiments and Results in Multi-modal, Distributed, Robotic Perception. In: DARS 4, L. Parker *et al.*, (Eds.), Springer-Verlag, Tokyo, 283–292
8. Kasiński A., Skrzypczyński P. (2001) Perception Network for the Team of Indoor Mobile Robots. Concept. Architecture. Implementation. Engineering Applications of Artificial Intelligence, **14(2)**, 125–137
9. Matia F., Moraleda E. et al. (1998) Distributed Task Planner for a Set of Holonic Mobile Robots. In: DARS 3, T. Lueth *et al.*, (Eds.), Springer-Verlag, Berlin, 35–44
10. Ozaki K., Asama H. et al. (1994) Negotiation Method for Collaborating Team Organization among Multiple Robots. In: DARS, Springer-Verlag, Tokyo, 199–210
11. Parker L. E. (1994) Heterogeneous Multi-Robot Cooperation, PhD Thesis, MIT.
12. Schwartz D. (1995) Cooperating Heterogenous Systems. Kluwer Acad. Publ.
13. Simmons R., Apfelbaum D. et al. (2000) Coordination for Multi-Robot Exploration and Mapping. Proc. AAAI Conf. on Artificial Intelligence, Austin.
14. Smith R. G., Davis P. (1981) Frameworks for Cooperation in Distributed Problem Solving. IEEE Trans. on SMC, **11(1)**, 61–70
15. Werger B. B., Mataric, M. J. (2000) Broadcast of Local Eligibility for Multi-Target Observation. In: DARS 4, L. Parker *et al.*, (Eds.), Springer-Verlag, Tokyo, 347–356

A Geometric Arbiter Selection Algorithm on Infrared Wireless Inter-robot Communication

Hiroyuki Takai*, Hirotoshi Hara*, Keisuke Hirano*,Gen'ichi Yasuda** and Keihachiro Tachibana*

* Department of Information Machines and Interfaces, Hiroshima City University, 3-4-1 Ozuka-Higashi, Asa-minami, Hiroshima 731-3194 Japan

**Department of Mechanical Engineering, Nagasaki Institute of Applied Science, 536 Aba-machi, Nagasaki 851-0193 Japan

Abstract

Recently, robots that can perform group operations have been developed. A space-division infrared wireless communication system has been designed, that is capable of detecting and locating other robots even if they are independently mobile. In addition, this communication system decides an arbiter that is a local temporary controller, taking advantage of the geometrical nature of the triangle that is formed by communication links between communicating partners. A geometric arbiter selection algorithm was proposed and the performance of the infrared receiver was tested. The result of the experiments agreed with all expectations of the space-division wireless communication system.

Keywords: Inter-robot communication, Infrared wireless communication, Space-division wireless communication, Ad hoc communication network

1. Introduction

Recently, autonomous mobile robots, which perform group operations, have been developed [1]. These multi-agent robots are capable of carrying baggage, and/or supporting rescue operations, among other tasks.

These multi-agent robots have to communicate to perform their tasks smoothly. These robots must arbitrate among themselves in order to avoid collisions when using radio waves or infrared rays in a working area. Usually radio waves are used for omni-directional long distance communication. On the other hand, Infrared rays are restricted in direction, so have limited communication applications.

When the number of robots increases to accomplish complicated tasks, congestion often occurs. This can lead to communication interference in the working area. If robots send communication signals from within the same area at the same time, communication interference often occurs. Therefore, many wireless communication systems have been proposed to reduce communication interference on wireless communication networks.

One of the methods of reducing communication interference is to limit the size the communication area, which usually decrease the number of robots within a given communication area [2]. This method limits transmitting signal strength and width. If the communication system controls each communication area suitably, and is able to reduce communication interference, parallel transmission can take place. This means that the same communication channel can be utilized in the same working area by one or more robots. However, suitable signal strength and/or width depend on the working environment. In addition, when a robot runs and/or rotates, the communication link can be easily lost, as narrow infrared beams require accurate location information.

Another method of reducing communication interference divides time for communication links among robots. This method of reducing communication interference uses media access control (MAC) algorithms, which are a part of the open system interconnect (OSI) standard, such as TDMA (Time Division Multiple Access). In this method, all robots are synchronized by one controller. However, if the controller malfunctions, all robots stop and no tasks can be accomplished.

An inter-robot wireless communication system using infrared rays has been designed [3-5]. Infrared rays have strong directivity and limited beam width. As a consequence, communication interference can be reduced. In addition, this communication system can detect the direction of signals from communicating partners. One robot is chosen as an arbiter geometrically by measuring the angles of the triangle formed by the communication links. The arbiter is a local temporary controller and is intended to manage a part of the communication network.

This paper describes a geometric arbiter selection algorithm on space-division infrared wireless communication systems for multiple autonomous mobile robots.

2. Space-division infrared wireless communication system for multiple autonomous mobile robots

Infrared rays have strong directivity and limited beam width. Accordingly, infrared wireless communication systems suffer from hardly any interference. This is especially true when the infrared receiver faces a different direction to the transmission signal. Therefore, infrared wireless communication systems reduce communication interference and allow parallel transmissions to be utilized.

However, most infrared wireless communication systems lose connections when a robot runs and/or rotates due to the limited beam width of infrared rays. Also, when there are a number of robots in the same communication area, these robots are synchronized in order to reduce communication interference using the media access control (MAC) algorithm.

An infrared wireless communication system, that is suitable for multi-agent mobile robots, is shown as follows.

2.1 Hardware design of a space-division infrared wireless communication system

Infrared rays have strong directivity and limited beam width. Because of this feature of infrared rays, communication links are lost when a robot runs and/or rotates. The communication link can be preserved by extending communication signals in a wide area in all directions or by tracking the communication signals of communicating partner robots. However, when the communication signal spreads out in a wide area in all directions, communication interference increases. Accordingly, the most desirable method for maintaining communication links is the tracking method.

We have designed an infrared wireless communication system for mobile robots, which is capable of detecting the angle of incidence of infrared rays and tracking communicating partner robots to maintain communication links [3-5]. In this infrared wireless communication system, a number of infrared transceivers are arranged on the circumference of the robot body. Figure 1 shows an arrangement of the infrared transceiver modules.

Eight transceivers are arranged at 45 degree intervals on the circumference of the robot body. Each transceiver has a reception angle of 60 degrees overlapping on each side with the adjacent transceiver by 15 degrees. This is necessary to maintain the communication link when switching transceivers to face the communicating partner when the robot runs and/or rotates.

The values of these parameters were decided temporarily and in order to optimize communication performance. Figure 2 shows the process of communication.

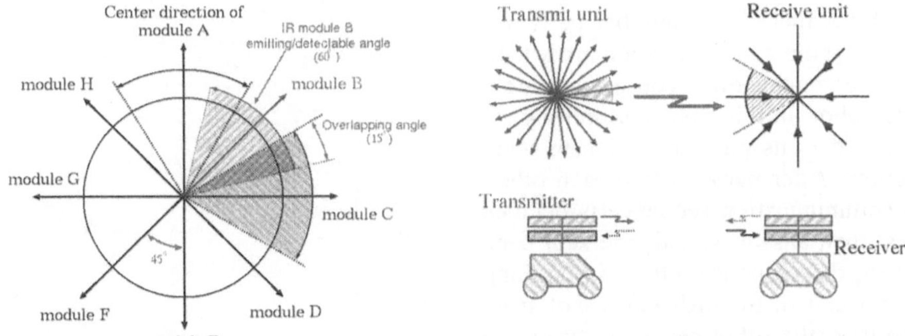

Fig.1 An arrangement of transceivers

Fig.2 The process of the communication

Each transceiver communicates independently as well as with adjacent transceivers. Each transceiver communicates with a partner only in the direction in which it faces. As a result, a space-division infrared wireless communication system is composed. This communication system suffers from hardly any interference, if the transceiver faces a different direction to other communicating partners.

2.2 Building a infrared wireless communication networks

This communication system uses only one communication carrier for all robots. Therefore, simplex or half-duplex communication is possible. Simplex communication systems make stable communication links for transmission or reception. On the other hand, Half-duplex communication systems require synchronization for quickly switching transmission and reception functions on a communication link.

Normally, this communication system creates half-duplex communication links. After communication signals are received transmission signals are sent. Point-to-point communication links are established among a number of robots and they become part of the communication network.

The building process of the communication network shows as follows:

1. At first, all robots are isolated from each other. Each robot repeatedly sends out search signals in all directions in order to find partner robots.
2. Each partner robot receives a search signal and detects the direction from which it came. The partner robot sends back an answer signal in that direction.
3. After the reception of the answer signal from the partner robot, the robot sends a conformation signal back to establish a communication link.
4. A communication link is established between two mobile robots. These robots keep sending search signals in the directions where a partner has not yet been found.
5. Robots spread communication links among themselves and expand the communication network.

When there are a number of robots in the same communication area, their answer signals often confuse a searching robot. In this case, a searching robot orders its partners to locate other robots. After partners find each other, a communication sequence is decided and then answer signals are sent back along the sequence. Once a searching robot can distinguish one robot from another, the robot sends confirmation signals back to each robot to establish communication links. These functions are sequenced so as not to suffer from interference from other robots' transmissions.

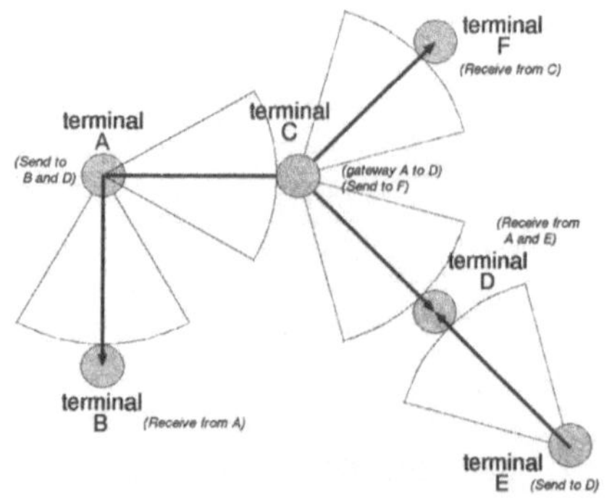

Fig3.Inter-robot communication network

Figure 3 shows the inter-robot wireless communication network using the infrared wireless communication system.

In this communication network, some robots have several communication links with partners in different positions. Each robot relays information from/to the different links. It is an ad hoc communication network because these robots are independently mobile and may change position depending on the task.

2.3 Utilizing media access control (MAC) algorithm on wireless communication network

The media access control (MAC) algorithm, that is in the data-link layer of the open system interconnect (OSI) standard reference model, controls signal power, timing and direction. It is used to reduce communication interference. Algorithms can be categorized into two groups, distributed algorithms and centralized algorithms. Many media access control (MAC) algorithms have been proposed to reduce communication interference.

In distributed algorithms such as CSMA (Carrier Sense Multiple Access) that are used in the Ethernet LAN (Local Area Network), all robots communicate randomly. Each robot transmits communication signals independently. However when robot numbers increase in a working area, communication interference increases.

On the other hand, in centralized algorithms such as TDMA (Time Division Multiple Access), all robots communicate synchronously. Generally, in such networks only one controller synchronizes all inter-robot communication and its location is fixed. Therefore, if the controller malfunctions, all robots stop and no tasks are accomplished.

However, if all robots are capable of communication control, the controller role can be switched to another robot when malfunction or damage occurs. In this case, most of tasks can still be accomplished.

An arbiter is a temporary local controller in a small communication network. In infrared wireless communication networks, an arbiter is selected geometrically because of the nature of infrared rays and the function of transceivers in this communication system.

2.4 Selecting arbiters for the network

When two or more transmitters send signals at the same time inside the reception angle of the receiver, it cannot separate the signals and confusion results. This can occur when two or more transmitting robots are in close proximity and are both facing a receiving robot. Therefore, the direction in which transmitters and receivers face, the beam width of the transmitter and the reception angle of the receiver can affect the level of communication interference.

This infrared wireless communication system detects the direction of partner robots based on the angle of incidence of the infrared rays and calculates the interior angles that are formed by the communication links. These communication links form a number of triangles with robots on their apexes.

This infrared wireless communication system reduces communication interference by calculating the angle of incidence of the infrared rays and takes advantage of the geometrical nature of the triangle that forms from the communication links.

In a triangle, if an interior angle becomes narrow, the others become wide, because the sum of the interior angles of a triangle is 180 degrees. Each robot reports its interior angle to two partner robots. The robot with the widest angle becomes an arbiter for this communication network. Figure 4 shows the selection of the arbiter.

Normally, the arbiter has an interior angle that is equal to 60 degrees or greater, because each interior angle in an equilateral triangle is 60 degrees. If the reception angle of the receiver is smaller than 60 degrees, an arbiter can separate communication signals, because each communication signal is detected by separate receivers.

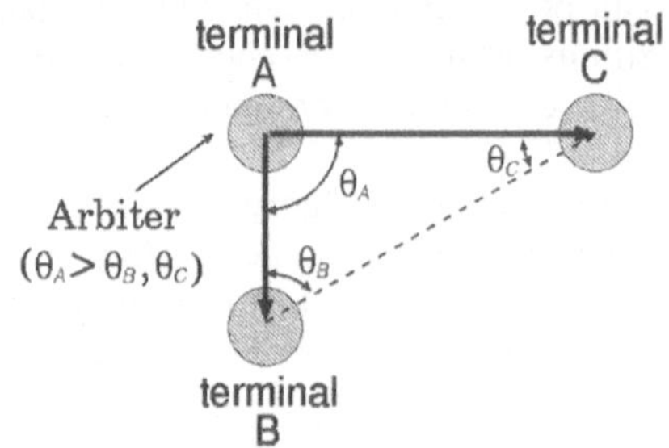

Fig.4 The selection of the arbiter

An arbiter gives instructions to all partner robots in not only transmission timing but also transmission direction. This ensures that only one partner can send a signal inside the reception angle of the receiver at a time.

3.Hardware realization and performance measurements

This infrared wireless communication system finds the direction of a communicating partner and gets location information on each robot by determining the angle of incidence of the infrared rays.

The direction detection function was successfully tested in an experiment.

3.1 The detection of the angle of incidence of the infrared rays

We conducted an experiment, which tested the detection of the angle of incidence of infrared rays. We used the PIN photo diode (HAMAMATSU S6560) to detect infrared rays in this experiment. Figure 5 shows a schematic view of the PIN photo diode. The PIN photo diode can detect the angle of incidence of infrared rays. The PIN photo diode has two electric current outputs 'a' and 'b'. The electric current outputs 'a' and 'b' of the PIN photo diode have the relation of the equation 1 with the angle of incidence of the infrared rays θ.

$$\theta = (a - b)/(a + b) \qquad (1)$$

A source of infrared rays was placed in front of the angle of the incidence detector. Then the source was moved from the left 50 degrees to the right 50 degrees in 0.5 degree increments, and the distance between the source and the detector was moved from 15cm to 30cm in 5cm increments.

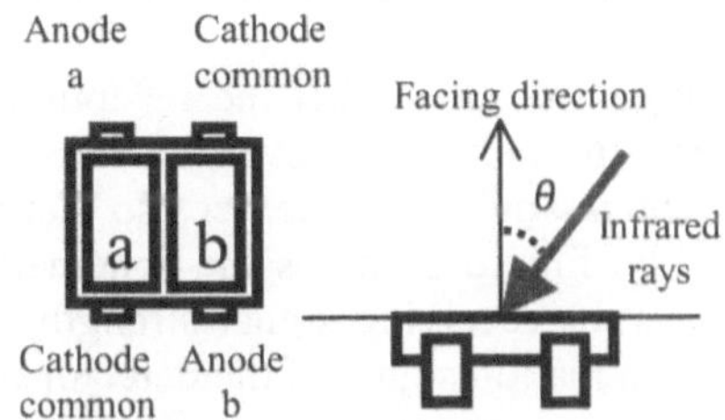

Fig.5 Schematic view of photo detector

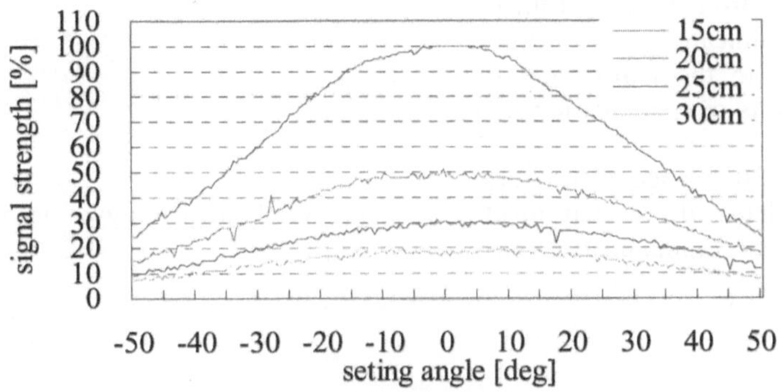

Fig.6 Receiving signal strength

Figure 6 shows the receiving signal strength using a PIN photo diode (HAMAMATSU S6560) and the analog signal processor (HAMAMATSU C3683-01). It shows the summation signal of the detector output 'a' and 'b'.

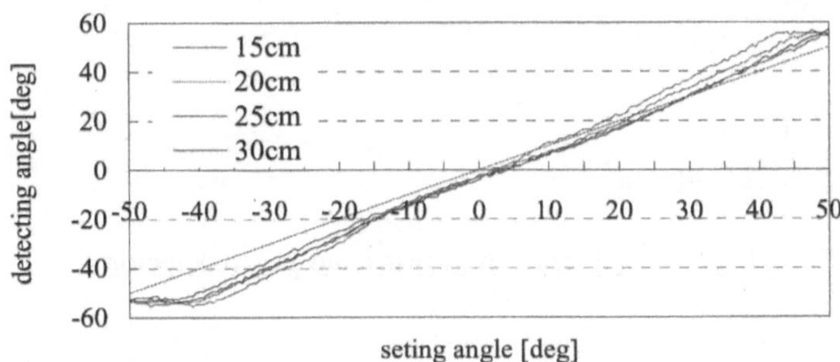

Fig.7 Angle detection error of photo detector (S6560)

Figure 7 shows the angle detection error of this infrared detector, and it shows the angles at which infrared beams were detected compared with the angles at which the infrared source was set. In this figure, the gradient of results relates to the distance from the detector to the signal source and/or the pivoting center.

3.2 Angle restriction for the detector

Figure 6 shows that the reception angle of the receiver is bigger than 60 degrees. In practice however, barriers are put in front of the detector to restrict the reception angle to 60 degrees so that it may separate communication signals completely. Figure 8 shows the schematic view of barriers and a detector. Figure 9 shows the receiving signal strength of the photo detector with barriers. It shows the summation signal of the detector output 'a' and 'b'.

Figure 9 shows the receiving signal outside the 60 degrees detection area being eliminated by barriers. Barriers shut out signals from outside of the 60 degrees communication area. This is necessary due to the hardware configuration of this communication system.

The angle of incidence of the infrared rays was detected precisely inside of the 60 degrees communication area.

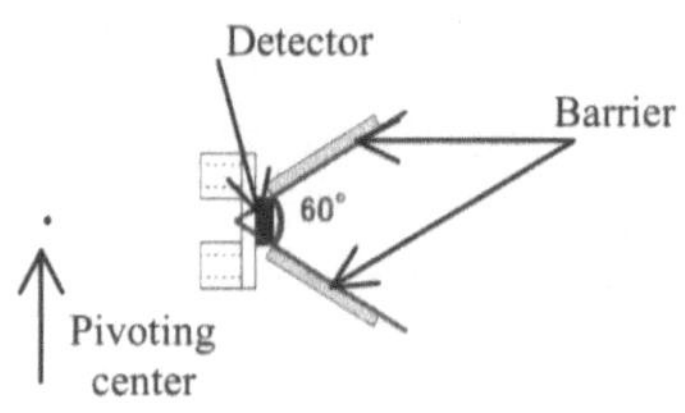

Fig.8 Schematic view of barriers

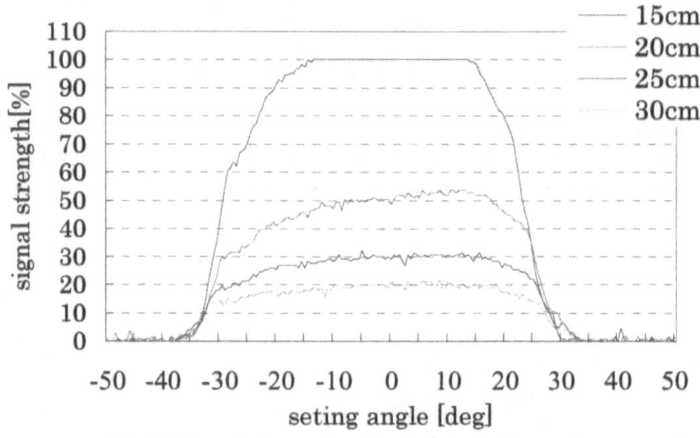

Fig.9 Signal strength of restricted photo detector

3.3 A simulation of directional receivers

In this communication system, eight transceivers are arranged at 45 degrees intervals on the circumference on the robot body. Each transceiver has a limited reception angle of 60 degrees overlapped by 15 degrees on each side by adjacent transceivers.

In this experiments, two detectors are placed on the same center pivot with a gap of 45 degrees. This is in order to simulate part of this communication system. Figure 10 shows a schematic view of the detector arrangement. Figure 11 shows the signal strength of each detector.

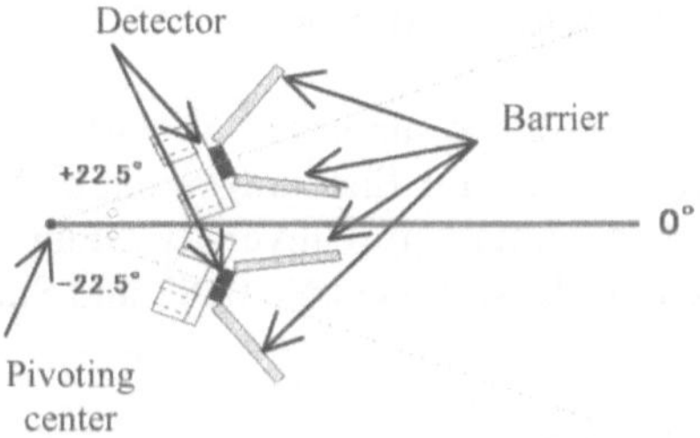

Fig.10 Arrangement of two detectors

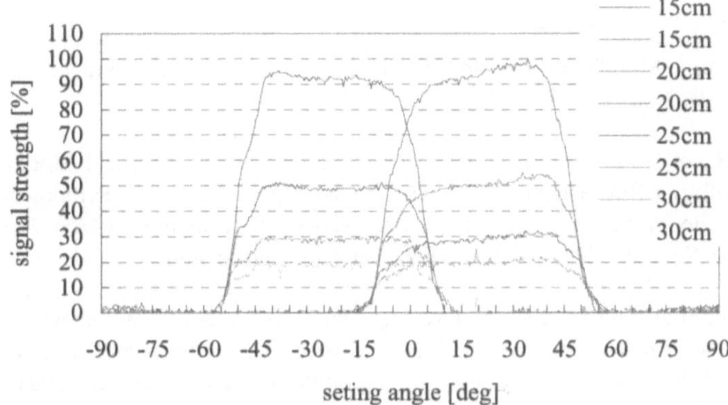

Fig.11 Receive signal strength and receiving area of restricted detector

Figure 11 shows the receiving signal strength and receiving area of each receiver. Each communication area is restricted to a width of 60 degrees and overlaps with the adjacent receivers by 15 degrees. This result confirms the feasibility of the layout of figure 1.

4.Conclusions

We proposed a space-division infrared wireless communication system for mobile robots and a geometric arbiter selection algorithm in this paper.

The angle detection experiment was successful and shows the system is able to detect angles of incidence of infrared beams precisely.

The result of the simulation using directional receivers proved the feasibility of the space-division wireless communication system that we proposed.

A Geometric arbiter selection algorithm that we proposed will also be realized from these results.

Acknowledgements

This work is supported, in part, by Japan Society for the Promotion of Science Grant-in-Aid for Encouragement of Young Scientist (No. 13750365) and, in part, by Hiroshima City University Grant for Special Academic Research (Encouragement for Researchers No.0087 and Support for Researchers No.1611).

References

1. ASAMA H., et al (Eds.) (1994) Distributed Autonomous Robotic Systems, Springer, Tokyo

2. YOSHIDA E., and ARAI T., (2000) Performance Analysis of Local Communication by Cooperating Mobile Robots. In: IEICE Transactions of communication, Series EB, Vol.E83-B, No.5, pp.1048-1059

3. TAKAI H., et al (2000) Wireless Communication System for Local Mobile Terminals using Infrared Rays (in Japanese). In: Proceedings of The 23rd Symposium on Information Theory and Its Applications (SITA2000) at Kumamoto (Aso), Japan, Oct.10-13, 2000. Society of Information Theory and Its Applications, pp.181-184

4. TAKAI H., et al. (2001) A Space-Division Optical Wireless Communication System for Fully Distributed Multiple Autonomous Mobile Robots. In: K.SCHILLING, H.ROTH (Eds.). Telematics Applications in Automation and Robotics 2001 (TA2001). PERGAMON press, pp.333-338

5. TAKAI H., et al. (2001) A Space-Division Optical Wireless Inter-ROBOT Communication System with Mutual Localization Ability for Multiple Autonomous Mobile ROBOTS. In: INOUE H., ASAMA H., (Eds.). Preprints of The 4th IFAC Symposium on Intelligent Autonomous Vehicles (IAV2001) at Sapporo, Japan, Sep.5-7, 2001. IFAC, pp.338-343

6.SUZUKI S., et al. (1996) Development of an Infrared Sensory System with Local Communication Facility for Collision Avoidance of Multiple Mobile Robots. (In Japanese) In: Transactions of JSME, Series C (Japanese Edition) Vol.62 No.602, pp.14-20

7.ARAI Y., et al. (2001) Collision Avoidance in Multi-Robot Environment based on Local Communication. (In Japanese) In: Journal of RSJ, Vol.19, No.1, pp.45-58

8.MINAMI T., et al. (2000) Essential Effect of Directional Antenna on Packet Radio Network. (In Japanese) In: IEICE Transactions of communication (Japanese Edition), Series B, Vol.J83-B, No.8, pp.1148-1155

9. AKIZUKI O., et al. (1997) TDMA with Multiple Parallel Transmission using Directional Antenna. (In Japanese) In: IEICE Transactions of communication (Japanese Edition), Series B-I, Vol.J80-B-I, No.10, pp.719-728

A Study of Communication Emergence among Mobile Robots : Simulations of Intention Transmission

Kuniaki Kawabata[1], Hajime Asama[1], and Masayuki Tanaka[2]

[1] Advanced Engineering Center, RIKEN, 2-1, Hirosawa, Wako, Saitama, 351-0198, Japan
[2] Dept. of Mechanical Engineering, Toyo University, 2100, Kujirai, Kawagoe, Saitama, 350-8585, Japan

Abstract. Purpose of this study is to realize communication emergence among multiple robots. An important role of communication in multi-agent system is to make it possible to control the other agents based on intention transmission. We consider that multiple robots system can be more and more adaptable by treating communication as one of behavior. We discuss the method of how the robots can learn appropriate actions including communication to adapt the environment without giving communication manner. In this paper, we attempt computer simulations of collision avoidance as an example of cooperative task and discuss the results.

1 Introduction

In multiple robots system, the agents can multiply their capabilities and effectiveness, or work in parallel as occasion arises. In such system, each agent must work on own purpose and also consider the common purpose of whole system. Namely, communication is a necessary skill for autonomous robots engaged in the cooperative task. The works were already done on communication among multiple robots as a means of signal transmission to the other ones to achieve tasks cooperatively[1][2]. However, these studies largely set rules to communicate. By these methods, communication may not be useful and adapt in a dynamic and complex environment. On the point of view, we should discuss about adaptable communication scheme.

Yanco et al. tried to develop the method to acquire an adaptive communication for cooperation of two robots[3]. In this system, un-interpreted vocabulary is given and the robots acquire the usage of the words. Billard et al. proposed a learning method of communication through imitation [4]. This is an interesting approach but the system needs a teacher robot. In these methods, the communication is treated as special function for the robotic system. However, we consider that the communication should not be treated with physical behaviors, separately. We discuss how the robot acquires the selection of communication and actions for collaborative task execution.

On the other hand, in recent years, reinforcement learning [5][6] is remarkable method as a machine learning for robots to acquire adaptive behavior.

This method is a sort of unsupervised learning and doesn't need explicit or implicit knowledge about working environment. Also, the agents can adapt to dynamic environments when it is hard to design accurate model of the environment. Such unsupervised learning method is suitable for emerging function among the agents. In this paper, we describe the communication emergence as robots learn what to do through interactions under dynamic environment. Here, we discuss collision avoidance between two mobile robots as an example of cooperative task. The robot learns behavior including communication, so as to emerge appropriate actions or functions, autonomously.

2 Communication as a Behavior

In the field of multiple robots cooperation, most of researches consider that only physical action (for example, just moving) means behavior of the robot. However, on the condition of the communication among the robots, the transmission of the intention to the other robot should be treated as a behavior and it is required to consider receiving the other one's intention as a perception. Therefore, we attempt to adopt these transmissions of intention to the robot's control architecture. On the point of control, one of important characteristics of communication is to realize to control the other's behavior utilizing the intention transmission, If the robot's behaviors are limited by number of the actuators, D.O.F. of the robot is equal to the number. However, if the robot can transmit its own intention and the other one do it, it means that the robot could control the other's behavior, too. It indicates the expansion of the robot's D.O.F. Here, we consider that it realizes to improve flexibility and adaptability of multiple robot system with this expansion of D.O.F. In this paper, we apply this concept to collision avoidance task of mobile robots. Using computer simulations, we discuss emerged communication scheme among the robots and its effect. Here, we assume to utilize omnidirectional mobile robot as a platform. As an external sensor, we also assume that the robot equips an omnidirectional vision system. In this paper, especially we attempt two types of communication emergence among two mobile robots. One is to acquire how to use pre-defined meaning messages (Case 1). Another one is to acquire a common protcol for cooreration using no meaning symbols (Case2) . As an example of cooperation task, we set collision avoidance among mobile robots. In each case, reinforcement learning scheme is utilized.

3 Reinforcement Learning

3.1 Q-learning Algorithm for SMDP

Emerging functions for adaptation to dynamic environment, we utilize reinforcement learning scheme. Utilizing this method, the robots realize to acquire reactive and adaptive behavior. In this learning mechanism, the agents don't need priori knowledge, and are able to adapt dynamic environment. Especially, we apply Q-learning algorithm for robots' learning. In Q-learning algorithm, the evaluation of the state and actions are given. When the agent act to get reward, the evaluation value for the state and the action would increase. On the contrary, the agent acts to get punishment, the evaluation value for the state and the action decrease. The robot acquires behavior adapted to the condition by selecting actions that makes the evaluation value large. We assume that given information to the robots is from visual sensor and constant communication. In the projective sensor space like visual information, the same physical action causes different transition on the image plane. When a robot takes actions at some situation, but it returns almost the same situation on the image plane[5]. Therefore, we use reinforecement learning for Semi Markov Decision Processes (SMDP) to solve such problems. The algorithm for SMDP is as follows.

1. observe state s_t at time t in the environment.
2. perform the behavior a_t that is selected by prospecting strategy.
3. keep receiving the reward until an event occurs (for N sampling period), and calculates the sum of the reward R_{sum} using following equation.

$$R_{sum} = r_t + \gamma r_{t+1} + \gamma^2 r_{t+2} + \cdots + \gamma^{N-1} r_{t+N-1} \tag{1}$$

4. observe the environment state s_{t+N} after the event occurring.
5. renew the Q-value using the following equation using state s_{t+N}.

$$Q(s_t, a_t) \leftarrow (1-\alpha)Q(s_t, a_t) + \alpha \left[R_{sum} + \gamma^N \max_{a'} Q(s_{t+N}, a') \right] \tag{2}$$

   where α is the learning factor ($0 < \alpha \leq 1$), γ is the discount factor ($0 \leq \gamma < 1$).
6. clear $R_{sum} = 0$
7. Set time step $t + N$, back to 1.

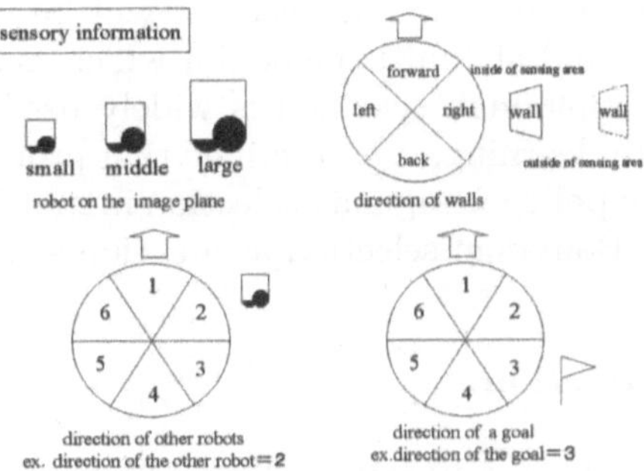

Fig. 1. State Space

3.2 State Space Construction

To apply Q-learning as a learning algorithm, we must design state space, appoint action selection method, and set up robots' possible actions and reward. We assumed that each robot employs omni-directional visual system. Also, robots can perceive the environment from constant communication with the other. It is better to divide state space into small for selecting the most suitable action at high precision. But, convergence time is order of exponential function depends on the size of state space, state space should be divided roughly. Therefore, we construct state space as follows (see Fig.1);

- the other robot's motion velocity(high, middle, low) .
- direction which can be recognized from partner robot (front, back, right, left) as information given by constant communication
- the other robot's size on the image plane (large, middle, small)
- directions of the other robots on the image plane(six area)
- directions of the destination on the image plane(six area)
- information about walls (five area : outside of sensing area and front, back, right, left in the sensing area) on the image plane
- motion velocity of itself (high, middle, low).

3.3 Prospecting Strategy

There are some strategies to choose actions, such as roulette selection, and combination of max selection and random selection. Also, there is a probability to select behavior based on Boltzman distribution, in terms of improving behavior gradually according to learning level. The probability $p(.)$ of state x to select behavior l is as follows,

$$p(l|x) = \frac{\exp(Q(x,l)/T)}{\sum_{m \in possible \quad actions} \exp(Q(x,m)/T)} \tag{3}$$

where T is temperature constant. If T is large, the agent become positive in search. If T is nearly to 0, action selection will be influenced by a little difference of Q value, and at limit, the action which maximizes the Q value is chosen. Generally, Boltzman selection is widely used, and as a means of selecting behavior after learning, it is described that probability policy is more effective than decisive policy using max selection in multi-agent environment. Therefore, we utilize Boltzman selection as a action selection method.

3.4 Evaluation Function

The purpose of the learning is to avoid collision in complex environment with speech. We define following evaluate function r_t as reward. j is a subscript that specify the robot. The rewards depend on the distance between robot and the goal.

$$r_{jt} = \kappa \Delta dg_j \tag{4}$$
$$\Delta dg_j = dg_j(t - \Delta t) - dg_j(t) \tag{5}$$

where, $dg_j(t)$ and Δdg_j indicate distance to the distination at time t and differencial value. This equation is utilized as reinforcement signal to Q-learning trials.

4 Behaviors

Here, we define the robot's basic behaviors for collision avoidance task. We attempt to observe two types of emerging communication and consider two types of behavior set. For case 1, the robot understands the meaning of speech but does not know how to use them.

Case 1

- move forward
- change own moving direction ($\pm\frac{\pi}{4}$ [deg])
- change own moving velocity
  - slow down (-0.002[m/sec])
  - speed up (+0.002[m/sec])
- message
  - change moving direction(+$\frac{\pi}{4}$ [rad]) and slow down(-0.002 [m/sec])
  - change moving direction(-$\frac{\pi}{4}$ [rad]) and slow down(-0.002 [m/sec])
  - change moving direction(+$\frac{\pi}{4}$ [rad]) and speed up(+0.002 [m/sec])
  - change moving direction(-$\frac{\pi}{4}$ [rad]) and speed up(+0.002 [m/sec])

For case 2, the speech behaviors are just symbols and there are no predifined means.

Case 2

- move forward
- change own moving direction ($\pm\frac{\pi}{4}$ [deg])
- change own moving velocity
  - slow down (-0.002[m/sec])
  - speed up (+0.002[m/sec])
- message
  – A – B – C – D

Here, we just give the symbols without meaning as messages. We utilize these two sets for learning system and observe the results.

5 Simulation

5.1 Setup

The robot's diameter is 0.36[m]. Each robot's speed is under 0.04 [m/sec], and over 0.01[m/sec]. The field is 3.0[m] × 3.0[m] and two robots(Robot A and B) are running in the environment.

Case 1
The experimental conditions in Case 1 are as follows.

- The destination for Robot A is given
- Robot B learns learns adaptive behavior generation.
- Robot B moves randomly. (It doesn't learn behavior.)
- Non-learning robot must follow the speech from Robot A

In this case, we attempt to confirm that the robot can acquire intention transmission manner on common protocol.

Case 2
The conditions for Case 2 are as follows

- Both of the destinations for the robots are given
- Both of the robot learns adaptive behavior generation.

In this case, we attempt to confirm that the robots can acquire intention meaning manner and its protocol.
One trial finishes when the learning robot reaches its destination, when the robot collides with the other, a wall, or when the learning loop is over 3000. After the learning trial, the position of robots and goal get new positions when every learnig trial starts. Here, sampling period is 1.0[sec]

5.2 Results

For collision avoidance simulation, we set the learning constant $\alpha = 0.04$, the discount factor $\gamma = 0.9$, the temperature constant $T = 0.2$ for Boltzman

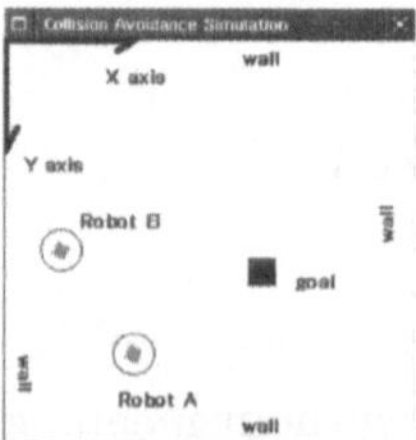

Fig. 2. Actual Simulation environment for Case 1

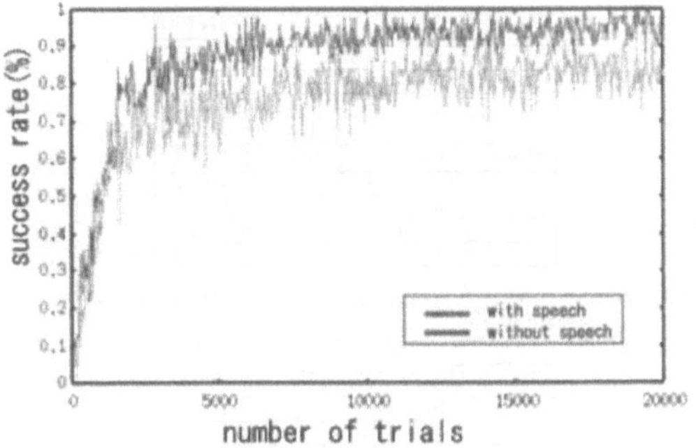

Fig. 3. Success rate

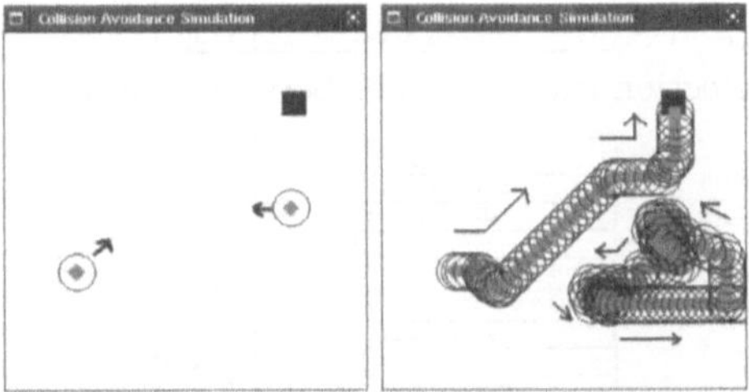

Fig. 4. An Example of Simulated Collision Avoidance Task

selection, and the parameters of the evaluation function $\kappa = 1.0$. As for velocity, we define over 0.03[m/sec] as high speed, less than 0.03[m/sec] and over 0.01[m/sec] as middle, and less than 0.01 [m/sec] as low. Regarding the size on the image plane, it is defined as large when the distance between robots is over 1.2[m], middle when it is less than 1.2[m] and over 0.6[m], and small when it is less than 0.6[m].

Case 1

We attempt learning simulations with speech and without speech for 20000 trials under Case 1 setup. Figure 3 show the success rate when the robot learns behavior with speech act and without speech act. Success rate depends on the times robot reached its goal in every 100 trials. Figure 4 shows examples of simulated collision avoidance task. Table 1, 2 and 3 show the behavior selection rate on the distance between the robots. Speech act 1 is the request "change course to $+\frac{\pi}{4}$ [rad] and slow down", speech act 2 is "change course to $+\frac{\pi}{4}$ [rad] and speed up", speech act 3 is "change course to -$\frac{\pi}{4}$ [rad] and slow down", and speech act 4 is "change course to -$\frac{\pi}{4}$ [rad] and speed up". It is clear that the case of learning with speech act is more adaptable to the task that requires reaching the goal without colliding with the other robot or obstacles. This shows that the robot learned the good

Table 1. Behavior selection rate (distance between robots is under 0.6[m])

action	rate[%]
move forward	7.95
slow down	6.59
speed up	8.31
change course ($+\frac{\pi}{4}$ [rad])	8.58
change course ($-\frac{\pi}{4}$ [rad])	10.56
speech act 1	13.89
speech act 2	15.42
speech act 3	11.98
speech act 4	16.69

Table 2. Behavior selection rate (distance between robots is from 0.6[m] to 1.2[m])

action	rate%]
move forward	17.52
slow down	15.02
speed up	15.07
change course ($+\frac{\pi}{4}$ [rad])	10.35
change course ($-\frac{\pi}{4}$ [rad])	10.96
speech act 1	6.62
speech act 2	5.48
speech act 3	7.29
speech act 4	11.65

Table 3. Behavior selection rate (distance between robots is over 1.2[m])

action	rate%]
move forward	13.67
slow down	13.40
speed up	10.49
change course ($+\frac{\pi}{4}$ [rad])	14.61
change course ($-\frac{\pi}{4}$ [rad])	13.95
speech act 1	7.52
speech act 2	9.91
speech act 3	7.71
speech act 4	8.69

usage of speech act, namely, faculty emerged. Moreover, the probability to select speech act is 57.98 [%] when the interval between robots is under 0.6[m], 31.04 [%] when it is from 0.6[m] to 1.2[m], and 33.83 [%] when it is over 1.2[m]. The robot selects actions according to the distance between robots, and speech is effective when the distance to the other robot is short. Thus, the robot learns when to transmit the intention through interactions with the environment, and communication facility emerged. Fig. 4 shows the examples of simulation result. Since this result, learning robot can acquire usage of message (intention) transmission under pre-defined protocol.

Case 2

We set the environment on Figure 5. The parameters and almost conditions are similar with Case 1. In this case, the initial point and the destination of each robot are fixed. The number of learning trials is 20000 using setup of

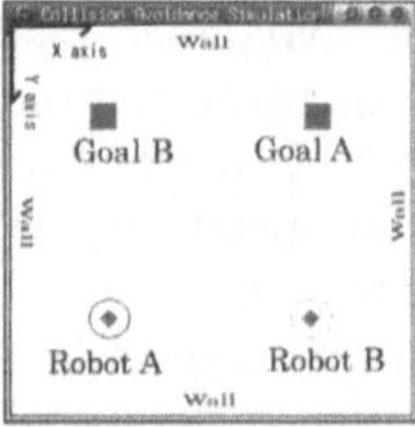

Fig. 5. Simulation Environment for Case 2

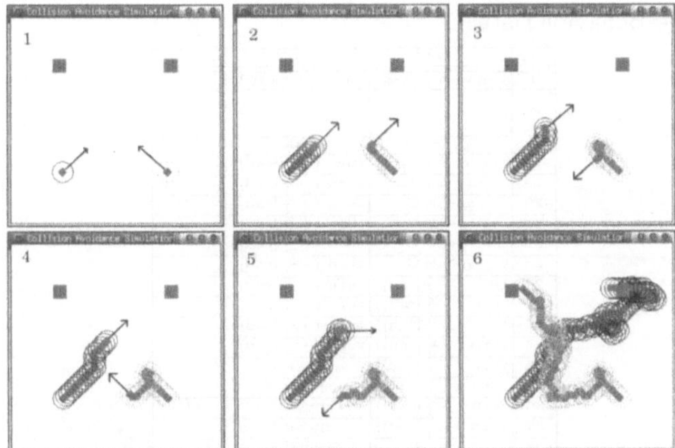

Fig. 6. A Trajectory of Each Robot

Case 2. An example of simulation result is shown on Figure 6. Detail procedure is also shown on Table. 4. By this result, it could be observed that each robot changes the behavior of the partner and realizes the collision avoidance, effectively. As the result, the robots realize to acquire the protocol and intention transmission to the partner.

6 Conclusion

In this paper, we have addressed communication emergence as the robot adapting the environment without giving communication manner, previously. Experimental results show that the action probability changes depends on the distance between robots, and the robot can learn the availability of communication where required. Thus, communication procedure emerges properly, and proposed system works effectively.

References

1. Y. Ishida, H. Asama, K. Ozaki, A. Matsumoto, I. Endo, Design of Communication System and Development of a Simulator for an Autonomous and Decentrized Robot System, *Journal of Robotics Society of Japan*, 10(4), 1992, 544-551 (in Japanese)
2. N. Hutin, C. Pegard, E. Brassart, A Communication Strategy for Cooperative Robots, *Proc. of IEEE/RSJ Intl.Conference on Inteligent Robots and Systems*, 1998, 114-119
3. H. Ynaco, L. A. Stein, An Adaptive Communication Protocol for Cooperating Mobile Robots, *From Animals to Animats 2*, 1993, 478-485
4. A. Billard, G. Hayes, Learning to Communicate Through Imitation in Autonomous Robots, *Artificial Neural Networks - ICANN'97*, 1997, 763-768

Table 4. State-Behavior Data

step	robotA	robotB	number
1	move forward	move forward	1
⋮	⋮	⋮	
17	speed up	speak A	
18	move forward	turn ($+\frac{\pi}{4}$ [rad])	
19	move forward	turn ($+\frac{\pi}{4}$ [rad])	
20	speak C	turn ($+\frac{\pi}{4}$ [rad])	2
21	speak C	turn ($+\frac{\pi}{4}$ [rad])	
22	speak C	turn ($+\frac{\pi}{4}$ [rad])	
23	speak C	turn ($+\frac{\pi}{4}$ [rad])	
24	turn ($-\frac{\pi}{4}$ [rad])	turn ($-\frac{\pi}{4}$ [rad])	
25	turn ($-\frac{\pi}{4}$ [rad])	turn ($-\frac{\pi}{4}$ [rad])	
26	turn ($+\frac{\pi}{4}$ [rad])	turn ($+\frac{\pi}{4}$ [rad])	
27	turn ($+\frac{\pi}{4}$ [rad])	turn ($+\frac{\pi}{4}$ [rad])	
28	turn ($-\frac{\pi}{4}$ [rad])	speak C	
29	turn ($-\frac{\pi}{4}$ [rad])	speak C	
30	turn ($+\frac{\pi}{4}$ [rad])	speak C	3
31	turn ($+\frac{\pi}{4}$ [rad])	speak C	
32	speak B	speak C	
33	speak B	speak C	
34	speak B	speak C	
35	speak D	speak C	
36	speak D	speak C	
37	speak D	speak C	
38	speak D	speak C	
39	speak C	speed up	
40	speak C	speed up	4
41	speak C	turn ($-\frac{\pi}{4}$ [rad])	
42	speak C	turn ($-\frac{\pi}{4}$ [rad])	
43	turn ($-\frac{\pi}{4}$ [rad])	slow down	
44	turn ($-\frac{\pi}{4}$ [rad])	slow down	
45	turn ($+\frac{\pi}{4}$ [rad])	turn ($+\frac{\pi}{4}$ [rad])	
46	turn ($+\frac{\pi}{4}$ [rad])	turn ($+\frac{\pi}{4}$ [rad])	
47	turn ($+\frac{\pi}{4}$ [rad])	turn ($-\frac{\pi}{4}$ [rad])	
48	turn ($+\frac{\pi}{4}$ [rad])	turn ($-\frac{\pi}{4}$ [rad])	
49	turn ($-\frac{\pi}{4}$ [rad])	speak C	
50	turn ($-\frac{\pi}{4}$ [rad])	speak C	5
⋮	⋮	⋮	
121	move forward	slow down	6

5. M. Asada, A. Noda, K. Tawaratsumida, K. Hosoda, Purposive Behavior Aquisition for a Robot by Vision-Based Reinforcement Learning, *Journal of Robotics Society of Japan*, 13(1), 1995, 68-74 (in Japanese)
6. R. S. Sutton, A. G. Barto, *Reinforcement Learning -An Introduction-* (MIT Press, Cambridge, Massachusetts,1998)
7. H. Asama, M. Sato, L. Bogoni, H. Kaetsu, I. Endo, Development of an Omni-Directional Mobile Robot with 3 DOF Decoupling Drive Mechanism, *Proc. of International Conference on Robotics and Automation*, 1995,1925-1930
8. K. Yokota, K. Ozaki, A. Matsumoto, K. Kawabata, H. Kaetsu, H. Asama, Omni-directional Autonomous Robots Cooperatinng for Team Play, *RoboCup-97 : Robot Soccer World Cup I* (Springer Verlag, Tokyo, 1998), 333-347
9. M. Hoshino, H. Asama, K. Kawabata, Y. Kunii, I. Endo, Communication Learning for Cooperation among Autonomous Robots, *Proc. of IEEE International Conference on Industrial Electronics, Control and Instrumentation*, 2000, SS38-SOR-4

Communication by Datagram Circulation among Multiple Autonomous Soccer Robots

Takuya Sugimoto[1], Takeshi Matsuoka[2], Toshihiro Kiriki[3], and Tsutomu Hasegawa[3]

[1] Hitachi Information & Control Systems, Inc.
[2] Fukuoka University
[3] Kyushu University
sugimoto@hasegawa.is.kyushu-u.ac.jp, matsuoka@ctrl.tec.fukuoka-u.ac.jp, kiriki@hasegawa.is.kyushu-u.ac.jp, hasegawa@hasegawa.is.kyushu-u.ac.jp

Abstract. This paper describes a communication method for multiple autonomous soccer robots. This method has advantages as compared with the conventional broadcast communication; the rate of loss of packets is low, it is guaranteed that communication data has reached other robots, and processing load of communication does not concentrate in a short period. Precise cooperation among robots requires sharing global information of the field in real-time. The proposed method is suitable for this kind of work. This method consists of two elements. One is the method of building a ring to move packets among robots. Another is the method of reconstructing a ring dynamically corresponding to temporary interruption of communication of a certain robot, exclusion of a robot, and inclusion of a robot. These two methods can substitute for the conventional broadcast communication.
keywords: communication, datagram circulation, robot soccer

1 Introduction

Communication is needed in order to perform obstacle avoidance or a task assignment etc., in a multiple autonomous robots system. Various communication methods corresponding to the characteristics of the various works are proposed[1][2][3][4]. In the work like exploration of environment, the communication is allowed to be done occasionally, because each robots mostly behave independently and don't need precise cooperation[2][3]. On the other hand, in the work like bearing a common load[5] or robot soccer[6][7][8][9], a real-time and close communication is required to achieve precise cooperation[4].

We propose a communication method among multiple autonomous soccer robots playing cooperatively in a dynamically changing soccer field. Our soccer robot is equipped with an omni-directional camera[10] as the vision sensor. Within a limited field of view of the vision sensor, each robot locally observes goal posts, other robots, and a ball. Then the robot calculates its own pose, the positions of colleague robots, the positions of opponent robots, the position of a ball, etc.. The results of the measurement always contains errors. In proportion as the distance between the camera and the observed object becomes further, the error will increases. In addition, the ball and the landmarks like goal posts are often occluded partly or entirely by other robots. The occlusion often disables the measurement. Especially the ball is very often occluded in the game. This causes difficulties in planning of the behavior of the robot. Coping with these problems, the robots communicate with each others to obtain the global and precise information of the field as much as possible. Based on this global information, the robots plan effective behavior of cooperation. The following two requirements have to be satisfied in communication of the robot soccer.

1. The communication must be done very fast and frequently to cope with the dynamic changes in the field. It must also be very reliable to assure successful cooperated behaviors.
2. The communication must be maintained even if the number of robots involved in the communication changes due to the nature of the robot soccer game. Some robot may be excluded because of the foul play during the game. Some robot may be excluded because of breakdown or malfunction of components and then included again after the repair or reboot of the system.

We have developed a communication protocol, which is a variation of token ring protocol[1]. This protocol achieves following two functions to satisfy the above conditions.

1. the function of constituting a ring of communication which provides efficient throughput of global information to be shared
2. the function of reconstructing the ring dynamically corresponding to the change of the number of the robots

2 Datagram circulation method

We deal with the robot soccer of the RoboCup Middle-size robot league[6]. In the rules for the league, using wireless LAN communication based on IEEE-802.11 standard is allowed. In case that a small number of robot share global information of the field, conventional broadcast method based on the UDP/IP protocol is often used. TCP/IP protocol is not used when a real-time communication is required. Because TCP/IP protocol does not guarantee that the datagram is send immediately when the user program commands the OS to send it. The following problems occur in the use of UDP/IP broadcast method.

1. In IEEE-802.11 standard, there is no re-sending function when a broadcasting packet is lost. Therefore, if many robots communicate simultaneously, the communication tends to be unstable because of lost of packets.
2. In the actual game, robots leave and rejoin the game frequently. When a robot doesn't receive a packet from a colleague robot for a long time

period, the robot can not judge whether the temporary interruption of communication occurs or the colleague robot is excluded from the game. Therefore, the robot may fail planning of the adequate behavior of cooperation.

3. Communications load may exceed the robot's processing capability when communication data concentrates in a short period.

To overcome these problems, we propose the datagram circulation method. The method uses unicast communication based on the UDP/IP protocol. An unique identification number is assigned to each robot. We call this number "Robot Number". We define and use the "Global Information datagram". This consists of local information each robots have. The information of one robot is its own pose, the positions of opponent robots, the position of a ball, the distance between the robot and the ball, the distance between the robot and the opponent goal, the number of opponent robots the robot can recognize, and some flags. "Global Information datagram" is relayed in order of the number from the small-numbered robot to the large-numbered robot. The largest-numbered robot sends the "Global Information datagram" to the smallest-numbered robot. In this way the ring is formed. We call the ring "Communication Ring"(Fig. 1).

Each robot has the number of the preceding robot and the number of the succeeding robot. We call the former the "Preceding Number" and call the latter the "Succeeding Number". When a robot relays the "Global Information datagram", the robot renews its own information and reads information of the other robots to plan an adequate behavior of cooperation. The robot which has received the "Global Information datagram" checks the renewal of information of the other robots and recognizes that the "Global Information datagram" has reached the other robots. Since the possible velocity

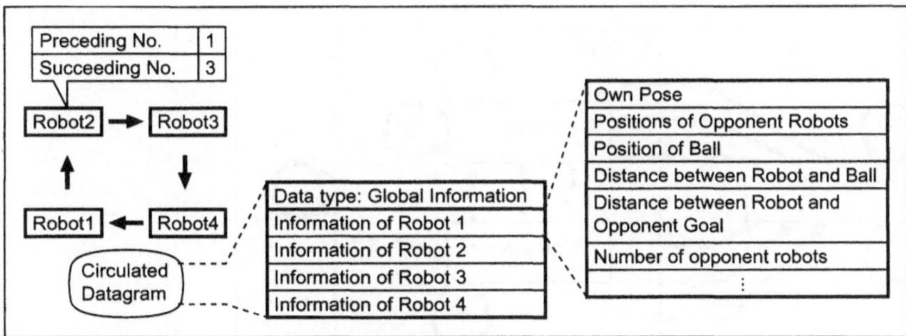

Fig. 1. Datagram circulation method

of the motion of the ball and the robots is limited to be rather slow, the robot does not need to communicate more frequently than the frequency of change of the state. In addition, the robot has to ensure sufficient CPU resource for real-time processes, for example, the vision process or the servo process, etc. Hence, the communication cycle needs to be maintained within the fixed range. To satisfy this condition, the robot delays the relay of received "Global Information datagram" if the receiving interval does not exceed the fixed time.

3 Reconstruction of the "Communication Ring"

The "Communication Ring" must be maintained coping with the temporary interruption of communication of a certain robot, the exclusion of a robot, and the inclusion of a robot. To satisfy this requirement, the communication process is designed to have three different states of function.

1. Stable communication state
   The process relays the "Global Information datagram". The "Communication Ring" is maintained well.
2. Confirmation state
   The process checks the soundness of function of the "Communication Ring". The process is triggered to enter this state when it doesn't receive the "Global Information datagram" within a fixed time. Once the soundness of the ring is confirmed, the process reconstructs the "Global Information datagram" and returns to the "Stable communication state". Otherwise the process issues an event to enter the "Reconstruction state".
3. Reconstruction state
   The process tries to reconstruct the "Communication Ring". If succeeded, the process reconstructs the "Global Information datagram" and resumes the "Stable communication state".

Nine different events trigger the transitions among the states(Fig. 2). "Ring

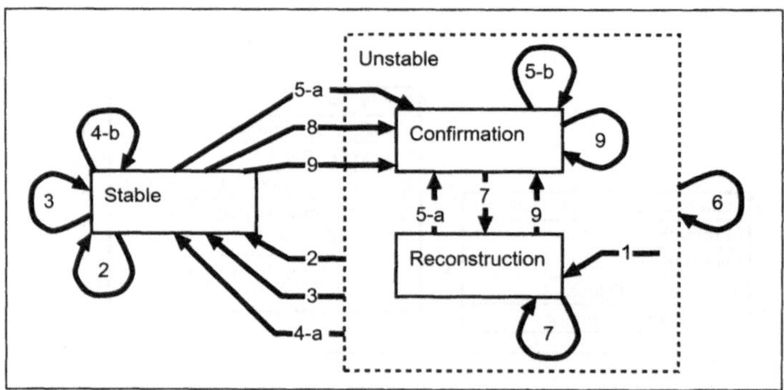

Fig. 2. Events and state transitions

Confirmation datagram" and "Ring Reconstruction datagram" are defined for ring reconstruction process. In addition to identification symbol entity, these datagrams have entities for "Sender" and "Info". "Sender" is the "Robot Number" indicating the robot which sent the datagram. "Info" is one of the

"Robot Number" to be used in various way for reconstruction of the "Communication Ring". Table 1 shows the state transitions of the communication process. In the table "Unstable" is used to indicate either "Confirmation" or "Reconstruction".

Table 1. Events and state transitions

event1:	Activation of the communication process
transaction	Whenever a robot try to be included in the "Communication Ring", it constructs the "Ring Reconstruction datagram". "Info" is the "Robot Number" of the robot. The process sends it to the "Preceding Number" robot.
transition	The process enters into "Reconstruction" state.
event2:	Receiving the "Global Information datagram"
transaction	The robot relays the "Global Information datagram" to the "Succeeding Number" robot.
transition	The state changes to "Stable" if the state is "Unstable". The state remains if the state is "Stable".
event3:	Receiving the "Ring Confirmation datagram" "Info" of the datagram > robot's own "Robot Number"
transaction	The robot relays the "Ring Confirmation datagram" to the "Succeeding Number" robot as it is.
transition	The state changes to "Stable" if the state is "Unstable". The state remains if the state is "Stable".
event 4–a:	Receiving the "Ring Confirmation datagram" "Info" = robot's own "Robot Number" (in "Unstable" state)
transaction	The robot relays the "Ring Confirmation datagram" to the "Succeeding Number" robot as it is.
transition	The state changes to "Stable" if the state is "Unstable". The state remains if the state is "Stable".
event 4–b:	Receiving the "Ring Confirmation datagram" "Info" = robot's own "Robot Number" (in "Stable" state)
transaction	The process ignores this event.
transition	The state remains as it is.
event 5–a:	Receiving the "Ring Confirmation datagram" "Info" < robot's own "Robot Number" (in "Stable" or "Reconstruction" state)
transaction	The process replaces the "Info" of the "Ring Confirmation datagram" by the robot's own "Robot Number". Then it sends the "Ring Confirmation datagram" to the "Succeeding Number" robot.
transition	The state changes to "Confirmation".

event 5–b:	Receiving the "Ring Confirmation datagram" "Info" < its own "Robot Number" (in "Confirmation" state)
transaction transition	The process ignores this event. The state remains as it is.
event 6:	1st timeout of receiving the "Ring Confirmation datagram"
transaction transition	The process sends the "Ring Reconstruction datagram" to the "Preceding Number" robot. The state remains as it is.
event 7:	2nd timeout of receiving the "Ring Confirmation datagram"
transaction transition	The process changes the "Preceding Number" to [Preceding Number - 1]. Then the process sends "Ring Reconstruction datagram" to the "Preceding Number" robot. The state changes to "Reconstruction" if the state is "Confirmation". The state remains if the state is "Reconstruction".
event8:	timeout of receiving the "Global Information datagram"
transaction transition	The process sends the "Ring Confirmation datagram" to the "Succeeding Number" robot. Info is the "Robot Number" of the sender. The state remains.
event9:	Receiving the "Ring Reconstruction datagram"
transaction transition	The process changes its "Succeeding Number" to the number indicated by the "Ring Reconstruction datagram". Then the process send the "Ring Confirmation datagram" to the "Succeeding Number" robot. The state changes to "Confirmation" if the state is "Reconstruction" or "Stable". The state remains if the state is "Confirmation".

3.1 Temporary interruption of communication

When a temporary interruption of communication occurs by interference of wireless LAN etc., the "Global Information datagram" will be lost. In this case, each robot sends "Ring Confirmation datagram" to confirm the function of "Communication Ring". Each robot recognizes the recovery of "Communication Ring" by detecting that the data has make a round in the "Communication Ring".

An example is shown in Fig. 3(i) and (ii). When a "Global Information datagram" sent from the robot 3 to the robot 4 is lost, each robot sends "Ring Confirmation datagram" almost simultaneously (Fig. 3(i)). After that, the "Ring Confirmation datagram" whose "Info" is 4 makes a round. When the "Ring Confirmation datagram" whose "Info" is 4 has arrived to the robot 4, the robot 4 recognizes that the "Communication Ring" is recovered. Fig. 3(ii) shows the circulation of the "Ring Confirmation datagram" sent from robot 1. Then the robot 4 constructs the "Global Information datagram" newly and sends it to the robot 1. Thus the "Communication Ring" is recovered.

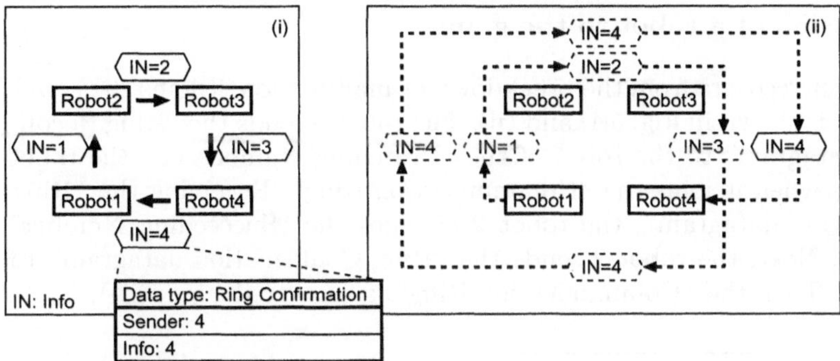

Fig. 3. Sending of "Ring Confirmation datagram"

3.2 Exclusion of a robot from the game

When a robot is excluded from the game because of foul, breakdown or malfunction etc., neighboring robot changes the "Preceding Number" or the "Succeeding Number" dynamically. Thus, the "Communication Ring" is maintained among remained robots.

An example is shown in Fig. 4(i),(ii) and (iii). When the robot 3 leaves the "Communication Ring", the "Ring Confirmation datagram" is lost. Each robot sends "Ring Confirmation datagram"(Fig. 4(i)). Then the robot 1 and the robot 2 receive "Ring Confirmation datagram". However, the robot 4 can not receive the "Ring Confirmation datagram" which the robot 3 is to send. The robot 4 changes the "Preceding Number" from 3 to 2. Then the robot 4 sends the "Ring Reconstruction datagram" to the robot 2(Fig. 4(ii)). The robot 2 receives the "Ring Reconstruction datagram" and changes its "Succeeding Number" to 4. Then the robot 2 sends the "Ring Confirmation datagram" to the robot 4 (Fig. 4(iii)). Thus the "Communication Ring" is recovered.

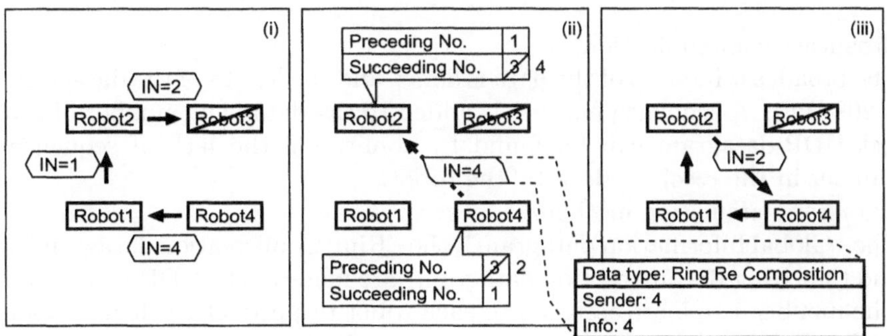

Fig. 4. Exclusion of a robot

3.3 Inclusion of a robot to the game

When a robot recovers from the breakdown or malfunction, it joins the game. An example is shown in Fig. 5(i) and (ii). The robot 3 sends the "Ring Reconstruction datagram" to the robot of the "Preceding Number"i.e., the robot 2 (Fig. 5(i)) when it joins the "Communication Ring". Receiving the "Ring Reconstruction datagram", the robot 2 changes the "Succeeding Number" from 4 to 3. Next, the robot 2 sends the "Ring Confirmation datagram" to the robot 3. Then the "Communication Ring" is recovered(Fig. 5(ii)).

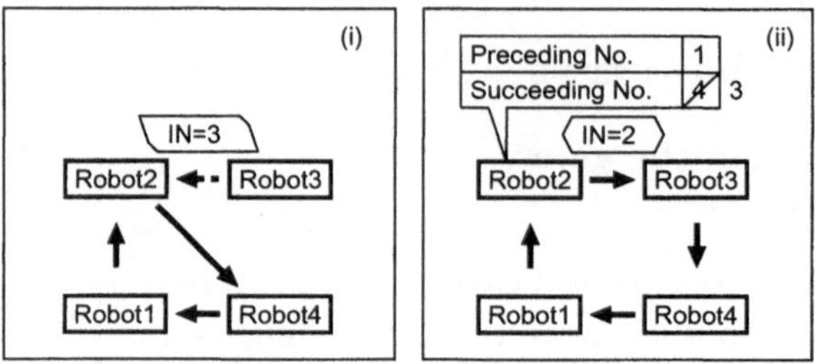

Fig. 5. Inclusion of a robot

4 experiment

4.1 The rate of loss of packets

We show an experiment of the measurement of the rate of loss of packets. In case that a robot joins the "Communication Ring", we compare the rate of loss of packets in the conventional broadcast method with that in the proposed method. We use ART-Linux on the celeron 800MHz CPU board with 128Mbyte RAM. The whole communication band of wireless LAN is used for the experiment. The experimental setup is as follows in consideration of receiving the data by 50ms of an average.

- Broadcast communication
  The broadcast interval of the data is 50ms, and the length of the datagram is 200-Byte. A sequence number is added to the data. The number of the lost UDP-datagram can be found by confirming the lack of sequence number in the received data by the robot.
- Datagram circulation method
  The "Global Information datagram", the "Ring Confirmation datagram", and the "Ring Reconstruction datagram" are made as the UDP-datagram. The number 1 to 5 are assigned to each robot respectively. When a robot received the "Global Information datagram", the robot sends it to the succeeding robot in 20ms. The length of the "Global Information datagram" is 1000-Byte composed of information of the 5 robots. The number

of the lost UDP-datagram can be found by detection of the receiving interval of the "Global Information datagram" exceeding 120ms. The rate of lost of communication data is defined as;

$$loss\ rate = \frac{number\ of\ lost\ packets}{number\ of\ lost\ packets\ and\ received\ packets} \quad (1)$$

Table 2 shows the results of the broadcasting and the datagram circulation method. The rate of loss of communication data of the latter is lower than that of the former.

Table 2. Data loss rate

Number of robots	Broadcasting (%)	Datagram circulation (%)
2	0.0	0.0
3	0.2	0.0
4	24.7	0.2
5	38.2	0.3

4.2 The recovery of the communication

We show an experiment of the recovery of the communication with our datagram circulation method. The dynamic reconstruction of the "Communication Ring" is confirmed in the following three events.

- Temporary interruption of communication
  When 5 robots are normally communicating, we erase the "Global Information datagram" on the robot 3. Then the time of the reconstruction of the "Communication Ring" among 5 robots is measured.
- Exclusion of a robot
  When 5 robots are normally communicating, we stop the communication process of the robot 3. Then the time of the reconstruction of the "Communication Ring" among 4 robots except for the robot 3 is measured.
- Inclusion of a robot
  When the robot 1, 2, 4, and 5 are normally communicating, we spawn the communication process of the robot 3. Then the time of the reconstruction of the "Communication Ring" among 5 robots is measured.

The experimental setup is similar to the setup of the preceding experiment with datagram circulation method.

Table 3 shows the experimental result. In each case, the dynamic reconstruction of the "Communication Ring" was achieved.

5 Conclusion

We proposed the communication system by datagram circulation, and evaluated the effectiveness by experiment. First, we described the method of building the ring which datagram moves by sending information to another robot

Table 3. Time for recovery of communication

Problem	Time for recovery(ms)
Temporary interruption of communication	245
Exclusion of a robot	410
Inclusion of a robot	300

in succession with using unicasting communication. Decline of the rate of loss of packets, the guarantee of information having reached other robots, and the effect of evasion of communications processing load concentrating in a short period are acquired by this method. Next, the method of dynamically reconstructing a ring coping with temporary interruption of communication of a certain robot, exclusion of a robot, and inclusion of a robot was described. This method can substitute for the conventional method using broadcast communication.

References

1. Bhargav P. Upender, Philip J. Koopman, Jr. (1994) Communication Protocols for Embedded Systems. Embedded Systems Programming, 7(11), 46–58.
2. Jing wang. (1994) On Sign-board Based Inter-Robot Communication in Distributed Robot Systems. Proc. IEEE Int. Conf. Robotics and Automation, 1045–1050
3. Shoji SUZUKI, Hajime ASAMA, Akira UEGAKI, Shin'ya KOTOSAKA. (1995) An Infra-Red Sensory System with Local Communication for Cooperative Multiple Mobile Robots. Proc. IEEE Int. Conf. Robotics and Automation, 220–225
4. Michael Mock, Edgar Nett and Stefan. Schemmer. (1999) Efficient Reliable Real-Time Group Communication for Wireless Local Area Networks. Proc. 3rd European Dependable Computing Conf., Prague, Czech Republic, 1999, 380–397.
5. Daniel J. Stilwell. (1994) Optimal Control for Cooperating Mobile Robots Bearing a Common Load. Proc. IEEE Int. Conf. Robotics and Automation, 58–63
6. M. Asada, S. Suzuki, M. Veloso, G. K. Kraetzschmar, H.Kitano. (1999) What We Learned from RoboCup-97 and RoboCup-98. Proc. IEEE/RSJ Int. Conf. Intelligent Robots and Systems, 1426–1431.
7. R. Polesel, R. Rosati, A. Speranzon, C. Ferrari, E. Pagello. (2000) Using Collision Avoidance Algorithms for Designing Multi-robot Emergent Behaviors. Proc. IEEE/RSJ Int. Conf. Intelligent Robots and Systems
8. Ashley Tews, Gordon Wyeth. (2000) Thinking as One: Coordination of Multiple Mobile Robots by Shared Representations. Proc. IEEE/RSJ Int. Conf. Intelligent Robots and Systems
9. C. Castelpietra, L. Iocchi, D. Nardi, M. Piaggio, A. Scalzo, A. Sgorbissa. (2000) Coordination among Heterogeneous Robotic Soccer Players. Proc. IEEE/RSJ Int. Conf. Intelligent Robots and Systems
10. Y. Yagi, H. Nagai, K. Yamazawa and M. Yachida. (1999) Reactive Visual Navigation based on Omnidirectional Sensing –Path Following and Collision Avoidance–. Proc. IEEE/RSJ Int. Conf. Intelligent Robots and Systems, 58-63.

Autonomous Robots Sharing a Charging Station with no Communication: a Case Study

Francois Sempé[1], Angelica Muñoz[2], and Alexis Drogoul[2]

[1] France Telecom R et D, 38 du Gal Leclerc, 92131 Issy-les-Moulineaux, France. francois.sempe@rd.francetelecom.com
[2] Laboratoire d'Informatique de Paris 6, 4 place Jussieu, 75252 Paris Cedex 05, France. angelica.munoz - alexis.drogoul@lip6.fr

Abstract. This research focuses on the design of a group of self-sufficient mobile robots, such that a group of three robots can remain in operation and efficiently share a charging station, using simple mechanisms. An experimental bottom-up approach has been adopted in order to test various strategies to manage collective self-sufficiency, which rely upon low-level mechanisms such as non-direct communication and non-complex decision making.
keywords: collective robotics, mobile robot, self-sufficiency, charging station, sharing strategy.

1 Introduction

This research focuses on the design of a group of self-sufficient mobile robots. **Self-sufficiency** denotes the ability of a system to maintain itself in a viable state for long periods of time [1], which means that a self-sufficient robot has to be able to ensure its power supply by itself. For that, it has at its disposal certain recharging facilities, *i.e.* rechargeable batteries and a self-recharge device, and relies on several mechanisms so as to be able to examine its power supply constantly and to locate and use a charging station. The design of a group of self-sufficient robots introduces new challenges, because a robot that is part of a group not only has to ensure its own operation, but also has to cope with its partners and share common, usually essential resources with them.

Self-sufficiency is at the core of the design of autonomous robots but surprisingly this topic has not been much addressed in the literature. Maybe it is not considered to be a *noble* issue but merely a technical one. Or maybe the specialized skills required to study the question of recharging batteries are not commonly found in robotic research teams. There is even less work on the self-sufficiency of a group of robots.

The research described in this paper is part of the MICRobES[1] project, which aims at implementing a group of mobile robots which will be able

[1] MICRobES is an acronym in French for Implementation of Robot Collectivities in a Social Environment.

to operate, or *survive*, permanently in the corridors of our laboratory. The paper describes the research to enable a group of three MICRobES robots to remain in operation and efficiently share a charging station, using simple mechanisms. An experimental bottom-up approach has been adopted in order to test various startegies to manage collective self-sufficiency which rely upon low-level mechanisms such as non-direct communication and non-complex decision making.

The paper is organized as follows: section 2 adresses the problem of sharing a station by a group of robots and considers related work. Section 3 describes our proposal to design self-sufficient robots, and gives details of the hardware and the control of our robots, as well as the framework of the experiments. Section 4 discusses the results and section 5 presents conclusions and future lines of research.

2 Sharing a charging station

2.1 The problem: to survive and to be useful

What is it expected of a group of robots sharing a charging station? There are two main goals. First, all robots must *survive* for long periods of time; second, they need to share the station efficiently, in order to carry out certain tasks.

What does efficient sharing mean? Let us reuse the concept of basic cycles presented by McFarland and Spier [2]. A basic cycle corresponds to the three stages a self-sufficient and useful robot has to repeat: *to work, to reach and connect to the station* and *to recharge the battery.* Roughly speaking, an efficient cycle maximizes the time spent working. How can that be done? Charging time is irrelevant here, as it cannot be modified by robot behaviors and depends mostly on charger speed. Supposing that it is constant, increasing working time means decreasing the time spent in reaching the station. For a single robot the best strategy will probably be to reduce the number of recharges by only trying to reach the station when the battery is low. For a group of robots, however another stage has to be considered: once a robot has reached the recharge area, it may have to wait because a partner has already occupied the station. Waiting time is a problem for two reasons: first, it represents a danger as it may lead to a flat battery and second, it is a waste of time that must be reduced.

What does a group of robots sharing a charging station need? In order to make sharing possible, the charging station must be sufficiently fast with respect to the number of robots. *Recharge speed* refers to the autonomy provided by, let us say, one minute of recharge. For example, a recharge speed of 5 means that one minute of recharge will give a robot 5 minutes of autonomy. Such a charging station can theoretically supply 6 robots with the same autonomy: once the first robot has gained five minutes of autonomy, five one-minute slots are available for five more robots. But in practice, a charging robot cannot change places instantaneously with another one and the station must be free for a while to let the robots switch. In fact, the longer this free time, the easier the sharing, and it should not be very difficult for three robots to share a station with a recharge speed of fifty. On the other hand it is difficult, not to say impossible, to build such a station.

What decisions do robots have to make? For the station-sharing problem, a robot has to take two major decisions: when to go and recharge and when to leave the station. There are many possible strategies, based on individual state, partner state, priority rules, negociation and so on.

2.2 Related work

The station-sharing problem has rarely been addressed in research and, as far as we know, no paper has been written just on this subject. Two research projects do exist, however. Steels [3] built an ecosystem where rivals have to compete for power supply. Two mobile robots share a station but no detailed results are presented. Michaud [4] suggested using artificial emotions in order to organize long term activity for a group of robots. However, no experiments involving physical robots have been presented yet.

The questions of recharge and self-sufficiency for one robot are a bit more common. Birk [5] points out the problem of batteries and shows that cell chemistry may constrain robot behavior. Last, a kind of sport record has been established by Yuta ans Hada [6]. They made a robot that ran continuously for a week recharging its battery every ten minutes.

3 The proposal

This paper proposes a bottom-up experimental approach to the station-sharing problem. It has been shown that a group of robots exhibiting simple behaviors can achieve non-trivial tasks without the help of communication[7], and we want to explore this further and to identify the conditions for such a sharing method.

3.1 A basic strategy for sharing

What could be a basic strategy for a robot team to share a station? As already mentioned above, two decisions have to be taken by each robot: when to go and recharge and when to leave the station. One of the first ideas that comes to mind is to set a go-and-recharge threshold that triggers the equivalent behavior. This threshold refers to an energy level. For instance, a 4-minute go-and-recharge threshold means that a robot tries to reach the station when less than four minutes of autonomy are left. But when should a robot leave the station? When it has reached a given level of energy. Thus all robots will leave the station with the same remaining autonomy.

This is a basic strategy but why should it work? The assumption is that robots will alternate at the station because their power supply will reach the go-and-recharge threshold at different times. In our case, since there is only one station to share and each robot leaves the station with the same amount of power supply, their autonomy at a given time is always different.

However, there is a bootstrap problem. What if the robots start with the same autonomy? They are going to rush to the station at the same time, which is probably a bad idea. We carried out two series of experiments on alternation: the first with robots starting with different energy levels, the second with robots starting with the same energy level.

3.2 Hardware

The experiments have been caried out using three Pioneer 2-DX mobile robots from ActivMedia©, provided with odometers, bumpers, sonars, radio modems and video cameras. We also have a charging station made by the LIP6 laboratory and France Telecom R & D. The robots have been modified in order to use this station. Both an electronic card to control the temperature and the current, and charging plates at the rear of the robots have been adapted. The original batteries have been replaced by lead batteries that accept high currents up to 60 Amps. The charging station has been designed to accept an approximate connection between it and the robots[2]. Robots take from 20 to 30 seconds to reach and connect to the charging station that they have just located and the time needed to recharge depends on the current supply of a robot and the age of its battery; on average 15 minutes are sufficient to provide 2 hours of autonomy.

The current supplied by the charger is very irregular. In order for experiments to be reproduced, the recharge speed percieved by the robot is constant, and below the actual but unpredictible one.

3.3 Control

Robots have a repertoire of basic behaviors that are activated by the external stimuli they perceive. Basic behaviors such as avoiding obstacles, wandering, navigation and localization have been designed for our research team. These behaviors have been successfully implemented in and tested on our robots, but discussion is beyond the scope of this paper (some details can be found in [8]).

In order to be self-sufficient, a robot has to examine its power supply constantly. When it perceives that it is below the go-and-recharge threshold,

[2] Patent pending

it goes towards the area where it knows that there is a charging station. If the robot recognizes that it is within this area, it starts to search for the exact position of the charging station by revolving around itself and goes in that direction when it perceives the visual landmark that identifies it. Then the robot adjusts its position to connect to the charging station, stays there while its battery is being recharged, and finally leaves the station to restart its loop. If a robot finds the station occupied, it wanders for 15 seconds and tries to connect again. Figure 1 illustrates the behavior described, for a robot whose main task is to wander.

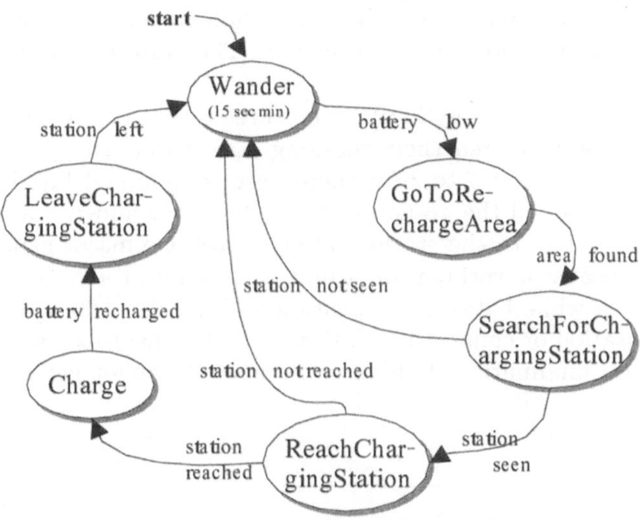

Fig. 1. Automaton that summarizes the behavior of an individual robot.

3.4 Experimental settings

The environment, about 20 square meter in area, is L-shaped *i.e.* since the station is not always in sight, navigation is necessary. Three Pioneers start at the same time. Each run lasts one hour at most or until a robot *dies*. The maximum autonomy has been set to 15 minutes in order to have more cycles. Two go-and-recharge thresholds are tested: 4 and 7.5 minutes. These values are discussed later. The recharge speed represents a critical parameter, as underlined earlier. In our experiments, the charger gives approximately 5 minutes of autonomy for 1 minute of recharge. As three robots work together, the station will be free more or less half of the time.

4 Results and discussion

4.1 Maintaining alternation

In all the experiments presented in this section, the three robots start with 5, 10 and 15 minutes of autonomy respectively. Thus, at the beginning alternation at the station should occur; the questions being the maintenance of alternation and its efficiency.

The basic strategy. Two experiments were done with a go-and-recharge threshold set at 7.5 minutes. In both cases, robots managed to survive for one hour before we interrupted the experiments. They alternated at the station but irregularly.

Figure 2 shows the energy level of the three robots during run 1. We can see that the robots started their charging stages (increasing functions) at very different levels, from 7 to 2 *i.e.* there were often long delays between the moment they triggered the go-and-recharge behavior and the moment they connected themselves to the station. Delays have one major cause: interference. This is because a working robot may wander into the recharge area and may stand for a while between the station and another robot that is trying to reach the station or connect itself. Consequently this robot will fail, either because station landmark is hidden or because it cannot get access to the station.

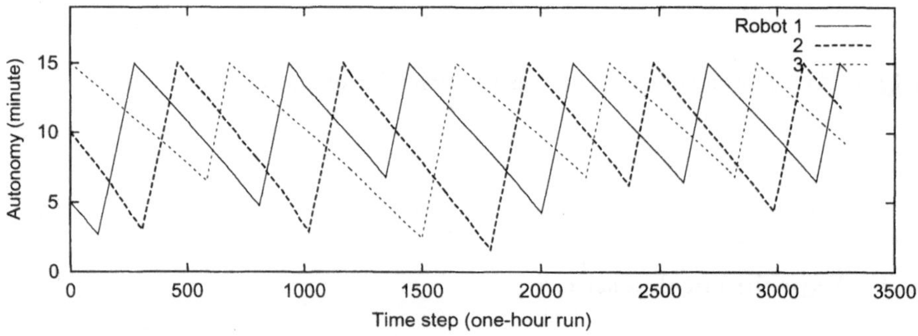

Fig. 2. The autonomy of the 3 robots with a 7.5-minute go-and-recharge threshold.

Sanctuary strategy. The interference problem is not surprising and it has already been pointed out [9]. In order to reduce failures caused by this phenomenon, we have chosen to prevent working robots from entering the station area. The benefit is clearly demonstrated if we compare the number of recharge tries shown in Table 1. Using the sanctuary strategy, the number of failures to reach the station has dropped drastically, though not yet entirely. There are three reasons for this persistence. First a robot may try to reach a busy station. Second, interference still occurs with a robot leaving the station after its recharge stage. Third, identification of the station using the camera fails from time to time.

The following strategies are based on sanctuary: working robots avoid the station area.

Table 1. Total numbers of attempts and success in connecting to the station. Basic and sanctuary strategies are compared.

strategy		*Total tries*	*Successes*
Basic	run1	28	15
	run2	30	15
Sanctuary	run1	18	15
	run2	21	17

Sharp strategy. This strategy is used to improve the efficiency of sanctuary strategy by decreasing the go-and-recharge threshold. Two experiments have been caried out with a 4-minute threshold. The first important point is the success of these new runs; 4 minutes are enough to reach the station and connect to it despite remaining connection failures.

Second, when sanctuary and sharp strategies are compared, we see that robots spend noticeably more time working if a sharp strategy is adopted, with a 5 pecentage point increase from 66.6% to 71.6%, on average. The explanation is simple: robots recharge themselves more rarely - approximately one time fewer in a one-hour run - and in consequence spend less time in reaching and connecting to the station. Another benefit must be mentioned: the average duration of working periods increased too.

4.2 Creating alternation

The failure of the sanctuary strategy. It has been shown above that the robots alternated efficiently at the station, thanks to the different initial energy levels. But what would happen if robots started with the same autonomy? In other words, how can the alternation be created?

Let us now consider the case of robots that all start with 10 minutes of autonomy. No experiment is necessary to show that a 4-minute recharge threshold will lead to a disaster as more than two minutes are necessary to recharge one robot. The 7.5-minute threshold has been tested twice. In both cases one of the robots *died* before its first recharge: three robots cannot recharge themselves within a period of 7.5 minutes. A negative side effect makes the situation worse: as the robots reached the threshold approximately at the same time, they all rushed together to the station, thus creating a lot of interferences. Increasing the threshold is not the right solution, since the robots has to spend most of their time working.

Opportunism : a way to create alternation. The new strategy has to exhibit two contradictory properties: greed and priority for a starving robot. On the one hand, when they start robots must try to go and recharge as soon as possible in order to use all their remaining autonomy for a vital purpose. On the other hand, a starving robot must access the station more easily than the others. But remember that no robot knows anything about the other robots' state.

Opportunism is the behavior that makes a robot recharge itself when it sees the station, whatever its remaining autonomy. In theory, associating opportunism with a go-and-recharge threshold regroups the two properties. First, greed, because robots may use the station before the go-and-recharge threshold is reached, *i.e.* before they really need energy. Secondly, starving robots have a better chance of reaching the station because their behavior is deterministic: even if they cannot see the station they stay around and check for it regularly.

Figure 3 shows the power supply of an opportunistic strategy experiment. Opportunistic behavior may be triggered only when remaining autonomy is less than 10 minutes: we do not want too greedy robots. Alternation was successfully created and maintained. On this run, no robot failed to connect to the station: nineteen connection attempts led to nineteen recharges. None had to wait, since a robot charging its battery hides the station landmark, thus inhibiting opportunistic behavior. On the other hand, robots went and recharged very early, reducing the duration of working period (7.2 minutes on average).

Opportunism is necessary only at the first stage, to create the alternation, after which the sharp strategy proved to be efficient. In order to take advantage of both strategies, a temporary opportunism was implemented. Robots are opportunistic for two cycles, and then adopt a pure sharp strategy. Two successful experiments have shown the creation of alternation and long working periods (more than 10 minutes).

Last, figure 4 illustrates how opportunism can cope with robots starting with different energy levels. We can see at the beginning that the most starving robot has been overtaken by another one because of an opportunistic behavior. However the first one waited until the end of the recharge and connected the station in time.

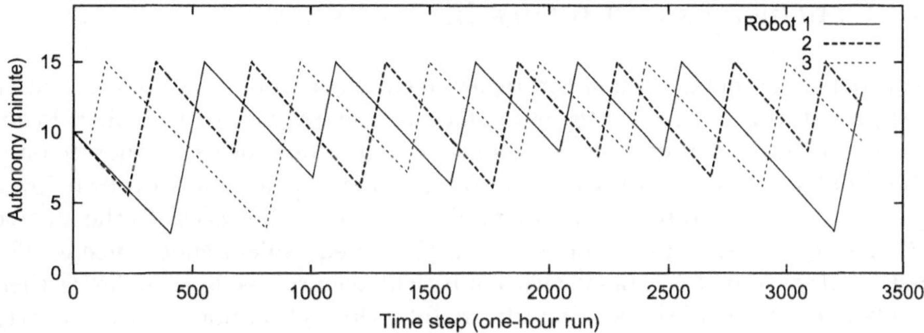

Fig. 3. The autonomy of the 3 robots with an opportunistic strategy.

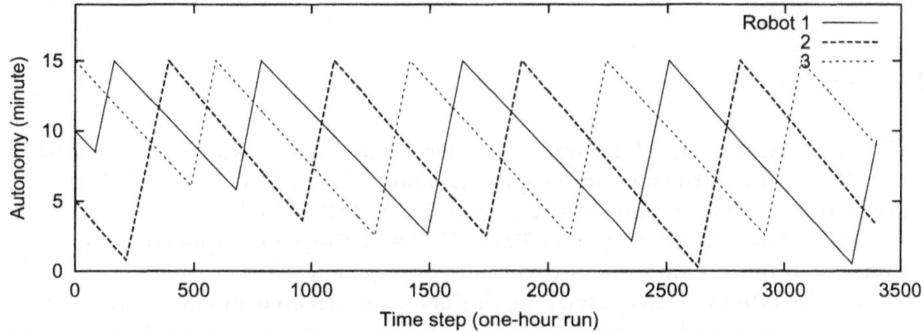

Fig. 4. The autonomy of the 3 robots with a temporary opportunistic strategy.

4.3 Discussion about the go-and-recharge threshold

The go-and-recharge threshold is at the core of the strategies presented above. This section discusses its properties in relation to two other parameters: the time necessary to reach the station and the average recharge time of one robot.

Obviously, the go-and-recharge threshold must be set above the maximum duration of the navigation and connection phases. In an extensive environment, it should depend on the distance between the location of the robot and that of the charging station. Interference with other robots, navigation and connection problems must be taken into account; with our experimental settings a 4-minute threshold was enough.

In our case, the navigation stage rarely exceeded 1 minute and the connection stage a mere 30 seconds. With a 4-minute threshold, this left more than 2 minutes of autonomy, which is the approximate duration of a recharge. Thus a robot can wait for another one to finish recharging, which is why the opportunistic strategy works.

5 Conclusions and future lines of research

This paper presented various strategies to enable a group of robots to share a charging station. These strategies are based on simple mechanisms that do not suppose communication between robots. The first ones use a go-and-recharge threshold: a robot tries to reach the station when its autonomy drops below a given value. The strategies are oriented to maintain the access to the station alternately and were tested on robots with non-equivalent energy needs. The last strategy enables robots with equivalent energy needs to balance their needs and access to the station alternately, through opportunistic recharge behavior. Results of the different strategies are presented and compared.

Future work will focus on experiments with groups of more robots in more complex situations, in order to analyze the influence of the size of the group on the strategies, and to improve mechanisms to share a charging station.

References

1. D. McFarland.(1995) Autonomy and Self-Sufficiency in Robots. The Artificial Life Route To Artificial Intelligence: Building Embodied, Situated Agents. L. Steels (ed). Lawrence Erlbaum Ass. Pub. USA, 187–213.
2. D. McFarland, E. Spier. (1997) Basic Cycles, Utility and Opportunism in Self-sufficient Robots. Robotics and Autonomous System, 20, 179–190.
3. L. Steels. (1994) A case study in the behavior-oriented design of autonomous agents. From animals to animats 3: Proceedings of the 3rd International Conference on Simulation of Adaptive Behavior, 445–452
4. F. Michaud, E. Robichaud, J. Audet. Using Motives and Artificial Emotions for Prolonged Activity of a Group of Autonomous Robots. Proceedings of the AAAI Fall Symposium on Emotions. Cape Code Massachussetts.
5. A. Birk.(1997) Autonomous Recharging of Mobile Robots. Proceedings of the 30th International Symposium on Automative Technology and Automation. Isata Press.
6. S. Yuta, Y. Hada. (2000) First Stage Experiments of Long Term Activity of Autonomous Mobile Robot: Result of Repetitive Base Docking over a Week. Proceedings of ISER'00, Seventh International Symposium on Experimental Robotics, 235–244.
7. R. Beckers, O.E. Holland, J.L. Deneubourg. (1994) From Local Actions to Global Tasks: Stigmergy and Collective Robotics. Artificial Life IV, 181–189.
8. P.E. Viel., A. Drogoul A., M. Milgram. (2001) Localization and Identification: a Robust Landmark Recognition Systems for Mobile Robots. Technical report LIP6, UPMC, France.
9. D. Goldberg, M. Mataric.(1997) Interference as a Tool for Designing and Evaluating Multi-Robot Controllers. Proceedings of the AAAI-97 Conference, 637–642.

Chapter 4
Human-Machine Cooperative Interaction

A Scalable Command and Control System for Human-Machine Work Systems

Aaron C. Morris, Charles K. Smart, and Scott M. Thayer

Robotics Institute
5000 Forbes Avenue
Carnegie Mellon University
Pittsburgh, PA 15213
email: acmorr@ri.cmu.edu, cks@andrew.cmu.edu, sthayer@ri.cmu.edu

Abstract. The emergence of complex work systems has yielded new challenges for efficient and reliable collaboration between humans and machines. Robots are now working autonomously beside human counterparts to accomplish critical tasks; however fully autonomous robot action is still considered unreliable. This paper examines an approach to increasing the robustness, reliability, and efficiency of human-machine work systems by dynamically establishing dynamic control relationships between humans and robots as well as altering the effective autonomy manifested by each robot. The process involves **workload estimation** to determine the parameters of the system, **workload optimization** to analyze and modify system parameters, and **workload mitigation** to enact these modifications in a non-intrusive manner. Furthermore, heuristic approaches to approximating an optimal system configuration for real-time environments are also addressed and simulated.

Key words: human-machine work systems; workload estimation, optimization, and mitigation

1 Introduction

Unstructured and hostile environments impose risk to exposed humans and present ideal domains for robotic applications; however, the conditions of these environments present significant challenges (i.e. complex obstacles, lack of sufficient prior knowledge, etc.) for autonomous robot operation and can impede a robot from achieving a desired goal. The emerging use of multiple robots has compounded these problems since the success of one robot may depend upon the performance of other robots within the collective.

To mitigate these challenges, an approach that integrates human cognition into the robot control structure is proposed. As first discussed in [1], researchers from a spectrum of fields have developed technologies aligned with this theme. Some recent examples of this research include human-computer interfaces [2], distributed software agents [3], human factors modeling [4], and adjustable autonomy [5]. Even components of human thought have been modeled [6] [7] in order to create a more seamless connection between human and computer.

This paper discusses the enabling merger of computational cognitive estimation with robot control theory to produce a system enabling a small group of humans to manage a larger group of semi-autonomous robots. The bulk of the work outlined herein examines the mathematical model used to create this linkage between human cognition and robot control. This work is presented as a three-part process (**workload estimation**, **workload optimization**, and **workload mitigation**) detailing the system representation. Finally, an analysis of simulations is presented that portends improvements in human-machine control.

2 The Human-Machine Work System

A human-machine work system is a collaboration of multiple heterogeneous "natural and artificial cognitive systems" [2] engaged in the execution of shared tasks. The corresponding notation for representing this system is $\mathbf{H} = \{h_1, h_2, ..., h_M\}$ for M humans, $\mathbf{R} = \{r_1, r_2, ..., r_N\}$ for N robots, and $\mathbf{T} = \{t_1, t_2, ..., t_P\}$ for P tasks. Robots are also decomposed into two sub-layers identified as subsystems and functions. A subsystem describes a particular interface, behavior, or capability of a robot and is denoted as $r = \{s_1, ..., s_i, ..., s_K\} \Rightarrow r \subseteq \mathbf{S}$ where s_i is a subsystem (i.e. *navigation*, *target recognition*, *manipulation*, etc) within the set of all subsystems **S**. A function represents a particular action composing the robotic subsystem and is denoted as $s_i = \{f_1, ..., f_j, ..., f_J\} \Rightarrow s_i \subseteq \mathbf{F}$ where f_j is a particular function (i.e. *sensor feedback, path planning,* and *steering*) within the set of all functions **F**. Furthermore, **T**, **H**, **R**, **S**, and **F** create a system that operates by estimating workload parameters, optimizing these parameters through analysis of system configuration, and enacting an optimized configuration through alteration of robot autonomy and human-function paring (Figure 1).

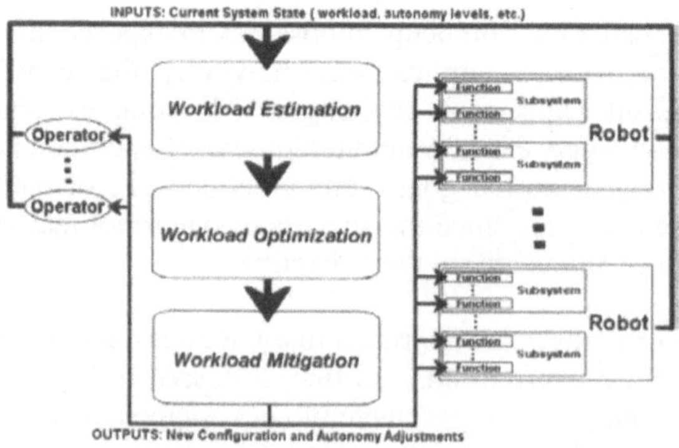

Fig. 1. Architecture of the human-machine work system.

From these baseline definitions, human-machine work systems can be evolved into complex structures that consist of multiple tasks distributed over multiple robots and managed by multiple operators. To formalize and constrain system organization, the following axioms will be used.

(1) A function can be controlled by only one operator
(2) An operator can control multiple functions
(3) Tasks can occupy multiple subsystems
(4) A single subsystem can concurrently execute multiple tasks

3 Workload Estimation

The human's role as an administrator requires the injection of control signals at the robot functional level. Human cognition, however, is an exhaustible resource that restricts the number of functions they can manage. Additional activities beyond a given threshold will sacrifice the performance of some (or all) concurrent activities [4]. To quantitatively monitor this threshold, the ACT-R (Adaptive Character of Thought – Rational) cognitive architecture [6] is utilized. According to ACT-R, human cognition can be separated into six distinct, quantifiable areas consisting of working memory, long-term memory, vision, speech, motor, and audition. With this representation, human cognitive capacity can be modeled in a vector form, $\vec{\Lambda} = [\lambda_{wm}, \lambda_{ltm}, \lambda_v, \lambda_s, \lambda_m, \lambda_a]^T$, to quantify the capacity within each area of the human brain.

In a similar manner, a formal means of representing the induced cognitive load of human interaction with functional components is derived. The cognitive loading resulting from human interaction with robotic functions is defined as $\vec{\Gamma} = [\varphi_{wm}, \varphi_{ltm}, \varphi_v, \varphi_s, \varphi_m, \varphi_a]^T$ corresponding to the cognitive load induced upon a particular region of the human brain as measured by ACT-R.

It is important to note that the level of autonomy governing the behavior of a robot function will have a profound impact upon the induced cognitive load [7]. For example, robot functions operating under full autonomy may require little or no attention from the operator and consequentially, the induced cognitive load will be relatively small. The approach to modeling this effect is discussed in the following section.

4 Adjustable Autonomy and Workload Mitigation

To enable environmental adaptation, this system requires that robot autonomy dynamically vary as the robot encounters uncertainties. Adjustable autonomy ([8] and [5]) allows robot functions to be controlled at various levels of human interaction so that the shared workload of humans and robots may be mitigated to the constraints of the work system. For example, a robot may require only minimal supervision during a navigation task until complex terrain obstructs its path. A lower level of autonomy may then be invoked so that human control assists the robot in avoiding obstacles. Adapted from [3], we define four general autonomy levels for robot functions:

1. **Fully autonomous:** the highest level of autonomy where control is determined by intrinsic functional capabilities. Operators may periodically monitor and asynchronously alter the progress of a unit, but the robot is otherwise independent.
2. **Semi-Autonomous:** the autonomy level that maintains the independence of robot functions with minimal intervention from the human operator. Interaction between operator and function is infrequent, yet sometimes critical, causing small or moderate cognitive loadings.
3. **Indirect Manual Control:** the level of autonomy that requires human intervention for discrete yet frequent time intervals. The cognitive demand resulting from function control can vary from moderate to relatively high.
4. **Direct Manual Control:** the lowest level of autonomy that requires continuous and direct functional control. The cognitive demand under this level of autonomy is consistently high (and perhaps maximal).

Any of these four levels of autonomy may be invoked during functional execution of a task. Both robot functions and human operators have the capacity to request an autonomy adjustment. Adjustment can impact the cognitive load applied to the human operator: either amplifying or attenuating the load depending upon the direction of transition. To represent adjustable autonomy $\vec{\Gamma}$ must be to vary as the level of autonomy varies. By this definition, an autonomy alteration can be represented as

$$\vec{\Gamma} \xrightarrow{\text{autonomy adjustment}} \vec{\Gamma}' \text{ or } T(\vec{\Gamma})_{\substack{autonomy\\ adjustment}} \rightarrow \vec{\Gamma}' \text{ where } T \text{ is a linear transform.}$$

To compute T, the four autonomy levels are assumed to be qualitatively consistent across the entire set of functions. This assumption is reasonable since the same joystick and monitor set-up used to control a *steering* function for the *navigation* subsystem will also control a *guidance* function for the *manipulation* subsystem. Thus, for a given function under a base autonomy level, the resulting cognitive load under an alternative autonomy level can be modeled as a scaled multiple of the base level or $\varphi_i' \propto \varphi \Rightarrow \varphi_i' = \delta_{ik}\varphi_i$ where each $i \in \{wm, ltm, ...a\}$ is a cognitive component at a default autonomy level and δ_{ik} is the constant of proportionality for the desired level of autonomy $k \in \{level1, ..., level4\}$. A diagonal scaling matrix Δ_k is used to map functional cognition vectors at the default autonomy level to a new level k so that $\Delta_k\vec{\Gamma} = \vec{\Gamma}'$

Furthermore, by selecting $\mathbf{\Delta_k}$ to be isomorphic, a transform from level *k* with $\mathbf{\Delta_k}$ to level *k'* with $\mathbf{\Delta_{k'}}$ can be expressed as $\mathbf{\Delta_{k'}}\mathbf{\Delta_k^{-1}}$. Hence, the transform from any autonomy level to another is determined by multiplying the cognitive loading vector by $\mathbf{\Delta_{k'}}\mathbf{\Delta_k^{-1}}$. The existence of an inverse matrix $\mathbf{\Delta_k^{-1}}$ to perform level jumping is a key benefit for modeling autonomy adjustment in the form of a linear transformation. In addition, the use of linear transformations (based upon the properties of these transformations to act on the basis of a vector space [10]) also minimizes:

1. Prior knowledge of the system
2. The data representation of the system
3. The computational time required obtaining a new level of autonomy.

In fact, the construction of each $\mathbf{\Delta_k}$ would require prior knowledge of cognitive loading values for only one function across all autonomy levels. The subsequent scaling components can then be acquired by taking ratios of components from these known cognitive loading vectors.

5 Workload Representation

Given $\vec{\Lambda}$ for each operator within the human population and the $\vec{\Gamma}$ for each robotic function within the set **F**, the relationship binding humans to robot functions can be stated as: for each $h_i \in \boldsymbol{H} \rightarrow h_i \Leftrightarrow \vec{\Lambda}_i$ where i = 1 to M, a matrix of the from $[\vec{\Lambda}_1^T, \vec{\Lambda}_2^T, ..., \vec{\Lambda}_M^T] = \mathbf{H_C}$ is created. This 6×*M* matrix represents the entire human cognitive capacity for a human-machine work system. In addition, the previous representation can be extended to functions, subsystems, and robots by stating for each $f_i \in \boldsymbol{F} \rightarrow f_i \Leftrightarrow \vec{\Gamma}_i$ where i = 1..J, a matrix of the form $[\vec{\boldsymbol{\Gamma}}_1^T, \vec{\boldsymbol{\Gamma}}_2^T, ..., \vec{\boldsymbol{\Gamma}}_J^T] = \mathbf{F_C}$ is created. This 6×*J* matrix represents the functional cognitive loading for the all system functions. Additionally, using the definition for robots, subsystems, and functions, the following is derived:

$$r = \{\{f_{11},..,f_{1j},..,f_{1Y}\},...,\{f_{i1},..,f_{ij},..,f_{iY}\},...,\{f_{X1},..,f_{Xj},..,f_{XY}\}\}.$$
$$r = \{\{\vec{\boldsymbol{\Gamma}}_{11}^T,..,\vec{\boldsymbol{\Gamma}}_{1j}^T,..,\vec{\boldsymbol{\Gamma}}_{1Y}^T\},...,\{\vec{\boldsymbol{\Gamma}}_{i1}^T,..,\vec{\boldsymbol{\Gamma}}_{ij}^T,..,\vec{\boldsymbol{\Gamma}}_{iY}^T\},...,\{\vec{\boldsymbol{\Gamma}}_{X1}^T,..,\vec{\boldsymbol{\Gamma}}_{Xj}^T,..,\vec{\boldsymbol{\Gamma}}_{XY}^T\}\}$$

Each robot *r* in the population of *N* robots is composed in this manner. No explicit association among functions, subsystems, and robots is enforced since axioms **(1)** and **(2)** state the existence of a direct relationship between humans and functions; however, system designers are free to enforce any required constraints. In this manner, an operator constrained to control a particular robot can equivalently be constrained to control only those functions associated with that robot. This $6\times(J{\cdot}K{\cdot}N)$ matrix (denoted $\mathbf{R_c}$) can be used to calculate the cognitive loading of a work system (denoted $\mathbf{H}_\mathbf{L}$) with the equation $\mathbf{R_C M} = \mathbf{H_L}$. $\mathbf{M}$ is a $(J{\cdot}K{\cdot}N)\times M$ Boolean configuration matrix that has the effect of summing all $\vec{\Gamma}$s in $\mathbf{R_c}$ that are controlled by a particular operator.

5.1 Workload Optimization

The next objective of human-machine work systems is to maximize the number of functions under human supervision while concurrently minimizing the cognitive load induced on each operator. These constraints act in opposition to each other: maximization seeks to consume human cognition while minimization tends to withhold it. To resolve the conflicting constraints, an optimization procedure must create a reasonable balance based upon the current state of the system.

Minimization: Recall that every column in $\mathbf{H_C}$ is a cognition capacity vector $\vec{\Lambda}$ and every column in $\mathbf{H_L}$ is cognitive loading vector defined as

$$\vec{\Gamma}_j = \sum_{k=1}^{J\cdot K\cdot N} \vec{\Gamma}_{kj} m_{kj}$$

This definition implies that the jth operator having a cognitive capacity of $\vec{\Lambda}_j$ must command a set of functions inducing a cognitive load of $\vec{\Gamma}_j$. Thus, the minimizing constraint requires $\lambda_{ij} \geq \varphi_{ij}$ for each element λ_{ij} in $\vec{\Lambda}_j$ and φ_{ij} in $\vec{\Gamma}_j$ for all $i \in \{wm, ltm, ...a\}$ ensuring that operators do not become cognitively overloaded.

Maximization: Let Z be number of all-zero rows in the matrix $\mathbf{M}$ (i.e. functions assigned to no operator). Recall M is the number of columns in $\mathbf{M}$ and $(J{\cdot}K{\cdot}N)$ is the number of rows in $\mathbf{M}$. The maximizing constraint desires that the number of managed functions be greater than the number operators in the system or $(J \cdot K \cdot N) - Z > M$ ensuring the maximum number of functions is controlled.

Ideally, all functions should be operator supervised; however, complete functional coverage for a relatively large system may not be achievable, especially during periods of heavy system activity. This complication creates a trade-off between the number of functions obtaining operator attention and the number of functions each operator can adequately manage. Therefore, a cost metric must be established in order to determine the optimal functional assignment given the cognitive capacity of each operator and the optimal level of autonomy governing each function.

5.2 Cost Metric

The cost metric encodes a number of systemic parameters that reflect the situational significance of a function. These parameters assist in determining an optimal configuration of humans, functions, and autonomy levels.

- **Functional Priority**: encodes the relative priority of a function. It provides precedence to critical functions when cognitive loading becomes excessive.
- **Initial Configuration**: is the initial configuration matrix $\mathbf{M}_0$ allowing the system to preserve the default grouping structure whenever possible.
- **Pairing Authorization**: encodes the permission granting or forbidding the existence of a pairing so that certain robot functions (i.e. those functions controlling weaponry or sensitive sensory equipment) are available to only authorized operators.
- **Request Wait Time**: represents the elapsed time of a function's or operator's request to adjust autonomy in an attempt to prevent starvation: the act denying functions from receiving operator attention.
- **Control Inertia**: records the time elapsed between a linked operator and function (provided the function is not fully autonomous) to prevent unnecessary context switching between operator-to-function assignments

By quantifying these parameters, the cost function can be tailored to suit the requirements of any work system and allow efficiency to be determined by comparing relative system cost. For example, the minimum system cost can be cast as the goal for searching the configuration space. The resulting configuration defined at the goal will be the system's **M**.

5.3 The Workload Optimization Algorithm

The Transportation Algorithm is a well-documented problem of optimization that parallels the intentions of workload optimization. In summary, the Transportation Algorithm involves the determination of an optimal shipping network for groups of suppliers and consumers. Similarly, workload optimization involves the determination of an optimal configuration of operators (suppliers of cognitive capacity) and functions (consumers of cognitive capacity). Despite the similarities, the baseline Transportation Algorithm cannot be directly applied to workload optimization due to the following complications:

1. The Transportation Algorithm optimizes scalar quantities whereas scalable command and control relies upon vector quantities
2. The Transportation Algorithm produces a shipment matrix consisting of real-number values with multiple suppliers connected to a single consumer whereas scalable command and control produces a configuration matrix of Boolean values with a single operator per function

Complication (1). This situation requires modification to the vector components. As such, the cognitive capacity and loading vectors are collapsed into scalar components by selecting the smallest component from each $\vec{\Lambda}$ for operator representation and the largest component from each $\vec{\Gamma}$ for function representation. This reduction of dimensionality does affect the optimality of the configuration; however, the approximation drastically reduces the configuration space to improve the speed of computation.

Complication (2). This situation requires manipulation of continuous flow into discrete containers. To approximate an optimal configuration, the maximal flow component from each column of the configuration matrix is selected as the human-function match. Occasionally, selecting the maximal component can lead to cognitive overloading; however, if this case occurs, the work system varies the level of autonomy for any overloading functions and reprocesses the configuration.

Finally, these solutions are only reasonable when the cognitive loading vectors are small relative to cognitive capacity vectors. When these vectors are on the same order of magnitude, the accuracy of approximation algorithm will degrade resulting in a systemic tendency to unnecessarily increase levels of functional autonomy.

6 Simulation Results and Conclusions

To demonstrate the capabilities of scalable command and control in human-machine work systems, a simulation of the system described in sections 3, 4, and 5 was implemented. The cost function used during the simulation was of the form $\mathbf{C} = f(\mathbf{M}_0, i, j) \cdot e^{g(\mathbf{p}_i, \mathbf{q}_i, \mathbf{Z}_{ij})} \cdot (1 + \alpha \mathbf{A}_{ij})$ where **C** is cost; $\mathbf{M}_0$ is the preferred configuration; **p** is functional priority; **q** is request wait time; **Z** is task inertia; and **A** is the authority. The function *f*(*) returns a constant that reflects the grouping assignment while *g*(*) returns a composition of the priority, request time, and control inertia. This composition is placed into an exponential operator to give numerical importance to the mentioned parameters. Finally, α is selected to be significantly large such that when operator-robot parings are prohibited, $\mathbf{C}_{ij}$ becomes extremely costly thereby effectively prohibiting the potential pairing.

To obtain the results in Table 1, 10 trials were simulated at each human-robot configuration. Each simulation lasted 100 iterations; each robot had between one and two subsystems (randomly selected); and each subsystem was allowed to have between one and three functions (randomly selected). To serve as a comparison, a second system was created using a fixed-assignment configuration where operators were only permitted to control pre-assigned functions up to their cognitive capacity. Autonomy alterations were requested on a Gaussian distribution (Figure 3). Finally, 10% noise was added to simulate random disturbance in workload estimation and communication errors. The results include the average number of functions (AF), the unassigned function per iteration (UF), the unassigned functions with high priority per iteration (UFHP), total autonomy switches (TAS), and total context switches (TCS).

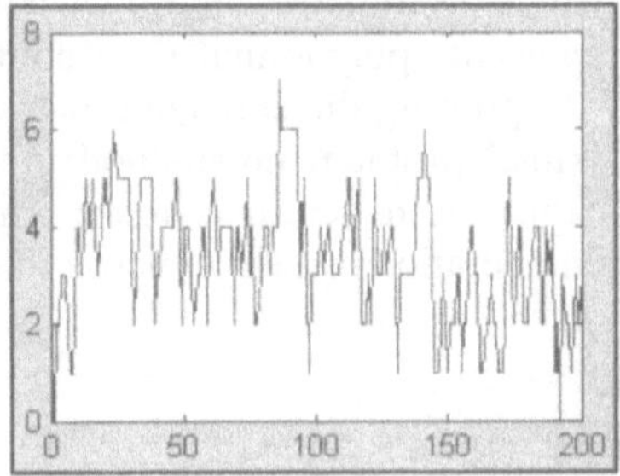

Fig. 3. Autonomy-transition intensity characteristic of our simulation runs.

Table 1. Simulation Results

H:R*	AF	Method	UF	UFHP	TAS	TCS
3:1	2.5	Optimized	0.00	0.00	54.00	0
		Static	0.00	0.00	54.00	NA
3:3	6.9	Optimized	0.15	0.05	145.10	10.8
		Static	0.34	0.11	145.10	NA
3:9	20.8	Optimized	9.90	0.95	240.20	4.8
		Static	12.46	2.90	240.20	NA
8:2	47.0	Optimized	23.66	1.06	232.00	54
		Static	27.06	4.72	232.00	NA

*H:R is the number of humans to robots used for a given run.

The results reflect the intentions of scalable command and control. Both the fixed and optimized procedures produced identical results when ample cognitive capacity was available to meet all cognitive loading conditions (resulting in no unnecessary assignment switches). When system complexity was increased, the system still performed quite well by consistently maintaining a lower number of unmanaged functions than static control. The last set of trials (those using 8 humans and 18 robots) demonstrates the true strength of the optimization procedure. The average number of unassigned functions per iteration is nearly half of what the fixed procedure generated. Furthermore, of those unassigned functions, an average of 1.06 of the top priority functions were left unassigned while the fixed method missed an average of 4.72. Clearly, if these top-priority functions were absolutely critical to the successful completion of a task, then the optimizing procedure would be the preferred implementation.

Scalable Command and Control demonstrates potential as a multi-human, multi-robot control mechanism. The simulations estimate that optimization consistently out-performs static control structures as system complexity increases. This fact is most apparent when the static control paradigm left 300% to 470% more top-priority functions unassigned. The use of optimization also maintains a ratio of approximately 3.33 functions per operator for complex systems (i.e. the simulations with more than one function per operator) whereas static assignment produced only 2.7. This 21% difference, when accumulated across the operator population, results in a marked loss of efficiency. The time complexity of the approximation algorithm can be estimated as Kn^{α} where K is the number of autonomy levels and n^{α} represents polynomial time complexity of the standard Transportation Algorithm. By pruning the shipment matrix, the problem becomes analogous to the "bin packing" problem understood to be NP-hard [12]. The polynomial time approximation used herein is more amenable to the real-time requirements of unstructured operations, at the expense of optimality.

Reference

1. N. Jordan. (1963) Allocation of functions between men and machines in automated systems. Journal of applied psychology, Vol. 47, No. 3.
2. E. Hutchins. (1995). Cognition in the wild. Cambridge, MA
3. T. Fong, C. Thorpe, C. Baur. (2001) Collaboration, Dialogue, and Human-Robot Interaction. 10th International Symposium of Robotics Research, November 2001, Lorne, Victoria, Australia.
4. J. Ferber. (1999) Multi-Agent Systems. Addison-Wesley, London.
5. C. Lebiere, J. R. Anderson, D. Bothell. (2001) Multi-tasking and cognitive workload in an ACT-R model of a simplified air traffic control task. Proceedings of the 10th Conference on Computer Generated Forces and Behavior Representation. Norfolk, Va.
6. D. Kortenkamp, Keirn-Schreckenghost, R. Peter Bonasso. (1991). Adjustable Control Autonomy for Manned Space Flight Systems. Proc. of IEEE.
7. J.R. Anderson, C. Lebiere. (1998) The Atomic Components of Thought. Mahwah, NJ: Erlbaum.
8. C. Lebiere, J. R. Anderson, D. Bothell. (2001) Multi-tasking and cognitive workload in an ACT-R model of a simplified air traffic control task. In Proceedings of the 10th Conference on Computer Generated Forces and Behavior Representation. Norfolk, Va.
9. K. S. Barber, C. E. Martin. (1999) Specification, Measurement, and Adjustment of Agent Autonomy: Theory and Implementation. Autonomous Agents and Multi-Agent Systems.
10. R. Larson, (2000) Elementary Linear Algebra. 4th Ed. Houghton Mifflin Company. Boston, MA.
11. M H. Sohn, S. Ursu, J. R. Anderson, V. A. Stenger, C. S. Carter. (2000) The role of prefrontal cortex and posterior parietal cortex in task switching. Proceedings of the National Academy of Sciences, vol. 97, no. 24, pp. 13448-13453.
12. G. Strang. (1986) Introduction to Applied Mathematics. Wellesley-Cambridge Press. Wellesley, MA.

Implementing a Face Detection System Practically Robust against Size Variation and Brightness Fluctuation for Distributed Autonomous Human Supporting Robotic Environment

Hiroshi Mizoguchi, Ken-ichi Hidai*, Kazuyuki Hiraoka, Masaru Tanaka, Takaomi Shigehara, and Taketoshi Mishima
Saitama University, Saitama 338-8570, JAPAN (* Current affiliation is SONY Corp.)
Email: {hm | hidai | hira | mtanaka | sigehara | mishima}@me.ics.saitama-u.ac.jp

Abstract This paper presents a robust face detection system intended to be used for practical human interactive distributed robotic environment. Towards future aging society, there are much expectation and social demands for such human interactive environment that is possible to collaborate and support humans. Typical examples are Intelligent Room at MIT, Intelligent Space at University of Tokyo, Easy Living at Microsoft Research, and so forth. For these distributed robotic environment, face detecting function of the each agents are very crucial. However, in the real situation, it cannot be easy to realize the robust detection, because position, size, and brightness of face image are much changeable.

To solve these problems the authors develop such system that can detect the face robustly in the practical situation. Since the system has wide dynamic range of detectable size and brightness of the face image, it is robust against size variation and brightness fluctuation. The dynamic range of the maximum and minimum face size is 7:1. The range of the brightness is 8:1, where maximum illumination is 1290 lx and minimum is 160 lx. By combining several techniques, such as skin color extraction, correlation-based pattern matching, multi-scale, and histogram equalization, the authors succeed to realize these robustness

Key words: correlation-based pattern matching, multi-scale, histogram equalization

1 Introduction

Towards future aging society, there are much expectation and social demand for distributed autonomous robotic environment that is possible to interactively collaborate and support human. In such environment, each element or agent of the system must properly recognize and interact with the human. Although these recognition and interaction by each elemental agents are partially and incomplete, the distributed system as a whole is possible to provide adequate support to the human.

Research and development activities on these type distributed autonomous robotic environments have increased in recent years. Typical examples are Smart Rooms and Intelligent Room both at MIT, Robotic Room and Intelligent Space both at University of Tokyo, SELF at AIST(former ETL), Aware Home at Georgia Tech., Easy Living at Microsoft Research, and so forth[1]-[15].

For these distributed robotic environment to support humans properly, each elemental part or agent of the environment must recognize human face. If the robotic environment could recognize human face, quality of its interaction with human would be improved more properly. In other words, distributed face detection systems should robustly detect and track the target human face for wide range as a whole. In other words, face detecting function of the each agents are very crucial.

Conventional face recognition technology [15]-[20] is possible to identify face image at accuracy over 99%, if the face is located at specified position and scaled at predefined size in a scene. That is the case in which frontal face image can be taken with simple or no background under properly controlled lighting condition and the face area occupies large amount part in the scene. But in case of the real situation, such good condition cannot be obtained. Background is often complex and cluttered. There is a possibility that the face may not be included in the scene. Even if included, position of the face is not fixed. Brightness of the image may be so fluctuant. Size and orientation of the face may vary quite much. Thus, to use the current face recognition technology in the real situation practically, a robust method to detect and extract the face in a scene with cluttered and complex background is keenly required.

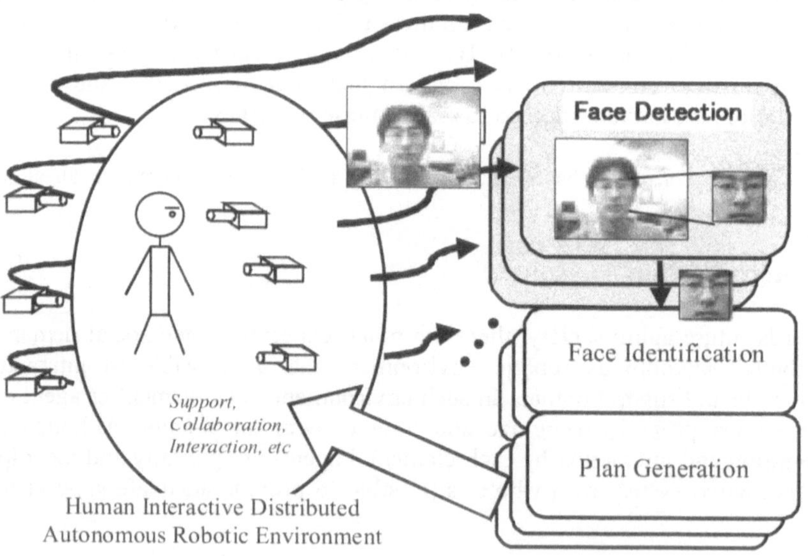

Fig. 1 Role of face detection in distributed autonomous robotic environment

Therefore the authors have implemented a practically robust face detection system to be used for distributed autonomous human support robotic environment. Among various problems, the implemented system tackles the problems of position variation, size variation, and brightness fluctuation of the face image. Figure 1 illustrates its role in the whole distributed system and relationship to the face identification system. Since the implemented system has wide dynamic range of detectable size and brightness of the face image, it is robust against size variation and brightness fluctuation. The dynamic range of the face size is 7:1 and brightness is 8:1. Key idea of the system is adoption and combination of several techniques, such as skin color extraction, correlation-based pattern matching, multi-scale image generation, and histogram equalization. This combination contributes the robustness mentioned above.

In the following, from section 2 to 4, we discuss the above mentioned problems and our solutions to them. Not only speculative description, but quantitative discussion based on experimental results are also presented. Section 2 treats position variation, section 3 does size variation and section 4 does brightness fluctuation. Section 5 is concluding remarks.

2 Position Variation

In case there is one or more face images in a scene, the system must detect the face images even though they locate at arbitrary location within the scene. On the contrary, in case there is no face within the scene, the system must output "no face" surely without any confusion of non-face area with face area. Thus the implemented face detection system solves this problem by utilizing correlation-based pattern matching. In other words, the system judges whether there is at least one face image in the scene. And the system also locates the position of the face if it exists. An averaged face image of ten persons is used as the template for the matching.

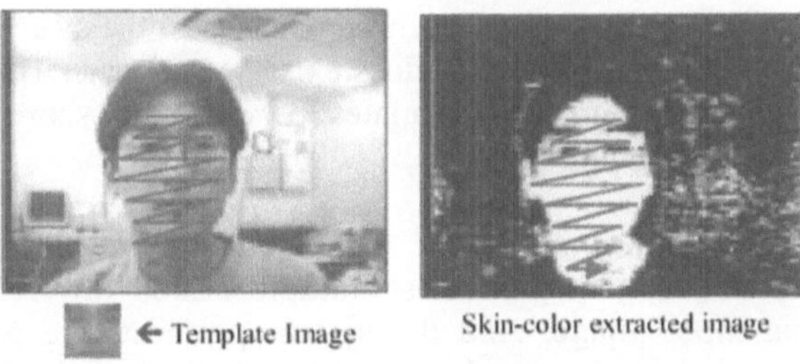

Fig. 2 Limiting search area for pattern matching to skin-color area

Although the correlation based pattern matching is effective to detect an a face image at arbitrary location in the scene, the matching process has a demerit that it consumes much computation and takes long execution time. To solve the problem, the authors utilize skin color extraction method. The implemented system reduces the execution time by limiting search area for the matching to the skin color area. The basic idea is that the skin color area must be a candidate of face area.

Figure 2 illustrates the limited search area for the matching. Since the detection does not rely only on the skin color extraction, it can be free from confusion of cardboard with skin color. Figure 3 shows an example of the execution time reduction. These results are measured with/without the skin color extraction on Pentium II 450MHz PC. Size of the input image is 240 by 180 pixels and the template image size is 24 by 24. In this case, about 4.5 times speed up can be gained.

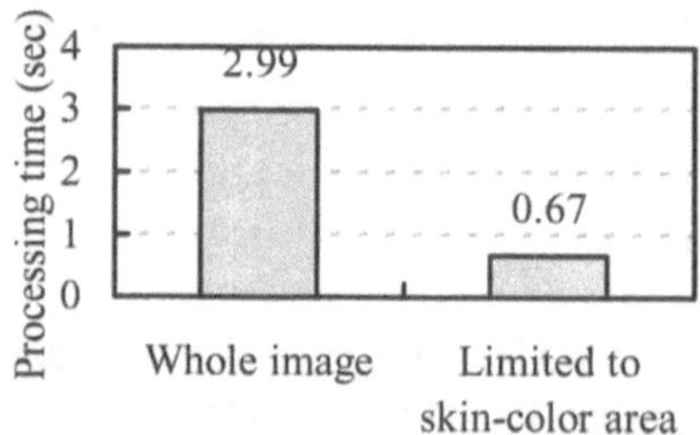

Fig. 3 Effect of search are limitation to skin-color area

3 Size Variation

If the size of a face to be detected is so different from the size of the template, the pattern matching fails to detect the face. In other words, the system reports wrong judgment as "no face", even if the face does exist in the scene. This is because the correlation based matching process calculates low similarity score and thus cannot detect the face candidate properly.

In order to overcome this problem, the implemented system utilizes multi-scale technique. Where the system generates multiple scaled low resolution images from the original input image. And the system performs the pattern matching process in each scaled images as shown in fig. 4. Size of the template for the matching is fixed. If at least one face exists in the input image, there must be a generated image in which the face size is similar to the template size. To this generated image, the matching process calculates high similarity score and succeeds to detect the face.

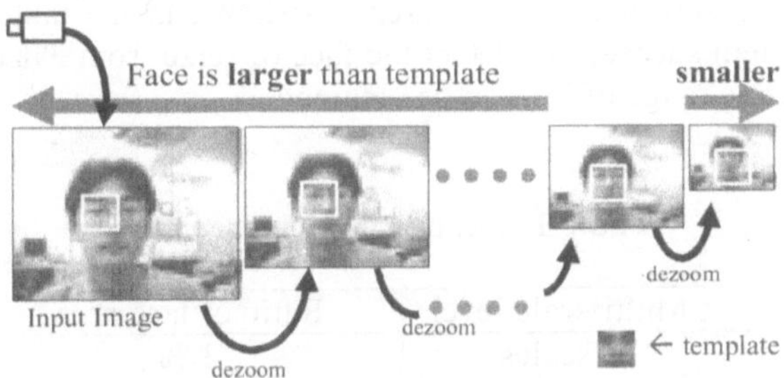

Fig. 4 Multi-scale

To confirm the effectiveness of the multi-scale, the authors conduct an experiment to measure success rate of face detection. In this experiment the area of the presented face is not fixed. Percentage of the area of the face to that of the whole image varies from 2% to 32%. Other conditions, such as CPU type, clock frequency, input image size, template image size and so forth, are the same as the experiment described in the previous section.

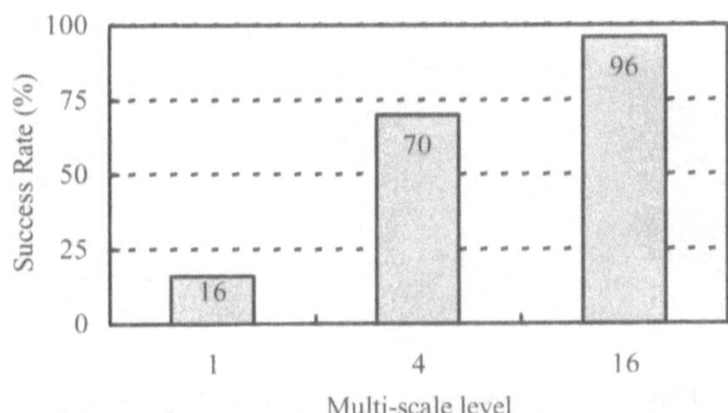

Fig. 5 Effect of Multi-scale

The experimental results are shown in fig. 5 and table 1. Figure 5 shows the success rates of the detection at different scales, 1, 4, and 16. The scale 1 denotes no multi-scale, i.e. the original input image itself. In case of the scale 1, since the system can only detect such face that its size is similar to that of the template, the system fails to detect most of the presented faces. As shown in fig. 5, the success rate rises as the number of scales increases.

Table 1 lists up the dynamic range of detectable face size. In this table the size is represented as the percentage of the area. As shown in the table, in case of the scale 16, the system succeeds to detect the face of seize from 3% to 22%. In other words, the dynamic range of 7:1 can be obtained.

Table 1 Range of face area size

Multi-scale level	Ratio of face area
1 scales	3 %
4 scales	3 ... 16 %
16 scales	3 ... 22 %

4 Brightness Fluctuation

Figure 6 shows a typical example of brightness fluctuation. In the left image, face part is well lighted and has high brightness. On the contrary, in the right image, face part is back-lighted and has quite low brightness. In the middle, there is little difference of brightness between face and back images. The middle image has low contrast comparing to other two images.

Fig. 6 Example of brightness fluctuation

To overcome these problems on the brightness fluctuation, the authors adopt histogram equalization method as a preprocess for the template matching. The method stretches the range of brightness and makes the distribution flat as shown in fig. 7. As the result, contrast enhanced image can be obtained. Fig. 8 and fig. 9 show the effectiveness of the histogram equalization. Figure 8 shows results of the template matching with/without the equalization. The images in fig. 6 are used as input. The figure plots the correlation value distribution. In case of the matching with the equalization, a steeple peak of the distribution can be observed at around face position. On the contrary, in case of the matching without the equalization, the steeple peak does not appear and high correlation value is found even at no face region.

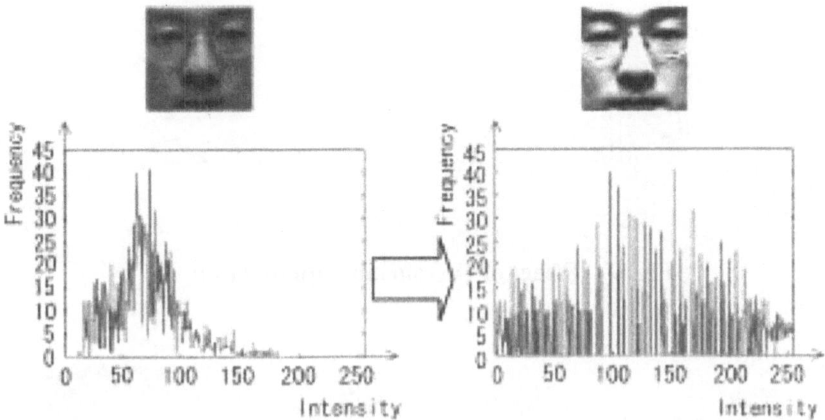

Fig. 7 Histogram equalization

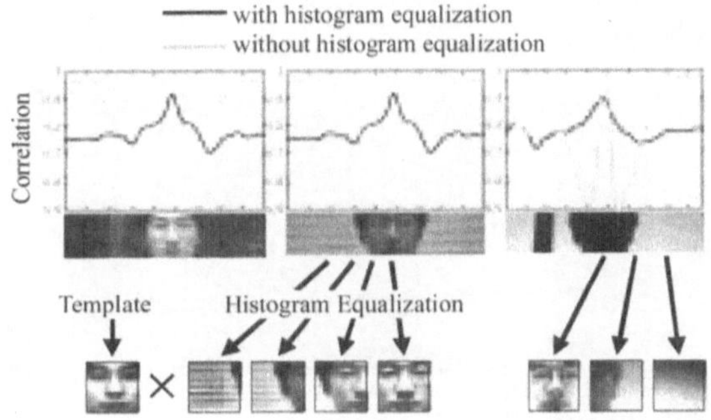

Fig. 8 Effect of histogram equalization

Figure 9 shows success rate of the face detection with/without the equalization. Experimental condition is the same to that of the previous sections. The scale 16 is used as the multi-scale. The success rate of 96.8% can be obtained with the equalization. On the contrary, in case of the detection without the equation, the rate is only 57.9%. In input images, maximum illumination is 1290 lx and minimum is 160 lx. Thus the dynamic range of 8:1 can be obtained against brightness fluctuation.

Figure 10 shows several examples of the face detection. These results are obtained by combining several techniques described so far. You can find that the system is possible to detect more than one faces without confusion.

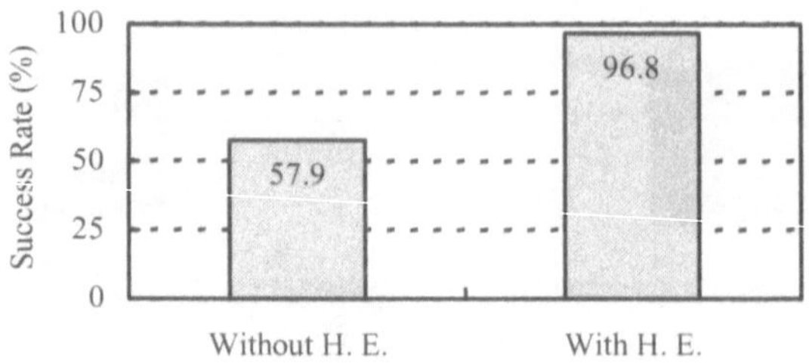

Fig. 9 Effect of Histogram Equalization

Fig.10 Examples of face detection

5 Concluding Remarks

This paper presents a practically robust face detection system that is implemented by the authors. The system is intended to be used as an elemental function for human interactive distributed robotic environment. Since the implemented system has wide dynamic range of detectable face size and brightness, the system is robust against position variation, size variation and brightness fluctuation of the face. The dynamic range of the face size is 7:1 and brightness is 8:1. The system demonstrates over 96% success rate of the face detection, even though the background is cluttered. Combination of correlation-based template matching, skin-color extraction, multi-scale, and histogram equalization realizes the above mentioned robustness. Effectiveness of the system is shown by quantitative discussion based on experimental results. Performance evaluation, combination with the face identification system, application to the distributed environment are future works.

Acknowledgments

This work has been partly supported by CREST of JST(Japan Science and Technology) 279102. The work also has been supported in part by Grant-in-Aid for Scientific Research (C) of JSPS and in part by Kayamori Foundation.

References

1. J. H. Lee, G. Appenzeller, H. Hashimoto. (1998) Physical Agent for Sensored, Networked and Thinking Space, Proc. of ICRA'98, pp.838-843.
2. A. Pentland. (2000) Looking at People: Sensing for Ubiquitous and Wearable Computing, Trans. on PAMI, Vol.22, No.1, pp. 107-119.
3. T. Sato, et al. (1996) Robotic room: Symbiosis with human through behavior media, Robotics and Autonomous Systems, No.18, pp.185-194.
4. H. Mizoguchi, et al. (1996) Robotic office room to support office work by human behavior understanding function with networked machines, IEEE/ASME Trans. on Mechatronics, Vol.1, No.3, pp.237-244.
5. M. C. Torrance. (1995) Advances in Human-Computer Interaction: The Intelligent Room, Working Notes of the CHI95 Research Symposium.
6. A. Pentland. (1996) Smart Rooms, Scientific American, pp.54-62.
7. H. Asada et al. (1996) Total Home Automation and Health Care/Elder Care, Tech. Rep., Dept. of Mech. Eng., MIT.
8. C. D. Kidd, R. Corr, G. D. Abowd, C. G. Atkeson, I. MacIntyre, E. Mynatt, T. E. Starner, and W. Newstetter. (1999) The Aware Home: A Living Laboratory for Ubiquitous Computing Research, Proc. of CoBuild'99, Position paper.
9. I. Essa. (1999) Computers Seeing People, AI Magazine, Vol.20(1), pp.69-82.
10. A. K. Dey, D. Salber and G. D. Abowd. (1999) A Context-based Infrastructure for Smart Environments, Proc. of MANSE'99.
11. B. Brumitt, B. Meyers, J. Krumm, A. Kern, and S. Shafer. (2000) EasyLiving: Technologies for Intelligent Environments, Proc. of Int'l Sympo. Handheld and Ubiquitous Computing , 2000.
12. Y. Nishida, T. Hori, T. Suehiro, and S. Hirai. (2000) Sensorized Environment for Self-communication Based on Observation of Daily Human Behavior, Proc. of IROS2000, pp.1364-1372.
13 Y. Nishida, T. Hori, T. Suehiro, and S. Hirai. (2000) Monitoring of Breath Sound under Daily Environment by Ceiling Dome Microphone, Proc. of SMC2000, pp.1822-1829.
14. R. P. Picard. (1997) Affective Computing, MIT Press.
15. T. Mori et al. (1997) Action Recognition System based on Human Finder and Human Tracker, Proc. of IROS'97, pp.1334-1341, 1997.

16. R. Chellappa, C. Wilson, and S. Sirohev. (1995) Human and Machine Recognition of Faces: A Survey, Proc.IEEE, Vol. 83, No. 5, pp.705-740.
17. T. Kurita, K. Hotta, and T. Mishima. (1998) Scale and Rotation Invariant Recognition Method Using Higher-Order Local Autocorrelation Features of Log-Polar Image, Proc. of ACCV'98, Vol. II, pp.89-96.
18. H. A. Rowley, S. Baluja, and T. Kanade. (1998) Neural Network-Based Face Detection, IEEE Trans. on PAMI, Vol. 20, No.1, pp.23-38.
19. Q. B. Sun, W. M. Huang, and J. K. Wu. (1998) Face Detection Based on Color and Local Symmetry Information, IEEE Proc. of FG'98, pp.130-135.
20. K. Hotta, T. Kurita, and T. Mishima. (1998) Scale Invariant Face Detection Method using Higher-Order Local Auto-correlation Features extracted from Log-Polar Image, Proc. of FG'98, pp.70-75.

Voice Communication in Performing a Cooperative Task with a Robot

Koliya Pulasinghe[1], Keigo Watanabe[2], Kazuo Kiguchi[2], and Kiyotaka Izumi[2]

[1] Faculty of Engineering Systems and Technology,
[2] Department of Advanced Systems Control Engineering,
Graduate School of Science and Engineering, Saga University,
1-Honjomachi, Saga 840-8502, Japan.
†E-mail: koliya@ieee.org, {watanabe, kiguchi, izumi}@me.saga-u.ac.jp

Abstract. This paper investigates the credibility of voice (especially natural language commands) as a communication medium in sharing advanced sensory capacity and knowledge of the human with a robot to perform a cooperative task. Identification of the machine sensitive words in the unconstrained speech signal and interpretation of the imprecise natural language commands for the machine has been considered. The system constituents include a hidden Markov model (HMM) based continuous automatic speech recognizer (ASR) to identify the lexical content of the user's speech signal, a fuzzy neural network (FNN) to comprehend the natural language (NL) contained in identified lexical content, an artificial neural network (ANN) to activate the desired functional ability, and control modules to generate output signals to the actuators of the machine. The characteristic features have been tested experimentally by utilizing them to navigate a Khepera® in real time using the user's visual information transferred by speech signals.

1 Introduction

Use of robots as human assistants has been taken much attention in research community since they are very far from human dexterity to use them in place of human. In this context, robots are taken out from the manufacturing floor and used with humans in the human environment for the tasks like nursing and aiding where people can use them as assistants with limited functionality. A flexible communication medium is a must, where voice has the most plausible features among the others. Figure 1 describes the nature of the voice commands that we can use in performing a cooperative task. As described in the Fig. 1, the words used in natural conversations consist of lot of particles to maintain the grammatical structure of the uttered sentence and words having imprecise meaning. These two features can be emphasized as: 1) identifying the keywords which lead to activate the robot's functions (action words like lift) and 2) imprecise nature of words which describes how smoothly the robot should perform the particular action (fuzzy predicates like little, very slow). This paper proposes a methodology, which is capable to represent above mensioned properties in voice based man-machine communication in performing a cooperative task.

Fig. 1. Voice as the communication medium in a cooperative task

With the advent of fuzzy reasoning most researchers used it in the design of controller for a machine. But still fuzzy reasoning is not very popular in voice based machine control, [1,2,4], where it is an innate feature [3]. Because voice signal consists of natural language sentences which is inherently composed of imprecise words. Furthermore, in some implementations, speech recognizer is designed to identify a particular set of words which describe the machine functions where users restrain from natural conversation [1,4,5]. Therefore we are encouraged to design an FNN running on unrestricted speech, which can anticipate the above difficulties. The proposed system used a keyword spotting system to identify the machine sensitive words [6–9] and interpret them for machine using FNN. The new concept called "significance of words" is discussed to equate the system output to the users desire.

In the reminder of this paper, in Section 2, the system overview is briefed with its major components: the speech recognizer (SR), action selection network (ASN), action modification network (AMN) and significance of the imprecise words with their characteristics. We discuss experimental setup and results in Section 3 and give conclusions and future directions in Section 4.

2 System Overview

The system functionality should cater for the major demands, i.e., identify the machine sensitive words from the running speech and interpret them to the machine in its identifiable form. Consequently, the system should be capable to give a response similar to user's desire, which should be the output of any control system, contained in the speech signal.

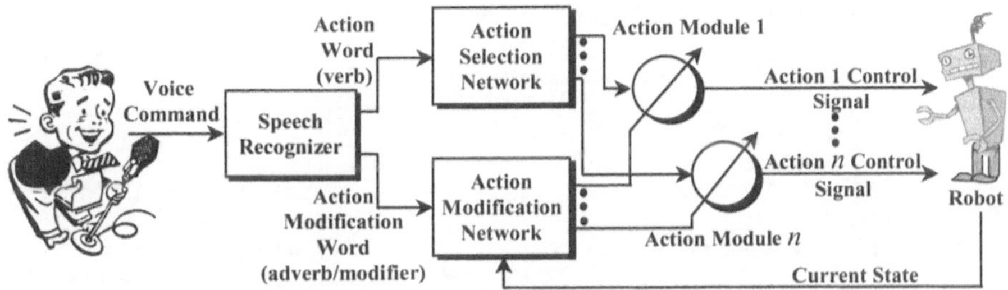

Fig. 2. The system overview

As illustrated in Fig. 2, an SR captures the user's utterances by means of a microphone. Captured voice signal is processed to recognize the machine sensitive words (pseudo sentence) and pass this pseudo sentence to an artificial neural network (ANN) to decode it into an action word (the verb) and several action modification words (the adverb). The nature of the action word is definite or precise but the action modification words are not definite, or imprecise [5]. Therefore, the system handles them in a different manner. The action word is fed into Action Selection Network (ASN) to fire the prospective action. Action modification words together with current machine status are fed into Action Modification Network (AMN), which modifies the operating behavior of the action fired by the ASN. Actions are implemented as modules. The robot may have N different functionalities; as an example, turning capability where it can be activated by turning module, and moving capability where it can be activated by moving module, etc. Each module emits the activation signals for the robot. The ASN fires one of these modules at a time.

2.1 Speech recognizer

The SR is developed using HMM Toolkit[1], which is an integrated suite of software designed for building and manipulating continuous density HMMs to develop speech recognition systems [10].

The SR consists of two parts as shown in Fig. 3. The keyword spotting module has two parallel networks as in Fig. 4. The outline of the work carried out by the keyword spotter can be explained by means of the following conversation:

User	Robot, Can you **go very fast**
Robot	You ask me to **go very fast**
User	Yes/No

[1] Hidden Markov Model Toolkit used in this design was developed at the Speech Vision and Robotics Group of the Cambridge University Engineering Department.

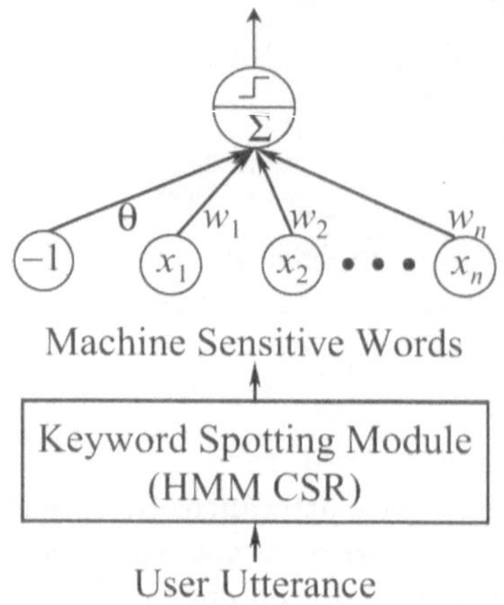

Fig. 3. The speech recognizer

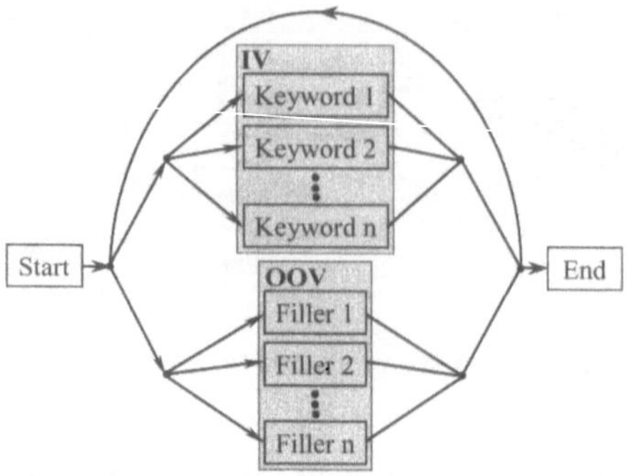

Fig. 4. The architecture of the keyword spotter

Keyword spotter filters the machine sensitive words "**go very fast**" from the utterance. This has been achieved by implementing filler network to identify the out-of-vocabulary (OOV) words as an alternative network to the baseline speech recognition (in-vocabulary or IV) network as in Fig. 4.

Once the utterance has completed, the HMM continuous speech recognizer (HMM CSR) recognizes the lexical contents. Then, the built-in keyword spotter separates out the pseudo sentence and forwards it into ANNs for the identification and classification [11]. Each perceptron identifies a particular word. If the word exists in the user command, output of the perceptron is set to one. Otherwise it is set to zero. The convergence procedure of the perceptron is explained below by using the notations used in Fig. 3. In learning phase, initial weights $w_i(t), (0 \leq i \leq n)$, and the threshold at the output node θ, are initialized to small random values. Here $w_i(t)$ is the weight from input i at time t. Then new continuous valued inputs $x_1, x_2, \ldots, x_n$ along with the desired outputs $d(t)$ are presented to compute the output as

$$y(t) = f_h\left(\sum_{i=1}^{n} w_i(t)\, x_i(t) - \theta\right) \tag{1}$$

where

$$f_h(\alpha) = \begin{cases} +1 & \text{if } 0 \leq \alpha \\ -1 & \text{otherwise.} \end{cases}$$

The weights are updated by

$$w_i(t+1) = w_i(t) + \eta\left[d(t) - y(t)\right] x_i(t) \tag{2}$$

with

$$d(t) = \begin{cases} +1 & \text{if input is a desired word} \\ -1 & \text{otherwise} \end{cases}$$

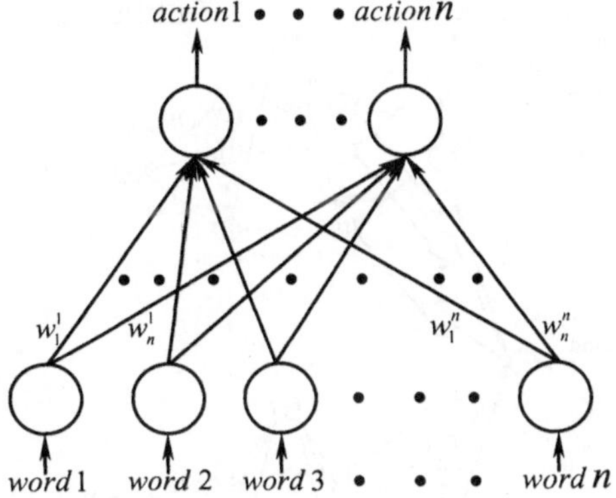

Fig. 5. The action selection network

where η is a positive gain fraction less than 1. Weights are unchanged if the net makes the correct decision.

2.2 Action selection network

Action words recognized at the SR are fed into the ASN illustrated in Fig. 5, to fire the desired action. It also generates a binary output as SR, which switches on the desired output module to trigger the desired action. Namely, if user wants to turn the mobile robot then the turning module is triggered by suppressing the other and vice versa. The ASN is also an ANN consisting of perceptrons, where it is trained using the same methodology explained under the ANN at SR.

2.3 Action modification network

Inherent properties of FNN controllers, i.e. their ability to manipulate imprecise data, naturally persuade us to select it as the controller for designing a machine control system driven by NL [12]. The system proposed here uses the linguistic rules of fuzzy algorithms to seize the user's desire coming in the form of adverbs/fuzzy predicates of the NL.

The proposed AMN shown in Fig. 6, consists of FNN for each and every actions which have been modified by fuzzy predicates (adverbs). The antecedent part takes the current value of the particular action as well as command input of the user for output value calculation at the consequent part. Every part for the each action is learned separately by the gradient descent algorithm, which modifies the consequent part of the particular action in the FNN. The adaptation process is illustrated in Fig. 7.

As shown in Fig. 6, at the layer 1, i.e., at the linguistic labels, every node computes a membership function of $\mu_{A_i}(x_i)$, where x is a crisp input to the

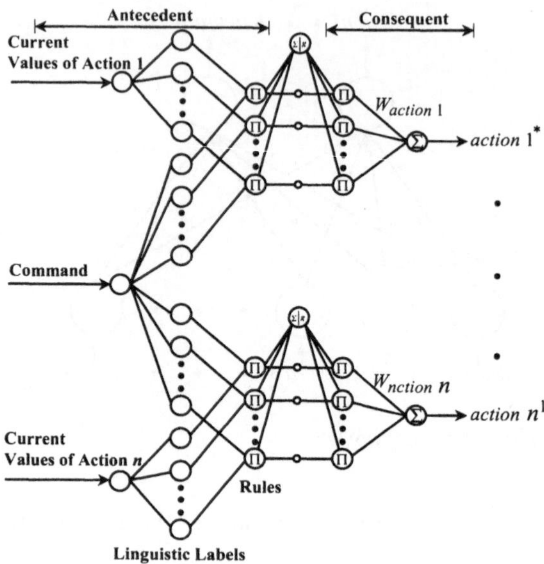

Fig. 6. The action modification network

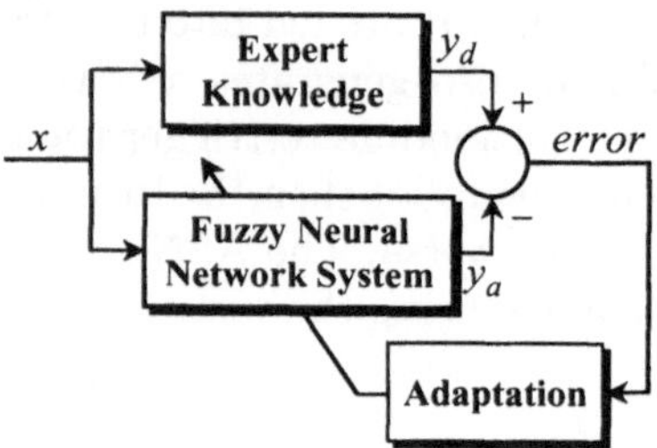

Fig. 7. Scheme for adaptation of AMN

node and A_i is the linguistic label associated with this node. In other words, the first layer is the fuzzification layer and its outputs are the membership values. The shape of these functions is triangular except command nodes. At the each rule node, layer performs the T-norm operator and it is usually the multiplication of the incoming signals:

$$h_i = \mu_{A_i}(x_i) * \mu_{B_i}(y_i) . \tag{3}$$

Here h_i is the confidence in the antecedent, $\mu_{A_i}(x_i)$ and $\mu_{B_i}(y_i)$ are the confidences in the linguistic labels, and "$*$" is the algebraic product. Then the output consequent, *action* $k^*(k = 1, ..., n)$, can be calculated as the following weighted mean of w_i with respect to the weight h_i:

$$action\ k^* = \frac{\sum_{i=1}^{r} h_i w_i}{\sum_{j=1}^{r} h_j} \tag{4}$$

where r represents the number of rules.

The connection weights at the consequent part are trained off-line by the information gathered for the particular action. When w_i represents an element of the weight vector W_x, where x means any action, it is updated by using the following equation:

$$w_i(t+1) = w_i(t) + \gamma [y_d - y_a] \frac{h_i}{\sum_{j=1}^{r} h_j} \tag{5}$$

where γ represents the learning rate, and y_d and y_a represent the desired output and actual output respectively for the action selected for the training.

2.4 Significance

We came across a new phenomena called as significance here, while interpreting imprecise words for the machine. It describes the contextual meaning of the word, i.e., transformation of words' meaning according to the current state of the machine. In natural language based commanding, this is related with the machine functions, which has limitations. As an example if machine has limited range in velocity in performing its actions the significance can be described as illustrated in Fig. 8. The significance of the command "go very fast" diminishes when machine arrives to its maximum speed, similar effect can be seen with the command "go very slow" at low speeds. This concept is taken into consideration in the adaptation process to represent the user desire more closely in the implementation.

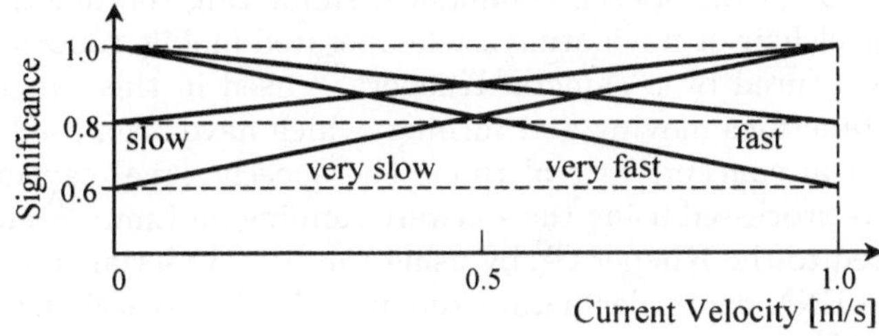

Fig. 8. Significance of fuzzy predicates of velocity command

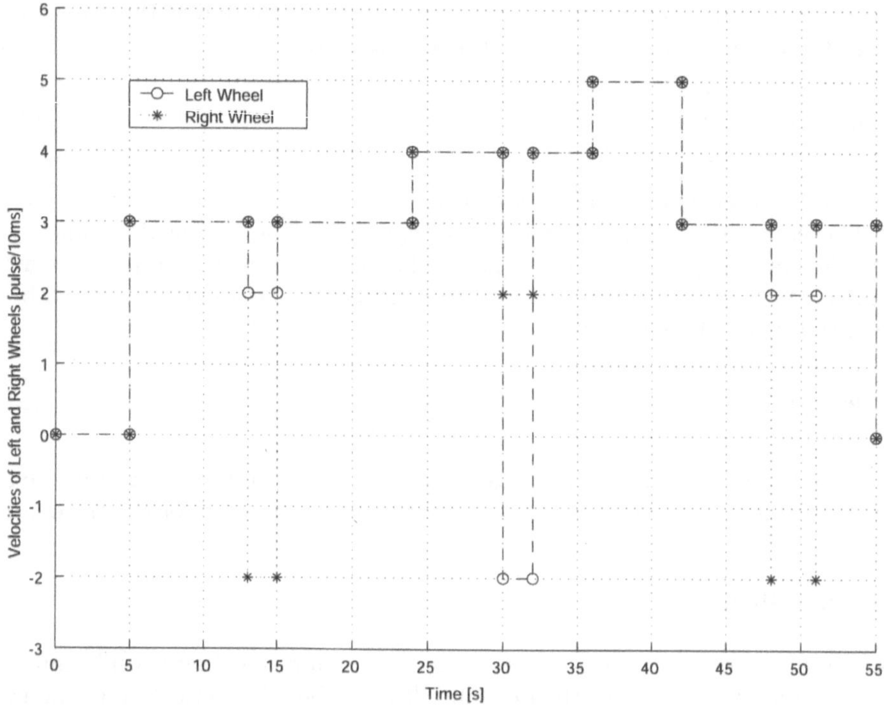

Fig. 9. Velocity profile of the two wheels of the Khepera®

3 Experimental Setup and Results

The concepts described in the above section have been applied to navigate the miniature robot, Khepera®, in real time using the user's vision information transferred through the speech commands, simulating the function of voice controlled wheelchair or navigating a tele-operated mobile robot using vision information captured by a camera. Khepera® used in this experiment can perform two functions moving and turning, which have been used to test the above theoretical concepts in real time. The speech signal captured by the microphone, is processed using the software running on Linux® environment and transferred to the Khepera® by using the RS-232 serial communication protocol. The FNN controller parameters used in this experiment were kept as same as in [5].

At first, the output of the speech recognition module is described here to show the machines capability to capture the machine sensitive words from unrestricted user utterances.

User	robot, can you go very fast
Recognizer	`FILLER FILLER FILLER GO VERY FAST`
User	robot turn right
Recognizer	`FILLER TURN RIGHT`
User	please turn left
Recognizer	`FILLER TURN LEFT`

The above "FILLER" occurrences have been filtered out for further processing. The keywords and out-of-vocabulary (OOV) words used in training are shown in Table 1. The alternative network used in the experiment has one filler node.
The velocity profile shown in Fig. 9, illustrates the velocities of left and right wheels of its 55 [sec] navigation for the commands: "robot, go very fast," "please turn right," "can you go fast," "turn left," "please go very fast," "robot, go very slow," and "please turn right." Velocity values evaluated at the FNN are truncated to nearest integer value since velocity commands to the Khepera® should be in integer format.

Table 1. Keywords and OOV words used for the training of the SR.

Keywords	OOV words
go	please
turn	robot
very	can
little	you
fast	I
left	want
right	
forward	
backward	
to	

4 Conclusions and Future Directions

The two major requirements, i.e., interpreting imprecise words in natural language commands and filtering machine sensitive words from the running utterance, for natural and flexible conversation with machines have been implemented in the experiment. This has immense help in controlling the nature of performing the task assigned by the human counterpart in the cooperation with the robot. Khepera® has two functional capabilities like in wheelchair, but there are no restrictions in using these concepts with the machines having several functional capacities. The design suffers from lack of parsing, where it couldn't identify the context grammar. Hence it couldn't differentiate the commands, "I want to go fast" from "I **do not** want to go fast." We are focusing to include context grammar identification unit in future works. In addition to that we plan to investigate the construction of filler elements for robust elimination of out-of-vocabulary words including background noise and apply these elements to a robot, which has high degrees of freedom.

References

1. Mazo M., Rodrìguez F. J. et al. (1995) Electronic control of a wheelchair guided by voice commands. Control Eng. Practice **3(5)**:665–674
2. Sugisaka M., Fan X. (2001) Control of a welfare life robot guided by voice commands. In: Proc. of the ICCAS 2001, Cheju Korea, 390–393
3. Lin C. T., Kan M. C. (1998) Adaptive fuzzy command acquisition with reinforcement learning. IEEE Transaction on Fuzzy Systems **6(1)**:102–121
4. Komiya K., Morita K. et al. (2000) Guidence of a wheelchair by voice. In: IECON 2000, Nagoya Japan, 102–107
5. Pulasinghe K., Watanabe K. et al. (2001) Modular fuzzy neural controller driven by voice commands. In: Proc. of the ICCAS 2001, Cheju Korea, 194–197
6. Rose R. C., Paul D. B. (1990) A hidden Markov model based keyword recognition system. In: Proc. of the IEEE ICASSP '90, Albuquerque New Mexico, 129–132
7. Jeanrenaud P., Ng K. et al. (1993) Phonetic-based word spotter: Various configurations and application to event spotting. In: Proc. of the EUROSPEECH '93, Berlin Germany, 1057–1060
8. Bazzi I., Glass J. R. (2000) Modeling out-of-vocabulary words for robust speech recognition. In: Proc. of the ICSLP 2000, Beijing China.
9. Leeuwen D. A. V., Kraaij W. et al. (1999) Prediction of keywords spotting performance based on phonemic contents. In: Proc. of the ESCA/ETRW, Cambridge UK, 73–77
10. Young S. J. (1993) The HTK hidden Markov model toolkit: Design and philosophy. Technical Report TR.153, Department of Engineering, Cambridge University UK
11. Lippmann R. P. (1987) An introduction to computing with neural nets. IEEE Magazine on Acoustics, Signal, and Speech Processing **4**:4–22
12. Jang J. S. R., Sun C. T. (1995) Neuro-fuzzy modeling and control. In: Proc. of the IEEE **83(3)**:378–406

Chapter 5
Multi-Robot Coordination

Position Control of Load by Traction of Two Mobile Robots

Hiroshi Hashimoto[1], Shinya Nakagawa[1], Naoki Amano[1], and Chiharu Ishii[2]

[1] Tokyo University of Technology, 1404-1 Katakura Hachioji Tokyo 192-0982,
hasimoto@cc.teu.ac.jp,nakagawa@hiha.mech.teu.ac.jp,amaono@media.teu.ac.jp
[2] Ashikaga Institute of Technology, 268-1 Omae Ashikaga Tochigi 326-8558,
c-ishii@ashitech.ac.jp

Abstract. This paper discusses the cooperative traction work on a load by two mobile robots. The reference trajectory of the load is given as a function of time, and the robots pull the load such that the load tracks the given reference trajectory. To achieve this, the virtual traction force vector generated by the PID controller is divided between the two mobile robots. Two force division methods of the virtual traction force vector are proposed, and their effectivenesses are examined through the simulation experiments.

1 Introduction

Recently, in many publications, the cooperative transfer work by multiple mobile robots has been studied. Reference [1] discussed the cooperative transfer work of handling the load by robots with grasp mechanisms at the tips of the robots. References [2] and [3] discussed the tracking problem such that the load tracks the reference trajectory given on the x-y plane. Those studies hardly consider the inertia of the load and exchange of absolute positions of the two mobile robots. Reference [4] discussed the cooperative transfer work of pushing the load from its back. In this study, the center of gravity of the load always comes in front of the action point of the pushing force. This may give rise to rotational movement of the load. Therefore, high-level and complex strategy is required in this approach. References [5]- [7] discussed a dynamic control method to transfer a common object by the coordinate of multiple wheeled mobile robots with a passive rotaly under the condition the relative positions of robots be restrained.

On the other hand, the cooperative traction work is considered as one of the methods of cooperative transfer work. It is the one such as pulling the vessel by two or more tugboats. The traction has an advantage that it has self-regulation. This means that, in the case of pulling the load by traction force, the center of gravity of the load can automatically approach the direction of the action vector of the traction force due to the influences of the traction force and viscous friction force even if the moment of inertia arises. Therefore, the operation of the traction work is superior from a view-point of transfer work. Thus, it is required to establish the division strategy of the traction forces for the practical multiple mobile robots. To the best of our knowledge, however, argument about the strategy to divide traction force even into two mobile robots has not seen.

This paper considers the cooperative traction work on a load by two mobile robots, and presents two division methods of the traction force on the load for two mobile robots. The reference trajectory of the load is given as a function of time, and robots must pull the load such that the load tracks the given reference trajectory. To achieve this, firstly assuming the operation of traction work by one mobile robot, the virtual traction force vector is determined using a PID controller so that the load tracks the given reference trajectory. Secondly, the virtual traction force vector is divided between the two mobile robots. Two force division methods are proposed: (i) the symmetrical division method and (ii) the unsymmetrical division method. In the former, the traction force of each robot is determined symmetrically centering around the virtual traction force vector. This method has an advantage that the algorithm becomes simple. On the other hand, in the latter, based on a priori information of the reference trajectory, the traction force of each robot is determined unsymmetrically centering around the virtual traction force vector. The effectiveness of those two force division methods are examined through the simulation experiments for a circular arc reference trajectory, and features of those methods are discussed.

The effectiveness of those two force division methods are examined through the simulation experiments for a circular arc reference trajectory, and features of those methods are discussed.

2 Virtual Traction Model

In this section, under the assumption that one mobile robot pulls a load, modeling of the load and construction of the position control system to track the reference trajectory are explained.

The model of the load is regarded as a mass system, which has inertia and the friction force of velocity. It is defined as follows, independently for x and y axes.

$$G_i(s) = \frac{1}{s^2M + sD} \quad (i = x, y) \tag{1}$$

where $G_i(s)$ is a transfer function of the model of the load, M is the mass of the load [kg] and D is the coefficient of the friction force [N·s/m]. Let us define a reference trajectory at time t to be $p_r(t) = [x_r(t), y_r(t)]$, and a position vector of the load at time t to be $p(t) = [x(t), y(t)]$. The load is pulled by virtual traction force of one robot toward the target point $p_r(t+\Delta t)$, which is the point at Δt after time t on the reference trajectory $p_r(\cdot)$ as shown in Fig.1. The reason for this is explained as follows. If there exists a deviation for $p(t)$ from its reference $p_r(t)$ at time t, by aiming at the point $p_r(t+\Delta t)$, the load can follow the reference trajectory earlier than aiming at the point $p_r(t)$. The control objective is to reduce the deviation between the point $p(t)$ and the point $p_r(t+\Delta t)$ as much as possible. Then the load follows the reference trajectory. To this end, in this paper we consider the PID control system shown in Fig.2.

The feedback loop system is constructed independently for x and y axes respectively, and the output $\mathbf{f}(t)$ of the controller is a traction force vector in the case where

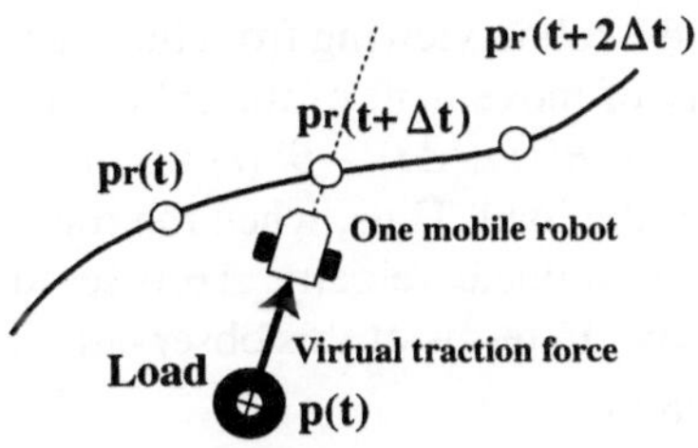

Fig. 1. Schematic diagram of $\mathbf{p}(t)$ and $\mathbf{p_r}(t)$

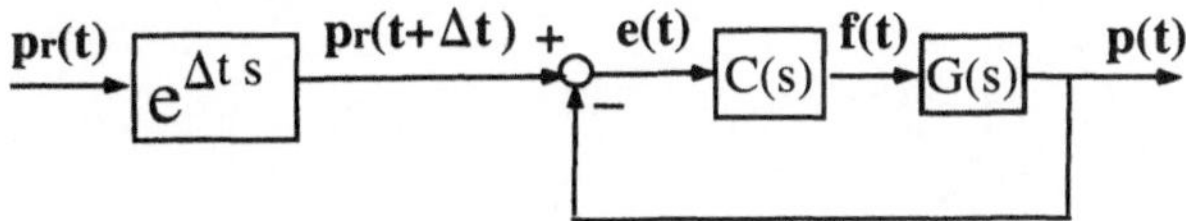

Fig. 2. Feedback control system

one mobile robot pulls the load to track the given reference trajectory. We call this vector "the virtual traction force vector" throughout this paper. $\mathbf{C}(s)$ is a PID controller, and $\mathbf{e}(t)$ is a position error. They are defined as follows.

$$e_i(t) = p_{ri}(t+\Delta t) - p_i(t) \quad (i = x, y) \tag{2}$$

$$C_i(s) = k_{Pi} + \frac{k_{Ii}}{s} + k_{Di}s \quad (i = x, y) \tag{3}$$

where k_{Pi}, k_{Ii}, k_{Di} are design parameters.

Considering the actual motion of the load, there exists some constraint in the relationship between the straight line velocity of the load $\mathbf{v}_{load}$[m/s] and the rate of change for the rotation angle of the robot. The physical meaning of this is illustrated in Fig.3,

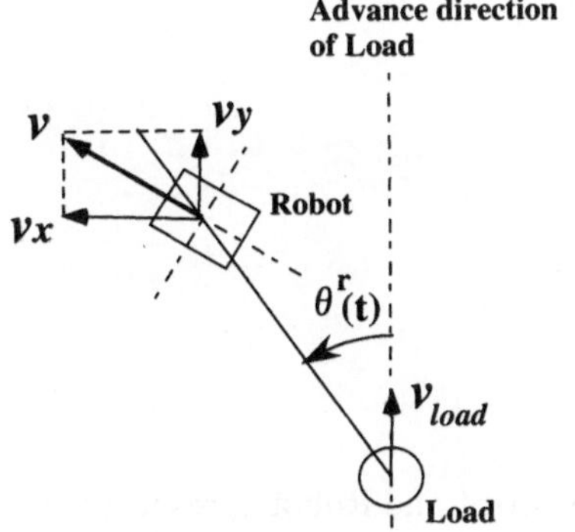

Fig. 3. Relation of $\mathbf{v}_{load}$ and rotation angle

where $\mathbf{v}$ is the linear velocity of movement of the robot, and $\theta^r(t)$ is the relative angle from the direction of $\mathbf{v}_{load}$ to the direction of the robot, viewing from the load. Since there exists a limitation in the linear velocity of movement of the robot, the difference of the relative angles $\Delta\theta$, defined by $\Delta\theta = \theta^r(t+\Delta t) - \theta^r(t)$ becomes small if $\mathbf{v}_{load}$ is enlarged in the advance direction of the load. Thus, when the robot moves to the advance direction of the load at the highest linear velocity, $\Delta\theta$ is set to zero. From this, the model is not pure holonomic type. In terms of this observation, the constraint between $\Delta\theta$ and $\mathbf{v}_{load}$ is shown in Fig.4.

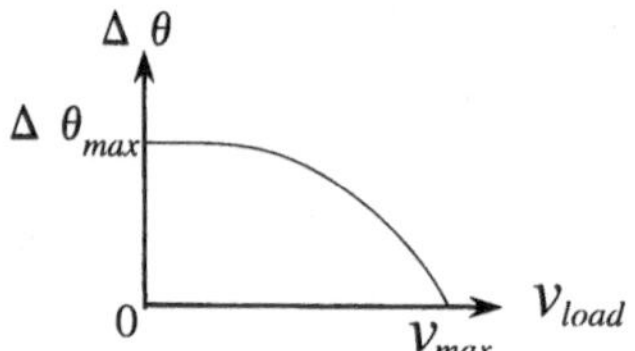

Fig. 4. $\mathbf{v}_{load}$ v.s. $\Delta\theta$

3 Division Method

In this section, we describe the division method of the virtual traction force vector $\mathbf{f}(t)$ between two mobile robots. This is illustrated in Fig.5,

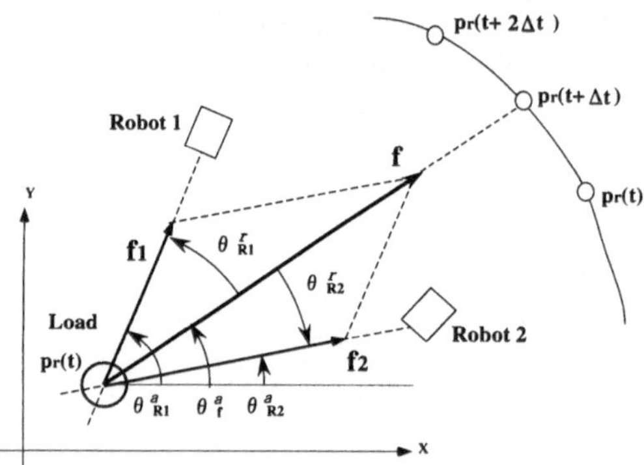

Fig. 5. Division of traction force

where $\mathbf{f_1}$ and $\mathbf{f_2}$ are real traction force vectors of Robot 1 and Robot 2, respectively, θ^a_{R1}, θ^a_{R2} and θ^a_f are absolute angles of the vectors $\mathbf{f_1}$, $\mathbf{f}$ and $\mathbf{f_2}$ from the x axis, and θ^r_{R1} and θ^r_{R2} are relative angles from the direction of $\mathbf{f}$ to the directions of $\mathbf{f_1}$ and $\mathbf{f_2}$, respectively.

From Fig.5, the following equations hold.

$$|\mathbf{f_1}(t)| = \frac{\sin(\theta^r_{R2})}{\sin(\theta^r_{R1}+\theta^r_{R2})}|\mathbf{f}(t)|$$
$$|\mathbf{f_2}(t)| = \frac{\sin(\theta^r_{R1})}{\sin(\theta^r_{R1}+\theta^r_{R2})}|\mathbf{f}(t)| \tag{4}$$

In order to obtain the values of $\mathbf{f_1}(t)$ and $\mathbf{f_2}(t)$, several parameters must be determined. Two force division methods to determine these two vectors are explained as follows. Those are the symmetric division method and the unsymmetrical division method.

Symmetrical division method

By splitting $\mathbf{f}(t)$ into $\mathbf{f_1}(t)$ and $\mathbf{f_2}(t)$ symmetrically, the algorithm to obtain the values of $\mathbf{f_1}(t)$ and $\mathbf{f_2}(t)$ becomes simple. Thus, we assume that the mobile robots pull the load keeping the relative angles θ^r_{R1} and θ^r_{R2} symmetric and constant. In this case, from (4), the norms of the traction forces are obtained as follows.

$$|\mathbf{f_i}(t)| = \frac{|\mathbf{f}(t)|}{2\cos\theta^r_{Ri}} \quad (i=1,2) \tag{5}$$

where $\theta^r_{R1}(=\theta^r_{R2})$ are given parameters.

It is found that this algorithm is simple comparing with the unsymmetrical division method stated later.

Unsymmetrical division method

From (4), it is found that the norms of the traction forces $\mathbf{f_i}(t)$ $(i=1,2)$ and relative angles θ^r_{Ri} $(i=1,2)$ are required to determine $\mathbf{f_1}(t)$ and $\mathbf{f_2}(t)$. Therefore, we determine them in the following way. First, as shown in Fig.6, the angle θ^r_{R1} is determined as the relative angle from the direction of $\mathbf{p_r}(t+\Delta t)$, viewing from $\mathbf{p}(t)$ to the direction of $\mathbf{p_r}(t+2\Delta t)$, also viewing from $\mathbf{p}(t)$. Next, determine the value of $|\mathbf{f_1}(t)|$ adequately, then $\mathbf{f_1}(t)$ and $\mathbf{f_2}(t)$ are obtained from (4) in turn. According to this method, $|\theta^r_{R1}|$ and $|\theta^r_{R2}|$ become generally unsymmetric.

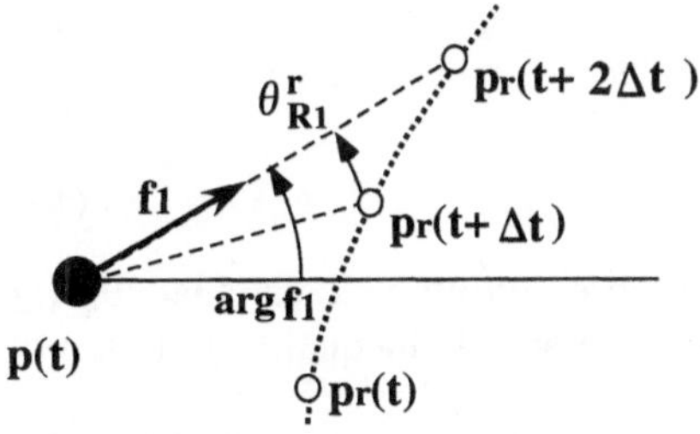

Fig. 6. Setting of f_1

Here, in the case where θ^r_{Ri} is variable, the following two problems must be taken into consideration. The first one is that the values of $\arg\mathbf{f_1}(t)$ and $\arg\mathbf{f_2}(t)$ may reverse, depending on the direction of $\mathbf{f}(t)$. As a result, the robots will intersect (Intersection Problem). The second one is that there exists a limitation on the minimum distance that the robots can come close together (in other words, $min|\theta^r_{R1} + \theta^r_{R2}|$) since an actual robot has width (Minimum Angle Problem).

Taking the above problems into account, the calculation procedure for $\mathbf{f_1}(t)$ and $\mathbf{f_2}(t)$ is described as follows.

Step 1 At time t, calculate θ'_{R1} so as to satisfy the following equation.

$$\begin{aligned}\theta'_{R1}(t) &= \arg \mathbf{f_1}(t) \\ &= \arg(\mathbf{p_r}(t+2\Delta t) - \mathbf{p}(t))\end{aligned} \tag{6}$$

Step 2 Let i be 1. For the given line velocity $\mathbf{v}_{load}$ at time t, refering the $\Delta\theta(\mathbf{v}_{load})$ shown in Fig.4, if

$$|\theta'_{Ri}(t) - \theta^a_{Ri}(t-\Delta t)| \leq \Delta\theta(\mathbf{v}_{load}) \tag{7}$$

holds, then set $\theta^a_{Ri}(t)$ as

$$\theta^a_{Ri}(t) = \theta'_{Ri}(t). \tag{8}$$

Otherwise, set $\theta^a_{Ri}(t)$ as

$$\theta^a_{Ri}(t) = \theta^a_{Ri}(t-\Delta t) + \mathrm{sgn}(\theta'_{Ri}(t) - \theta^a_{Ri}(t-\Delta t))\Delta\theta(\mathbf{v}_{load}). \tag{9}$$

Step 3 Define $|\mathbf{f_1}(t)|$ using the output $\mathbf{f}(t)$ of $\mathbf{C}(s)$ as follows.

$$|\mathbf{f_1}(t)| = \alpha|\mathbf{f}(t)| \tag{10}$$

where α $(0 < \alpha \leq 1.0)$ is a discrete variable given as

$$\alpha = m \cdot \Delta\alpha \ \ (m = 1, 2, \cdots) \tag{11}$$

using the given constant step $\Delta\alpha$ $(0 < \alpha < 1)$. Thus, $\mathbf{f_1}(t)$ is obtained.

Step 4 Using the obtained $\mathbf{f_1}(t)$, calculate $|\mathbf{f_2}'(t)|$ and θ'_{R2} as follows.

$$|\mathbf{f_2}'(t)| = |\mathbf{f}(t) - \mathbf{f_1}(t)| \tag{12}$$

$$\theta'_{R2}(t) = \arg(\mathbf{f}(t) - \mathbf{f_1}(t)) \tag{13}$$

With regard to the $\mathbf{f_2}'(t)$, if

$$|\mathbf{f_2}'(t)| \leq f_{2max} \tag{14}$$

is satidfied, then set $|\mathbf{f_2}(t)| = |\mathbf{f_2}'(t)|$. Furthermore, setting $i = 2$, calculate $\theta^a_{R2}(t)$ in the same way of Step 2. Thus, $\mathbf{f_2}(t)$ is obtained. If inequality (14) is not satisfied, update α and return to Step 3.

If there exists an α satisfying inequalities (10) and (14), go to Step 5. Otherwise the traction forces are calculated as follows.

$$|\theta^r_{R1}| = |\theta^r_{R2}| \tag{15}$$

$$|\mathbf{f_i}(t)| = \min(f_{1max}, f_{2max}) \ \ (i = 1, 2) \tag{16}$$

Step 5 If the following inequality is satisfied,

$$\theta^a_{R1} < \theta^a_{R2} \tag{17}$$

then, subscript "1" and "2" in the symbols are exchanged, and the positions of Robot 1 and Robot 2 are adapted to the actual relation of position.

Step 6 if

$$|\theta^a_{R1} - \theta^a_{R2}| < \theta_{min} \tag{18}$$

then

$$\theta^a_{R1} = \theta^a_f + \frac{\theta_{min}}{2}, \quad \theta^a_{R2} = \theta^a_f - \frac{\theta_{min}}{2} \tag{19}$$

where θ_{min} means the relative angle $(\theta^r_{R1} + \theta^r_{R2})$ when the two robots come closest together.

In this algorithm, the Intersection Problem is considered in Step 5, and the Minimum Angle Problem is considered in Step 6.

4 Simulation

The proposed two force division methods are examined through simulation experiments. A circular arc reference trajectory given by

$$\mathbf{p_r}(t) = \mathbf{p_c} + \begin{bmatrix} R\cos(\omega t + \varphi) \\ R\sin(\omega t + \varphi) \end{bmatrix} \tag{20}$$

is used, where $\mathbf{p_c}$ is the position vector of the center of the circular arc, R is the radius of the circular arc, and φ and ω are parameters related to the starting point and angular velocity of the circular arc trajectory. The parameters used in the simulation are listed in Table 1. The radius of each robot is 10 [cm], and the relative angle in the case of the symmetrical method is $\theta^r_{Ri} = \pm 40$ [deg], while the minimum relative angle in the case of the unsymmetrical method is $\theta_{min} = 5.7$ [deg].

Table 1. Parameters of circular arc , load and controller

$\mathbf{p_c}$	R [m]	ω [rad/s]	φ [rad]
$[1.5\ \ 0]^T$	1	$-\frac{\pi}{20}$	π

Mass [kg]	Dumping Coefficient [N·s/m]	$[k_{Pi}, k_{Ii}, k_{Di}\ (i = x, y)]$
10	5	[10, 0.1, 10]

Figs.7 and 8 show the results in the cases of the symmetrical method and the unsymmetrical method respectively. Comparing these results, it is found that the intersection problem and the minimum angle problem have not occurred in the case

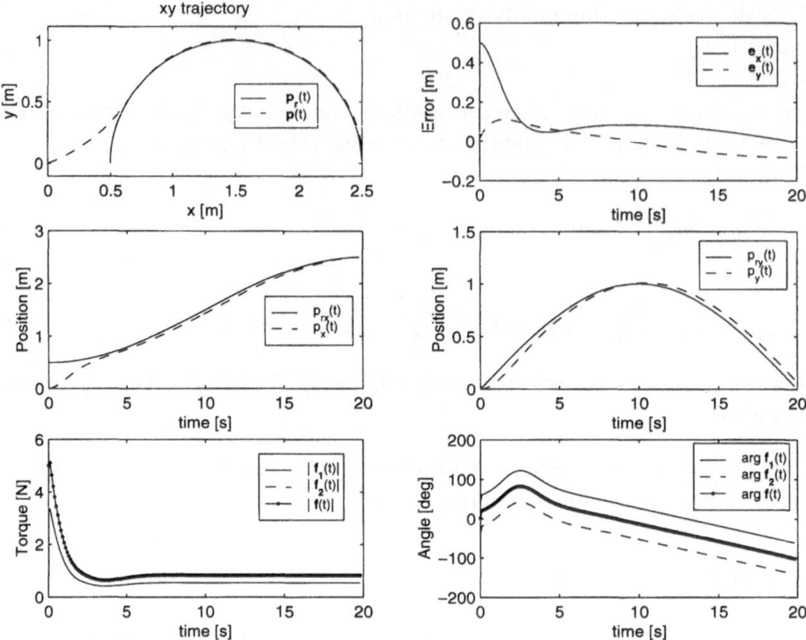

Fig. 7. Simulation results. (symmetrical method)

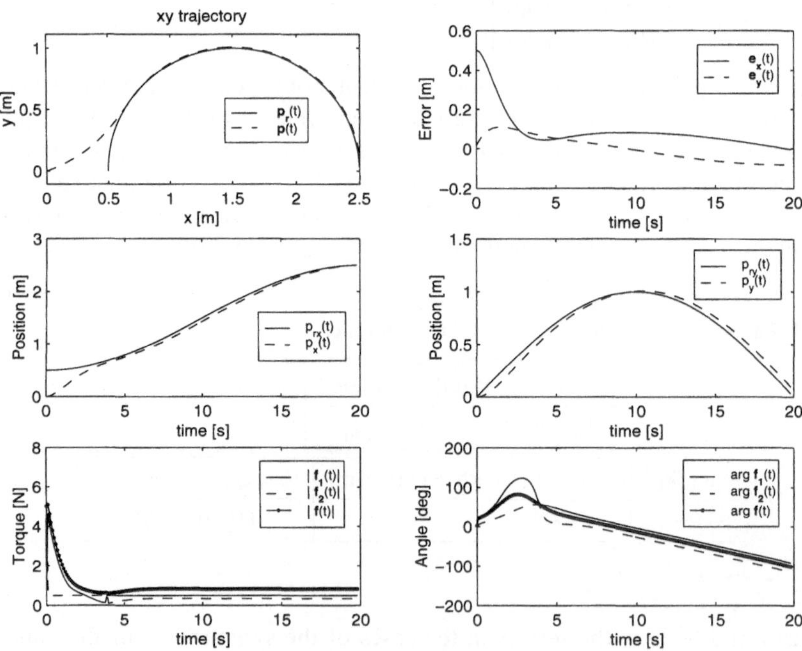

Fig. 8. Simulation results. (unsymmetrical method)

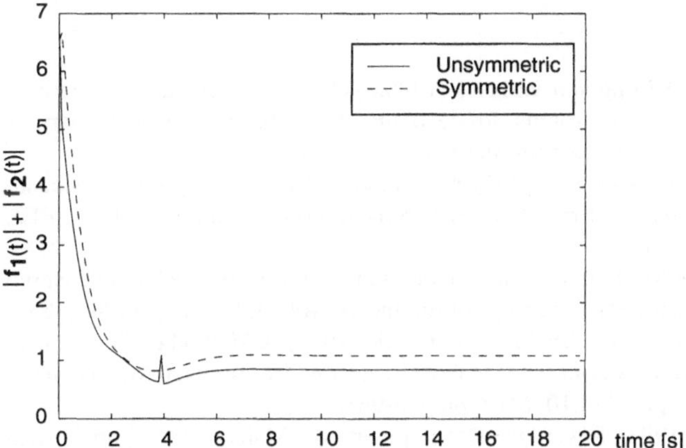

Fig. 9. Total force for each method

of the symmetrical method, although the intersection has been seen in the case of the unsymmetrical method. However, in the graph at the bottom of the right in Fig.8, $\arg \mathbf{f}_2(t)$ becomes negative in spite of $\arg \mathbf{f}(t)$ being positive during the period $0 \le t \le 1$. This means that the divided force $\mathbf{f}_2(t)$ prevents the load advancing toward the target point. To investigate this in detail and compare with the case of the unsymmetrical method, the total forces $|\mathbf{f}_1(t)| + |\mathbf{f}_2(t)|$ in both methods are shown in Fig.9. From this result, it is found that the total force by the unsymmetrical method is smaller than the one by the symmetrical method during most of the time. This means that the unsymmetrical method can save energy to pull the load for the robots compared with the symmetrical method. On the other hand, a steep change of force is shown at the time when the intersection occurred (around four second) in the case of the unsymmetrical method. In order to generate such a force, high power is required for the actuator.

5 Conclusions

In this paper, the two force division methods for the virtual traction force vector are proposed, and simulation experiments were carried out to examine the effectiveness of the proposed methods. Discussing the simulation results, the following features of these two methods are pointed out.

The symmetrical method has features such as the simplicity of the algorithm, the smoothness of the change of force, and no problems of intersection and minimum angle in comparison with the other one. On the other hand, the unsymmetrical method has the feature that smaller force is enough to pull the load compared with the other one. The problem which concerns the determining of parameter α remains unanswered.

References

1. C.Kanamori, M.Kajitani, A.Ming and V.Masek (1995.6) Obstacle Avoidance of the transporting system using two mobile robots, Proceedings of the '95 The Japan Society of Mechanical Engineers Robotics Mechatronics Conference, pp.804-807.
2. M.Nakamura, T.Maeda and N.Kyura (1997) A Method of Contour Control for Mechatronic Servo System by use of Error Feedback Synchronous Position Control, IEEJ, Vol.117-D, No.5, pp.544-551.
3. T.Egami and K.Yoda (1998) Optimal Synchronous Path Control of Mechatronic Servo System, The Society of Instrument and Control Engineers, Vol.34, No.9, pp.1178-1183.
4. H.Asama, K.Ozaki, A.Matsumoto, Y.Ishida and I.Endo (1992) A Method of Cooperative Task Assignment by Multiple Autonomous Robots Based on Decentralized Management, The Robotics Society of Japan, Vol.10, No.7, pp.955-963.
5. D.J.Stilwell and J.S.Bay (1993) Toward the Development of a Material Transport System using Swarms of Ant-like Robots, Proc. IEEE Int. Conf. Robotics and Automation, Vol.1, pp.766-771.
6. D.J.Stilwell and J.S.Bay (1994) Optimal Control for Cooperating Mobile Robots Bearing a Common Load, Proc. IEEE Int. Conf. Robotics and Automation, Vol.2, pp.58-63.
7. H.Hashimoto, F.Oba and T.Eguchi (1995) Dynamic Object-Tranportation Control Method by Multiple Mobile Robots, J. of the Robotics Soc. of Japan, Vol.13, No.6, pp.152-159.

Cooperative Motion Coordination Amidst Dynamic Obstacles

Stefano Carpin*[1] and Lynne E. Parker[2]

[1] Intelligent Autonomous Systems Laboratory
Department of Electronics and Informatics
The University of Padova, ITALY

[2] Center for Engineering Science Advanced Research
Computer Science and Mathematics Division
Oak Ridge National Laboratory, Oak Ridge, Tennessee, U.S.A.

Abstract. The cooperative leader following task for multi-robot teams is introduced and discussed. We describe the design and implementation of a distributed technique to coordinate team level and robot level behaviors for this task, as well as a multi-threaded framework for the implementation of a heterogeneous multi-robot system. This approach enables robots to remain in formation as they deal with other obstacles that may appear within the formation. We describe how the robot behaviors are realized and scheduled. The proposed approach has been run and validated on a team of robots performing in both indoor and outdoor environments.

1 Introduction

In this paper we address the problem of *leader following* in the case of a heterogeneous multi-robot team. This task requires the robots to move in a *linear* pattern, each following the previous robot, with the first robot either following a human operator, being teleoperated, or going through a predetermined path. While others have previously studied this leader following behavior, our focus is specifically on enabling the team to perform in a *robust* way, so that the team is able to correctly operate even if some external undesired or unforeseen event occurs. Examples of unexpected events include an obstacle in the way or one or more of the robots failing.

We propose a distributed policy based on explicit communication that allows this goal to be achieved at the team level. We introduce a multi-threaded structure employed on different robots, and we also give the details about the tracking techniques. This framework allows us to abstract the coordinated tracking process from the low-level sensor details. This paper is organized as follows: related work is discussed in Section 2, while our approach in described in Section 3. Section 4 briefly outlines how the various sensors are used for leader following and obstacle avoidance. Finally, Section 5 presents experimental results, with conclusions offered in Section 6.

* Research performed while visiting Oak Ridge National Laboratory.

2 Related Work

The task of pattern formation and formation marching has gained a lot of attention in the last years, and is one of the challenging issues in multi-robot research [10]. In [2] a real implementation on a team of outdoor robots is discussed. A class of reactive behaviors that implement formations is introduced and tested on a team of military unmanned ground vehicles performing in an outdoor terrain. While this work is similar to ours in terms of the behavior based approach and the outdoor operating scenario, the main differences are that we do not use a global positioning system (e.g., GPS) and we also deal with a heterogeneous system in which robots are equipped with different sets of sensors.

Potential field approaches are also widely used (see for example [11,12,2]). Robots move by being attracted to their desired position in the pattern and being repulsed from obstacles and other robots. In this context one of the main problems is dealing with systems that operate in environments that exhibit significant dynamics, since the group has to promptly react to unforeseeable circumstances that can emerge inside or outside the team itself [4]. In [5] the idea of moving a team by means of a leader that conducts the rest of the robots is discussed and some simulation results are shown. The problem of designing control laws for this kind of problem is discussed in [9]. One of the main issues is related to the design of *distributed* control schemes – i.e., schemes where each robotic agent acts on the basis of local decisions [3,7,13] as opposed to *centralized* schemes where decisions are made by a unique subject, which has a global view of the situation, and are then communicated to robotic agents [1]. In constrast to much of this previous research, our research explicitly addresses issues of maintaining formation in a leader-following application while also avoiding obstacles that may unexpectedly appear within the formation.

3 Distributed Control Architecture

We address the following problem: given a set of possibly heterogeneous robots arranged in a linear pattern, design a strategy so that if the first moves in an unknown environment, the others follow while at the same time avoiding obstacles that may appear within the formation. We call this task *Cooperative Leader Following*. In our approach, we distinguish between *team-level* behaviors and *robot-level* behaviors. At the team level we use a situated automaton [8] approach, with the team seen as a finite state automaton whose inputs come from the environment. The transitions between the three group-level behaviors are shown in Figure 1. The behaviors at the team level are *Team-Follow*, *Team-Wait*, and *Team-Recover*. When the team is in *Team-Follow*, every robot (except the first) will follow its local leader. The *Team-Wait* behavior is executed when the team is waiting for some event to happen. This is the case when, for example, a moving object approaches a robot, so that it is not safe to keep moving. In this case all robots stop to avoid breaking the formation. The *Team-Recover* behavior is executed when the team is trying to recover from a *wait* condition. Since not all robots are involved in a collision danger situation, when the team is performing this behavior single robots will execute different behaviors, with some robots trying to go around obstacles, and others simply performing regular following at a reduced speed.

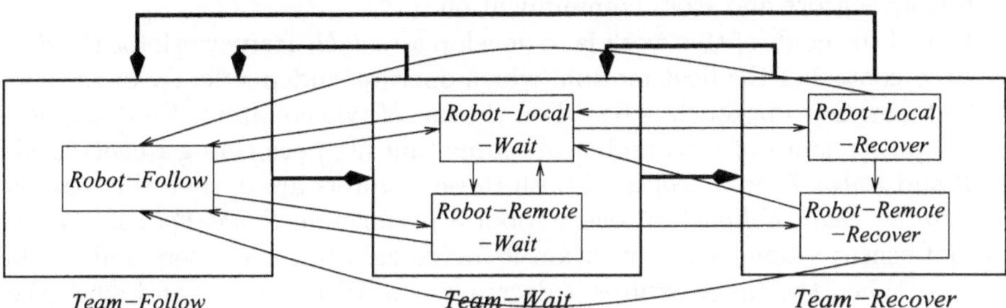

Fig. 1. Team-Level behaviors (bold lines), Robot-Level behaviors (thin lines) and their transitions.

The introduction of the *Team-Wait* behavior is motivated by the assumption that the team will not operate in an interference-free environment, but rather in an unknown, possibly unstructured, environment shared with other moving entities, such as humans or other robots. In this scenario, it could happen that a moving object approaches the robot, so that it is necessary for it to stop to avoid a collision. The robot first *waits* for a certain amount of time for the obstacle to go away. If this time is exceeded, then the robot attempts to circumnavigate the obstacle – i.e., to *recover* from the situation and to resume the *follow* behavior. The wait stage is introduced because recovering is a difficult task, and it is preferred to execute it only when there is no alternative. Of course, when a robot is waiting, other robots should wait too, to keep the formation together. From the above discussion it is evident that some sort of *explicit* communication is necessary to gain the desired team level behavior, especially for large team sizes.

At the single robot level we design a set of five behaviors that, locally executed, give the team level behaviors previously described. They are:

- **Robot-Follow**. The robot is following its leader.
- **Robot-Local-Wait**. The robot is waiting because an obstacle does not let it move safely.
- **Robot-Remote-Wait**. The robot is waiting because one or more other robots are in *Robot-Local-Wait.*
- **Robot-Local-Recover**. The robot is trying to recover from a *Robot-Local-Wait* situation. This means that it is trying to overtake an obstacle while keeping track of its leader.

- **Robot-Remote-Recover**. The robot is following its leader but at a reduced speed, so that if its follower is doing a *Robot-Local-Recover* behavior, it will be easier for the robot to keep tracking while overtaking the obstacle.

At the team level the switching between different behaviors is triggered by explicit communication, while at the single robot level the triggering comes both from sensors and from communication.

One of the goals of this work is to develop a *scalable* framework for the distributed control of the fleet motion, which operates independently of the size of the team. Our approach is based on the use of two counters, *Wait* and *Recover*, to keep track of the number of robots that are performing *Robot-Local-Wait* and *Robot-Local-Recover*. If both those counters are 0, then the team is performing *Team-Follow* (i.e., every robot is performing *Robot-Follow*). When a robot senses a dangerous situation it increments both counters and starts a timer. When the timer expires it decreases the Wait counter and when the dangerous situation does not hold anymore it decreases the Recover counter. Other robots decide what to do on the basis not only of the information they sense, but also on the basis of counters' values. If Wait is greater than zero and no local dangerous situation is detected the behavior is *Robot-Local-Wait*. If Wait is zero but Recover is positive *Robot-Local-Recover* is executed. Table 1 illustrates this approach. S_n is the state of the automata at time n, while W_n and R_n are the values of the counters at time n. The input of the automata, I_n, can be one of the following:

- **LWB** (Local Warn Begin): sensors provide information that something is too close to keep following
- **LWE** (Local Warn End): sensors provide information that a previously detected dangerous situation does not hold anymore
- **LTO** (Local Time Out): the timer started with LWB expired so it is necessary to switch from waiting to recovering

In order to have a distributed implementation, the Warn and Recover counters are not allotted to a unique entity, but are rather shared among all the robots – i.e., every robot has its own copy. For this reason four different messages are sent, namely two for updating Warn (increase and decrease) and two for updating Recover. Messages are anonymously sent in a broadcast fashion. In this way, for both sending and receiving, robots do not need to know the number of members of the team. Thus the approach is independent of the size of the team itself.

4 Sensor Processing Techniques

To test the proposed framework, we implemented this approach on a heterogeneous team of three to five mobile robots. The robots are heterogeneous

S_n	W_n	R_n	I_n	S_{n+1}	W_{n+1}	R_{n+1}
Follow	0	0	LWB	Local Wait	1	1
Follow	1	1	-	Remote Wait	1	1
Local Wait	n	1	LWE	Remote Recover	n-1	0
Local Wait	1	1	LWE	Follow	0	0
Local Wait	n	m	LWE	Remote Wait	n-1	m-1
Local Wait	n	m	LTO	Timer Elapsed	n-1	m
Timer Elapsed	0	m	-	Local Recover	0	m
Timer Elapsed	n	m	LWE	Remote Wait	n	m-1
Local Recover	0	m	LWE	Remote Recover	0	m-1
Local Recover	0	1	LWE	Follow	0	0
Local Recover	1	m	-	Timer Elapsed	1	m
Remote Wait	n	m	LWB	Local Wait	n+1	m+1
Remote Wait	0	m	-	Remote Recover	0	m
Remote Wait	0	0	-	Follow	0	0
Remote Recover	n	m	LWB	Local Wait	n+1	m+1
Remote Recover	n	m	-	Remote Wait	n	m
Remote Recover	0	0	-	Follow	0	0

Table 1. Transitions of the automaton based on messages coming from sensors and on the value of Warn and Recover counters. The state *Timer Elapsed* is used when a robot's timer elapsed, but the robot has to wait because of W_n is greater than 0. In that case its behavior is *Robot-Local-Wait.*

in that they are equipped with different sensor suites and have different mobile platforms. We developed a multi-threaded software architecture in which each sensor is handled by a separate thread that uses its data to obtain information about tracking and navigation. Each thread then sends its output (e.g., the polar coordinates of the point to track) to a *Decisor* thread which, on the basis of the desired behavior and sensory input, drives the robot. The Decisor first merges the provided points to track and then decides how to move on the basis of a set of *fuzzy rules.* A separate thread handles explicit communcation. Each robot is equipped with sonar sensors that provide distance readings within four meters. Three of the robots are equipped with a SICK PLS (Proximity Laser Scanner) laser range finder. Also, three robots have a Sony camera mounted on a pan-tilt unit. Each robot is equipped with wireless Ethernet, which allows them to communicate over the standard TCP/IP protocol. Finally, we note that even though these robots also have a Differential Global Positioning System (DGPS), we did not use it in this research because of its frequent unavailability (e.g., due to indoor operations, satellite obstructions, etc.).

In our approach, we use the available sensors that are most appropriate for each navigation subtask. Our experimental trials involving sonar-based versus laser-based (or camera-based) tracking have clearly shown that the laser and camera sensors are much more accurate than the sonar beyond a short distance. Thus, the laser scanners and cameras are used for robot tracking. However, the laser and camera sensors do not have 360 degree coverage of the space around the robot, whereas the sonars are positioned around the entire robot periphery. Thus, the sonars are used to detect obstacles that approach the robot from directions unseen by the laser and camera sensors, and for short-range obstacle detection.

4.1 Using laser PLS sensor to track the leader

The SICK PLS sensor provides readings with a scan angle of 180 degrees and an angular resolution of 0.5 degrees. Distances returned under 15 meters are considered reliable. Starting from this sensor data, the tracking routine removes spikes and averages across samples to obtain a smoother sequence (see Figure 2). From the smoothed sequence, the routine then extracts all the local minima in the sequence and tries to match each one with the previous tracked minima[1]. The one which is closest is considered to be the new local minima to track and will be tracked in the next step. The routine returns the polar coordinates (distance, ρ, and angle, θ) of the minima to track. This is the information that the thread dealing with the laser will provide to the Decisor module. The chosen local minima is the one which minimizes the quantity:

$$d = \sqrt{(\rho_c \cos\theta_c - \rho_t \cos\theta_t)^2 + (\rho_c \sin\theta_c - \rho_t \sin\theta_t)^2}$$

where (ρ_c, θ_c) are the polar coordinates of the candidate local minima and (ρ_t, θ_t) are the coordinates of the minima currently being tracked. This formula gives the distance on the cartesian plane between the candiate local minima and the currently tracked minima. In addition, during the local minima search, the routine verifies that no obstacle is too close to the robot. If this is not the case, the LWB condition is raised and the corresponding LWE condition is issued when the obstacle moves away.

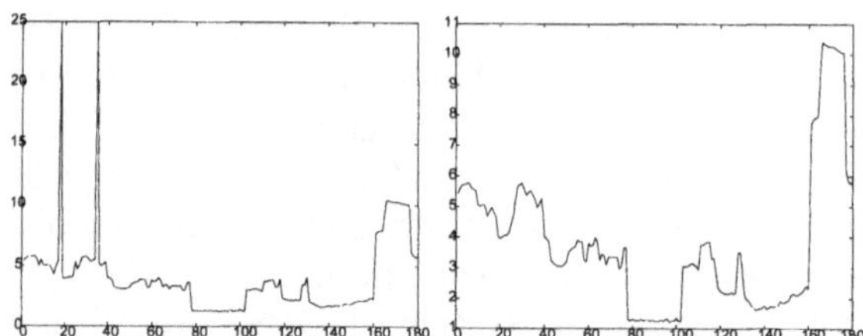

Fig. 2. Raw readings from the PLS sensor and a smoothed sequence (note the difference in the two scales).

[1] It is assumed that at the beginning all the robots are arranged in a line pattern, so that at the first scan cycle the current position of the leader is roughly known.

4.2 Using CCD camera to track the leader

Three of our robots are equipped with a CCD camera mounted on a pan-tilt unit. The associated framegrabber returns a color image in the RGB coordinate space of 120 × 160 pixels. Robots equipped with a camera have to detect and follow in real time a similar robot in their field of view. The designed approach for detecting the position of the leader robot is as follows:

- *Color segmentation* is accomplished by defining a range of colors that match the red color of our robots, discarding all other pixels.
- *Averaging (Smoothing)* is achieved using a neighborhood averaging technique in which pixel color is updated by the eight surrounding pixels. If four or more are red it is set to red, otherwise it is set to white (white pixels are ignored in the following steps).
- *Blob detection* is done by checking the boundaries of twenty-five pixels. When a red pixel along the boundary of a region is found, the region is flagged as a possible hit.
- *Object assignment* gives a different label to each connected component, using the iterative algorithm by Haralick [6].
- *Object selection* decides which of the objects should be tracked. This is done by comparing the center of mass of every distinct object with the position in the image plane of the previous object being tracked (in a similar way to what is done in the laser routine, as outlined in Section 4.1).
- *Proximity estimation* gives a rough estimation of the distance to the robot, based on the dimension of the blob being tracked, and of its position on the image plane.

Fig. 3. Results of visual tracking approach. Small white circles indicate locations of robots detected by the visual tracking algorithm.

Thus, as for the laser, the camera-handling thread gives a position to track in polar coordinates (i.e., distance and direction). Figure 3 shows examples of the detection of robot position using this approach.

4.3 Using sonar to detect obstacles

Every robot is equipped with sonar sensors. In our experiments they have been used for obstacle avoidance only, and not for tracking, because from a number of trials it became evident that crosstalk and environmental conditions were such that tracking could be obtained only within a small distance range. Instead, they proved to be highly effective for obstacle avoidance. The thread which deals with sonars identifies the LWB and LWE conditions, as for the laser, and in addition it produces a vector whose direction and intensity indicate the direction and distance to obstacles. Since readings from sonars come in polar coordinates (ρ_i, θ_i), the cartesian components x_r and y_r of the resulting vector are calculated as:

$$x_r = \sum_{i=1}^{n} f(\rho_i) \cos \theta_i \qquad y_r = \sum_{i=1}^{n} f(\rho_i) \sin \theta_i$$

where n is the number of samples returned by the sonar sensor and f is the function plotted in Figure 4.

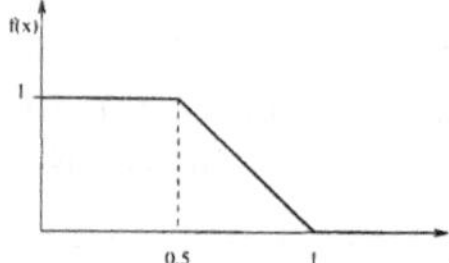

Fig. 4. Sonar weight function, where x is in meters. The x axis is directed along the heading of the robot and the y axis is perpendicular and positive to the left.

5 Robot Team Experiments

The proposed framework has been implemented and tested in both indoor and outdoor environments using teams of three to five robots. Figure 5 shows the robots performing these behaviors in an outdoor grassy environment, while Figure 6 shows the robots performing these behaviors in an outdoor gravel environment. Figure 7 gives an example of the robot state changes that occur to maintain formations when the robots encounter obstacles within their formation. In this figure, all three robots are initially in the *Robot-Follow* behavior. Then, at time $T0$, Robot 1 encounters an obstacle that puts it into the *Robot-Local-Wait* behavior, causing the other two robots to enter the *Robot-Remote-Wait* behavior. At time $T1$, after waiting a period of time for the obstacle to leave but with the obstacle still in the way, Robot 1 enters the *Robot-Remote-Wait* behavior, causing the other two robots to enter the *Robot-Remote-Recover* behavior. At time $T2$, Robot 3 itself then encounters an obstacle, causing it to go into the *Robot-Local-Wait* behavior. When that obstacle does not move, Robot 3 enters the *Robot-Local-Recover* behavior at time $T3$. However, since Robot 1 had not yet completed moving around its obstacle, it also enters the *Robot-Local-Recover* behavior at this time. Robot 2 enters the *Robot-Remote-Recover* behavior. Then, at time $T4$,

Robot 1 successfully passes its obstacle and moves to the *Robot-Remote-Recover* behavior to wait on Robot 3 to complete the bypass around its obstacle. At time $T5$, Robot 3 completes its obstacle bypass, and all robots return to the *Robot-Follow* behavior. The introduction of the *wait* behavior proved to be effective in reducing the number of recover stages, where it is more difficult to both go around an obstacle and to keep track of the leader.

Fig. 5. Results of approach implemented on robots operating in an outdoor grassy environment.

Fig. 6. Results of approach implemented on robots operating in an outdoor gravel environment.

6 Conclusions

A distributed technique for the coordinated motion in a linear pattern of a multi-robot team has been illustrated. The framework, based on explicit anonymous broadcast communcation, is fully scalable with the size of the team and deals with communcation failures. This approach allows a team of robots to remain in a linear formation even when dynamic obstacles appear in the path. The proposed approach has been implemented and validated on a heterogeneous multi-robot team performing in both indoor and outdoor environments.

Acknowledgments

The authors thank Hunter Brown for his work in developing the visual tracking algorithm. This research was supported in part by the Italian Ministry of Education Scientific Research and in part by the Engineering Research Program of the Office of Basic Energy Sciences, U. S. Department of Energy. Accordingly, the U.S. Government retains a nonexclusive, royalty-free license to publish or reproduce the published form of this contribution, or allow others to do so, for U. S. Government

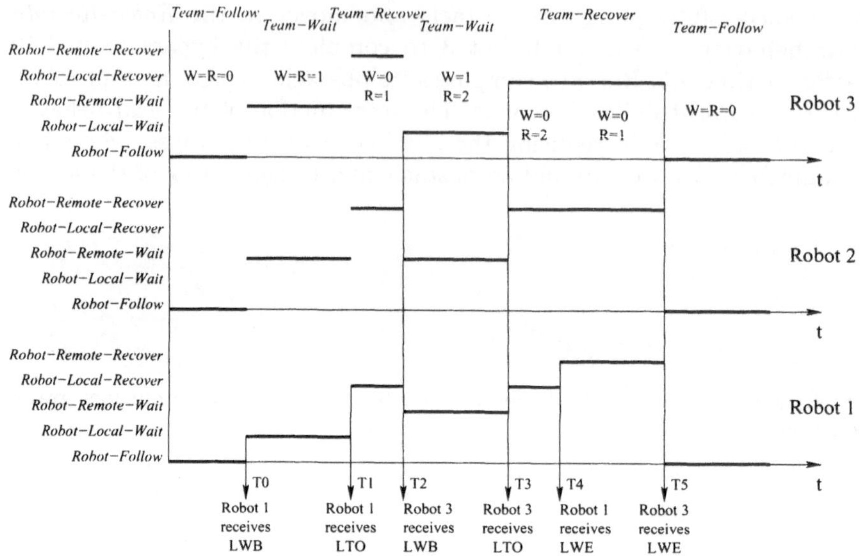

Fig. 7. An example of local behavior scheduling in the case of a three robot team.

purposes. Oak Ridge National Laboratory is managed by UT-Battelle, LLC for the U.S. Dept. of Energy under contract DE-AC05-00OR22725.

References

1. T. Arai, J. Ota. "Dwarf intelligence - A large object carried by seven dwarves". *Robotics and Autonomous Systems*, 18(1-2):149-155, 1996.
2. T. Balch, R.C. Arkin. "Behavior-based formation control for multirobot teams". *IEEE Transactions on Robotics and Automation*, 14(6):926-939, 1998.
3. G.Beni, S. Hackwood. "Coherent Swarm Motion under Distributed Control". *Proc. 1992 Symp. on Distributed Autonomous Robotic Systems*, 39-52.
4. D.C. Brogan, J.K. Hodgins. "Group Behaviors for Systems with Significant Dynamics". *Autonomous Robots*, 4(1):137-153,1997.
5. J.P. Desai, V. Kumar, J.P. Ostrowski. "Control of changes in formation for a team of mobile robots". *Proc. of the 1999 IEEE International Conference on Robotics and Automation*, pp. 1556-1561.
6. R.M. Haralick. "Some Neighborhood Operators". In *Real-Time Parallel Computing Image Analysis*. M. Onoe *et al.* (Eds.), Plenum, New York, 1981.
7. P.J. Johnson, J.S. Bay. "Distributed Control of simulated autonomous mobile robot collectives in payload transp.". *Autonomous Robots*, 2(1):43-63, 1995.
8. L. Kalebling, S. Rosenschein. "Action and Planning in Embedded Agents". In *Designing Autonomous Agents*, ed. P. Maes, MIT Press, Cambridge, MA, 1990.
9. L.E. Parker. "Designing Control Laws for Cooperative Agent Teams". *Proc. 1993 IEEE Int'l. Conf. on Robotics and Automation*, Vol.3, pp. 582-587.
10. L.E. Parker. "Current State of the Art in Distributed Autonomous Mobile Robots". In L.E. Parker, G. Bekey, J. Barhen (Eds.), *Distributed Autonomous Robotics Systems 4*, pp. 3-12, Springer, 2000.
11. J.H. Reif, H. Wang. "Social Potential Fields: A Distributed Behavioral Control for Autonomous Robots". K. Goldberg (Ed.) *Algorithmic Foundations of Robotics*, AK Peters, 1995, pp. 331-336.
12. F.E. Schneider, D. Windermulth, H.L. Wolf. "Motion coordination in formations of multiple mobile robots using a potential field approach". In L.E. Parker, G. Bekey, J. Barhen (Eds.), *Distributed Autonomous Robotics Systems 4*, Springer, 2000, pp. 305-314.
13. H. Yamaguchi, T. Arai, G. Beni. "A distributed control scheme for multiple robotic vehicles to make group formations". *Robotics and Autonomous Systems*, 36(4):125-147, 2001.

Collision Avoidance Algorithm for Two Tracked Mobile Robots Transporting a Single Object in Coordination Based on Function Allocation Concept

Hiroki Takeda, Yasuhisa Hirata, Zhi-Dong Wang, and Kazuhiro Kosuge

Department of Machine Intelligence and Systems Engineering, Tohoku University, Aoba-yama 01, Sendai 980-8579, Japan

Abstract. In this paper, we propose a collision avoidance algorithm for two tracked mobile robots transporting a single object based on a function-allocation concept. In this algorithm, the desired trajectory of the object is given to the leader robot, and the follower robot estimates the desired trajectory of the leader along the heading direction of the follower and generates the motion of the object for avoiding obstacles. We experimentally implement the proposed algorithm in the nonholonomic tracked mobile robots, and illustrate the validity of the proposed control algorithm.

1 Introduction

When a human being handles a large or a heavy object, we usually carry it in cooperation with other people. It is natural to extend this behavior to robot system. Based on different concept on function distributed and information management, many researchers have been proposed various motion control algorithms for the multiple mobile robots to handle a single object in coordination [1]–[9] etc.

We have proposed a leader-follower type motion control algorithm for nonholonomic tracked mobile robots to transport a single object [4]. In this algorithm, a motion command of the object is given to one of the robots referred to as a leader, and the rest of the robots referred to as followers estimate the motion of the leader through the motion of the object and transport it together with the leader in coordination. Because the robots do not use explicit communication, the execution of more reliable transport task is expected.

However, this control algorithm is designed under the assumption that each robot would not collide with obstacles. Then we could not apply the same control algorithm to robots system directly, which are used in an environment with obstacles. In this paper, based on the control algorithm presented in [4], we propose a collision avoidance method for two tracked mobile robots transporting a single object in coordination based on a function allocation concept. We implement the proposed algorithm in two tracked mobile robots and illustrate the validity of the algorithm.

2 Conventional Algorithm

In this section, we briefly explain the decentralized control algorithm proposed in [4]. In case of the mobile robots under a nonholonomic constraint, the robots could not move on the ground freely if each robot held the transported object firmly. To solve this problem, we assume that each robot holds the object through a free rotational joint, which is located in the middle of both wheels, as shown in Fig. 1. In this case, the motion of the robot is characterized by two kinds of motion. One is the translational motion along the heading direction of the robot, and the other is the rotational motion around the free rotational joint.

In the algorithm proposed in [4], each robot was controlled so as to have a following dynamics along the heading direction.

$$D\Delta^l \dot{x}_l + K\Delta^l x_l = {}^l F_{lx} \tag{1}$$

$$D\Delta^1 \dot{x}_1 + K\Delta^1 x_1 = {}^1 F_{1x} \tag{2}$$

where D is a positive definite damping coefficient and K is a positive definite stiffness coefficient. $\Delta^l x_l$ and $\Delta^1 x_1$ are the motion deviations of the leader and the follower according to the forces applied to the each robot ${}^l F_{lx}$, ${}^1 F_{1x}$, respectively. A desired trajectory of the object is given to the leader and the follower estimates its desired trajectory along the heading direction of the follower by using the estimation algorithm proposed in [1]. For the rotational motion of the robot, each robot is controlled as if it has a caster-like dynamics as shown in Fig. 3 and the follower makes its orientation rotate to the heading direction of the object based on the caster-like dynamics. Let us briefly review this caster-like dynamics.

The real caster has an offset C_{off} between the axis of the wheel and the free rotational joint as shown in Fig. 2(a). The caster turns to the direction of the force applied to the caster by this offset. If the robot is controlled so as to imitate the motion of the real caster, the robot rotates more than 90 degrees as shown in Fig. 2(c), when the robot is pulled backward. The rotational motion more than 90 degrees could apply the excessive force to the transported object. To avoid this problem, we consider that the follower is controlled to have two different caster dynamics as shown in Fig. 3(a). That is, we consider changing the position of the free rotational joint according to the force direction applied to the robot. When the robot is pushed forward, the offset is set equal to $C_{off}(> 0)$(Fig. 3(b)), and when the robot is pulled

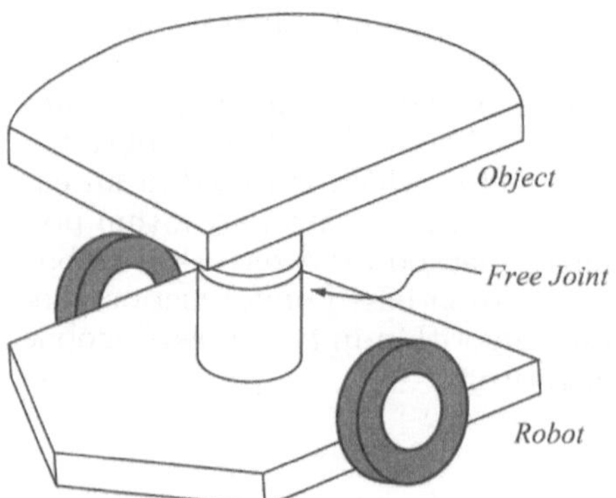

Fig. 1. Holding mechanism applied to the robot, which has nonholonomic constraints to transport a single object in coordination

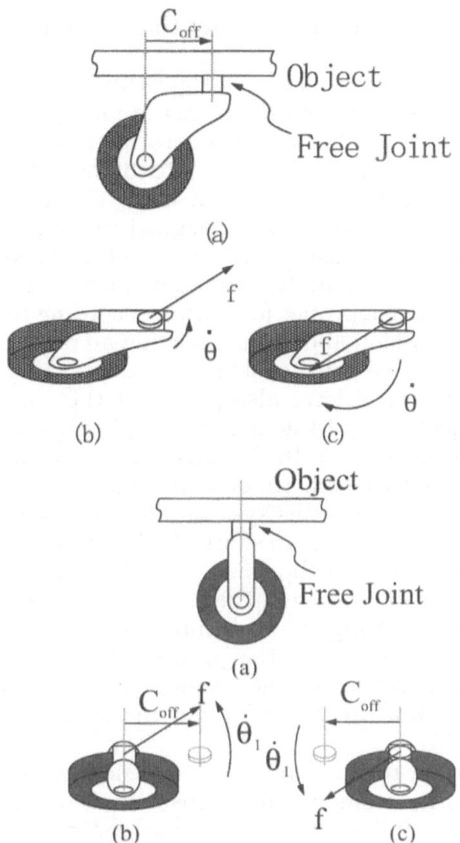

Fig. 2. Real-caster action: (a)General view of the real caster (b)Caster can rotate minimal angle to align the wheel direction with the force direction (c)Caster has to rotate more than 90 degrees to align the wheel direction with the force direction.

Fig. 3. Dual-caster action: (a)General view of the dual caster (b)(c)The direction of the offset of the caster C_{off} changes so as to minimize the rotational angle.

backward, the offset is set equal to $-C_{off}$(Fig. 3(c)).We refer to this caster-like dynamics as dual-caster action.

3 Control Algorithm for Collision Avoidance

3.1 Function Allocation Concept

In the control algorithm explained in the previous section, the leader could transport a single object in coordination with the follower by applying the desired position of the object to the leader. However, the leader could not manipulate the orientation of the object actively during the transportation of the object, because the orientation of the object depend on the position of the follower, which is controlled based on the force applied to the follower itself by using the dual-caster action. That is, the orientation of the object is changed passively based on the motion of the follower, which is affected by the force applied to the follower.

In this case, if we use this system in an environment with obstacles, the object supported by two robots or the follower might collide with the obstacles, although the leader could avoid the obstacles by applying the desired trajectory appropriately to the leader. To realize the transportation of an object in an environment with obstacles, we have to consider the problem of collision avoidance for multiple mobile robots.

Many researchers have proposed the motion planning algorithm for the collision avoidance [11] etc. However, these control algorithms are designed for a single robot, so that we could not apply the same control method to the multiple robots system directly. K.Inoue et. al have proposed the collision avoidance algorithm for the leader-follower type system using the multiple mobile robots [5]. However, this control system is designed for omni-directional mobile robot under the holonomic constraints. J.Desai et. al have proposed the formation control algorithm of the multiple nonholonomic car-like robots [10]. In this system, robots achieve a proper formation according to an environment. However, there is no discussion about transporting an object by these robots in this paper. In another word, the constraints of supporting the object are not considered. N.Miyata et. al have also proposed the control algorithm transporting a single object by multiple nonholonomic car-like robots based on the function allocation concept [6]. In this algorithm, several functions for achieving tasks are allocated to each robot and multiple mobile robots realize the tasks effectively. To realize this system, however, each robot is controlled in the centralized control system to share the information of other robots. M.N.Ahmadabadi et. al proposed the constraint-move based cooperation strategy for multiple mobile robots system[7][8]. In this method, the cooperative task is executed by decomposing the task into two different sub tasks. As robots are sharing the information of the path of the object specially in constraint group, some consideration for the accumulation of the position and orientation error of each robot are needed when the length of the path becomes longer.

In this paper, we restrict our attention to the problem of collision avoidance for multiple mobile robots under the nonholonomic constraints, which are controlled in the decentralized way, and propose a motion control algorithm of multiple tracked mobile robots transporting a single object in coordination without colliding with the obstacles. To realize this method, we consider the function allocation concept to realize the transportation of an object effectively by multiple nonholonomic mobile robots. In its algorithm, we allocate the two types of function to each robot. One is to control the position of the object and the other is to control the orientation of the object and to transport the object in coordination.

In this paper, we allocate the position control function of the object to the leader, so that the leader transport the object based on the desired position of the object applied to its leader. We also allocate the orientation control function of the object and the transport function with the other robot to the follower, so that the follower transport a single object with the leader without colliding with obstacles by controlling the orientation of the object.

3.2 Control Algorithm of Robots

When each robot grasps the object through a free rotational joint, the relative position of the follower with respect to the position of the leader is restricted on the circle as shown in Fig. 4, whose radius is the size of the object L. In this paper, the destinate path of the object given to the leader is based on path-planning in the C-space of the leader. In this section, we

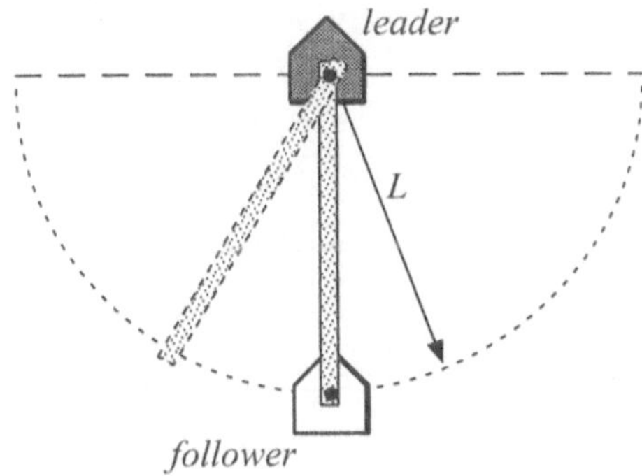

Fig. 4. The orientation of the object is constrained by the position of the leader and the follower

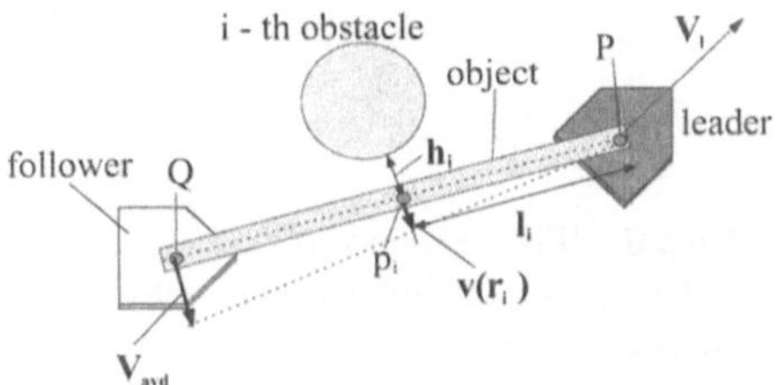

Fig. 5. The avoidance velocity vector of the follower V_{avd} is calculated based on the minimal distance between the obstacle and the object h_i to avoid the obstacle / obstacles.

consider that how the follower generates the motion of the object for the collision avoidance by moving on the circle as shown in Fig. 4.

First, we consider the avoidance velocity vector of the follower V_{avd}, which is generated by the follower to avoid obstacles. Let p_i be the point on the object closest to the i-th obstacle as shown in Fig. 5. Let r_i be a vector perpendicular to the heading direction of the object defined as follows.

$$r_i = h_i y_o \tag{3}$$

Where h_i is the minimum distance between i-th obstacle and the object and y_o is a unit vector of the object coordinate frame. The avoidance velocity vector V_{avd} is generated as follows.

$$V_{avd} = \sum_{i=1}^{n} v(r_i) \cdot \frac{L}{l_i} \tag{4}$$

$$v(r_i) = \frac{k}{|r_i|^2} r_i \tag{5}$$

Where l_i is the distance from the grasping point of the leader to the point p_i, and k is a constant.

From eqs.(4) and (5), the follower could avoid obstacles by specifying the avoidance velocity vector V_{avd} perpendicular to the heading direction of the object. However, the follower could not always generate the velocity perpendicular to the heading direction of the object, because of the motion constraints of the follower under the nonholonomic constraints.

The motion of the follower is characterized by two kinds of motion. One is the translational motion along the heading direction of the follower and the

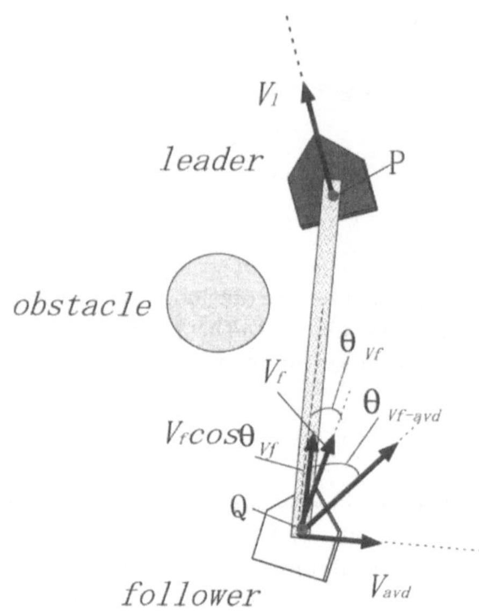

Fig. 6. The motion direction of the follower is derived by the velocity of the object and the avoidance velocity.

other is the rotational motion around the free rotational joint of the follower. For the translational motion along the heading direction of the follower, the follower generate the velocity along its heading direction V_f by estimating the desired trajectory of the leader using the estimation algorithm proposed in [1].

For the rotational motion around the free rotational joint of the follower, we consider that how the follower generates the angular velocity of the follower around its free rotational joint by using the velocity along the heading direction of the follower V_f and the avoidance velocity V_{avd}. Let P and Q be the positions of the rotational joints of the leader and the follower respectively as shown in Fig. 6. Let V_l be the velocity of the leader at the point P, and V_f be the velocity of the follower at the point Q. Let θ_{Vf} be the angle between the vector from P to Q and V_f. From Fig. 6, the velocity along the heading direction of the object is $V_f \cos\theta_{Vf}$. The desired orientation of the follower is also calculated as follows.

$$\theta_{Vf-avd} = \tan^{-1}\frac{|V_{avd}|}{|V_f \cdot \cos\theta_{Vf}|} \tag{6}$$

Let us consider how the follower aligns its orientation to the direction calculated in eqs.(6). We calculate the force $\tilde{f}_{1x}$ along the direction of θ_{Vf-avd} and the force $\tilde{f}_{1y}$ perpendicular to the direction of θ_{Vf-avd} as following equation based on the magnitude of the force F applied to the follower.

$$\tilde{f}_{1x} = |F|\cos(\theta_{Vf} - \theta_{Vf-avd}) \tag{7}$$

$$\tilde{f}_{1y} = |F|\sin(\theta_{Vf} - \theta_{Vf-avd}) \tag{8}$$

Then we apply $\tilde{f}_{1x}$ and $\tilde{f}_{1y}$ to the dual-caster action as follows;

$$D_{cast}\dot{\theta}_1 = C\tilde{f}_{1y}sgn(\tilde{f}_{1x}) \tag{9}$$

$$sgn(\tilde{f}_{1x}) = \begin{cases} 1, & \text{for } \tilde{f}_{1x} \geq 0 \\ -1, & \text{for } \tilde{f}_{1x} < 0 \end{cases} \tag{10}$$

where D_{cast} is the damping coefficient of the dual caster and C is the offset of the dual caster as shown in Fig. 3. By specifying the velocity along the heading direction of the follower calculated by the estimation algorithm proposed in [1] and the angular velocity of the follower around its free rotational joint expressed in this section, the follower could transport a single object in coordination without colliding with obstacles.

3.3 Effect of Misalignment

To estimate the desired trajectory of the leader along the heading direction of the follower, we utilize the control algorithm proposed in [4]. This algorithm was designed under the assumption that the relative orientation between the object and the follower is small. In the control algorithm proposed in this paper, however, the follower generates the avoidance motion to avoid obstacles, so that the relative orientation between the object and the follower might be large.

In this section, we investigate the effect of misalignment between the orientation of the object and the orientation of the follower. In the algorithm proposed in [4], we design the estimator based on the estimation error expressed by the following equation.

$$\Delta^1 x_{d1} = 2U_1 \Delta^1 x_1 \tag{11}$$

$$U_1 = \frac{1}{2}\left(1 + \frac{\cos(\theta_l - \theta_1)\cos(\theta_l - \theta_s)}{\cos(\theta_1 - \theta_s)}\right) \tag{12}$$

Where $\Delta^1 x_{d1}$ is the estimation error of the trajectory of the follower and $\theta_l, \theta_s, \theta_1$ are the orientation of the leader, object and follower respectively.

If the follower calculate U_1, the follower could estimate the trajectory precisely based on the estimation algorithm proposed in [1] as shown in Fig. 7(a). However, the follower could not calculate the estimation error because the follower does not know the orientation of the leader and the orientation of the object. We design the estimator based on the estimation error by the following equation instead of using eqs.(11).

$$\Delta^1 \tilde{x}_{d1} = \frac{\Delta^1 x_{d1}}{U_1} = 2\Delta^1 x_1 \tag{13}$$

From this relationship, the resultant estimator is expressed by Fig.7(b). By designing G_1 to keep the stability of this system for any U_1, the follower can estimate the trajectory using Fig.7(b). In the algorithm proposed in this paper, the stability of the system is guaranteed as long as $U_1 > 0$. When the relationships expressed as following equations are satisfied by the dual-caster action, the stability of the system is guaranteed.

$$-\frac{\pi}{2} < \theta_l - \theta_s < \frac{\pi}{2} \tag{14}$$

$$-\cos^{-1}\left(\frac{1}{3}\right) < \theta_1 - \theta_s < \cos^{-1}\left(\frac{1}{3}\right) \tag{15}$$

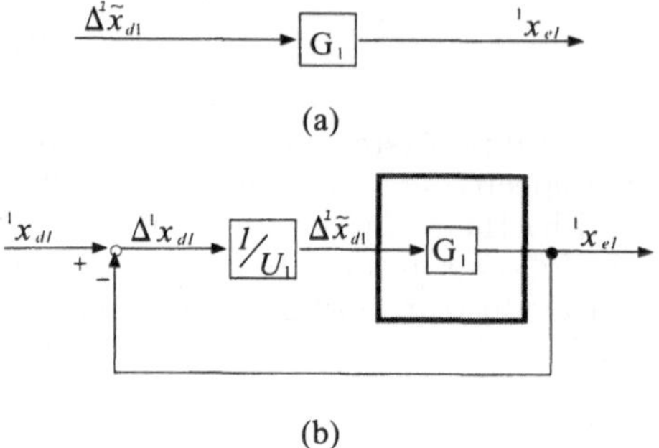

Fig. 7. Estimator of the follower to estimate the trajectory of the leader along the heading direction of the follower.

As long as the resultant relative orientation between the object and the follower satisfies the eqs.(15), the follower could estimate the desired trajectory of the leader along the heading direction of the follower and avoid obstacles, so that two tracked mobile robots transported a single object in coordination without colliding with obstacles.

4 Experiments

The proposed control algorithm was implemented in the experimental system, which consists of two tracked mobile robots with force sensor at the rotational joint of each robot as shown in Fig. 8. Each robot is controlled by its on-board controller, powered by rechargeable battery and connected to the network system of our laboratory through a wireless Ethernet. The desired trajectory was given only to the leader, and the follower was controlled using the proposed algorithm in the previous section. Since vision sensors, such as a camera, is under construction, the position of obstacles was given to the follower as the map information in this experiment.

In the experiments, first, the leader was commanded to move along the heading direction of the leader with its orientation constant. Then the leader changed its orientation, and moved along its heading direction with its orientation constant again. To illustrate the validity of the proposed control algorithm, we did two experiments. One is the experiment using the proposed algorithm, and the other is the experiment using the conventional algorithm proposed in [4].

The results are shown in Figs. 9–11. Fig. 9 shows the path of the leader, the path of the follower using the proposed algorithm (follower-proposed) and the path of the follower using the conventional algorithm proposed in [4] (follower-conventional) respectively. As shown in Fig. 9, the path of "follower-proposed" follows the leader and avoids the obstacles successfully, although the path of "follower-conventional" collides with an obstacle. Snap shot of each experiment is shown in Figs. 10 and 11.

5 Conclusion

In this paper, we proposed a collision avoidance algorithm for two tracked mobile robots transporting a single object in coordination based on the function allocation concept. The follower transports a single object with the leader without colliding with obstacles by applying this method. We implemented

Fig. 8. Experimental system: the robot has a force sensor at the rotational joint of the robot and the robot was controlled by its on-board controller, powered by rechargeable battery and connected to the network system of our laboratory through a wireless Ethernet.

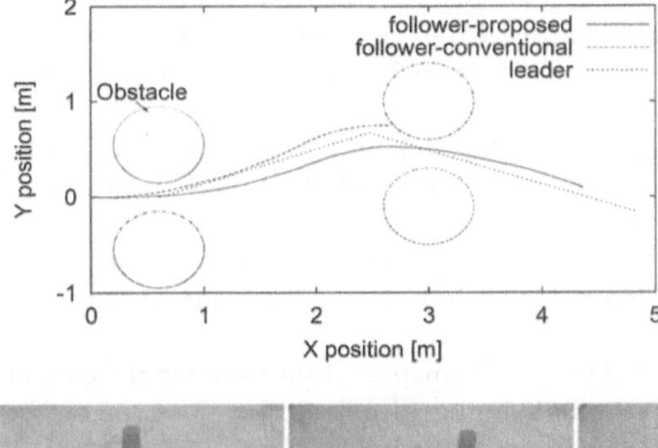

Fig. 9. Experimental results: the "follower-proposed" follows the leader and avoids the obstacles successfully, although the "follower-conventional" collides with an obstacle.

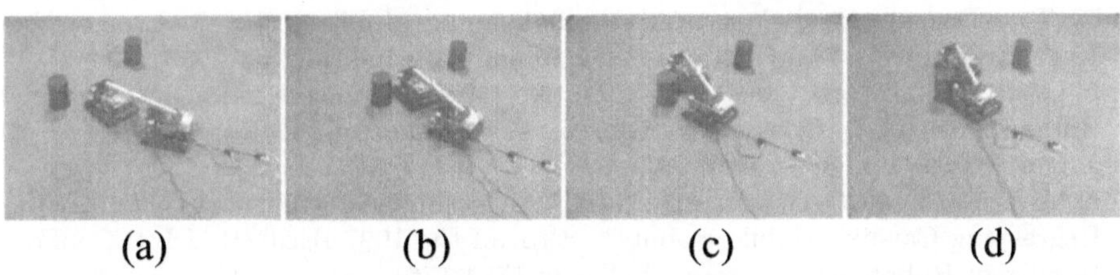

Fig. 10. Transportation using conventional algorithm proposed in [4]: the follower collides with the obstacle at (d) finally.

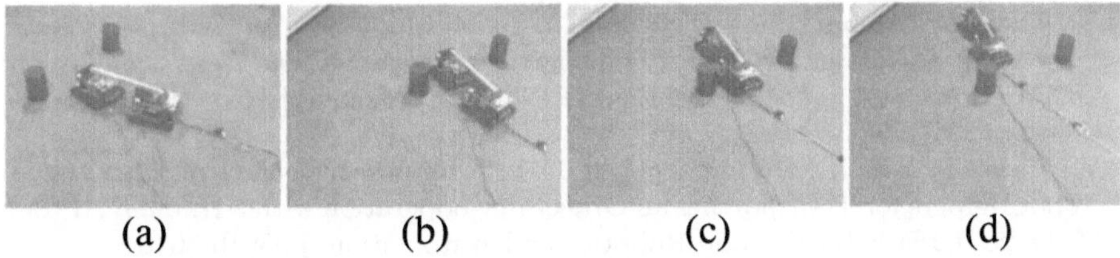

Fig. 11. Transportation using proposed algorithm: robots are transporting the object successfully without colliding with obstacles.

the proposed algorithm in the nonholonomic tracked mobile robots and illustrated the validity of the proposed algorithm. When transporting the object by more than 2 robots, the load of the object to be supported by each robot is reduced to 1/n. We will extend this system to three or more robots from now on.

Acknowledgments

This project has been partially supported by Grant-in-Aid for Scientific Research of Japan Society for the Promotion of Science (No. A-2-12305028).

References

1. K.Kosuge, T.Oosumi,(1996) "Decentralized Control of Multiple Robots Handling an Object", Proc.of 1996 IEEE Int.Conf.on Intelligent Robots and Systems, pp.318-323
2. D.J.Stilwell, J.S.Bay,(1993) "Toward the Development of a Material Transport System using Swarms of Ant-like Robots", Proc.of 1993 IEEE Int.Conf.on Robotics and Automation, Vol.1, pp766-771
3. L.Chaimowicz, T.Sugar,V.Kumar, et al,(2001) "An architecture for Tightly Coupled Multi-Robot Cooperation", Proc.of the 2001 IEEE Int.Conf.on Robotics and Automation, pp2992-2997
4. K.Kosuge, T.Oosumi, M.Sato, et al,(1998) "Transportation of a Single Object by Two Decentralized-Controlled Nonholonomic Mobile Robots", Proc.of the 1998 IEEE Int.Conf.on Intelligent Robots and Automation, pp.2989-2994
5. K.Inoue, T.Nakajima,(2001) "Cooperative Object Transportation by Multiple Robots with Their Own Objective Tasks", Journal of the Robots Society of Japan, Vol.19, pp.888-896.(In Japanese).
6. N.Miyata, J.Ota, Y.Aiyama, et al,(1997) "Cooperative Transport System with Regrasping Car-like Mobile Robots", Proc.of the 1997 IEEE/RSJ Intl.Conf.on Intelligent Robots and Systems, Vol.3, pp1754-1761
7. M.N.Ahmadabadi, E.Nakano,(1997) "Task allocation and distributed cooperation strategies in a group of object transferring robots", Proc.1997 IEEE Int.Conf.on Intel.Robots and Systems, pp.435-440
8. Z.Wang, M.N.Ahmadabadi, E.Nakano, T.Takahashi,(1999) "A Multiple Robots System for Cooperative Object Transportation with Various Requirements on Task Performing", Proc.of the 1999 IEEE Int.Conf.on Robotics and Automation, pp1226-1233
9. Y.Hirata, T.Takagi,K.Kosuge, et al,(2001) "Map-based Control of Distributed Robot Helpers for Transporting an Object in Cooperation with a Human", Proc. of the 2001 IEEE Int.Conf.on Robotics and Automation, pp3010-3015
10. J.P.Desai, V.Kumar, J.P.Ostrowski,(1999) "Control of changes in formation for a team of mobile robots", Proc. of the 1999 IEEE Int.Conf.on Robotics and Automation, pp1556-1561
11. M.Vendittelli, Jean-Paul Laumond, C.Nissoux,(1999) "Obstacle Distance for Car-Like Robots", IEEE Transactions on Robotics and Automation, Vol.15, No.4, pp678-691

A Decentralized Test Algorithm for Object Closure by Multiple Cooperating Mobile Robots

ZhiDong Wang[1] and Vijay Kumar[2]

[1] Intelligent Robotics Lab., Graduate School of Eng., Tohoku Univ.
Aobayama 01, Sendai 980-8579, JAPAN
[2] GRASP Lab., CIS, University of Pennsylvania,
3401 Walnut Street, Philadelphia, PA 19104, U.S.A

Abstract. We address the manipulation of planar objects by multiple cooperating mobile robots using the concept of *Object Closure.* In contrast to Form or Force Closure, *Object Closure* is a condition under which the object is trapped so that there is no feasible path for the object from the given position to any position that is beyond a specified threshold distance. Once *Object Closure* is achieved, the robots can cooperatively drag or flow the trapped object to the desired goal. In this paper, we define object closure and develop a set of decentralized algorithms that allow the robots to achieve and maintain *Object Closure.*

1 Introduction

By considering closure mechanisms, previous work on object manipulation by multiple robot mechanism can be categorized into three types: *grasping based, conditional closure and caging type manipulation* (Fig.1). The first type manipulation is the most popular one, specially in multiple finger or multiple arm manipulation [1][2][4][9][14][19]. All robots are arranged so that the total robots system is grasping the object during the manipulation(Fig.1-(a)). In this case, Form Closure or Force Closure condition should always be satisfied strictly. There are several research groups that have developed control strategies for coping with distributed control requirements and distributed sensing errors in such systems[5][11][18][20][21]. The second class of manipulation tasks requires *conditional closure manipulation.* This type does not guarantee Form Closure or Force Closure when we just consider robots in the system. By including gravity force, inertia, friction force, etc as an extra closure component, Force Closure is realized. In 2D manipulation, the most typical example of *conditional closure manipulation* is box pushing demonstration by two robots[12]. Lynch and Mason showed results on controllability in such manipulation tasks[7]. Tasks such as lifting objects[10] and throwing objects[8] could be viewed as examples of such *conditional closure*, but in a dynamic setting.

This paper is based on the notion of *caging* defined and studied in [13][16][17]. The key issue is to introduce a bounded movable area for the object. Then, the contact between object and robotics mechanism need not be maintained by robot's control. This makes motion planning and control of each robotic mechanism become simple and robust. We call this condition *Object Closure.*

When a group of robots can establish object closure, the object can be transported to the desired target set by simply flowing a rigid formation of robots [15]. However, for multirobot cooperation (in contrast to multifingered grasping), decentralized algorithms are essential on constructing a flexible system which be able to use distributed sensing ability, realize reconfiguration of robots system, and have other advantages of multiple robot system.

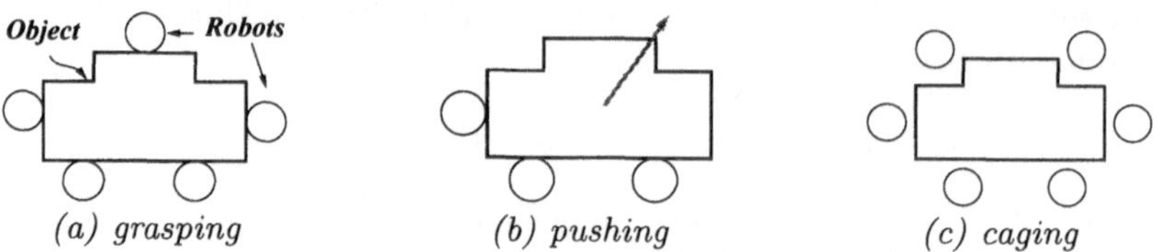

Fig. 1. Three type manipulations by multiple robotic mechanism

In this paper, we show a new approach for manipulation based on maintaining *Object Closure* with multiple mobile robots. The proposed method has distinct advantages, especially when the number of robots increases, and when the object geometry cannot be determined precisely.

2 Manipulation via Object Closure

The target of object manipulation is to generate some desired motion on the object so that it can reach its target position and orientation even under certain constraints. Then keeping contacts between robots and object in all the time is not a necessary condition for manipulation. In this paper, we study the problem of this type manipulation, and discuss properties and check conditions for *Object Closure.*

2.1 Assumptions

In this paper we will make the following assumptions about our task.

1. All robots have the same size circle shape body(discs) which be able to contact with the object in any direction. The object is star-shaped [1].
2. All robots know n, the number of robots attempting to maintain object closure, and can estimate the geometric properties (shape and center) of the object.
3. All robots can measure distance and direction toward any other robot and the object in its sensor range.
4. n is sufficiently large to guarantee object closure is feasible and each robot's sensor range is large enough to guarantee that each robot can see its closest neighbors while maintaining object closure.
5. All robots are holonomic and the controller dynamics can be reduced to a single integrator: $\dot{\boldsymbol{x}} = \boldsymbol{u}$.

[1] To star-shaped object, any half line connecting between the object's mass center and infinite point only has one intersection point with the outline of the object. It can be concave.

2.2 Approach of Object Closure based Manipulation

Our proposed approach to caging or trapping the object is realized in two stages: (a) All robots approach to the object independently (See Fig.2-(a)); and (b) The robots search for an inescapable formation (See Fig.2-(b)), based on two potential based distributed control components[22]. When the second step is successfully completed, a simple formation control strategy[3] can be used to drag the object to the designated goal destination. See Fig.2.

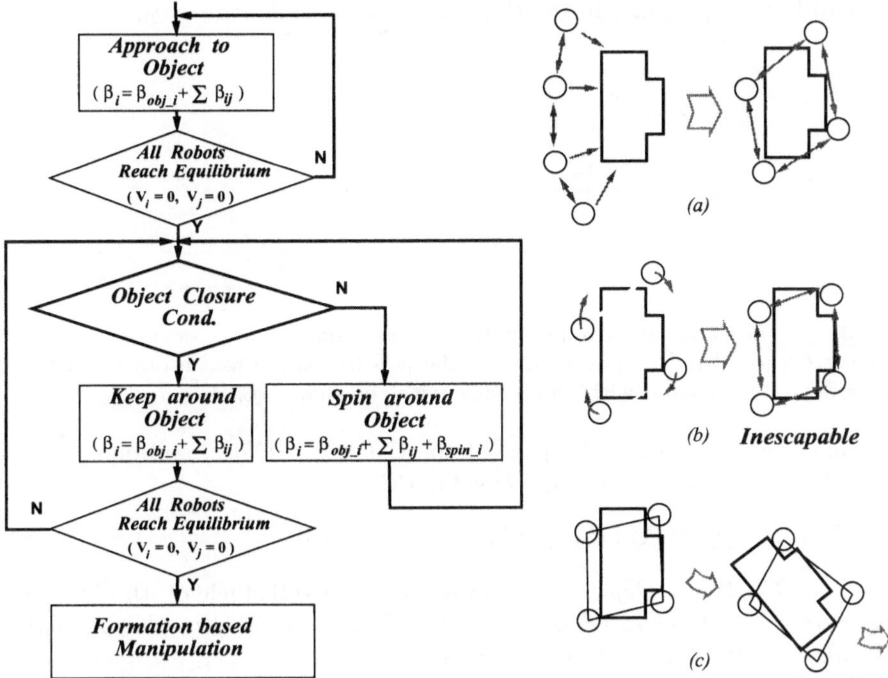

Fig. 2. The control flow chart in each robot based on two potential based control components and two stages on caging the object for formation based manipulation

2.3 Definitions

The problem of checking *Object Closure* condition is an inverse problem of finding a path from current configuration to a infinite point (free space). For solving this problem, we map the robots and the object in configuration space and consider the subset of configuration space which is connected to the object's current configuration.

Let $\mathcal{C}$ denote the C-space of the object and robots, $\mathcal{A}_{obj}$ denote the object and $\mathcal{A}_i$, $i = 1$ to n, denote the robot i in the working space. n is the total number of robots. A configuration $\boldsymbol{q}$ of $\mathcal{A}$ is a specification of robot's or object's position and orientation $(\boldsymbol{p}^T, \theta)^T$ in the working space.

First, we define a subset *C-Closure Object* as:

$$\mathcal{C}_{cls\_i} = \{\boldsymbol{q}_{obj} \in \mathcal{C} \mid \mathcal{A}_{obj}(\boldsymbol{q}_{obj}) \bigcap \mathcal{A}_i(\boldsymbol{q}_i) \neq \emptyset\} \tag{1}$$

which means that each robot $\mathcal{A}_i$ in the working space maps in $\mathcal{C}$ as a closure region to object's motion. This is shown schematically in Fig.3-(a) when the configuration space is limited to R^2. The union of all the *C-Closure Objects*:

$$\mathcal{C}_{cls} = \bigcup_{i=1}^{n} \mathcal{C}_{cls\_i} \tag{2}$$

is called the *C-Closure Object Region.* Also $\mathcal{C}_{cls\_i}(\theta_0)$ and $\mathcal{C}_{cls}(\theta_0)$ is the subset (a slice) in the $\mathcal{C}_{cls\_i}$ and in the $\mathcal{C}_{cls}$ respectively while $\theta_{obj} = \theta_0$.

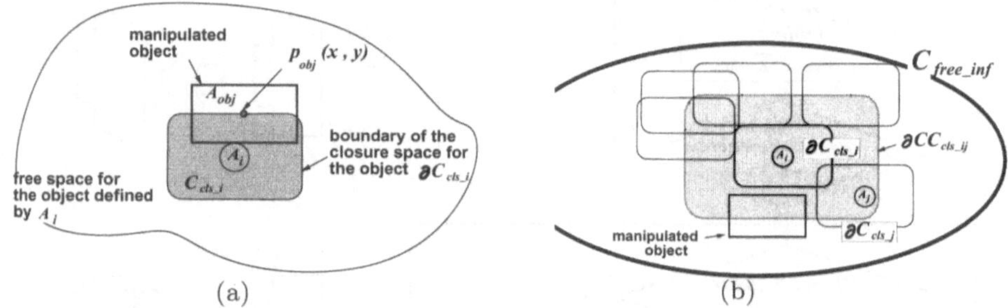

Fig. 3. (a)The constraints imposed by $\mathcal{A}_i$. The shaded area denotes $\mathcal{C}_{cls\_i}$, *C-Closure Object.* (b)$\mathcal{CC}_{cls\_ij}$ is defined as the non free space for an object with the shape $\partial\mathcal{C}_{cls\_j}$ associated with an obstacle with the shape $\partial\mathcal{C}_{cls\_i}$.

Next, we define a new C-space of the *C-Closure Object* and denote it as $\mathcal{CC}$. A C-Object of a *C-Closure Object* in $\mathcal{CC}$:

$$\mathcal{CC}_{cls\_ij} = \{\boldsymbol{q}_j \in \mathcal{CC} \mid \mathcal{C}_{cls\_i}(\boldsymbol{q}_i) \bigcap \mathcal{C}_{cls\_j}(\boldsymbol{q}_j) \neq \emptyset\} \tag{3}$$

is called *CC-Closure Object* which indicates the C-Obstacle of ith *C-Closure Object* to jth *C-Closure Object*(Fig.3-(b)). Here, $i \neq j$. Also, $CC_{cls\_ij}(\theta_0)$ is the subset in the CC_{cls_ij} while $\theta_j = \theta_0$.

Finally, the subset of configuration space in which the object does not contact or intersect any robot is:

$$\mathcal{C}_{free} = \mathcal{C} \backslash \mathcal{C}_{cls} = \{\boldsymbol{q}_{obj} \in \mathcal{C} \mid \mathcal{A}_{obj}(\boldsymbol{q}_{obj}) \bigcap (\bigcup_{i=1}^{n} \mathcal{A}_i(\boldsymbol{q}_i)) = \emptyset\} \tag{4}$$

the *Free Space* for the object.

3 Object Closure

The object and robots are in a condition of Object Closure if and only if there is no feasible path connecting the current object configuration to any point in $\mathcal{C}_{free\_inf}$, a pre-defined subset of the C-space. The minimum number of robots on realizing Object Closure is same with the number of robot to realize 2nd Order Form Closure[14]. In this paper, we will be concerned with the problem of keeping the position of the caged object contained but not the orientation. Thus we only consider Object Closure in R^2. Thus $\mathcal{C}_{free\_inf}$ will have the structure of a generalized cylinder.

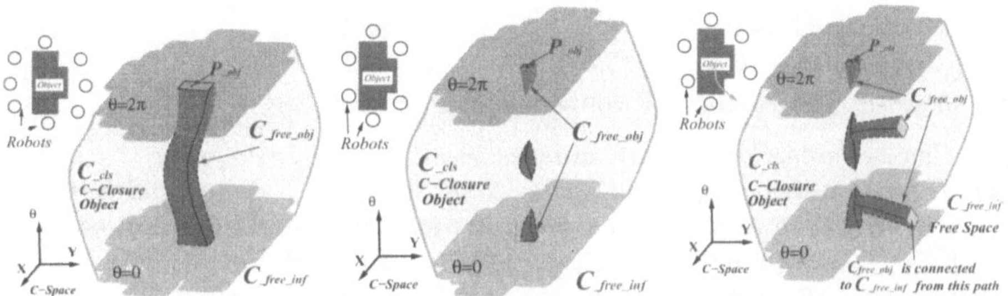

(a) (b) (c)

Fig. 4. $\mathcal{C}_{free\_obj}$ and $\mathcal{C}_{free\_inf}$: The object is closed in case (a) and (b) but is able to escape in case (c). The dark area in each is $\mathcal{C}_{free\_obj}$ which includes point $\boldsymbol{p}_{obj}$.

Let $\boldsymbol{q}_{obj} \notin \mathcal{C}_{cls}$ be a free initial configuration of the object and $\boldsymbol{q}_{inf}$ be the infinite point in $\mathcal{C}$. We define two sets $\mathcal{C}_{free\_obj}$ and $\mathcal{C}_{free\_inf}$ as follows:

$$\mathcal{C}_{free\_obj} = \{\boldsymbol{q} \in \mathcal{C}_{free} | \; connected(\boldsymbol{q}, \boldsymbol{q}_{obj})\} \tag{5}$$

Then *Object Closure* condition can be expressed as follow:

Proposition 1: *Let $\boldsymbol{q}_{obj}$ is the current configuration of the object. The object is in Object Closure iff the following conditions are satisfied.*

$$\begin{cases} \mathcal{C}_{free\_obj} \neq \emptyset, \;\; \{\boldsymbol{q}_{obj}\} \\ \mathcal{C}_{free\_obj} \bigcap \mathcal{C}_{free\_inf} = \emptyset \end{cases} \tag{6}$$

When Object Closure is achieved, there is a bounded free space($\mathcal{C}_{free\_obj}$) around the $\boldsymbol{q}_{obj}$, which is entirely kept inside of the $\mathcal{C}_{cls}$, as shown in Fig.4-a and Fig.4-b. On the other hand, Fig.4-c shows the case that there is a connection path between $\mathcal{C}_{free\_obj}$ and $\mathcal{C}_{free\_inf}$. In this case, the object is able to escape from robots' formation by this path.

As Fig.4 suggests, we must consider two cases of Object Closure. First, the free space around the $\boldsymbol{q}_{obj}$ is connected in the θ direction and the object can rotate while being completely contained (Fig.4-a). This case is an artifact of our loose definition of Object Closure, which emphasizes containment in R^2 and ignores orientations. The second case is shown in Fig.4-b. The free space $\mathcal{C}_{free\_obj}$ only has limited range in the θ direction from $\boldsymbol{q}_{obj}$, and consists of disjointed subsets.

We will check the conditions for *Proposition 1* by taking slices in the configuration space. Following the definitions in Eq.5 , we will define their slices along $\theta = \theta_0$ to be $\mathcal{C}_{free\_obj}(\theta_0)$ and $\mathcal{C}_{free\_inf}(\theta_0)$. The conditions for *Proposition 1* can now be expressed as follows:

Proposition 2: *Let θ_0 satisfy*

$$\begin{cases} \mathcal{C}_{free\_obj}(\theta_0) \neq \emptyset, \;\; \{\boldsymbol{q}_{obj}\} \\ \mathcal{C}_{free\_obj}(\theta_0) \bigcap \mathcal{C}_{free\_inf}(\theta_0) = \emptyset. \end{cases}$$

The Object Closure condition is satisfied iff for a small $\Delta\theta$, if θ_0 is replaced by $\theta_0 \pm \Delta\theta$,

$$\begin{cases} C_{free\_obj}(\theta_0) \neq \emptyset & (Con.2-A) \\ C_{free\_obj}(\theta_0) \bigcap C_{free\_inf}(\theta_0) = \emptyset & (Con.2-B) \end{cases}$$

are satisfied. The conditions can be satisfied in two ways:

(1) *Con.2-A and Con.2-B are satisfied in all* $\theta_0 \in [0, 2\pi)$

(2) *There is a pair of* θ_{0+} *and* θ_{0-} *so that conditions Con.2-A and Con.2-B are satisfied in all* $\theta_0 \in [\theta_{0+}\ ,\ \theta_{0-}]$ *and following conditions are satisfied.*

$$\begin{cases} C_{free\_obj}(\theta_{0+}) \neq \emptyset, & C_{free\_obj}(\theta_{0+} + \Delta\theta) = \emptyset \\ C_{free\_obj}(\theta_{0-}) \neq \emptyset, & C_{free\_obj}(\theta_{0-} - \Delta\theta) = \emptyset \end{cases}$$

4 Efficient Test for Object Closure

Because the check for Object Closure must run in real time, the computation cost should be low. But calculations involved in computing C_{cls_i} and checking the condition in Proposition 1 directly are hard, especially when n is large and the shape of the object is complicated. This is because of the geometrical complexity of C_{cls}. Based on the Proposition 2, we proposed a new algorithm which consists of two steps: (A) checking existence of object's configuration which is out of the set of *C-Closure Object Region* C_{cls} and (B) checking connected condition of each pair of neighbor *C-Closure Objects* so that all *C-Closure Objects* construct a close ring around the object's mass center.

The size and shape of C_{cls_i} and CC_{cls_ij} is not changed during the manipulation. Then they can be calculated in advanced for reducing the run time calculation cost even shape of the object is complicate. Also without losing the generality, we introduce the following two assumptions in the discussion.

- all robots' mass center $\boldsymbol{p}_i$ reach equilibrium in star-shape around the object
- all robots are numbered as $\theta_{i+1} \geq \theta_i$. Here θ_i denotes the angle of the vector from object's mass center to ith robot's mass center.

Checking Existence of C_{free_obj}: The check for Con.2-A is mostly used on searching boundary of C_{free_obj} in which θ_0 is shifted. Because the shape of C_{cls_i} is not always symmetrical, in general, not only the size and shape of $C_{free\_obj}(\theta_0)$ but also its center position is changed while θ_0 is shifted. Therefore it is necessary to check if there exists a neighbor $(\boldsymbol{p}_{obj} + \Delta\boldsymbol{p})$ which is not in $C_{cls}(\theta_0)$ when the $\boldsymbol{p}_{obj}$ is in $C_{cls}(\theta_0)$.

Checking Connection Condition of C_{cls_i}: In this check , we map *C-Closure Object* into CC space because the condition which a point lies inside a *CC-Closure Object* $(p_{i+1}(\theta_0) \in CC_{cls\_i(i+1)}(\theta_0))$ is the same with that two neighbor *C-Closure Objects* are connected $(C_{cls\_i}(\theta_0) \bigcup C_{cls\_i+1}(\theta_0) \neq \emptyset)$. This lets the checking problem be changed to a much simple problem.

Token-ring Type Decentralized Test: Because all robots are in a close ring around the object geometrically, we develop a token-ring type decentralized checking method which passes robot's local checking result to its neighbor. Then each robot only communicate with its two neighbor robots. To check Con.2-A procedure, because the two type checks(for $\boldsymbol{p}_{obj}$ and $\boldsymbol{p}_{obj}+\Delta\boldsymbol{p}$) exist, $(1+m)$ times "point in area checks" should be done and $(1+m)$ checking results should be passed to other robot with the token. In checking Con.2-B, the check procedure is based on information of its neighbor robot and itself only and each robot just needs one time "point in area checks" calculation. Totally, each robot should do $(m+2)$ times check of "point in area" on each testing step for Object Closure in each θ. Here, m is the number of neighbor points($\boldsymbol{p}_{obj}+\Delta\boldsymbol{p}$) around the $\boldsymbol{p}_{obj}$.

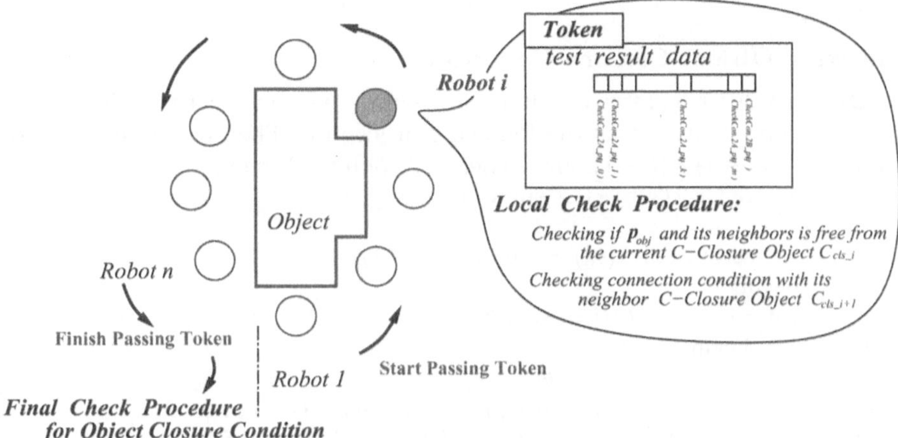

Fig. 5. A token-ring type local check procedure is designed for checking Con.2A and Con.2B. A final check procedure works on concluding testing results of a $\boldsymbol{p}_{obj}$ and its neighbors for Con.2A.

A simple protocol is designed to manage the token and decentralized test procedures(Fig.5). $(m+2)$ binary test results data is attached with token. The first $m+1$ bits are named as $CheckCon.2A\_p(\theta, k), k = 0, ..., m$ and the last bit is $CheckCon.2B\_p(\theta)$. In the beginning, all bits is set as true. Then the test is started from *Robot 1*. When the token is passed to *robot i*, the following procedures is done and the renewed test results will be passed to next robot with the token. It is clear that these procedures just use the local information of *robot i*, its neighbor *robot i+1* and the object.

procedure $LocalCheckCon.2-A\&B(\theta)$:

$$\begin{cases} CheckCon.2A\_p(\theta,0) = CheckCon.2A\_p(\theta,0) \bigwedge \{\boldsymbol{p}_{obj}(\theta) \notin C_{cls\_i}(\theta)\} \\ CheckCon.2A\_p(\theta,k) = CheckCon.2A\_p(\theta,k) \bigwedge \{\boldsymbol{p}_{obj}(\theta) + \Delta\boldsymbol{p}(k) \notin C_{cls\_i}(\theta)\} \\ \qquad\qquad\qquad\qquad\qquad\qquad\qquad\qquad k = 1, ..., m \\ CheckCon.2B\_p(\theta) \quad = CheckCon.2B\_p(\theta) \quad \bigwedge \{\boldsymbol{p}_{i+1}(\theta) \in CC_{cls\_i(i+1)}(\theta)\} \end{cases}$$

When the local check token is passed back to *robot 1* from *robot n*, the following final check procedures are used for concluding testing results:

1: procedure CheckCon.2-A(θ);
2: begin

```
3:   return ( ⋁_{k=0}^{m} CheckCon.2A_p(θ, k))
4:   if CheckCon.2A_p(θ, 0) = false  and  CheckCon.2A_p(θ, i) = true then
5:       p_obj = p_obj + Δp(i);
6: end;
```

and

```
1: procedure CheckCon.2-B(θ);
2: begin
3:   return CheckCon.2B_p(θ)
4: end;
```

Checking Object Closure Condition:

By using described two checking procedures, we construct the *Object Closure* checking algorithm as the following procedure. Then totally each robot should do $(m+2)*(l+1)$ times check of "point in area".

```
1: procedure CheckObjectClosureCon;
2: begin
3:   if ( CheckCon.2-A(θ0) = true ) and ( CheckCon.2-B(θ0) = true ) then
4:     begin
5:       for i → 1 to l  (l = 2π/Δθ)
6:         begin
7:           θ+ = θ0 + i * Δθ;
8:           if  CheckCon.2-A(θ+) = false then    go to Step 12;
9:           if  CheckCon.2-B(θ+) = false then    return false;
10:        end;
11:      return true;
12:      for i → 1 to l
13:        begin
14:          θ- = θ0 - i * Δθ;
15:          if  CheckCon.2-A(θ-) = false then    return true;
16:          if  CheckCon.2-B(θ-) = false then    return false;
17:        end;
18:      return true;
19:    end;
20:  else  return false;
21: end;
```

5 Simulation Results

A Simulation system is developed in Java2 package. The dynamics of the object and friction between object and supporting ground are implemented in the system. A compliance based contacting force model which includes friction and sliding model are introduced to simulate contacts between the robot and the object.

Fig.6 and Fig.7 show simulating results of the Object Closure by multiple mobile robots. In Fig.6, same controller which is based on the above mentioned algorithm is implemented in twenty robots. Without loosing generality, all robots are starting from random positions on the left side of the field. Because the number of robots is relatively large, the Object Closure is realized directly from approaching stage (Fig.6-IV), without the robots having to search around the perimeter. On the other hand, in the case which is shown in Fig.7, the four robots approach the object (Fig.7-II) and trap the object only for the current object orientation (Fig.7-III). The search by moving around the perimeter eventually yields Object Closure (Fig.7-IV).

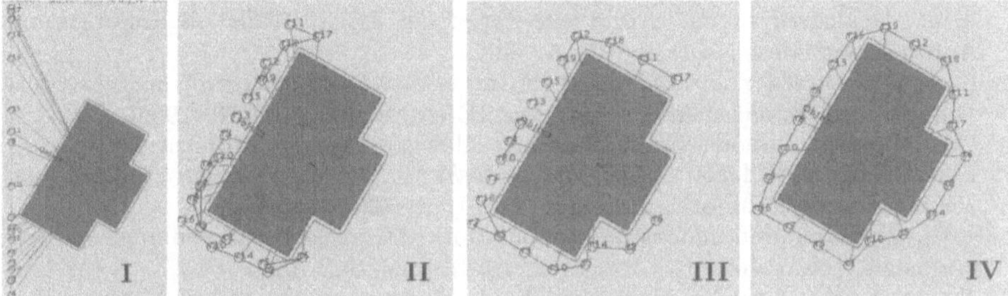

Fig. 6. A simulation result of object closure by twenty robots. Since the number of robots is large, the object closure is constructed immediately without spinning robots around the object.

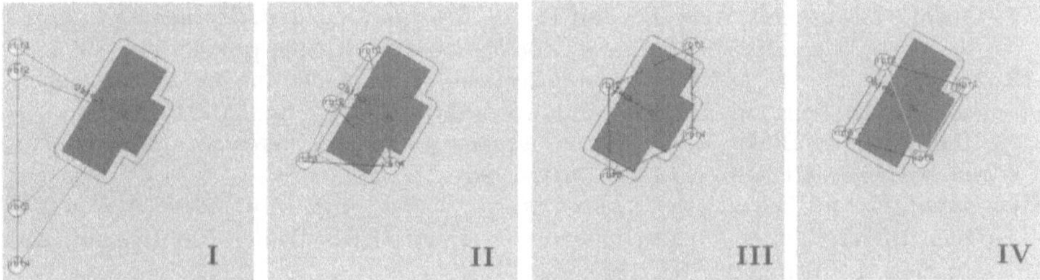

Fig. 7. A simulation result of four robots' object closure which need to search inescapable arrangement by spinning robots team(III,IV).

6 Conclusion

In this paper, we developed a novel approach to multi-robot manipulation based on the concept of Object Closure. While previous approaches have required form or force closure or variations on this theme, we use Object Closure, a condition where the object is trapped between the robots. We described algorithms for: (a) robot motion control that allow the robots to achieve and maintain object closure; and (b) testing for a condition of object closure. Our algorithms run in real-time on any browser with Java, with reasonably good performance. Also they can be implemented in decentralized token ring style. Therefore only local sensing and communication abilities should be equipped on each robot. The main limitations are our inability to handle heterogeneous robots and the lack of guarantees on completeness of the Object Closure Test.

References

1. Adams J., Bajcsy R., Kosecka J., Kumar V., and et.al, *Cooperative material handling by human and robotic agents: module development and system synthesis*, Expert System Applications, Vol.11, No.2, pp.89-97, 1996.
2. Bicchi, A. and Kumar, V, *Robotic Grasping and Contact*, ICRA2000, pp.348-353, 2000.
3. Fierro R., Das A., Kumar V., and Ostrowski J.P. *Hybrid Control of Formations of Robots*, ICRA2001, pp.157-162, 2001.
4. Koga M., Kosuge K., Furuta K., and Nosaki K., *Coordinated motion control of robot arms based on the virtual internal model*, IEEE Trans. on Robotics and Automation, Vol.1, No.1, pp. 77-85, 1992.
5. Kosuge K.,Hirata Y.,et.al, *Motion control of multiple autonomous mobile robots handling a large object in coordination*, ICRA99, pp.2666-2673, 1999.
6. Latombe J.C., Robot Motion Planning, Kluwer Academic Publishers Press,1991.
7. Lynch K.M. and Mason M.T., *Stable Pushing: Mechanics, Controllability, and Planning*, Int. J. Robotics Research, 16(6), pp.533-556, 1996.
8. ——, Dynamic nonprehensile manipulation: Controllability, planning, and experiments, Int. J. Robotics Research, 18(1), pp.64-92, 1999.
9. Nakamura Y., Nagai K., Yoshikawa T., *Dynamics and Stability in Coordination of Multiple Robotic Mechanisms*, Int.J.Robotics Research,8(2),pp.44-61,1989.
10. Ahmadabadi M.N., Rushan S.M., Wang Z.D., Nakano E., *A Constrain-Move based distributed cooperation strategy for four object lifting robots*, IROS2000, pp.2030-2035, 2000.
11. Ota J., Miyata N., Arai T., and et.al, *Transferring and Regrasping a Large Object by Cooperation of Multiple Mobile Robots*, IROS95, pp.543-548, 1995.
12. Parker L. E., *ALLIANCE: an architecture for fault tolerant multirobot cooperation*, IEEE Tran. on Robotics and Automation, 14(2), pp.220-240, 1998.
13. Rimon E. and Blake A., *Caging 2D bodies by one-parameter two-fingered gripping systems*, ICRA96, pp.1458-1464, 1996
14. Rimon E. and Burdick J.W., *Mobility of Bodies in in Contact -I: A New 2nd Order Mobility Index for Multiple-Finger Grasp.* IEEE Trans. Robotics and Automation, 14(5) pp.696-708, 1998
15. Spletzer J., Das A.K., and et.al., *Cooperative Localization and Control for Multi-Robot Manipulation*, IROS2001, pp.631-636, 2001.
16. Sudsand A. and Poncec J., *On Grasping and Manipulating Polygonal Objects with Disc-Shaped Robots in the Plane*, ICRA98, pp.2740-2746, 1998.
17. ——, *A New Approach to Motion Planning for Disc-Shaped Robots Manipulation a Polygonal Object in the Plane*, ICRA2000, pp.1068-1075, 2000.
18. Sugar T., and Kumar V., *Multiple Cooperating Mobile Manipulators*, ICRA99, pp. 1538-1543, 1999.
19. Uchiyama M. and Dauchez P., *A symmetric hybrid position/force control scheme for the coordination of two robots*, ICRA88, pp.350-355, 1988.
20. Wang Z.D., Nakano E., and Matsukawa T., A New Approach to Multiple Robots' Behavior Design for Cooperative Object Manipulation, *Distributed Autonomous Robotic Systems 2*, pp.350-361, 1996.
21. Wang Z.D., Ahmadabadi M.N., Nakano E., and Takahashi T., *A multiple robot system for cooperative object transportation with various requirements on task performing*, ICRA99, pp. 1226-1233, 1999.
22. Wang Z.D., Kumar V., *Object Closure and Manipulation by Multiple Cooperating Mobile Robots*, ICRA2002.

Coordinated Transportation of a Single Object by a Group of Nonholonomic Mobile Robots

Xin Yang[1], Keigo Watanabe[2], Kazuo Kiguchi[2], and Kiyotaka Izumi[2]

[1] Department of Production Control Technology,
Division of Engineering Systems and Technology,
[2] Department of Advanced Systems Control Engineering,
Graduate School of Science and Engineering,
Saga University, 1-Honjomachi, Saga 840-8502, Japan
†E-mail: xinyang@yahoo.co.jp, {watanabe, kiguchi, izumi}@me.saga-u.ac.jp

Abstract. In this paper, we propose a decentralized control system for transporting a single object by a group of nonholonomic mobile robots. One of these mobile robots acts as a leader, who is assumed to be able to plan and manipulate the omnidirectional motion of the object. Other robots, referred to as followers, cooperatively transport the object by keeping the initial relative position to the object. Different from the conventional leader-follower type system of transporting a signal object by multiple robots in coordination, the present follower can plan an action based on its local coordinate and does not need any absolute positional information. The simulation result exhibits a good performance of the proposed system.

1 Introduction

There has been plenty of researches concerning cooperative object manipulation by multiple robots [1][2]; the key driving forces for these researches are the industrial needs and the desire to understand human abilities and its intelligent cooperative behavior [3]. These existing researches covered a wide field from centralized control to decentralized control system. The centralized control is believed to be more effective for the motion control of manipulators handling an object than for multiple mobile robots cooperatively transporting a signal object, because huge computational power and communication capacity of controller are required in this structure and they must be varied with the number of individual agents. Many works have been done with decentralized motion control, especially with the leader-follower type in which the behavior that human does in cooperation is realized. But none of them has been used in industry mainly because the controller, which deals with not only the motion planning of each agent but also the interaction between agents, is too complex to execute in industrial environment.

As indicated in [4], any robot system should be developed in such a way that the robots are less dependent on the complicated sensor and explicit communication. Based on this principle, a leader-follower type decentralized control system for cooperatively transporting an object has been developed in our laboratory by using multiple robots. In this system, each agent is based on a nonholonomic mobile base with two independent driving wheels, which is the typical and simplest type of mobile platform. The follower robots cooperatively transport the object by keeping the initial relative position to the object, and therefore they plan their motions based on local coordinate and do not need any absolute position information, which means they do not need complex communication with each other or a supervisor.

In Section 2, we briefly describe the proposed system. The difference between the proposed system and the existing researches is explained in Section 3. After that, the control methods for the leader and followers are introduced in Section 4. In Section 5, the simulation results are shown and the conclusion of this paper is given in the last section.

2 System Description

Fig. 1 shows the sketch of a leader and three follower robots cooperatively transporting an object. All the agents are set on a horizontal plane and hold the object tightly using their arms.

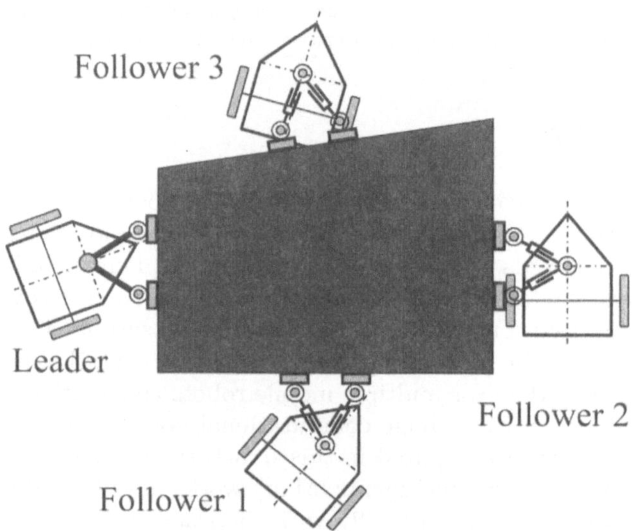

Fig. 1. Schematic view of the proposed system.

The leader has two rigid arms that are connected to the mobile base through an actuated joint and driven by a motor simultaneously. The other end of these arms connects with an end-effector through a passive joint. It is assumed that the end-effector is able to hold and grasp the object tightly.

When the leader robot holds an object, the two arms and the object construct a rigid body steering around the actuated joint. In this research, we intentionally design the steering joint offset from the central point of two driving wheels that enable this rigid body to perform an omnidirectional motion. In other words, the leader robot is able to manipulate the object to fulfil an omnidirectional motion.

The follower robot is similar to the leader in mechanical structure, i.e., the former is also equipped with two arms, one end of which connects with an end-effector through a passive joint and the other end is coupled with the mobile base at a point offset from the central point of two driving wheels. The difference is that these arms are compliant arms and link to the mobile base through a passive steering joint.

In actual design, clutch can be used to unify the design of the leader and follower robots. A clutch couples with the output shaft of the motor that drives two arms; when the clutch is open, two arms can steer freely. Other two clutches are mounted on two arms together with the linear compliant element so that they enable arms to switch between compliant and rigid structures.

It is assumed that the leader robot is able to observe the environment and plan the object motion, which is an omnidirectional motion manipulated by the leader. The followers estimate the motion of the object through the change of the arm length and control themselves based on their own local coordinates.

In the existing works about transporting an object in coordination, most researchers usually focus on the internal-force distribution among agents and the organization of behavior-based robots. In essential, the advantage of transporting a heavy object with multiple robots is keeping the stabilization of the whole system and reducing the load and output power of each agent. The idea of the proposed system is that all agents support the object together to share the load and keep the balance of object while each follower tries to minimize the inter-force with the object to reduce the driving power of the leader. When the stiffness coefficient of arms and the error of arm length are very small, the internal force that each follower acts to the object can be ignored. Then almost all the outputs of leader robot are used to manipulate the object. Therefore a small leader robot is able to move (accelerate and decelerate) a big and heavy object. The situation is almost like a person pushing a heavy box on a frictionless surface. The key point of the proposed concept depends on whether the error of arm length will be small enough or not; the simulation results give a satisfied answer and it should be noticed that in the proposed system the error of arm length is not related to the weight of the object. In the future work, a neural network will be developed for each follower to predict the motion of the object and a forward controller will be employed to control each follower. We hope, with the new control method, each follower robot not only reduces their internal force to the object but also shares some driving force/moment of the leader. In this paper, the omnidirectional motion of the object is manipulated by the leader; in fact, it can also be realized by two nonholonomic robots [5].

3 The Difference from the Existing Systems

The proposed system is different from the existing systems in two aspects: the control strategy and the mechanical structure.

From the view point of control strategy, the proposed system is a purely decentralized control system. In most of existing decentralized control structures, the motion of each agent is managed by itself, but the trajectory of each individual is usually planned previously or commanded by a supervisor, for example the leader [3][2]. Essentially in these systems, the low-level control is decentralized but the high-level control is still a centralized control. In the proposed system, the motion planning of a follower robot is based on its local coordinate without using any absolute position information provided by the leader or a supervisor; it is a completely decentralized system and it is much simpler and more realizable for practical applications, compared with the previous systems. The complete decentralized control system has also been applied in some research in which the advantages of such a system have been clearly proved [6].

More important feature of the proposed system lies on the mechanical design of each robot; we use a nonholonomic mobile robot as the individual agent, which is the typical and simplest mobile base. The feature of the follower mobile robot is that each robot has two arms equipped with a linear compliant structure and connected to mobile base through a passive steering joint offset from the central point of two driving wheels. In all of existing systems, the robot has only one arm and a "passive and active direction" concept is developed to suit the design [2], which means one robot acts to the object only in one direction. In order to manipulate an object on horizontal plane, it needs much more agents and more complex cooperation approach.

Some researchers use a force sensor fixed on the grasp point to estimate the desired motion of the object [6][7]. It is reasonable and effective, but it can not ensure that a big inter-force between agents and the slippage between ground and wheels of some agents will not be caused in high-speed operation.

In the proposed system, the passive steering joint of arms is designed to offset from the central point of two driving wheels, cooperating with the compliant structure to overcome the nonholonomic constraints of the mobile base and to make the omnidirectional motion of the object possible. It is a unique character different from the others [1][2][6].

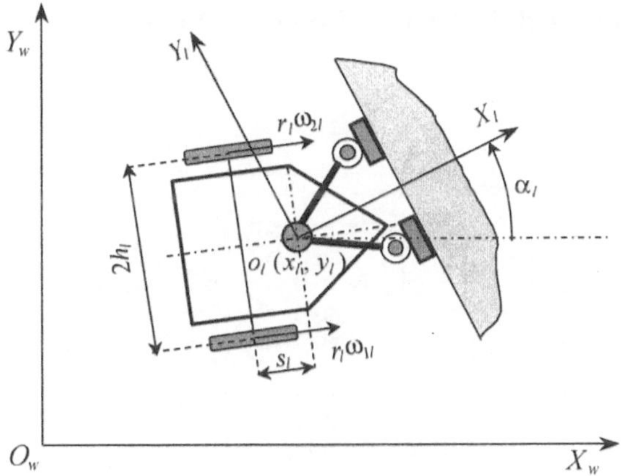

Fig. 2. Kinematic model of the leader robot.

4 Design of the Controller for Each Agent

4.1 Kinematic model of the leader robot

The leader robot holding an object is defined in the absolute world coordinate $\sum_w$, and a local coordinate system $\sum_l$ is fixed on it (as shown in Fig. 2). x_l and y_l are the position coordinates of the steering joint of leader's arm in $\sum_w$, the angle between X_w and X_l axes, α_l, denotes the posture of the rigid body constructed by two arms and the object. Therefore, the motion of the two arms or the object can be described by $\dot{\boldsymbol{x}}_l$ $(= [\dot{x}_l\ \dot{y}_l\ \dot{\alpha}_l]^T)$. The distance between two driving wheels is $2h_l$, s_l is the offset of arm's steering joint from the center of two driving wheels. Letting $\boldsymbol{\omega}_l = [\omega_{1l}\ \omega_{2l}\ \omega_{\alpha l}]^T$, the direct and inverse kinematic models of the leader holding an object are obtained by:

$$\dot{\boldsymbol{x}}_l = A_l \boldsymbol{\omega}_l \tag{1}$$

$$\boldsymbol{\omega}_l = A_l^{-1} \dot{\boldsymbol{x}}_l \tag{2}$$

where ω_{1l}, ω_{2l} and $\omega_{\alpha l}$ denote the angular velocities of two wheels and the steering velocity of arms, respectively, and

$$A_l = \begin{bmatrix} \frac{r_l}{2}\cos\alpha_l - \frac{s_l r_l}{2h_l}\sin\alpha_l & \frac{r_l}{2}\cos\alpha_l + \frac{s_l r_l}{2h_l}\sin\alpha_l & 0 \\ \frac{r_l}{2}\sin\alpha_l + \frac{s_l r_l}{2h_l}\cos\alpha_l & \frac{r_l}{2}\sin\alpha_l - \frac{s_l r_l}{2h_l}\cos\alpha_l & 0 \\ -\frac{r_l}{2h_l} & +\frac{r_l}{2h_l} & 1 \end{bmatrix}$$

$$A_l^{-1} = \begin{bmatrix} \frac{1}{r_l}\cos\alpha_l - \frac{h_l}{s_l r_l}\sin\alpha_l & \frac{1}{r_l}\sin\alpha_l + \frac{h_l}{s_l r_l}\cos\alpha_l & 0 \\ \frac{1}{r_l}\cos\alpha_l + \frac{h_l}{s_l r_l}\sin\alpha_l & \frac{1}{r_l}\sin\alpha_l - \frac{h_l}{s_l r_l}\cos\alpha_l & 0 \\ -\frac{1}{s_l r_l}\sin\alpha_l & \frac{1}{s_l r_l}\cos\alpha_l & 1 \end{bmatrix}$$

It can be found that equations (1) and (2) actually are the kinematic model of a holonomic and omnidirectional mobile robot [8], which means the robot is able to manipulate the object to fulfil an omnidirectional motion.

4.2 Kinematic model of the follower robots

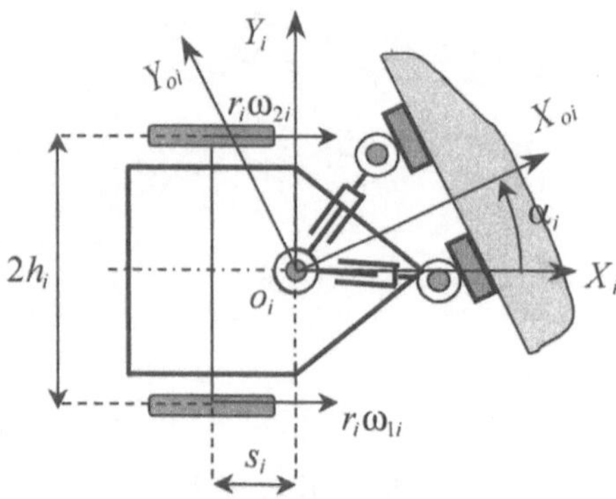

Fig. 3. Kinematic model of the follower robot.

Fig. 3 shows the kinematic model of the ith follower robot. The robot has two driving wheels with r_i radius, where the distance between them is $2h_i$ and the steering joint of arms (o_i) offsets from the center of two wheels with distance s_i. A local coordinate $\sum_i$ is fixed on the robot, where the X_i axis is along the central line of the robot and the origin locates at point o_i. We set another coordinate system $\sum_{oi}$ initiated from point o_i on the robot in such a way that the X_{oi} axis is always perpendicular to the edge of object held by two arms. This coordinate system will be used in the next paragraph for calculating the relative position of object to the follower robot. α_i is the angle from axis X_i to X_{oi}. The velocity of point o_i in space $\sum_{oi}$, ${}^{oi}\dot{\boldsymbol{x}}_{o_i}$, can be mapped into space $\sum_i$ by:

$$ {}^{i}\dot{\boldsymbol{x}}_{o_i} = {}^{i}_{oi}\boldsymbol{R}\; {}^{oi}\dot{\boldsymbol{x}}_{o_i} \tag{3} $$

where

$$ {}^{i}_{oi}\boldsymbol{R} = \begin{bmatrix} \cos\alpha_i & -\sin\alpha_i \\ \sin\alpha_i & \cos\alpha_i \end{bmatrix} $$

Letting ${}^{i}\dot{\boldsymbol{x}}_i = [{}^{i}\dot{x}_i \; {}^{i}\dot{y}_i]^T$ represent the motion of the follower in $\sum_i$ and defining $\boldsymbol{\omega}_i = [\omega_{1i}\; \omega_{2i}]^T$, the kinematic model of the robot is expressed as:

$$ \boldsymbol{\omega}_i = A_i^{-1}\; {}^{i}\dot{\boldsymbol{x}}_i \tag{4} $$

where

$$ A_i^{-1} = \frac{1}{r_i} \begin{bmatrix} 1 & \frac{h_i}{s_i} \\ 1 & -\frac{h_i}{s_i} \end{bmatrix} $$

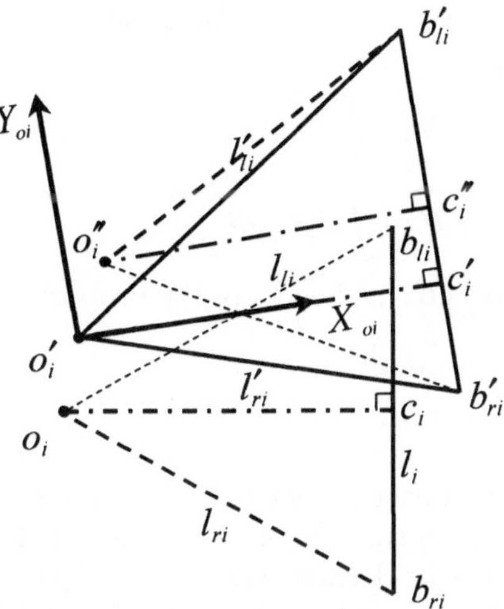

Fig. 4. Schematic drawing of the moving of two arms in a discrete time.

4.3 Motion control of the follower robots

For the follower robots, the motivation of the motion for keeping the initial relative position to the object is coming from the change of the arm length. In Fig. 4, points b_{li} and b_{ri} represent the two passive joints that connect end-effector to the arms. The length of line $b_{li}b_{ri}$ is l_i, which should be fixed while the object is moving because it is assumed that the end-effector can hold and grasp the object tightly. At first, points o_i, b_{li} and b_{ri} construct a triangle, $\overline{o_ib_{ri}} = l_{ri}$ and $\overline{o_ib_{li}} = l_{li}$. After a discrete time Δt, point o_i moves to o'_i and line $b_{li}b_{ri}$ moves to $b'_{li}b'_{ri}$, where $\overline{o'_ib'_{ri}} = l'_{ri}$ and $\overline{o'_ib'_{li}} = l'_{li}$. In order to recover the initial triangle, point o'_i should move to o''_i, where $\overline{o''_ic''_i} = \overline{o_ic_i}$ and $\overline{c''_ib'_{li}} = \overline{c_ib_{li}}$. Note that line $o''_ic''_i$ and $o'_ic'_i$ are perpendicular to line $b'_{ri}b'_{li}$; in coordinate system $\sum_{oi}$, the position differences of point o_i are ${}^{oi}e_{xi} = \overline{o'_ic'_i} - \overline{o''_ic''_i}$ in X_{oi} direction and ${}^{oi}e_{yi} = \overline{c'_ib'_{li}} - \overline{c''_ib'_{li}}$ in Y_{oi} direction. Therefore, it is easy to derive out:

$$\overline{c_ib_{li}} = \frac{{l_i}^2 + {l_{li}}^2 - {l_{ri}}^2}{2l_i} \tag{5}$$

$$\overline{o_ic_i} = \sqrt{{l_{li}}^2 - \overline{c_ib_{li}}^2} \tag{6}$$

$$\overline{c'_ib'_{ri}} = \frac{{l_i}^2 + {l'_{ri}}^2 - {l'_{li}}^2}{2l_i} \tag{7}$$

$$\overline{o'_ic'_i} = \sqrt{{l'_{ri}}^2 - \overline{c'_ib'_{ri}}^2} \tag{8}$$

$$\overline{c'_ib'_{li}} = l_i - \overline{c'_ib'_{ri}} \tag{9}$$

The velocities of steering joint o_i can be generated with the following PI controllers:

$$^{oi}\dot{x}_{oi} = K_{pxi} \ {}^{oi}e_{xi} + K_{ixi} \int_0^t {}^{oi}e_{xi} dt \tag{10}$$

$$^{oi}\dot{y}_{oi} = K_{pyi} \ {}^{oi}e_{yi} + K_{iyi} \int_0^t {}^{oi}e_{yi} dt \tag{11}$$

where $\{K_{pxi}, K_{pyi}\}$ and $\{K_{ixi}, K_{iyi}\}$ are the proportional and integral gains, respectively.

5 Simulation

A computer simulation was conducted to verify the proposed cooperative system and study the performance of each follower. In this simulation, one leader and two followers held each edge of an object shaped as a triangle together; the leader robot manipulated the object to fulfil a pre-planed motion in an absolute world coordinate. In fact, it is the motion of two actuated arms that was planed previously as they construct a rigid body with the object together. A resolved velocity control strategy with PI servo was adopted for the leader, so that:

$$\boldsymbol{\omega}_l = A_l^{-1} \dot{\boldsymbol{x}}_l^* \tag{12}$$

$$\dot{x}_l^* = \dot{x}_{dl} + K_{pxl} e_{xl} + K_{ixl} \int_0^t e_{xl} dt \tag{13}$$

$$\dot{y}_l^* = \dot{y}_{dl} + K_{pyl} e_{yl} + K_{iyl} \int_0^t e_{yl} dt \tag{14}$$

$$\dot{\alpha}_l^* = \dot{\alpha}_{dl} + K_{p\alpha l} e_{\alpha l} + K_{i\alpha l} \int_0^t e_{\alpha l} dt \tag{15}$$

where $\dot{\boldsymbol{x}}_l^* = [\dot{x}_l^* \ \dot{y}_l^* \ \dot{\alpha}_l^*]^T$ is the modified velocity, and $e_{xl} = x_{dl} - x_{actl}$, $e_{yl} = y_{dl} - y_{actl}$ and $e_{\alpha l} = \alpha_{dl} - \alpha_{actl}$ denote the errors between the desired trajectories $(x_{dl}, y_{dl}, \alpha_{dl})$ and actual positions $(x_{actl}, y_{actl}, \alpha_{actl})$ of the leader robot. $\{K_{px}, K_{pyl}, K_{p\alpha l}\}$ and $\{K_{ixl}, K_{iyl}, K_{i\alpha l}\}$ denote the proportional and integral gains, respectively.

In simulation, the object, handled by all agents cooperatively, moved along a straight line firstly, then turned left slowly and finally went straight along the Y_w axis; during the simulation, the object was kept with a constant posture. The initial position and the posture of each agent were shown in Fig. 5 and the parameters of each robot were listed in Table 1. In Figs. 6–8, the maximum velocity of leader robot in each direction was 0.2 [m/s] and the errors of arm length ($e_{l_{r1}}$ and $e_{l_{l1}}$ denote the length error of right and left arm of follower 1, and $e_{l_{r2}}$ and $e_{l_{l2}}$ are the arm length errors of follower 2) were less than 3.0 [mm], which means the inter-forces between the followers and the object were very small and the followers successfully transported the object in coordination.

Table 1. The parameters of each robot

$s_l,\ s_i$	[m]	0.075	$K_{pxl},\ K_{pyl},\ K_{pxi},\ K_{pyi}$	5.0
$h_l,\ h_i$	[m]	0.13	$K_{p\alpha l}$	3.0
$r_l,\ r_i$	[m]	0.05	$K_{ixl},\ K_{iyl},\ K_{ixi}, K_{iyi}$	3.0
Δt	[sec]	0.002	$K_{i\alpha l}$	2.0

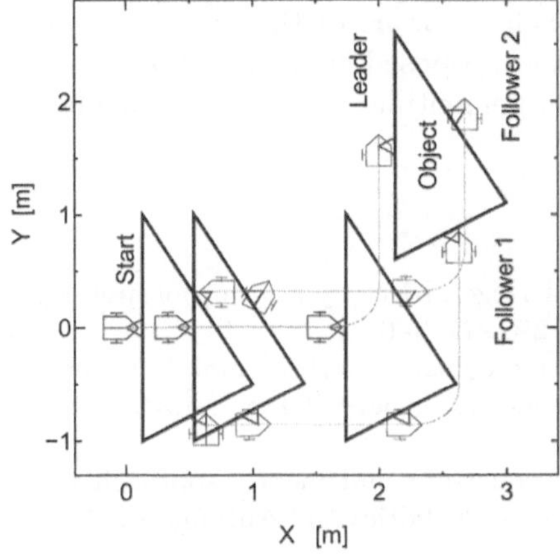

Fig. 5. Simulation result

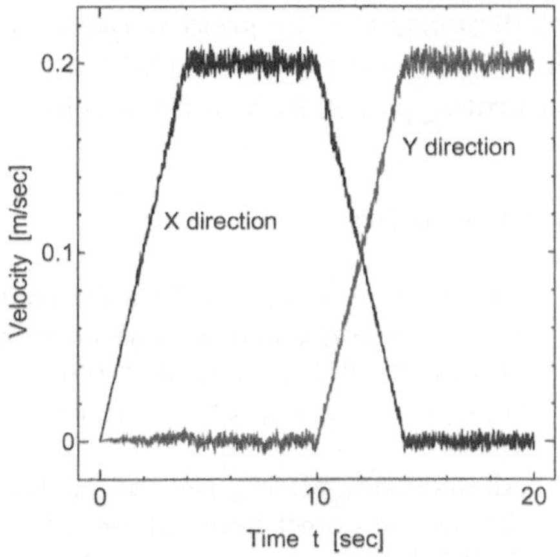

Fig. 6. Velocity of leader robot

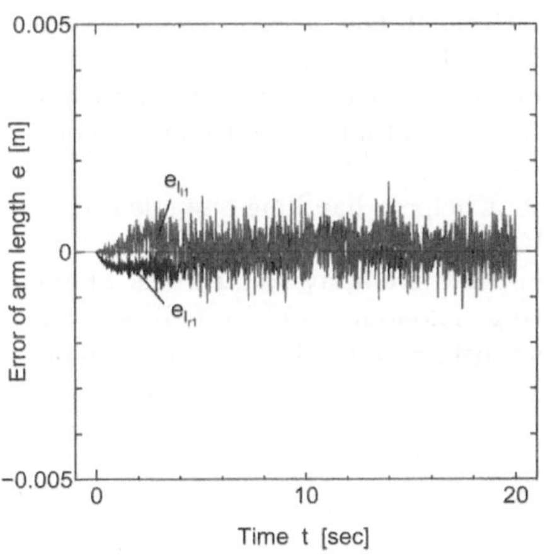

Fig. 7. Error of arm length of follower 1

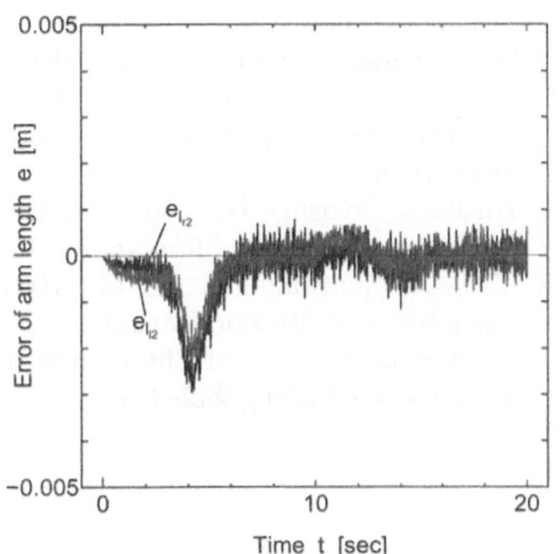

Fig. 8. Error of arm length of follower 2

6 Conclusion

In this paper, we have introduced a new coordinated control system for transporting a single object by a group of nonholonomic mobile robots. A leader-follower type decentralized control was adopted in the proposed system. The present follower robot was assumed to have two compliant arms that are connected with the nonholonomic mobile base through a passive steering joint offset from the central point of two driving wheels. The omnidirectional motion of the object could be manipulated by the leader robot and the followers were able to handle the object in coordination without any communication with each others. We just focused on the structure of the proposed system, and only some simple PI controllers were employed in the simulation to demonstrate the good performance of each follower. In the future, some intelligent control method will be used in the proposed system and a group of prototype robots will be developed for implementing several experiments.

References

1. Ota J., Arai T. (1999) Transfer control of a large object by a group of mobile robots. Robotics and Autonomous Systems **28**, 271–280
2. Osumi H., Nojiri H. et al. (2001) Cooperative control for three mobile robots transporting a large object. Journal of the Robotics Society of Japan **19-6**, 744–752
3. Ahamdabadi M. N., Nakano E. (2001) A "constrain and move" approach to distributed object manipulation. IEEE Trans. on Robotics and Automation **17-2**, 157–172
4. Donald B. R. (1995) Information invariants in robotics. Artif. Intell. **72**, 217–304
5. Yang X., Watanabe K., et al. (2002) Coordinated transportation of a single object by two nonholonomic mobile robots. In: Proc. of The Seventh International Symposium on ARTIFICIAL LIFE AND ROBOTICS (AROB 7th '02)
6. Kosuge K., Sato M. (1999) Transportation of a single object by multiple decentralized-controlled nonholonomic mobile robots. In: Proc. of IROS 99, Vol.3 1681–1686
7. Hirata Y., Kosuge K. (2000) Distributed robot helpers handling a single object in cooperation with a human. In: Proc. of ICRA 2000, 458–463
8. Yang X., Watanabe K., et al, (2001) An upper drive-active dual-wheel caster assembly and its application for constructing holonomic and omnidirectional plantform. In: Proc. of The Fourth IFAC Symposium on Intelligent Autonomous Vehicles (IAV2001), 442–447

Chapter 6
Robot Soccer

Coordinated Trajectory Planning and Formation Control of Soccer Robots to Pass a Ball Cyclically

Tzu-Chen Liang and Jing-Sin Liu

Institute of Information Science 20, Academia Sinica
Nankang, Taipei 115, Taiwan, R.O.C.
liu@iis.sinica.edu.tw

Abstract. A ball passing strategy for robot soccer games and the corresponded controller are presented in this paper. With this strategy, three mobile robots can pass ball in turn without holding while the global formation is moving at some specific direction. The moving trajectory of each robot is planned as an arc. After kicking, a robot predicts the next kick position and moves along the planned arc. Just like human soccer player, it corrects the prediction by getting accurate information, and changes its trajectory to achieve the next kick position on time. This converts the ball passing movement to a single objective trajectory planning and tracking control problem. A computer simulation is addressed to verify the strategy.

Key Words: Robot Soccer, Ball Passing, Trajectory Planning

1 Introduction

In recent years, robot soccer game enforces research issues such as multi-agent systems, multi-robot cooperative teams, autonomous navigation, sensor fusion, fast pattern recognition and vision-based real time control. It also has been proposed as a benchmark problem in the field of artificial intelligence and robotic systems. Since soccer game is a teamwork game, multiple soccer robots should realize some motion with collaboration to score and defend. However, communication between robots is limited and with noise, so a strategy for the motion decision of each robot by insufficient information should be carefully considered. Some delicate soccer skills, like incessant ball passing movements, requiring not only fast and accurate information getting and transferring but also high performance of controls and trajectory planning, still seldom appear in physical robot soccer games[6][7]. The "Benchmark Test of Robot Soccer Ball Skills" proposed by FIRA (1999)[9] includes a millennium benchmark challenge, which is to control three robots in passing the ball around them in turn forming a circuit. There are only two teams completed this exercise in FIRA Robot World Cup, 2000. This paper discusses an even more complicate ball-passing strategy for three robots. In this strategy, robots pass the ball with a direction intention and change their formation dynamically according to the ball position. We will also propose corresponded trajectory planning and control method.

The objective of the proposed passing strategy is for three robots to carry the ball to a specific direction without holding. This kind of passing strategy is common in human soccer games and is also a basic skill in practicing. Every player does not hold the ball so that opponents cannot intercept it easily. Furthermore, three players working with cooperation make the passing movement more dexterous.

The passing strategy includes three parts: the formation and role change scheme, the trajectory planning method and the tracking control. We will describe each of them in next three sections. Section 5 is a simulation to demonstrate its feasibility, and a conclusion will be addressed in section 6.

2 The Formation and Role Change Scheme

The ball locus and robot trajectories of our strategy are shown in figure 1. We design the zigzag type ball locus and on each direction change point, a robot kicks it. Three robots kick the ball in turn, as a result the relative positions of two kick points of the same robot are a to a' and b to b', etc. According to our strategy, the passing movement of each robot is a successive motion, i.e. after kicking, the robot runs to the next predicted kick position. Though the situation changes dynamically and the ball motion prediction has many uncertainties, the information is getting accurate while the next kick is impending. There are three times for a robot to correct its trajectory before arriving at the next kick position on time. All three robots do the same thing with different assigned roles, and then the passing strategy is achieved by single kind of controller and single trajectory planner.

For a passing movement, three roles are considered, named Passer, Previous Player and Next Player. Passer is the robot that prepares to kick the moving ball kicked by Previous Player, and Next Player is the robot that the Passer passes the ball to. Each robot acts as one of the three roles at a time and roles change after each kicking: Passer becomes the new Previous Player and Next Player becomes the new Passer. Besides, the Previous Player drives forward and becomes the new Next Player. Three robots do not have a fixed formation during the passing movements, but the formation dynamically changes according to the ball motion. We design the zigzag type ball locus so that the task of robot is to change ball direction and accelerate it to resist the natural deceleration caused by friction. For kick position prediction, we assume that the ball velocity after each kick can be observed by sensor. The ball locus is a straight line and the motion is pure slipping with friction force proportional to its velocity [8]. Therefore a ball motion predictor can predicts the ball location at some time after kicking. Since collision reaction is not considered, the motion predictor is trivial and not described here. By prediction, the ball locus is exactly known. The Passer chooses a point on this locus and run there to kick the ball on the specific time point. As for Previous Player and Next Player, because they cannot know their next kick positions exactly, an approximate prediction is performed. As shown in figure 2, the vector $\vec{v}_1$ and $\vec{v}_2$ are dependent on the zigzag locus, which is a function of kick direction and kick velocity. When Passer kicks the ball, two things will happen. First, roles change as described above; second, the kick position is observed by sensor and two vectors are added to this position for setting the goal position of new Next Player and new Previous Player, who are original Previous Player and Passer. Though vector $\vec{v}_1$ suggests the approximate next kick position for Previous Player and the robot is driven there, the trajectory will not be completed finally because as the robot's role changes to Next Player, a more accurate kick position prediction computed by $\vec{v}_2$ added by new kick position is applied to plan a new trajectory. The robot steers to track the new trajectory. Then when it becomes a Passer, the robot gets the exact ball information and tries to kick the ball.

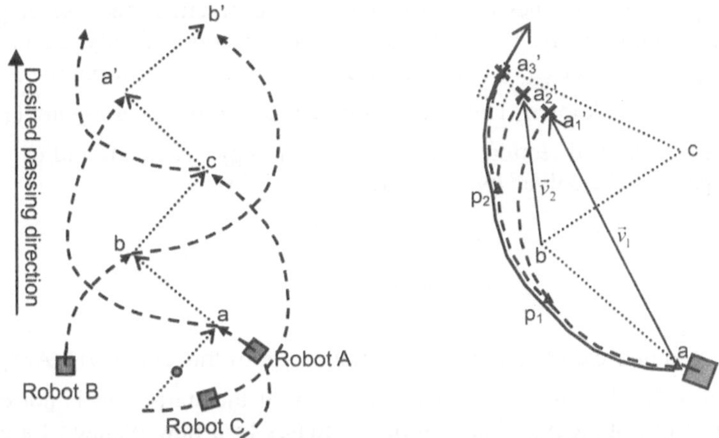

Fig.1.(LHS) The tactical plot of ball passing strategy. Dotted lines represent the ball locus. At each turning point (a, b, c, a' and b') a robot kicks the ball. Dashed lines are trajectories of robots. At this moment, Robot A is Passer, Robot B is Next Player and Robot C is Previous Player.

Fig.2.(RHS) This figure shows the three kick-position corrections of a robot. Dotted lines represent the ball locus; dashed lines are three trajectories generated by kick prediction, and thick solid line is the real trajectory that the robot finally performs. Vector $\vec{v}_1$ and $\vec{v}_2$ are applied to generate the trajectory a-a_1' and p_1-a_2' separately as robot's reference trajectory. Motion predictor tells the robot the exact kick position (a_3'), and trajectory p_2-a_3' is generated.

To summarize, each robot switches its trajectory three times between its two subsequent kicks and the corrections base on other robots' kick position and two prediction vectors. If the vectors $\vec{v}_1$ and $\vec{v}_2$ are chosen appropriately, the change of trajectory will be slight, and the robot motion will be fluent.

3 Trajectory Planning

Because all three robots in the passing strategy do the same thing: move to the next kick position, we have only one kind of trajectory planning problem. For convenience, we also let the time duration between two kicks be fixed and separate the trajectory planning problem to a path planning problem and a velocity planning problem. The path planning problem is to find a curve to connect the start-position, which is the robot position, and the end-position, which is the kick position. The selection of this curve should consider the kick direction, i.e. the robot orientation at the kick position. The velocity planning problem is to decide the velocity profile of a robot on the path.

3.1 Circular Path Planning: Computation of Center and Radius

Path planning problem for wheeled mobile robot has been widely discussed in many studies [2][3]. They show that a path synthesized by several basic curves, for example: straight lines or part of circles, are practical because of generation fast and tracking easily. For our strategy, the selection of only the circle as the planned path is enough. Referring to figure 3, the current-point (P_c) and the end-point (P_e) are known. The orientation vector of robot at end-point ($\vec{v}_{eb}$), which is the desired kick direction, is decided by strategy. Then the radius (R) of the circular path can be solved by geometric computation, as

$$\begin{aligned} &\beta = 2\alpha \\ &R = \frac{L}{\sqrt{2(1-\cos(\beta))}} \end{aligned} \tag{1}$$

Besides, geometric computation solves the center of this circle (O). Arc $\widehat{P_c P_e}$ is the proposed path. Note that the robot orientation at the start-point is not considered because in our strategy, when a robot switches to a new planned trajectory, the orientation error is slight and can be easily regulated by tracking control.

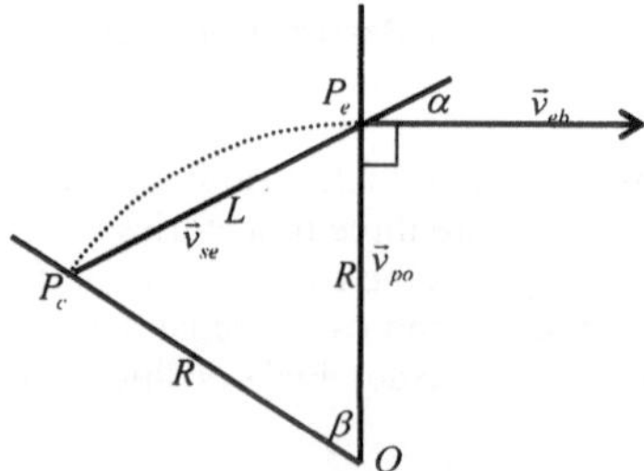

Fig.3. The geometric diagram for the computation of circular path. P_c is the current position of robot. P_e is the predicted kick position, which may be computed by $\vec{v}_{se}$ (either $\vec{v}_1$ or $\vec{v}_2$ in Fig. 2) or ball motion predictor. $\vec{v}_{eb}$ is the desired direction for robot to kick the ball. The dotted line shows the planned path, which is part of the circle centered at O with radius R.

3.2 Velocity Planning

For a selected arc path s_r, along the s_r direction, the velocity profile of the robot is planned by meeting four boundary conditions,

$$\begin{aligned} s_r(0) &= 0 \\ s_r(t_f) &= R \cdot \beta \equiv s_f \\ \dot{s}_r(0) &= V_{robot} \equiv \dot{s}_0 \\ \dot{s}_r(t_f) &= V_d \equiv \dot{s}_f \end{aligned} \tag{2}$$

where V_{robot} is the current velocity of robot; V_d is the desired kick velocity. The instant that the robot starts to execute the path is set to be 0, and the time duration allocated by strategy for the robot to traverse the path is assumed fixed as t_f. We formulate the time trajectory as a 3_{rd}-order polynomial,

$$s_r(t) = q_1 t^3 + q_2 t^2 + q_3 t + q_4 \tag{3}$$

Then the velocity profile $\dot{s}_r(t)$ is

$$\dot{s}_r(t) = 3q_1 t^2 + 2q_2 t + q_3 \tag{4}$$

Substitute the four B.C.s to above two equations, and the four coefficients of trajectory (3) can be solved as,

$$\begin{aligned} q_1 &= \frac{\dot{s}_f' t_f - 2s_f'}{t_f^3} \\ q_2 &= \frac{3s_f' - \dot{s}_f' t_f}{t_f^2} \\ q_3 &= \dot{s}_0 \\ q_4 &= 0 \end{aligned} \tag{5}$$

where $s_f' = s_f - \dot{s}_0 t_f$ and $\dot{s}_f' = \dot{s}_f - \dot{s}_0$.

We have found the reference of robot's mass center and its velocity profile along it now. Next section we realize the planed trajectory of soccer robot by designing suitable tracking controller.

4 Tracking Control

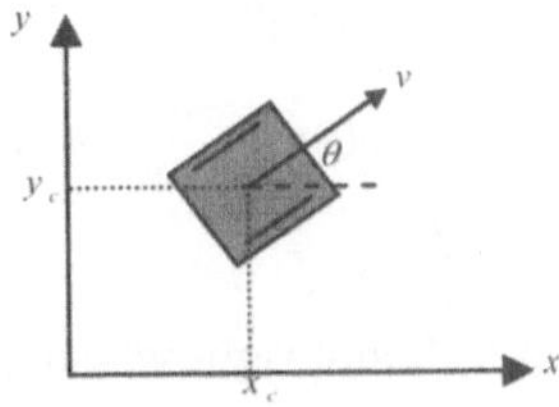

Fig.4. The schematic representation of the mobile robot.

The unicycle mobile robot is shown in figure 4, this kind of robot is mostly used in robot soccer games. Hence we choose it to perform the passing strategy and design controller. The vehicle position is described by the coordinate (x,y) of the mid point between the two driving wheels, and by the orientation angle θ with respect to a fixed frame. Under the hypothesis of "pure rolling" and "non slipping", the vehicle satisfies the nonholonomic constraint,

$$\dot{x}\sin(\theta) - \dot{y}\cos(\theta) = 0 \tag{6}$$

The motion of the robot can be described by the following kinematical model[4],

$$\begin{aligned} \dot{x} &= v\cos(\theta) \\ \dot{y} &= v\sin(\theta) \\ \dot{\theta} &= w \end{aligned} \tag{7}$$

where v is the linear velocity and w is the angular velocity of robot. Let $\dot{s}_L$ and $\dot{s}_R$ denote the velocities of the left driving wheel and the right driving wheel, respectively. Then v and w can be described as,

$$\begin{aligned} v &= (\dot{s}_R + \dot{s}_L)/2 \\ w &= (\dot{s}_R - \dot{s}_L)/l \end{aligned} \tag{8}$$

where l expresses the distance between two driving wheels.
We have computed the reference velocity profile along the planned trajectory in previous section, and the path radius R is also solved by equation (1). Therefore the reference wheel velocities are solved as

$$\begin{aligned} \dot{s}_{Mr}(t) &= \dot{s}_r(t)(1 + \frac{l}{2R}) \\ \dot{s}_{mr}(t) &= \dot{s}_r(t)(1 - \frac{l}{2R}) \end{aligned} \tag{9}$$

where $\dot{s}(t)_{Mr}$ and $\dot{s}(t)_{mr}$ denote the bigger and the smaller reference values of two wheel velocities. The sign of w decides which wheel has a larger or a smaller velocity.

The dynamical model of vehicle is described by the following equations [5],

$$\begin{aligned} \ddot{x} &= -\sin(\theta)\left[\dot{x}\cos(\theta) + \dot{y}\sin(\theta)\right]\dot{\theta} + \frac{\cos(\theta)}{mr}(\tau_R + \tau_L) \\ \ddot{y} &= \cos(\theta)\left[\dot{x}\cos(\theta) + \dot{y}\sin(\theta)\right]\dot{\theta} + \frac{\sin(\theta)}{mr}(\tau_R + \tau_L) \\ \ddot{\theta} &= \frac{l}{Ir}(\tau_R - \tau_L) \end{aligned} \tag{10}$$

Where τ_R and τ_L are driving torques of left and right wheels; m, I are the robot mass, moment of inertia, respectively; r is the wheel radius.
There are three errors in a tracking motion: e_s, e_d, and e_θ. As shown in figure 5, for some instant t', the signed distance between the robot and the reference path is

e_d, and distance between the projection, $s(t')$, and $s_r(t')$ is e_s. e_θ is the orientation error of the robot to the tangent direction of the reference path at $s(t')$. We define

$$\tilde{\tau}(t) = \ddot{s}_r(t) - k_2\dot{e}_s(t) - k_1 e_s(t) \quad (11)$$

where $e_s(t) = s_r(t) - s(t)$ and $\dot{e}_s(t) = \dot{s}_r(t) - \dot{s}(t)$. Furthermore, let

$$\tau_\Delta(t) = \frac{l}{R} k_2 \dot{s}_r(t) + k_3 e_d(t) + k_4 e_\theta(t) \quad (12)$$

The first term in the RHS of equation (12) is the compensation of differential velocity derived from equation (9); the second and third terms are feedbacks to let the robot ride along the reference path. The suggested tracking controller is

$$\tau_R(t) = \frac{1}{2}(\tilde{\tau}(t) - \tau_\Delta(t)) \quad (13)$$

$$\tau_L(t) = \frac{1}{2}(\tilde{\tau}(t) + \tau_\Delta(t))$$

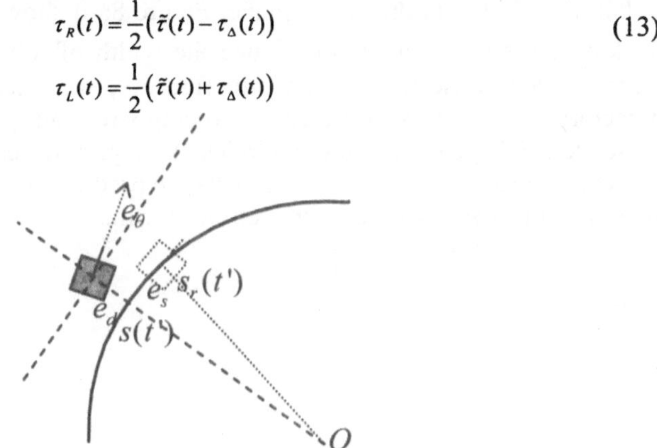

Fig.5. The diagram of three tracking errors at time t'. Solid line is the planned path. O is the center of the arc.

5 Simulations

The robot simulator is an imitation of Simurosot software of FIRA [1], except that the robot motion is simulated by dynamical model as in equation (10), not the kinematical one in Simurosot.

In the simulation, three identical mobile robots perform the proposed ball passing strategy. The robot mass, moment of inertia and length are $m = 1kg$, $I = 0.02kg \cdot m^2$, and $l = 0.08m$, respectively. Wheel radius is $r = 0.04m$. Maximum linear velocity is $1.5m/s$, and maximum torque is $0.4N \cdot m$. In each kick motion, the robot does not hold the ball. Instead, it kicks the ball by its front surface, which is orthogonal to the moving direction of robot. The desired kick velocity is $0.7m/s$, and the kick direction, is set as a $+/-0.25\pi$ angle apart from the desired passing direction. The two vectors $\vec{v}_1$ and $\vec{v}_2$ shown in figure 2 are represented by polar coordinate (l,ϕ) as $(2.34,0)$ and $(3.66,+/-0.145)$ respect to the passing direction. Therefore, when the passing direction rotates, these two vectors and desired kick direction also change accordingly. As a result an intention of change ball passing direction is realized without utilizing any more commands.

The initial relative positions of ball and robots are shown in figure 6. Robot "John" is the first Passer, "Mary" is the Next Player, and "Rosa" (Nominally, the Previous Player) will kick the ball after Mary. Then the ball returns to John and another passing cycle starts. Time duration between subsequent two kicks is set to be 1.6 sec. After first 5 kicks, desired passing direction is designed to rotate clock-wisely by an angle $0.2rad.$ after each kick.

Figure 7 is the simulation result. Three trajectories of robot motion and the locus of ball are plotted in the $x-y$ plane. Figure 8a-d show the velocity and position tracking error of robot "John". Since the width of robot is $80mm$, small tracking errors do not cause ball missing. In figure 9, we select a part of robot John's trajectory and the planned trajectory to compare. Path planning method described in section 3.1 generates these circles, and part of each circle is a reference trajectory of robot. As the robot corrects the prediction three times between its two subsequent kicks, in figure 9 three circles form a group.

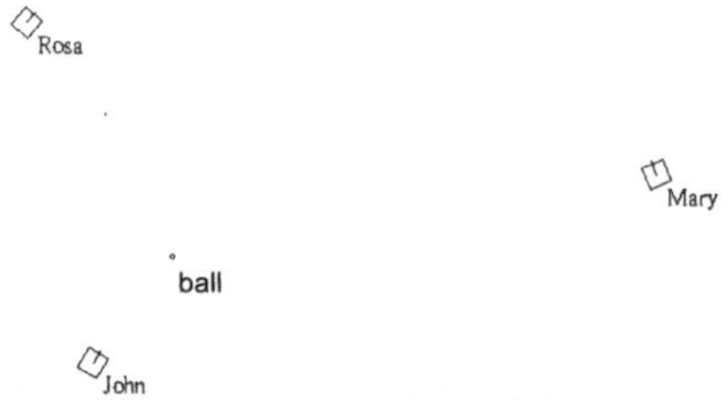

Fig.6. The initial positions of three robots and the ball.

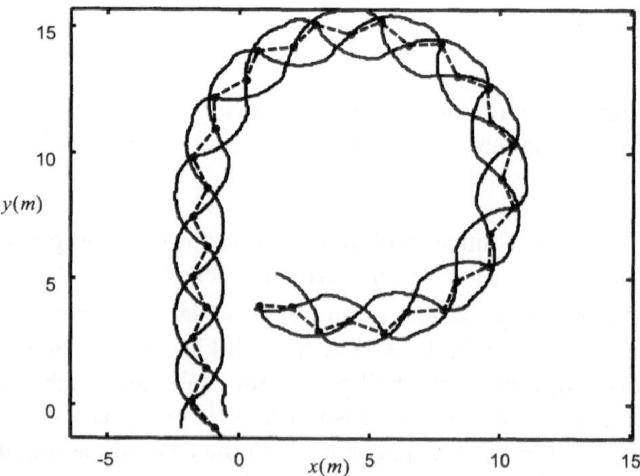

Fig.7. The simulation result of the ball passing strategy in first 36 kicks. Solid lines are trajectories of three robots; dashed line is the ball locus, and dots are the positions where a robot kick the ball.

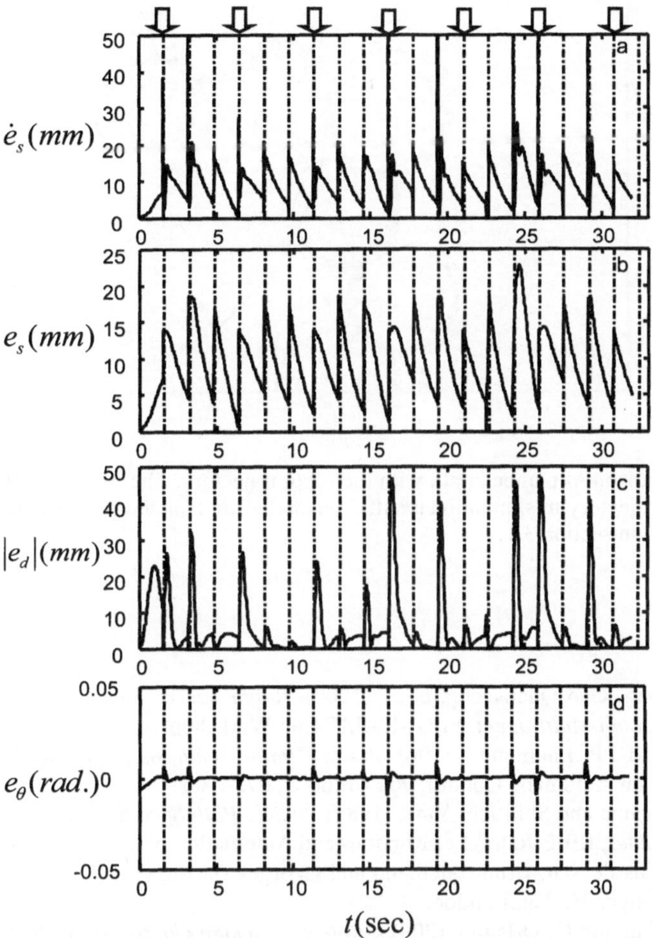

Fig.8. The error-time plots of selected robot "John". These four figures show the evolution of $\dot{e}_s, e_s, e_d, e_\theta$ in first 33 sec of simulation. Dot-dashed lines mark the time instant that a kick happens. Arrow signs point out that these kicks are completed by John. e_d is plot by its absolute value.

6 Conclusion

In this paper we proposed a ball passing strategy for robot soccer games, and verified it by computer simulation. The ball passing strategy makes three robots work with cooperation to carry the ball at a specific direction without holding. Instead of using complex trajectory planning method or long-term prediction, it is accomplished by appropriate role assignments and goal change. The simulation results evidence the practicability. Moreover, the idea that a soccer robot adjusts its trajectory by getting accurate information while executing a specific job hints that a human-like behavior pattern can be successfully applied in robot soccer games. To convert more soccer skills by the novel idea is the future work.

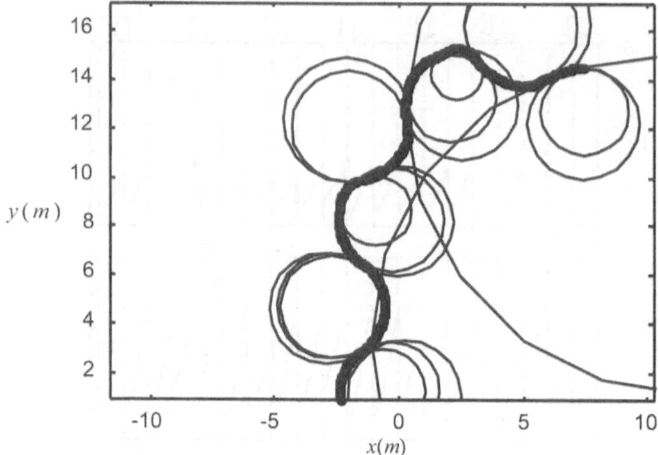

Fig.9. The comparison of planned path with the robot trajectory. The thick solid line is part of robot John's trajectory in simulation result. The circles are planned path generated by the method described in section 3.1.

References

[1] Hong Bingrong, Gao Quansheng, et al. (2000), *Robot Soccer Simulation Competition Platform Base on Multi-agent*, FIRA-KAIST Cup Workshop.

[2] P. Soueres and J.P. Laumond. (1996) *Shortest Paths Synthesis for a Car-Like Robot*, IEEE Trans. on Automatic Control, vol. 41, no. 5, 672-688.

[3] Yutaka Kanayama and Shin'ichi Yuta. (1988) *Vehicle Path Specification by a Sequence of Straight Lines*, IEEE Journal of Robotics and Automation, vol. 4, no. 3, 265-276.

[4] Carlos Canudas de Wit, Bruno Siciliano and Georges Bastin. (1997), *Theory of Robot Control*, Springer-Verlag, London.

[5] M. L. Corradini and G. Orlando. (2001), *Robust Tracking Control of Mobile Robots in the Presence of Uncertainties in the Dynamical Model*, Journal of Robotic Systems, vol. 18, no. 6, 317-323.

[6] Minoru Asada and Hiroaki Kitano. (eds.) (1999), *Robocup-98: Robot Soccer World Cup II*, Springer-Verlag, Berlin Heidelberg.

[7] Peter Stone, tucker Balch, Gerhard Kraetzschmar. (eds.) (2001), *Robocup 2000: Robot Soccer World Cup IV*, Springer-Verlag, Berlin Heidelberg.

[8] Byoung-Ju Lee and Gwi-Tae Park. (1999) *A Robot in Intelligent Environment: Soccer Robot*, Procceedings of the 1999 IEEE/ASME International Conference on Advanced Intelligent Mechatronics.

[9] Jeffery Johnson, Josep Lluis de la Rosa, et al. (1999), *Benchmark Tests of Robot Soccer Ball Control Skills*, The Federation of International

M-ROSE: A Multi Robot Simulation Environment for Learning Cooperative Behavior

Sebastian Buck, Michael Beetz, and Thorsten Schmitt

Munich University of Technology, Department of Computer Science IX, Orleansstr. 34, D-81667 Munich, Germany

Abstract. The development of high-performance autonomous multi robot control systems requires intensive experimentation in controllable, repeatable, and realistic robot settings. The need for experimentation is even higher in applications where the robots should automatically learn substantial parts of their controllers. We propose to solve such learning tasks as a three step process. First, we learn a simulator of the robots' dynamics. Second, we perform the learning tasks using the learned simulator. Third, we port the learned controller to the real robot and cross validate the performance gains obtained by the learned controllers. In this paper, we describe M-ROSE, our learning simulator, and provide empirical evidence that it is a powerful tool for learning of sophisticated control modules for real robots.

1 Introduction

The development of high-performance autonomous multi robot systems requires intensive experimentation in controllable, repeatable, and realistic robot settings. The need for experimentation is even higher in applications where the robots should automatically learn substantial parts of their controllers. For example, to improve the competence of our robot soccer team we have learned a model for predicting the time needed to perform a given single robot navigation task. To learn such a model with an expected inaccuracy of less than three percent we had to collect data from more than 4600 navigation episodes. Assuming that setting up and executing a navigation task takes only two minutes, we would have had to spend more than 150 hours of experimentation with the real robots. Obviously, this is not feasible.

We therefore propose to solve such learning tasks as a three step process. First, we learn a simulator of the robots' dynamics. Second, we perform the learning tasks using the learned simulator. Third, we port the learned controller to the real robot and cross validate the performance gains obtained by the learned controllers. Using this method we can perform the learning task mentioned above by collecting data from the real robots only for 6 hours and then perform the learning task in 3 hours of simulation.

In this paper, we describe M-ROSE, our learning simulator, and provide empirical evidence that it is a powerful tool for the learning of complex control modules for real robots.

Robot simulation in general is a powerful tool for the development of autonomous robot control systems because it allows for fast and cheap prediction and makes experiments controllable and repeatable. Simulation allows for the quick detection and diagnosis of software errors.

The use of robot simulators also yields a number of problems. For simulation purpose time must often be discretized. Also, simulators typically work in an abstract feature space and might therefore ignore key factors for the robot behavior [11]. Others argue that simulated controllers are doomed to succeed because of the design of the simulators [3]. As a consequence, software that succeeds in simulation may fail on a real robot [4]. Accurate and numeric simulations are typically extremely complex and expensive in terms of computational resources and are thus performed by parallel and distributed simulation [9,13]. Despite these problems a number of simulators have been proven to be valuable resources in the development of robot controllers [8].

Multi robot simulation includes the simulation of sensing as well as the simulation of motion dynamics. The simulation of motion maps the dynamic state of the robot and a low level robot command into a successor state. Sensor simulation maps the local surroundings and the sensor model into the sensed data. A substantial amount of work has been done on either one or both of these aspects: Lee et al. [12] generates an artificial neural network based model of the environment within a simulator of a Nomad robot, to learn an action model in a MDP framework simulators have been used [2], and many experiments of the successful well known rhino robot [18] were done in simulation. Furthermore the soccerserver [14] used in the RoboCup Simulation League mimics some sensory and motion abilities of a human-like soccer player. Another RoboCup simulator developed [10] is able to simulate a large number of different sensors (infrared, bumper, camera, and laser) while motion is simulated by directly using the values of translational and rotational control commands to compute a new state.

While most of the above work concentrates on the simulation of sensors while more or less neglecting the motion our main goal in this paper is an accurate simulation of the robots' dynamical behavior which becomes very important in high-speed environments such as autonomous robot soccer. In the RoboCup MidSize League abrupt changes in speed and direction are as common as they are in a real soccer game. Our approach to simulating changes in state in robotics involves neural learning from real robot data and is, as we will show, easy extendable and highly qualified because of its brilliant accuracy.

The remainder of this paper is organized as follows: In section 2 we will describe our motion (2.1) and sensory model (2.3) followed by statistics to document the accuracy (2.5). Section 3 shows the results of some successful experiments done with real soccer robots whose behavior has been implemented and learned in the simulator. Finally section 4 concludes.

2 Building a Multi Robot Simulation Environment

In addition to an accurate modeling of sensors and motion a multi robot simulator should be able to simulate different kinds of robots with their respective motion profile acting at the same time. A model of the robots' static environment as well as models of dynamic objects should be easy to integrate. Furthermore it is essential that a high number of learning data can be obtained in a reasonable period of time.

The main difference between our simulator and the RoboCup soccerserver [14] is that we simulate a team of real robots while soccerserver simulates a hybrid, in some aspects human-like, soccer player. Our team (the *AGILO RoboCuppers* [1]) consists of four Pioneer I robots [15] who obviously have several disadvantages compared with an agent of the soccerserver: They cannot hold the ball if it does not come directly into their ball-guiding device and path planning with real robots becomes more complicated than in the soccerserver where the field is about 105 times 70 meters and a player has a diameter of only 0.3 meters. Moreover the ball can go through a player in soccerserver (if it is fast enough) and can be given a velocity vector by a player having it in its kicking range. Teams like the *Karlsruhe Brainstormers* [16] have shown that in the soccerserver environment learning algorithms perform excellent and a lot better than a human being controlling a player with a joystick.
In addition to our Pioneer robots we started to extend our simulator by integrating a model of our B21 robots which we use for indoor exploration tasks. Before we describe the components of our simulator in detail we briefly point out the major advantages of M-ROSE:

- The individual dynamic motion profile of a robot can be learned.
- The simulator is easy extendable and can combine the simulation of robots with different dynamic motion profiles.
- Models of other objects (static and dynamic) or humans can easily be integrated.
- Powers are modeled according to their physical rules.
- Step by step analysis and monitoring supports the development of control software.
- Time lapse allows performing a great number of experiments in exiguous time.
- The control software can either be linked with the simulator core or the real robot.

2.1 Learning Models of Dynamic Behavior

In this section we describe how to learn a dynamic model of the motion behavior of a Pioneer I robot. Models can similarly be learned for other robots with the respective data at hand.

The dynamic state of the robot at a certain time i is given by the quintuple

$$\zeta_i = \langle x_i, y_i, \varphi_i, V_{tr_i}, V_{rot_i} \rangle \tag{1}$$

where x_i and y_i are coordinates in a global system, φ_i is the orientation of the robot and V_{tr_i} (V_{rot_i}) are the translational (rotational) velocities. The robot control system issues driving commands $\xi = \langle V_{tr}, V_{rot} \rangle$. The dynamic model for the change in state from the current state ζ_c to the successor state ζ_{succ} used by the simulator is acquired by learning the mapping

$$\Delta : \zeta_c \times \xi \mapsto \zeta_{succ} \tag{2}$$

from experience, that is recorded data from real robot runs. Our simulator learns this mapping using a standard multi layer neural network [7] with one hidden layer and supervised learning with the RPROP algorithm [17]. Considering the current state ζ_c at $x_c = 0, y_c = 0$, and $\varphi_c = 0$ in a local system we can reduce the input dimension to 4 by converting the successor state's (ζ_{succ}'s) $x_{succ}, y_{succ}, \varphi_{succ}$ into that local system (that means we regard $\Delta x, \Delta y, \Delta\varphi$ instead of $x_{succ}, y_{succ}, \varphi_{succ}$):

$$\Delta : \langle V_{tr_c}, V_{rot_c}, V_{tr}, V_{rot} \rangle \mapsto \langle \Delta x, \Delta y, \Delta\varphi, V_{tr_{succ}}, V_{rot_{succ}} \rangle \tag{3}$$

Using this mapping we have learned the dynamics of Pioneer I robots. During data acquisition we have executed a wide variety of navigation scenarios covering even abrupt changes in velocity and orientation to comply with the requirements of high-speed navigation. We have collected a total of more than 100.000 training patterns from runs with a real Pioneer I robot. Although the learning time took a few hours on a 800 MHz machine the computational amount of the neural network while running the simulation is infinitesimal.

2.2 Considering Physical Properties

All robots are dealing more or less with a dead time which means that at the moment a robot is performing a low level control command it has already sent control commands (e.g. translational and rotational target velocity) for further movements. In the case of Pioneer I robots the dead time is around 300 ms while the controller accepts ten commands per second. To consider the dead time M-ROSE writes low level control commands in a queue and waits a certain time before executing the commands.

Shapes of Objects and Environment The shapes of robots and any other objects are modeled as polygons or circles (see fig. 1). The corners of the polygon (radius of the circle) according to the objects reference point are read from a shape-file. A Pioneer I robot is described by a polygon while a B21 robot or a ball are regarded as a circle. The advantage of polygons and

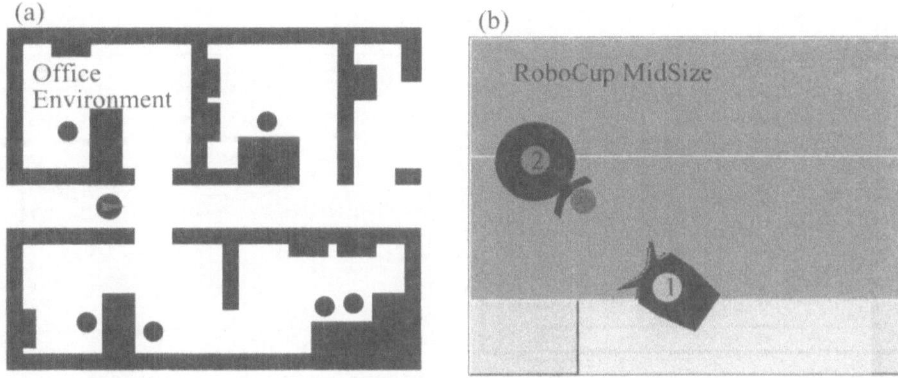

Fig. 1. Subfigure (a): Simulation of a B21-robot. All objects are drawn as filled polygons or circles. The direction of the robot is indicated by an arrow. Subfigure (b): Simulation of a RoboCup game with a Pioneer and a Nomad-shaped robot.

circles is that physical laws of nature can easily be applied to simulate for instance the collision with a ball. However, one can go far deep into physical simulation but we just consider the basic laws for collision and friction with a tunable factor of random.
Furthermore M-ROSE can simulate a manipulator on a robot interacting with an object. Currently this is done by analytic computation (position, direction, and power of the manipulator must be specified in the respective configuration file). But similar to the robots' dynamics the movements of the manipulator can be learned. For the description of the environment another configuration file similar to the shape-file for robots exists.

2.3 Simulation of Perception

The sensor model used in our RoboCup simulations so far is a very simple one: The robot receives his current state data with some simulated noise. Doing so we assume that the sensors of our robots are almost perfect which obviously is a wrong assumption. Therefore we are currently working on a more realistic approach: In several RoboCup games we have recorded the "*true*" state of a robot ζ (given by a camera mounted at the ceiling of the room) and the sensory data vector Υ in parallel. To achieve a realistic simulation of the sensors we will learn the mapping

$$\mathcal{I} : \zeta_c \mapsto \Upsilon_c \tag{4}$$

from a current robot state ζ_c in the simulator to the current sensory data vector Υ_c. Experiments will show if $\mathcal{I}$ is best represented by an interpolating lookup-table, decision trees or neural networks trained with the recorded data.

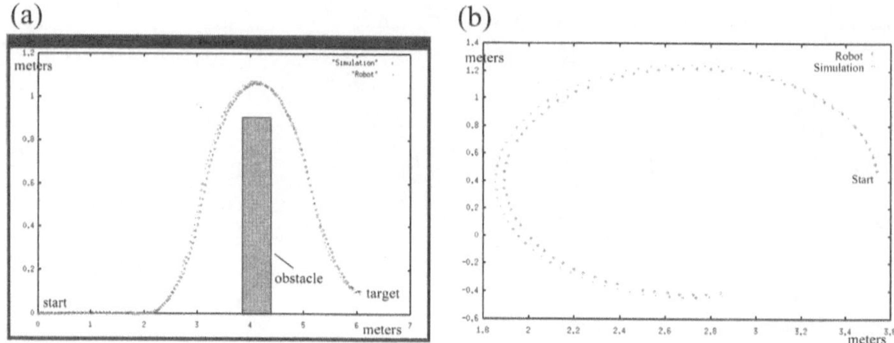

Fig. 2. A comparison between the position data of a real robot and the data simulated by M-ROSE. Subfigure(a): A robot driving a jink around an obstacle. Subfigure(b): A robot driving an ellipse.

2.4 Extensibility

The M-ROSE system in general is able to simulate different kinds of robots. Each robot to be simulated can be specified in a configuration file which contains information about the names of the shape-file and the network-file (which simulates the motion) of the robot. In future information about the robots sensors (and $\mathcal{I}$ in particular) will be included. The number of robots possible to be simulated is infinite but, however, a large number of robots (dependent on the computational power) will degrade the systems performance with respect to real time simulation and time lapse.

2.5 Accuracy

Evaluating around 100 trajectories each ten seconds in length driven with a real Pioneer I robot and *not* used for learning we found an average error of 2% in the simulation of position and orientation. We believe that this error largely results from the inherent indeterministic behavior of the robot controller. This indeterministic behavior itself usually leads to an average error close to one percent. After extreme navigation situations where the translational velocity was set from full to zero and the rotational velocity was set from zero to full (or both the other way around) sometimes an error of up to 10% in orientation (position) occurred. These cases are probably the hardest to predict at all and not very common in real robot navigation in general. Figure 2 shows real robot and predicted data of the simulator in two prevalent trajectories. Obviously the match of predicted and real robot data looks satisfying. The remaining small errors in simulation are balanced by the control software running at a frequency of 10Hz in real time.

3 Simulation Based Behavior Learning

After describing the simulation environment and regarding its accuracy we now want to present some demonstrations of successful robot behavior which was implemented in M-ROSE and works well on real robots. The examples cover cooperative tasks (coordination of actions in robot soccer, path planning) as well as a difficult manipulation task.
To demonstrate the successful behaviors we give statistical results and additionally provide robot videos on our web page. So far all examples are related to RoboCup.

3.1 Demonstration 1: Learning coordination

The first example shows how a cooperative behavior learned in M-ROSE works on real robots. One of the key issues in RoboCup is the cooperative coordination of the robots' actions. Part of it is the decision which robot of the team is to go for the ball. Considering the current dynamic states of all robots of one team as well as orientation and velocity of the target state near the ball, and additionally, the dynamic configuration of obstacles this is no simple decision. To support this decision we have implemented a neural projection $\mathcal{P}$ which maps start (ζ_s) and target state (ζ_t) of a robot to the estimated time need to reach the target state considering the configuration of obstacles and given a dynamic model of the robots motion (see [5] for details):

$$\mathcal{P} : \zeta_s \times \zeta_t \mapsto time \tag{5}$$

The robot r_f assumed to be the fastest at the ball (of all robots $\mathcal{R}$) is chosen to go there:

$$r_f = argmin_{r_i \in \mathcal{R}} \mathcal{P}(\zeta_s(r_i), \zeta_t) \tag{6}$$

This estimation is done by every single robot and consistent (1) in the simulator (because the robots got information about their *true* states) and (2) to a high extent in the real robot environment (because of our accurate visual localization system). As mentioned in the introduction the amount for data acquisition and learning was 9 hours while learning without simulation would have taken more than 150 hours.

Results in a real robot environment To demonstrate the quality of the coordination method implemented in M-ROSE we measure the number of robots going for the ball at the same time. Further we observe how long the same robot goes for the ball without any other robot going for the ball. The data was acquired from five real robot games against different opponent teams at the international robot soccer world cup 2001. The following table depicts in how much percent of the whole time played none, one, and more than one robot went for the ball.

#robots going for ball at the same time	quota in relation to the whole time played
0	00.34%
1	98.64%
>1	01.02%

The average time one robot goes for the ball or handles the ball without being interrupted by a decision of a fellow robot is 3.35 seconds. Videos showing typical decision situations are available at *http://www9.in.tum.de/people/buck/videos.html.*

3.2 Demonstration 2: Learning to get the ball

Another challenging task in RoboCup so far is to remove a ball from the wall. According to the rules of the RoboCup MidSize League it is not permitted to grip or fix the ball. Thus the ball has to be removed from the wall by carefully touching it without maneuvering the robot in a stuck situation at the wall. Most of the teams participating in the RoboCup MidSize League were not able to remove a ball from the wall in a controlled way by 2001. Our approach to that task is hand-coded and implemented in M-ROSE.

All experiments and tests to improve the performance and perfect the ball handling at the wall were done in the simulation environment. The software that we have developed in M-ROSE worked instantly on a real robot without any further adjustments. The hand-coded method to remove the ball from the wall is based on the simple idea to first approach the ball (see figure 3(a)) and then stop directly before it (see figure 3(b)) and turn rapidly towards the center of the field (see figure 3(c)). Thereby the ball should be touched by the outer part of the robots kicking device to move it inside the field.

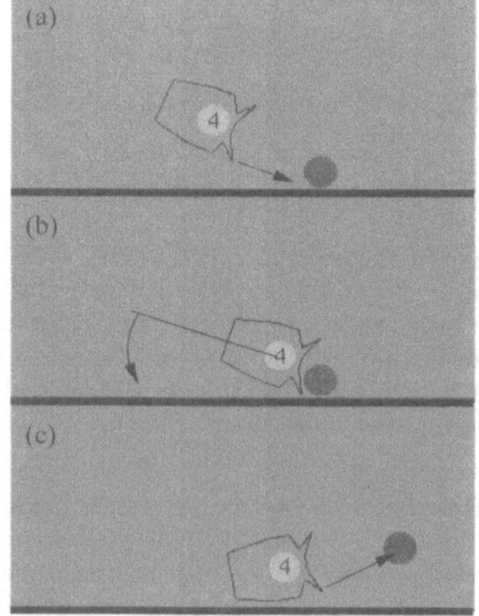

Fig. 3. A simple hand-coded method to remove a ball from a wall. The ball is moved by an impulse of a rapid turn.

Real robot Behavior So far there are no detailed statistics on removing the ball from the wall but in general it works on real robots. There are videos available at *http://www9.in.tum.de/people/buck/videos.html* showing typical situations where real robots take a ball from the wall in a controlled way.

3.3 Demonstration 3: Learning to choose the best planning method

Another software module developed in M-ROSE is the path planner of our RoboCup team. In this work, we propose to select problem-adequate navigation planning methods based on empirical investigations, that is the robots should learn by experimentation (which is done in the simulator) to use the fastest planning methods. The robots have learned predictive models for the performance of different navigation planning methods in a given application domain. A hybrid planning method that selects planning methods based on a learned predictive model outperforms the individual planning methods (statistics and detailed description can be found in [6]). All the learning data was generated in M-ROSE and the hybrid planning method was implemented in the simulator as well.

4 Conclusions

In this paper we propose a simulation environment for multi robot application domains. The main contributions are an accurate, robot specific neural simulation of robot motion and an easy extendable software environment. The dynamic behavior is simulated by a neural prediction of changes in state according to certain low level control commands.

The M-ROSE simulator addresses many key aspects of real robot behavior including accurate motion models, dead times, collisions of objects, powers of manipulators, shapes of objects, and extensibility. In contrast to most of the previous work the main focus is set on the simulation of motion which is most critical in high-speed scenarios such as RoboCup games. An accurate simulation of the sensors is considered in our ongoing research and is expected to greatly broaden the learning tasks that can be tackled and to improve the learning competence of M-ROSE for real robots. The examples in section 3.1, 3.2, and 3.3 have shown that control software hand-coded as well as learned in M-ROSE already leads to more than acceptable results on real robots which we believe is to a large extent founded in the accuracy of the simulator. While so far extensive simulations have been running with Pioneer I robots only the motion behavior of other robots can and will be learned analogously.

References

1. M. Beetz, S. Buck, R. Hanek, T. Schmitt, and B. Radig: *The Agilo Autonomous Robot Soccer Team: Computational Principles, Experiences, and Perspectives.*

International Joint Conference on Autonomous Agents and Multi Agent Systems (AAMAS) 2002.
2. T. Belker and M. Beetz: *Learning To Execute Navigation Plans* in F. Baader, G. Brewka and T. Eiter (eds): Lecture Notes in Artificial Intelligence, vol. 2174.
3. R.A. Brooks and M.J. Mataric: *Real Robots, Real Learning Problems*, in Robot Learning, Jonathan H. Connell and Sridhar Mahadevan, eds., Kluwer Academic Press, 193-213, 1993.
4. R.A. Brooks: *Artificial Life and Real Robots*, in F. J. Varela and P. Bourgine, editors, Proceedings of the First European Conference on Artificial Life, pp.3-10, 1992.
5. S. Buck, T. Schmitt, and M. Beetz: *Reliable Multi Robot Coordination Using Minimal Communication and Neural Prediction*, Seminar on Plan-based Control of Robotic Agents 2001, Schloss Dagstuhl, Lecture Notes in Artificial Intelligence, 2001, Springer Verlag.
6. S. Buck, U. Weber, M. Beetz, and T. Schmitt: *Multi Robot Path Planning for Dynamic Environments: A Case Study.* Proceedings of the IEEE International Conference on Intelligent Robots and Systems, 2001.
7. J. Hertz, A. Krogh, and R. G. Palmer: *Introduction to the Theory of Neural Computation.* Addison-Wesley, 1991.
8. N. Jakobi, P. Husbands and I. Harvey: *Noise and the Reality Gap: The Use of Simulation in Evolutionary Robotics*, Third European Conference on Artificial Life (ECAL95), pp.704-720, Springer Verlag, 1995.
9. M.L. Jugel and A. Sydow: *Parallelity in High-Level Simulation Architectures*, in Transaction of the Society for Computer Simulation International, Vol. 15, No. 3, pp.101-103, 1998.
10. H-Ul. Kobialka, P. Schoell, and A. Bredenfeld: *Tools for Assessing RoboCup Behavior* RoboCup Workshop, RoboCup-Euro 2000, Amsterdam, May 28th - June 2nd, 2000
11. T. Lee, U. Nehmzow, and R. Hubbold: *Computer Simulation of Learning Experiments with Autonomous Mobile Robots*, Proceedings of TIMR 99, Towards Intelligent Mobile Robots, Bristol, 1999.
12. T. Lee, U. Nehmzow, and R. Hubbold: *Mobile Robot Simulation by Means of Acquired Neural Network Models*, European Simulation Multiconference, Manchester 1998.
13. U. Mehlhaus and W. Rausch: *Distributed simulation of robot tasks*, ESS'93 European Simulation Symposium, pages 433–438, Delft, Holland, October 1993.
14. I. Noda, H. Matsubara, K. Hiraki, and I. Frank: *Soccer Server: A Tool for Research on Multiagent Systems.* Applied Artificial Intelligence, 12, 2–3, pp.233–250, 1998.
15. Pioneer Mobile Robots, Operation Manual, 2nd edition, Active Media, 1998.
16. M. Riedmiller and A. Merke: *Karlsruhe Brainstormers - a reinforcement learning approach to robotic soccer II.* In 5th International Workshop on RoboCup, Lecture Notes in Artificial Intelligence, 2001, Springer Verlag.
17. M. Riedmiller and H. Braun: *A direct adaptive method for faster backpropagation learning: the Rprop algorithm*, Proceedings of the ICNN, San Francisco, 1993.
18. S. Thrun, A. Bucken, W. Burgard, D. Fox, T. Frohlinghaus, D. Hennig, T. Hofmann, M. Krell, and T. Schimdt: *Map learning and high-speed navigation in RHINO* MIT/AAAI Press, Cambridge, MA, 1998.

Real-time Adaptive Learning from Observation for RoboCup Soccer Agents

Tomomi Kawarabayashi[1], Takenori Kubo[1], Takuya Morisita[1], Junji Nishino[2], Tomohiro Odaka[3], and Hisakazu Ogura[3]
tomomi@rook.fuis.fukui-u.ac.jp, kubo@rook.fuis.fukui-u.ac.jp, morisita@rook.fuis.fukui-u.ac.jp, nishino@fs.se.uec.ac.jp, odaka@rook.fuis.fukui-u.ac.jp, ogura@nqueen.fuis.fukui-u.ac.jp

[1] Graduate School of Engineering, Fukui University, Fukui City, Japan
[2] Faculty of Electro-Communications, The University of Electro-Communications, Choufu City, Japan
[3] Faculty of engineering, Fukui University, Fukui City, Japan

Abstract. In this paper, a real-time adaptive learning system from observation by teammates' behaviors and results from RoboCup soccer agents is proposed. The agent is required to adapt to its opponents in real time in a RoboCup simulated soccer game.

By only learning to adapt from its own behavior and results, the agent would limit its chances of learning during a game. Therefore, the proposed adapted learning system covers self and teammates' behaviors and results, which improves and increases its learning chances and learning ability about other opponents.

The agent tries to adapt to recognize possible scoring situations by learning and responding to opponents' intercept ability according to a learnt parameter. Compared to other systems (a "Non-Learning" system and a "Self-Behavior Learning" system), with the proposed system the score rate was improved from 0.04 (the non learning system) and 0.06 ("Self-Behavior Learning") to 0.10.

Keywords:Real-time adaptation, Observation, Multi-agent systems, Soccer agent, RoboCup.

1 Introduction

In this paper, a real-time adaptive learning system using observation of teammates' behaviors and results for RoboCup soccer agents is proposed. The learning system adapted in RoboCup Soccer Server System [1–4] was applied and its effectiveness was examined through some experimental games. The agents are required to adapt their behaviors to their opponents during game.

Some researches have addressed some basic acquiring skills, for example kick skills [5], ball interception [6] and other high-level decision making [5]. They have succeeded in automatically acquiring complex knowledge in parameter tuning and complex situations division (usually difficult and are of a high work cost to handle if coded). However, this approach is insufficient in a game with unknown opponents and agents' whose behaviors change during the game. Furthermore, when an agent tries to adapt in real time during a game, there is a problem that its learning chances is limited. In fact, real-time learning is limited by two factors; game time limit and learning chances (in terms of adapting to ball handling such as pass or shoot) in a multi-agent environment, twenty two agents.

Further, research in real-time adaptation covered agents' positions refinement (Reinforcement Learning[7]), improvement of behavior based on agents' roles (Q-learning), trees optimization for ball kicking behavior and positioning by Genetic Algorithm, [8–10]. The authors in [11] have indicated the possibility of real-time adaptation during a game to acquire cooperative knowledge for a pass in a small field.

Although the authors in [8–11] have addressed real time learning by gaining knowledge from results due to a previous self behavior. This imposed the problem of learning chances limitation. To improve on this learning limitation and to increase the learning efficiency, monitoring of other teammates' behaviors is proposed in addition to the above gaining knowledge technique.

In this paper, it is proposed that the agent learns a parameter in the "Shoot Behavior Rule" enabling it to recognize possible scoring situations. Experimental games evaluation showed that with the proposed learning system the agents were able to adapt their shoots to their opponents in a better way when compared to other systems during a game. Such improvement in agent adaptation have also reflected on the overall result by in increasing the score rate.

2 Architecture of a Learning Soccer Agent

The agent architecture consists of a "Decision Making System" and a "Learning System", Fig. 1. The "Decision Making System" is based on a three layered control model [12]; the "Strategic Layer"; the "Behavior Selection Layer", which includes a behavior selection and a behavior knowledge modules; and the "Executing Action Layer". The "Strategic Layer" decides on the strategy using the strategic knowledge and sensor information. The "Behavior Selection Layer" selects a particular behavior such as a pass or a shoot. The "Behavior Selection Module" resolves matched rules obtained from the "Behavior Rules" (stored in the "Behavior knowledge" module). A "Behavior Rule" may be represented as follows:

$$\textbf{IF}(Situation_1 \textbf{AND} Situation_2 \textbf{AND} ... Situation_N)\textbf{THEN}(Behavior_1)$$

Based on the information received from the server (through the sensor fusion module) the "Decision Making System" is updated in intervals depending on the vision angle and the resolution quality.

The "Learning System" contains two modules; "Self-Behavior Learning Module" and "Observation Learning Module" [13]. The first module learns from its own behavior, while the latter learns by observing other teammates'behavior. The "Observation Learning Module" also monitors teammates and opponent's behavior (action and response) and evaluates it before deciding on learning from this behavior (Learning Knowledge).

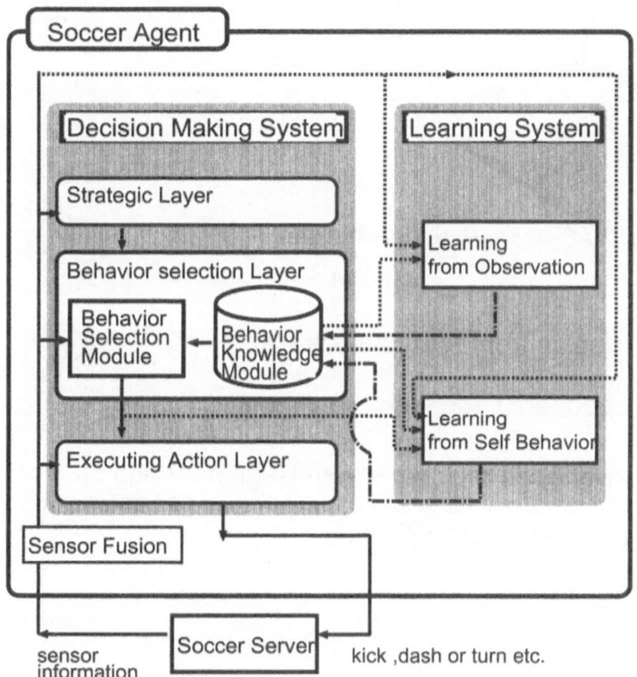

Fig. 1. The proposed learning agent architecture

3 Modeling Opponents' Interception Ability

The agent may recognizes a shoot situation, the opponents may intercept the ball and the agent will fail to score. This may be learnt as a not suitable shooting situation, however, if the agent is too strict with the results from this scenario, in the future, it may lose many possible scoring chances against opponents with low interception abilities. In this paper, learnt knowledge gained from games against previous opponents is not saved and the agent learns a parameter (opponents' interception ability) in the Shoot Behavior Rule from each new opponent.

At this stage it would be appropriate to describe both the Interception Ability Model and the Learning Algorithm. With a space for a shoot situation declared by the "Shoot Behavior Rule" the agent has to check opponents' arrangement and look for a shootable direction before continuing with the shooting option.

A shootable direction region is an area that partially covers the goal and falls outside the opponents' intercept-ability sectors. The shooting space recognition process may be explained with the aid of Fig. 2.

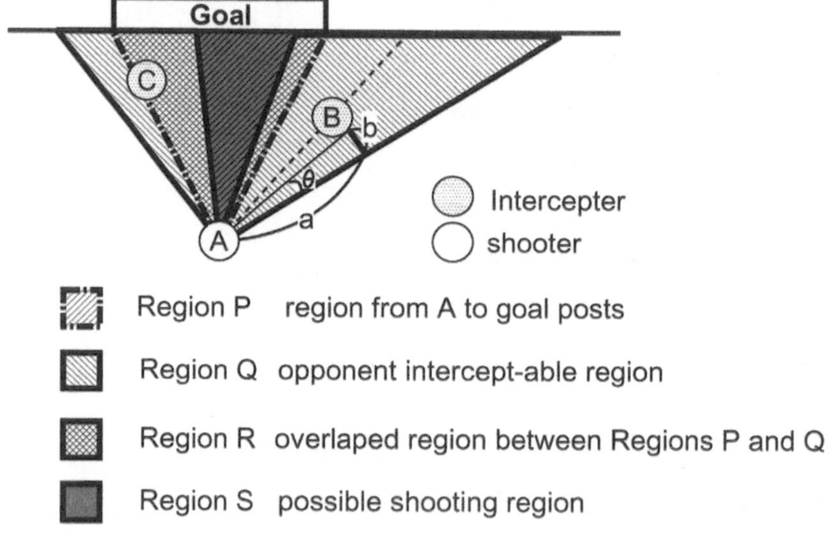

Fig. 2. Modeling of shooting space recognition

Let agent A be the shooter and agents B and C be the intercepting opponents. It is clear from the figure that Region S is the only possible shooting sector for Agent A. When a shoot situation is recognized, Agent A accordingly selects the "Shoot Behavior Rule".

Opponents' interception regions depend on their interception abilities and change in behaviors during a game. The degree of interception ability has to be learnt by the agent in real time during a game to achieve adaptive recognition. In fact the shooting Agent has to avoid two regions described by the opponent dimensions (as a stationary obstacle) and the other by the blocking scope described by its maneuverability.

The blocking region is described by the angle 2θ, where $\theta = \arctan(\alpha)$, $\alpha = b/a$. Therefore, parameter α represents opponent's interception ability and the agent recognizes that if α is large, then the opponent intercept-able region is large too, and vice versa.

Prediction of parameter α (through experience during a game) will enable the agent to recognize shootable situations without missing a chance. This will not only improve score rates, but is also considered to be more effective in selecting alternative behaviors to shooting, such as centering behavior.

4 Self-Behavior and Observation Learning Algorithm

4.1 Self-Behavior Learning algorithm

When the agent makes a successful shoot, α is decreased by Equation (1a). When the agent fails to shoot, α is increased by the Equation (1b).

$$\alpha = \alpha - \delta, \tag{1a}$$

$$\alpha = \alpha + \delta. \tag{1b}$$

Where α is a real number from α_{min} to α_{max}. and δ is a positive constant. Fig.3 shows changes in an opponent's interception regions as "Self-Behavior Learning" takes place. With an initially assumed value in α, subsequent shoot-success situations will result in decrementing α by δ. However, subsequent increments in α by δ follow shoot-failure situations.

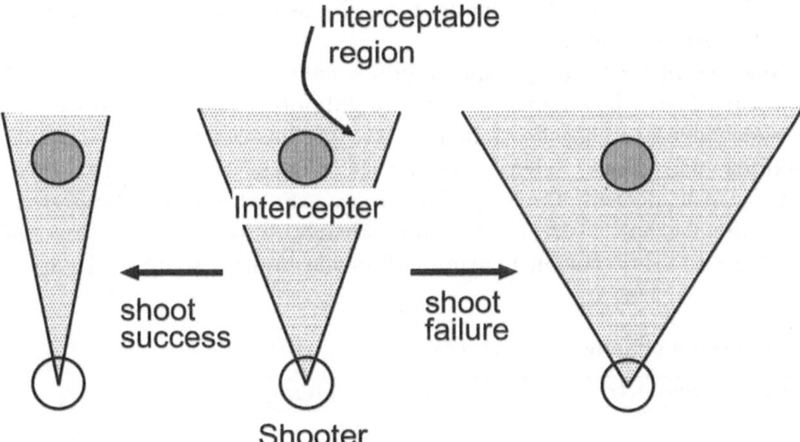

Fig. 3. Changes of an opponent(intercepter)'s interceptable region which a shooter assumes by "Self-Behavior Learning"

4.2 Observation Learning algorithm

As for "Observation Learning" algorithm, the agent assumes a teammate's-shoot-situation and evaluates a failure-situation or a success-situation based on a previously learnt parameter. It then compared its own evaluation with the actual outcome. If there is agreement between its knowledge and the actual outcome then agents knowledge is considered as a success, otherwise a failure in knowledge is declared. If failure in knowledge is declared; a wrongly predicted success-situation indicates underestimation of opponents'strength, therefore α is increased by δ; while, a wrongly predicted failure-situation indicates overestimation of opponents' strength, therefore α is decreased by δ.

5 Experimental Results from the Proposed Learning System

Experiments were performed to evaluate the effectiveness of "Self-Behavior Learning and Observation" system with "Self-Behavior Learning" and a non-learning system. In this experiments the competing teams are A, B, C and Neko. Each team had two Forward players and one Midfielder. Both Teams A and Neko employed agents without any learning capabilities. Team B employed only "Self-Behavior Learning" system. Team C is the only team that employed "Self-Behavior Learning and Observation" system. Teams B and C had an initial value of $\alpha_0 = 0$, with α between 0 and 1, and $\delta = 0.125$. Team A had α set to 0, as a constant. Teams A, B and C had a similar Decision Making System Architecture while Team Neko had a different Decision Making System Architecture.

Team Neko competed thirty times against each of Teams A, B and C, making a total of 90 played matches. Score, Shoot trials, Score Rate and Centering were calculated, as shown in Table1. The data represent the average of thirty games with Team C clearly having the highest average Score, Score Rate and Centering decisions among other teams. Furthermore, the Scoring Cumulative Sums graph over thirty games clearly shows, Fig. 4, that the "Self-Behavior and Observation Learning" has the highest score build up.

Table 1. The results of games against Team Neko (average of thity games)

Team	Score	Shoot trials	Score Rate	Centering
A	0.8	18.3	0.04	0.7
B	1.3	20.0	0.06	3.7
C	2.1	21.0	0.10	8.0

Score Rate = Score / Shoot trials

6 Conclusion

A learning system using "Self-Behavior and Observation Learning" was proposed for a real-time adaptive learning in simulated soccer games. A parameter representing opponent's interception ability was modeled in the "Shoot

Behavior Rule". The three methods of "Self-Behavior and Observation Learning", "Self-Behavior Learning" and "Non-Learning" were compared experimentally in a multi-agent environment.

The proposed "Self-Behavior and Observation Learning" had better abilities than the other methods. The effectiveness of its ability was reflected through the high Score, Score Rate and Centering decisions.

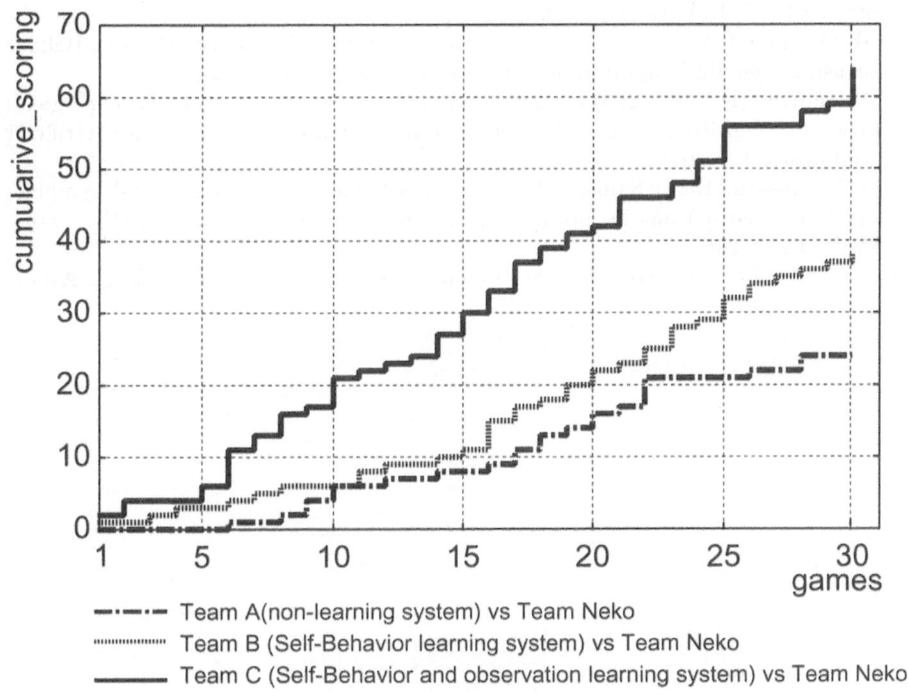

Fig. 4. Scoring cumulative sum over thirty games against Team Neko

References

1. H.Kitano, M.Asada, Y.Kuniyoshi, I.Noda, E.Ozawa.(1995) RoboCup:The Robot World Cup Initiative, Proceedings of IJCAI-95 Workshop on Entertainment and AI/Alife.
2. RoboCup official site. http://www.robocup.org/
3. I.Noda, H.Matsubara, K.Hiraki, I.Frank.(1998) Soccer server: A tool for research on multi-agent systems. Journal of Applied Artificial Intelligence,12,233–250.
4. Soccer Server web site. http://sserver.sourceforge.net/
5. P.Stone, M.Veloso.(1998) Layered approach to learning client behaviors in the robocup soccer server. Journal of Applied Artificial Intelligence,12,165–188.

6. M.Riedmiller, A.Merke, D.Meier, A.Hofmann, A.Sinner, O.Thate, and Ch.Kill.(2001).Kerlsruhe Brainstormers 2000.RoboCup2000:RobotSoccer WorldCup IV,Springer-Verlag, Berlin.
7. R.S.Sutton, A.G.Barto.(1998)Reinforcement Learning.The MIT Press.
8. T.Andou.(1998) Refinement of Soccer Agents' Positions Using Reinforcement Learning. RoboCup-97:RobotSoccer World Cup I,371–388.
9. K.Kostiadis, H.Hu(1999) Reinforcement learning and co-operation in a simulated multi-agent system. Proceedings of 1999 IEEE/RSJ Int. International Conference on Intel. Robots and Systems.
10. H.Akiyama,T.Nagao(2001)An Optimization Method of Soccer Player's Behavior using Genetic Programming,Proceeding of JSAI SIG-Challenge.
11. Y.Kumada, K.Ueda.(2001) Acquisition of Cooperative Tactics by Soccer Agents with Ability of Prediction and learning.Journal of Japanese Society for Artificial Intelligence,16,120–127.
12. J.Rasmussen.(1983) Skills, rules, and knowledge; signals, signs, and symbols, and other distinctions in human performance models. Journal of IEEE Trans. SMC,13,257–266
13. A.Bandura.(1971)Psychological modeling : Conflicting theories.Aldine Atherton, Chicago.

An Empirical Study of Coaching

Patrick Riley, Manuela Veloso, and Gal Kaminka*

Carnegie Mellon University, 5000 Forbes Ave., Pittsburgh, PA 15213-3891

Abstract. In simple terms, one can say that team coaching in adversarial domains consists of providing advice to distributed players to help the team to respond effectively to an adversary. We have been researching this problem to find that creating an autonomous coach is indeed a very challenging and fascinating endeavor. This paper reports on our extensive empirical study of coaching in simulated robotic soccer. We can view our coach as a special agent in our team. However, our coach is also capable of coaching other teams other than our own, as we use a recently developed universal coach language for simulated robotic soccer with a set of predefined primitives. We present three methods that extract models from past games and respond to an ongoing game: (i) formation learning, in which the coach captures a team's formation by analyzing logs of past play; (ii) set-play planning, in which the coach uses a model of the adversary to direct the players to execute a specific plan; (iii) passing rule learning, in which the coach learns clusters in space and conditions that define passing behaviors. We discuss these techniques within the context of experimental results with different teams. We show that the techniques can impact the performance of teams and our results further illustrate the complexity of the coaching problem.

1 Introduction

As multi-agent systems continue to grow more important, the types of relationships between agents continue to be studied. One important relationship that humans often exhibit is still largely lacking among our agents. This relationship is one of a coach or advisor who provides advice to others. We consider this to be the central feature of a coach relationship, and autonomous agents could from benefit from the development of this sort of relationship. One of the primary ways that advice can be generated is through an agent's observations of and experience with the world. Processing past and current observations into a form usable as advice is indeed a challenging problem.

We have implemented a coach for the Soccer Server System [10], a simulated robotic soccer environment. Notably, because of the creation of a standard language CLang [16], coaches and teams from researchers around the world are able to work together. We have worked towards this research goal of our coach working with a team for which it was not specifically designed. This was the basis for a small coach competition at RoboCup2001 [5] in which four

* This research was sponsored by United States Air Force Grants Nos. F30602-00-2-0549 and F30602-98-2-0135 and by an NSF Fellowship. The content of this publication reflects only the position of the authors.

teams competed. By exploring a few possible techniques for processing observations and providing advice and then evaluating their effects, we hope to further understand the challenges this problem poses. This paper reports on our coaching strategies implemented in simulated robotic soccer and presents the results of our focused experimentation. We believe these results provide a basis for future experimental work, as well as a grounding for more general explication of the coaching problem.

2 Environment

The Soccer Server System is a server-client system that simulates soccer between distributed agents. Clients communicate using a standard network protocol with well-defined actions. The server keeps track of the current state of the world, executes the actions which the clients request, and periodically sends each agent noisy, incomplete information about the world. Agents receive noisy information about the direction and distance of objects on the field (the ball, players, goals, etc.); information is provided only for objects in the field of vision of the agent.

There are 11 independent players on each side as well as a coach agent. The coach agent sees the position and velocity of all players and the ball, but does not directly observe the actions or the perceptions of the agents.

Actions must be selected in real-time, with each of the agents having an opportunity to act 10 times a second. Each of these action opportunities is known as a "cycle." Visual information is sent 6 or 7 times per second. Over a standard 10 minute game, this gives 6000 action opportunities and 4000 receipts of visual information. All units of distance discussed here are simulated meters, with the whole field measuring 105m x 68m.

The communication model between the coach and players was designed to require significant autonomy for the players, especially during the active parts of the games. Basically, the model permits the coach to say one message every 30 seconds (every 300 cycles). Messages are delayed 5 seconds (50 cycles) before being sent to the players.

The coach messages are in a standard coach language called CLang, which was developed by members of the simulated soccer community. Each message basically consists of a set of condition-action rules for the players. The conditions can include relative and absolute positions of the players and the ball as well as the play mode and the player currently controlling the ball. The actions include directions to pass or dribble, move to an area of the field, and "mark" (take a defensive position) against a player or region.

The exact communication model as well as further technical details can be found in [16].

3 Coaching Techniques

This section covers the techniques we use to coach simulated robotic soccer. All of these techniques are designed to learn information about the opponents and how to play effectively against them. Learning about the team to be coached the next research step, as discussed in the empirical results (Section 4).

3.1 Formations by Learning

One important concept in robotic soccer is that of the formation of the team [19]. The concept of formation used by CLang is embodied in the "home area" action. The home area specifies a region of the field in which the agent should generally be. It does *not* require that the agent never leave that area; it is just a general directive.

Our coach represents a formation as an axis aligned rectangle for each player on the team. From the home areas, agents can also a infer a role in the team, with the common soccer distinctions of defenders, midfielders, and forwards.

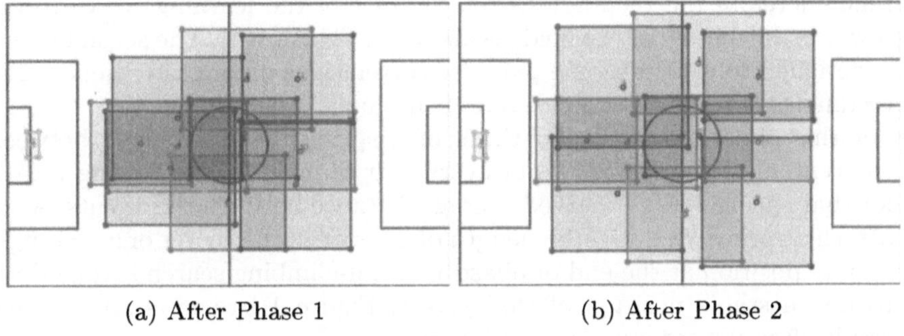

(a) After Phase 1 (b) After Phase 2

Fig. 1. The learning of the CMUnited99 formation from RoboCup2000 games.

All coaching based on formation uses an algorithm for learning the formation of a team based on observation of that team. The algorithm's input is the set of locations for each player on a team over one or more games. The learning then takes place in two phases.

1. The goal of the first phase is, for each agent, to find a rectangle which is not too big, yet encompasses the majority of the points of where the agent was during the observed games. The learning is done separately for each agent with no interaction between the data for each agent. First the mean position of the agent (c_x, c_y) is calculated, as well as the standard deviation (s_x, s_y). We then do a random search over possible rectangles

(σ is used a parameter for the search). The rectangles to evaluate are generated from the following distribution (for the left, right, top, and bottom of the rectangles), where $N(m, \sigma)$ represents a Gaussian with mean m and standard deviation σ (note that we use a coordinate frame where (0,0) is in the upper left):

$$(N(c_x - s_x, \sigma), N(c_x + s_x, \sigma), N(c_y - s_y, \sigma), N(c_y + s_y, \sigma)) \tag{1}$$

The evaluation function is a weighted sum (with parameter γ which we set to 0.95) of two quantities, both with maximum values of 1. The first involves f, the fraction of points where the agent was which are inside R. We simply use f^β to (where β is a parameter which we set to 1/3). The second quantity uses A (the area of R) and a scaling parameter M (which we set to 900). The evaluation function is then:

$$E(R) = \gamma f^\beta + (1 - \gamma)\left(1 - \frac{A}{M}\right) \tag{2}$$

2. The first phase of learning ignores correlation among the agents. In fact it quite common for all agents to shift one direction or another as the ball moves around the field. This tends to cause the average positions (and therefore the rectangles from phase 1 of the learning) to converge towards the middle of the field, as shown in Figure 1(a). The second phase is designed to capture some pairwise correlations among the agents. The rectangles are moved around, but their shape is not changed.
For this phase, conceptually think of a spring being attached between the centers of the rectangles of every pair of agents. The resting length for that spring is the observed average distance between the agents. Also, attach a spring with a resting length of 0 between the center of a rectangle and its position at the end of phase 1. A hill-climbing search is then done to find a stable position of the system. Figure 1(b) shows an example result after the second phase of learning.
Now we describe the details of the algorithm. First, the observed average distance t_{ij} between every two agents is calculated. Next, for each pair of agents, a value α_{ij} roughly corresponding the the tension of the spring in the above description is calculated as follows (w and m are parameters):

$$\alpha_{ij} = e^{m t_{ij}} \quad (i \neq j) \tag{3}$$

$$\alpha_{ii} = w \tag{4}$$

Eq. (3) provides a higher value (i.e. higher spring tension) between agents which are closer, reflecting the assumption that the correlated movement of nearby agents is more important than those of far away agents. The parameter m (which we set to -0.01) controls that exact weighting. Eq. (4) is used for the connection of an agent to its original position, since Eq. (3) would provide an extremely high weight since $t_{ii} = 0$. The constant w

is set to 0.5, which is the weight calculated by Eq. (3) for a distance of approximately 69 meters.
At each step of the hill-climbing search, a particular agent p is chosen at random to have its rectangle shifted. All other rectangles are held fixed. For all i, let o_i be the original position of rectangle i and let c_i be the vector of the center of current position of rectangle i. The evaluation function used is (where smaller is better):

$$\alpha_{pp} \left(\text{dist}(c_p, o_p)\right)^2 + \sum_{j \neq p} \alpha_{pj} \left(\text{dist}(c_p, c_j) - t_{pj}\right)^2 \tag{5}$$

This simply uses the α values computed in Eqs. (3) and (4) to compute the additive penalty for the imaginary springs not being at their resting length.
The gradient of the evaluation function as a function c_p is easily calculated and a small step is taken in the direction of the gradient (with learning rate 0.001).

Formation learning is used in two ways. The first is an instance of imitation where we imitate the formation of another team. This is especially important for the rule learning described in Section 3.3. The other technique we call "formation based marking." Here the coach observes the previous games of the opponent we will play and learns their formation. Each of the defenders is then assigned one of the forwards of the opponent to mark for the whole game. Ordinarily a team may change its assignment of which defenders mark which forwards. Sending static marking assignments may reduce flexibility, but it also reduces coordination problems and gives the player knowledge of the opponent they did not previously have.

3.2 Set plays

Set plays refer to times of the game when the ball is stopped (due to an out of bounds call, free kick, or kick off) and one team has time to prepare before kicking the ball. Our coach takes advantage of this time to make a plan for the movement of the ball and the agents. This plan is based on refinement of plan templates with a model of the opponent used in evaluating plan changes. Details about this process are described elsewhere [14].

An important difference to be noted is that the plans used in this work were described as a set of rules in CLang rather than as a Simple Temporal Network [7]. Compiling coordination constraints into rule-based systems can be difficult. **<I'd like to have a nice citation from this; maybe ask Gal>**

3.3 Rule Learning

The passing patterns of a team are an important component in a team's performance. Our coach observes the passes of teams in previous games in order to learn rules which capture some of these passing patterns. These rules can then be used either to imitate a team, or to predict the passes that an opponent will do.

The rule learning uses a combination of clustering (using Autoclass C [3]) to create regions on the field and C4.5 [12] to generate rules describing the passing behavior of a team. The attributes for the rules are the locations of the passer and receiver (using the regions learned from clustering) and the relative position of all teammates and opponents. The rules from C4.5 are then transformed into rules in CLang.

To illustrate, we now provide an example of an learned rule. The format here is almost the format of the CLang language. A few things have been renamed or left out for clarity.

```
1   ((and (play_mode play_on)
2         (bowner our)
3         (bpos "PLINCL0")
4         (ppos our {6} (arc (ball) 23 1000 -180 360))
5         (ppos opp {10} (arc (ball) 0 1000 151 29)))
6      (do our {2 - 11} (bto "PLOUTCL1" {p}))
7      (do our {11} (pos "PLOUTCL1")))
```

Lines 1–5 are the conditions for the rule and lines 6–7 are the actions. Line 2 says that some player on our team is controlling the ball. Line 3 says that that the ball in a particular cluster ("PLINCL0" is the name of the cluster). Lines 4 and 5 are on the position of particular players. Line 4 says that teammate number 6 is at least 23m away, while line 5 says that the angle of opponent number 10 is between 151 and 180 degrees. Note that we do use absolute player numbers here. This is one of the reasons we developed the formation learning techniques described in Section 3.1. As long as the opponent has not changed its formation, the absolute player numbers should be valid. Line 6 instructs all players on our team (except the goalie who is number 1) to pass the ball to the a particular cluster. Line 7 instructs a teammate number 11 (whose home formation position is closest to cluster "PLOUTCL1") to position itself in that region.

4 Experimental Setup and Results

The language CLang was adopted as a standard language for a coach competition at RoboCup2001. Four teams competed providing a unique opportunity to see the effects of a coach designed by one group on the team of another.

We participated in the coach competition, which consisted a single game in each test case. This section reports on our later thorough empirical evaluation of our coach and the techniques used. Each experimental condition was run for 30 games and the average score difference (as our score minus their score) is reported. Therefore a negative score difference represents losing the game and a positive score difference is winning. All significance values reported are for a two tailed t-test.

We use eight teams for our evaluation. We will use initials (denoted in parentheses here) for the teams. The teams that understand CLang are: the DirtyDozen (DD) from University of Osnabrück; and ChaMeleons (CM) from Carnegie Mellon University. Also from RoboCup2001, we use Gemini (GEM) from the Tokyo Institute of Technology and Brainstormers (B) from the University of Karlsruhe. Team descriptions for these teams are available in [5]. From the RoboCup2000 competition, we use the following teams: VirtualWerder (VW) from the University of Bremen; ATHumboldt (ATH) from Humboldt University; and FCPortugal (FCP) from the Universities of Aveiro/Porto (team descriptions can be found in [2]). We also use CMUnited99 (CMU99) from Carnegie Mellon [18], which competed at RoboCup99 and RoboCup2000. In order to run these experiments, we slowed the server down to 3-6 times normal speed so that all agents could run on one machine.

Our experiments aim to separate out the effects of the techniques of our coach. To do this, we ran a sequence of games with different combinations of the five techniques: formation (F) (Section 3.1), set plays (S) (Section 3.2), offensive and defensive rules (R) (Section 3.3), and formation based marking (M) (Section 3.1).

For playing against GEM, our coach observed one game of B playing against GEM. Advice was sent to imitate B's formation and formation based marking was used against GEM's formation. Rule learning was also done for those games. Similarly, our coach learned from 5 games of CMU99 playing against VW and from 10 games of FCP playing against ATH.

The results are shown in Figure 2. The CMvATH set is different from the others in several respects. No combination of the techniques resulted in an improvement for CM, and several combinations (F, FSR, FSRM) resulted in significantly worse performance ($p < .05$) compared to no coach.

For the other teams, the combination of all techniques (FSRM) is always significantly better than no coach ($p < .000002$). Looking at the individual techniques is also illustrative. Sending a formation sometimes helps the team (DD v GEM) and sometimes hurts the performance (CM v GEM), even though exactly the same formation is sent in each case. Even though the advice is the same, the effect on the team being coached is vastly different. From this, we conclude that the coach needs to learn something about the team being coached.

Except for the CM v ATH line, neither the rules nor the formation based marking make a significant impact on the score difference of the games. The formation based marking was a minor part of the coach and it is no great surprise that it's impact is small. The rule learning, however, was the most ambitious of the coaching techniques used. There are several reasons why the rules may have failed to have a large impact. The number of examples from which to learn varied considerably, from 51 to 1638) and so did the accuracy of

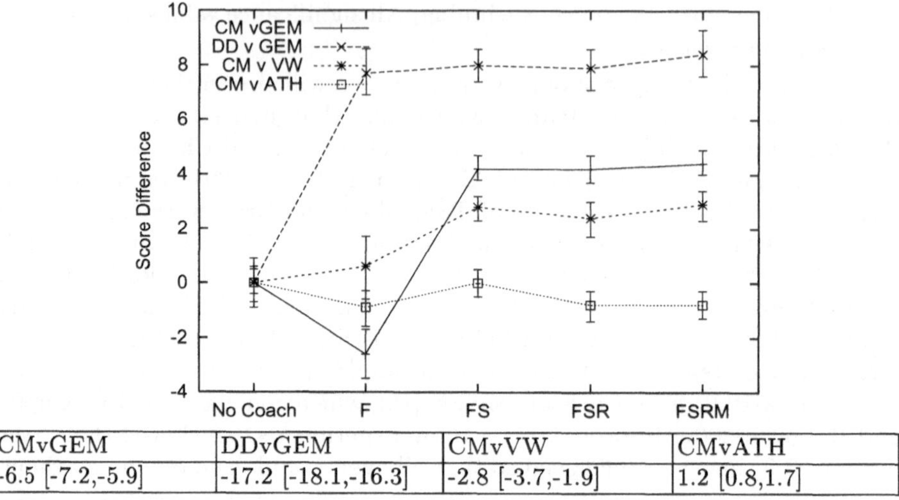

CMvGEM	DDvGEM	CMvVW	CMvATH
-6.5 [-7.2,-5.9]	-17.2 [-18.1,-16.3]	-2.8 [-3.7,-1.9]	1.2 [0.8,1.7]

Fig. 2. The score difference of teams coached by a random coach and various techniques of C-CM. The score differences have been additively normalized to the no coach values shown in the lower table. All error bars are 95% confidence intervals. Note that we do not have random coach results for all cases.

the rules on the reserved test set (35%–75%). Some preliminary experiments indicate that changing the input attributes could improve the performance. The attributes are currently based on the absolute player numbers, where sorting the player's by distance to the passer may be useful. This was done primarily because of the expressibility of the current version of CLang, and a new version is in progress (see [16] for details).

5 Related Work

The area of imitation has been studied under many different names. There has been extensive research in the robotics literature on learning a task by imitating a human being, called variously "teaching by guiding," "learning by watching," "programming by demonstration," and "imitation learning." Bakker and Kuniyoshi have a recent survey [1] and Dautenhahn emphasizes the biological connection [6]. Similarly, an area commonly called behavioral cloning deals with learning a control strategy for a task [17, 20]. Imitation is only one possible aspect of successful coaching. In particular for this work, we are imitating one aspect of agent interaction (passing), not simply agent interaction with the environment.

Some work has also been done in creating agents capable of receiving advice. For example, the RATLE system by Maclin and Shavlik [8] can incorporate advice generally specified as if-then rules (similar to the language we use here) into a reinforcement learning agent. Their results in Pengo, a grid and blocks world, also suggest that the learning agents need to be able to refine advice to achieve high performance. Clouse [4] find a similar results in a discrete driving task. They created an automated trainer to improve the learning speed of the learning agent. If the trainer gives too much advice, the learner can fail to converge.

Previous research in Intelligent Tutoring Systems (ITS) has examined how to give advice to human beings. For example, Miller, *et. al.* [9] consider how to give advice to students who are constructing arguments based on scientific data. The system works by comparing the structure of the student's argument (explicitly given by the student) to known patterns. The CAST system [11] trains humans to act in a team. Here, a coach agent provides advice based on tracking belief state of the human being coached. The primary difference between the ITS literature and this research is that tutoring systems generally rely on a fairly rigid and predefined task structure. Deviations from that structure are the focus of the advice. Here, we have no such predefined plan or structure.

The ISAAC system [13] is an automated game analysis tool for simulated robotic soccer. It does off-line analysis of games at several levels. It employs a local adjustment approach to suggest small changes (such as "shoot when closer to the goal") to a team's designer in order to improve performance. The suggestions are backed up by examples from the games analyzed and provided in a format useful for the designers to examine. However, ISAAC's suggestion are provided to the *designer* of the team, not to the agents themselves; there is no automated effect on the team.

6 Conclusion

We have presented several implemented coaching techniques for a simulated robotic soccer domain. We further presented the results from an extensive set of experiments to understand the effects of the coaching techniques presented here, using agents created at a variety of institutions. The experiments represent 630 games and over 20 days of computer time. The experiments justify that coaching can help teams improve in this domain. However, all of our coaching techniques are based on learning about the adversary and not on understanding the functioning of the team to be coached. Consequently, the effect of the advice on different teams varies greatly. Our results support the need for a coach to understand its team in order to achieve robust performance.

These empirical study is a first but significant step in the project of understanding an advice-based relationship between automated agents. We intend to use this experimental basis to aid in the understanding of the general coaching problem (see [15] for one characterization). This research raises many interesting question which we will continue to pursue.

References

1. P. Bakker and Y. Kuniyoshi. Robot see, robot do: An overview of robot imitation. In *AISB96 Workshop on Learning in Robots and Animals*, pages 3–11, Brighton,UK, 1996.
2. T. Balch, P. Stone, and G. Kraetzschmar, editors. *RoboCup-2000: Robot Soccer World Cup IV.* Springer Verlag, Berlin, 2001.
3. P. Cheeseman, J. Kelly, M. Self, J. Stutz, W. Taylor, and D. Freeman. Autoclass: A bayesian classification system. In *ICML-88*, pages 54–64, San Francisco, June 1988. Morgan Kaufmann.
4. J. Clouse. Learning from an automated training agent. In D. Gordon, editor, *Working Notes of the ICML '95 Workshop on Agents that Learn from Other Agents*, Tahoe City, CA, 1995.
5. A. B. S. Coradeschi and S. Tadokoro, editors. *RoboCup-2001: Robot Soccer World Cup V.* Springer Verlag, Berlin, 2002.
6. K. Dautenhahn. Getting to know each other—artificial social intelligence for autonomous robots. *Robotics and Autonomous Systems*, 16:333–356, 1995.
7. R. Dechter, I. Meiri, and J. Pearl. Temporal constraint networks. *Artificial Intelligence*, 49:61–95, 1991.
8. R. Maclin and J. W. Shavlik. Creating advice-taking reinforcement learners. *Machine Learning*, 22:251–282, 1996.
9. M. S. Miller, J. Yin, R. A. Volz, T. R. Ioerger, and J. Yen. Training teams with collaborative agents. In *ITS-2000*, pages 63–72, 2000.
10. I. Noda, H. Matsubara, K. Hiraki, and I. Frank. Soccer server: A tool for research on multiagent systems. *Applied Artificial Intelligence*, 12:233–250, 1998.
11. M. Paolucci, D. D. Suthers, and A. Weiner. Automated advice-giving strategies for scientific inquiry. In *ITS-96*, pages 372–381, 1996.
12. J. R. Quinlan. *C4.5: Programs for Machine Learning.* Morgan Kaufmann, San Mateo, CA, 1993.
13. T. Raines, M. Tambe, and S. Marsella. Automated assistant to aid humans in understanding team behaviors. In *Agents-2000*, 2000.
14. P. Riley and M. Veloso. Planning for distributed execution through use of probabilistic opponent models. In *AIPS-2002*, 2002. (to appear).
15. P. Riley, M. Veloso, and G. Kaminka. Towards any-team coaching in adversarial domains. In *AAMAS-02*, 2002. (extended abstract) (to appear).
16. RoboCup Federation, `http://sserver.sourceforge.net/`. *Soccer Server Manual*, 2001.
17. C. Sammut, S. Hurst, D. Kedzier, and D. Michie. Learning to fly. In *ICML-92*, Aberdeen, 1992. Morgan Kaufmann.
18. P. Stone, P. Riley, and M. Veloso. The CMUnited-99 champion simulator team. In Veloso, Pagello, and Kitano, editors, *RoboCup-99: Robot Soccer World Cup III*, pages 35–48. Springer, Berlin, 2000.
19. P. Stone and M. Veloso. Task decomposition, dynamic role assignment, and low-bandwidth communication for real-time strategic teamwork. *Artificial Intelligence*, 110(2):241–273, June 1999.
20. D. Šuc and I. Bratko. Skill reconstruction as induction of LQ controllers with subgoals. In *IJCAI-97*, pages 914–919, 1997.

Chapter 7
Distributed Control

Decentralized Control of Redundant Manipulator Based on the Analogy of Heat and Wave Equations

Kosei Ishimura[1], M.C. Natori[2], and Mitsuo Wada[1]

[1] Graduate School of Engineering, Hokkaido University, N13-W8, Kita-ku, Sapporo, Hokkaido 060-8628, Japan
E-Mail:{ishimura,wada}@complex.eng.hokudai.ac.jp
[2] The Institute of Space and Astronautical Science, 3-1-1 Yoshinodai, Sagamihara, Kanagawa 229-8510, Japan
E-Mail: natori@newslan.isas.ac.jp

Abstract. Based on the analogy of heat and wave equations, novel decentralized control methods of a redundant manipulator are proposed. Because these control methods are based on real physical phenomena, physical sense of control parameters can be understood intuitively. Some traditional control methods can be reconstructed from the viewpoint of proposed methods. Furthermore, stable characteristics of the heat conduction imply stability of the control method. As case studies, three subjects are discussed: position control of the end-effector of manipulator, obstacle avoidance, grasping behavior. Some numerical results are shown to illustrate capabilities of proposed control methods.

Keywords: redundant manipulator, heat equation, wave equation, decentralized control

1 Introduction

Robotic manipulators, which have more independent DOF than that are required to perform a specified task, are called kinematically redundant [1]. If manipulator's task is only to control position and posture of the end-effector, 6 DOF are enough. Redundant DOF is used to realize additional tasks such as obstacle avoidance, minimization of energy. Though applications of redundant manipulators are spread in various fields such as painting, inspecting and so on [2], calculation times for the control should be shortened for further uses: that is decentralization in the control. Decentralized parallel control would become a key to improve calculation times.

One of the most popular control methods, in which parallel computation is realized, is curve-fitting method [1][3]. In such methods, reference curve is not calculated in decentralized way. In this paper, we propose novel control methods based on analogy of the heat and wave equations. The heat equation is applied to average physical properties of each link. The wave equation is also applied to propagate them from one tip link to another tip link. The control methods can be also used in deployment control of modularized structures [4]. Because these control methods are based on real physical phenomena, physical sense of control parameters can be understood intuitively. Some traditional control methods can be reconstructed from the viewpoint of proposed methods. Furthermore, stable characteristics of heat conduction imply stability of the control method. As case studies, three subjects are discussed: position control of end-effector of the manipulator, obstacle avoidance, grasping behavior. As an example of redundant manipulator, serial link manipulator, which is simplified model of a Variable Geometry Trusses (VGT) [5], is studied.

2 Control Methods Based on the Heat and Wave Equations

2.1 The Control Input Based on the Heat Equation

The general heat equation can be written as

$$\frac{\partial T(\mathbf{x},t)}{\partial t}=\frac{\partial}{\partial x}\left(a\frac{\partial T(\mathbf{x},t)}{\partial x}\right)+\frac{\partial}{\partial y}\left(a\frac{\partial T(\mathbf{x},t)}{\partial y}\right)+\frac{\partial}{\partial z}\left(a\frac{\partial T(\mathbf{x},t)}{\partial z}\right) \tag{1}$$

where, $T(\mathbf{x},t)$ is temperature and a is thermal diffusivity. The thermal diffusivity a is a function of the temperature $T(\mathbf{x},t)$, and is written as $a(T)=\lambda/(\rho c)$ where, λ,ρ,c are heat conductivity, density and specific heat capacity respectively. When the heat equation is applied to the decentralized control, temperature $T(\mathbf{x},t)$ is replaced with a physical property, which wants to be averaged. In application, it is desirable that the effect of the control is only averaging the physical property. In other words, it is desirable that the volume integral of the physical property is conserved. Therefore, we derive boundary conditions such that the volume integral of temperature is conserved. The conservation equation of the volume integral of temperature can be written as

$$\frac{d}{dt}\int_V T(\mathbf{x},t)dv=0 \tag{2}$$

From the Eq.(1) and the theorem of Gauss, the left hand of Eq.(2) is

$$\frac{d}{dt}\int_V T(\mathbf{x},t)dv=\int_S a\,(n_x\frac{\partial T}{\partial x}+n_y\frac{\partial T}{\partial y}+n_z\frac{\partial T}{\partial z})\,ds \tag{3}$$

where, $[n_x\ n_y\ n_z]^T$ is a normal vector on the boundary S. Therefore, a boundary condition which satisfies Eq.(2) is described as follows

$$\frac{\partial T}{\partial x}=\frac{\partial T}{\partial y}=\frac{\partial T}{\partial z}=0 \qquad \text{(on the boundary } S\text{)} \tag{4}$$

Next, we derive the control input based on the heat equation. Physical properties of link No.i, which are aimed to be averaged, are denoted as $\mathbf{u}_i(t)$. In this section, the control input $\dot{\mathbf{u}}_i(t)$, which averages $\mathbf{u}_i(t)$ by using only local information, is derived from the heat equation. For simplicity, the derivation is carried out in the case of single variable $u_i(t)$. At first, the right hand of Eq.(1) is rewritten by using difference equations.

$$\frac{\partial T(x,t)}{\partial t}=\frac{a(T)(T(x+\Delta x,t)-T(x,t))-a(T)(T(x,t)-T(x-\Delta x,t))}{\Delta x^2} \tag{5}$$

Temperature $T(x,t)$ is replaced by the $u_i(t)$, and a_i is defined as $a(T(x,t))/\Delta x^2$. From the boundary condition(Eq.(4)), the control input $\dot{u}_{i\mathbf{heat}}(t)$ based on the heat equation can be derived as follows.

$$\begin{cases}\dot{u}_{1\mathbf{heat}}=a_1(u_2(t)-u_1(t))\\ \dot{u}_{i\mathbf{heat}}=a_i(u_{i+1}(t)-u_i(t))-a_{i-1}(u_i(t)-u_{i-1}(t)) \quad (i=2,\cdots,n-1)\\ \dot{u}_{n\mathbf{heat}}= \qquad -a_{n-1}(u_n(t)-u_{n-1}(t))\end{cases} \tag{6}$$

Here n is a total amount of links. Because the control input of each link is decided by only physical properties of adjoining links, the control input can be calculated in parallel. Furthermore, it is expected that the system reach a state of thermal equilibrium since the sum of control input $\dot{u}_i$ over the system equals zero. In the case of multivariable system $\dot{\mathbf{u}}_i$, control inputs can be derived in the same manner. If anisotropy of the thermal diffusivity is introduced, the proposed control input is expected to have further possibilities.

2.2 The Control Input Based on the Wave Equation

The control input based on the wave equation is derived. Because whole motions of VGT, manipulator and so on are mainly governed by the motion of the No.n link (end-effector), we suggest the propagation rule of local information based on the wave equation, in which shape information of each link propagates from end-effector to the base. There are some similar uses of wave equation in the curve fitting method of redundant manipulator [1]. As well as the derivation of Eq.(6), control input based on the wave equation for a variable $u_i(t)$ can be written by using a difference equation.

$$\begin{cases}\dot{u}_{1\mathbf{wave}}=c\cdot u_2(t)\\ \dot{u}_{i\mathbf{wave}}=c\cdot\left(u_{i+1}(t)-u_i(t)\right) \quad \left(i=2,\cdots,n-1\right)\\ \dot{u}_{n\mathbf{wave}}= \qquad -c\cdot u_n(t)\end{cases} \tag{7}$$

where, c is a constant which is corresponding to wave velocity.

3. Position Control of Redundant Manipulator

3.1 Redundant Manipulator

The VGT consists of repeated octet truss, and is one of redundant manipulators in broad sense. It can change its physical properties, such as its shape, rigidity and so on, by changing its member length. Because of the features, the VGT is a promising candidate as inspection robot in space and component of large space structure such as space station. Though it has multiplicity in its shapes because of its high redundancy, calculations for the position control of the end-effector require much times due to the redundancy. In previous works of the authors [5], a parallel control method was proposed from the viewpoint of the autonomous distributed system. However, the stability of the control method was not proved. In this paper, the realization of both parallel control and stable control is aimed. Our subject is to control the position of VGT's end-effector to a target.

To simplify the problem, we regard the VGT as serial link manipulator with high redundancy (Fig. 1). Each link needs at least 3DOF to control 3 dimensional position of the end-effector. Therefore, it is assumed that each link has 4DOF with additional 1DOF to keep independence of each link. Control input is time differential of link vector $\dot{\mathbf{r}}_i(t)$.

The movable range of serial link manipulator, which is corresponding to that of the VGT, is shown in Table 1. Here, we assume that the constant member length of VGT is 1, and variable member length can change from $1/\sqrt{3}$ to $2/\sqrt{3}$. This change in length can be realized by single telescopic mechanism. If the telescopic mechanism is multistage, the range of variable member becomes wider. A remarkable point is that minimum length of serial link manipulator corresponding to the VGT equals zero theoretically. The movable range in Table 1 is rough approximation in which coupled motion is ignored (ex. torsion and bending motion).

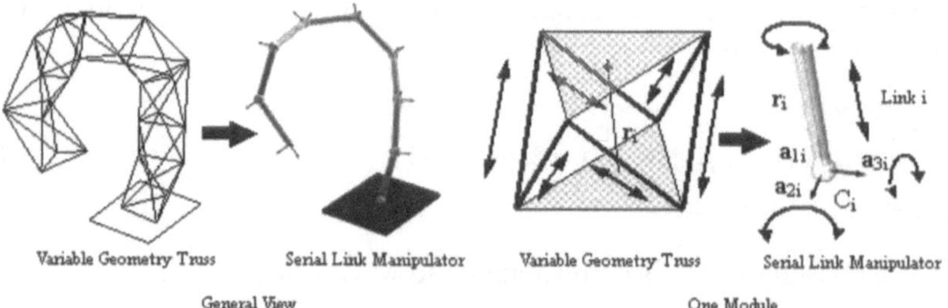

Fig. 1. Simplification of VGT to Serial Link Manipulator

Table 1. Movable Range

VGT		Serial Link Manipulator	
length of constant member	1	length of link : $\lvert\mathbf{r}\rvert$	from 0 to 1
length of variable members	from $1/\sqrt{3}$ to $2/\sqrt{3}$	torsion angle : θ_1	$\pm 11^\circ$
		bending angles : θ_2, θ_3	$\pm 33^\circ$

3.2 The Control Law for Position Control

The control input, which are the time differential of link vectors $\dot{\mathbf{r}}_i(t)$, is defined as the sum of following two kinds of inputs. First, control inputs by which only No.n link (end-effector) actively moves to the target can be written as

$$\begin{cases} \dot{\mathbf{r}}_{i\mathbf{inv}}(t) = 0 & (i \neq n) \\ \dot{\mathbf{r}}_{n\mathbf{inv}}(t) = K(\mathbf{x}_{\text{target}} - \mathbf{x}_{\text{e}}) \end{cases} \tag{8}$$

where, K is constants of feedback gain, and $\mathbf{x}_{\text{target}}$, $\mathbf{x}_{\text{e}}$ are position vectors of the target and end-effector, respectively. Then the control input based on the heat equation is applied to this problem to realize averaging shape changes of links because it is not desirable that local changes of link's shapes are large. $u_i(t)$ in Eq.(6) is replaced by the link vector $\mathbf{r}_i(t)$, which is representative property of link's shape. In the consider of Eq.(6) and (8), total control law based on the heat equation can be written as follows.

$$\begin{cases} \dot{\mathbf{r}}_1(t) = a_1(\mathbf{r}_2(t) - \mathbf{r}_1(t)) \\ \dot{\mathbf{r}}_i(t) = a_i(\mathbf{r}_{i+1}(t) - \mathbf{r}_i(t)) - a_{i-1}(\mathbf{r}_i(t) - \mathbf{r}_{i-1}(t)) \qquad (i = 2, \cdots, n-1) \\ \dot{\mathbf{r}}_1(t) = \qquad - a_{n-1}(\mathbf{r}_n(t) - \mathbf{r}_{n-1}(t)) + K(\mathbf{x}_{\text{target}} - \mathbf{x}_{\text{e}}) \end{cases} \tag{9}$$

Because each control input is decided by only physical properties of adjoining modules, it can be calculated in parallel.

As well as the case of heat equation, total control law based on the wave equation can be written as follows.

$$\dot{\mathbf{r}}_i = \dot{\mathbf{r}}_{i\mathbf{wave}} + \dot{\mathbf{r}}_{i\mathbf{inv}} \tag{10}$$

3.3 Numerical Results of Position Control

The converging behavior to the target is checked. Parameters and initial conditions, which are used in simulations, are shown in Table 2. The position of target is $\mathbf{x}_{\text{target}} = (4,2,2)^T$. In the control law based on the heat equation, $a_i = 4$ is used(Eq.(9)). A numerical result for the control based on the heat equation is shown in Fig.2(a). It shows that the position of end-effector (No.n link) converges to the target and the other modules follow after another. From this result, it is confirmed that proposed control law can realize the averaging interaction of link vectors $\mathbf{r}_i$ because link vectors show a tendency to make a straight line.

Next, numerical simulations for the control based on the wave equation are also carried out. The constant c, which is corresponding to the wave velocity (Eq.(10)), has to be decided in consideration of the movement of the end-effector because the constant would be relevant to the amount of link length which is sent out from the base link to the end-effector. Therefore, $c = 4|\dot{r}_{n\mathbf{inv}}|^2$ is used in this case. A numerical result for the control based on the wave equation is shown in Fig.2(b). To say nothing of the convergence to the target, it is shown that links (No.1 to n-1) trace locus of No.n link.

From the various numerical results about different constants a_i, c, it is qualitatively clarified that a_i controls the strength of averaging interaction, and c controls the amount of link length, which is sent out from the base link to the end-effector.

Table 2. Parameters of Simulations

constants	time step : Δt	0.1 (s)
	number of links : n	20
	feedback gain : K	0.1
initial conditions	link length	0.1
	joint angles : $\theta_{i1}, \theta_{i2}, \theta_{i3}$	0 (rad)

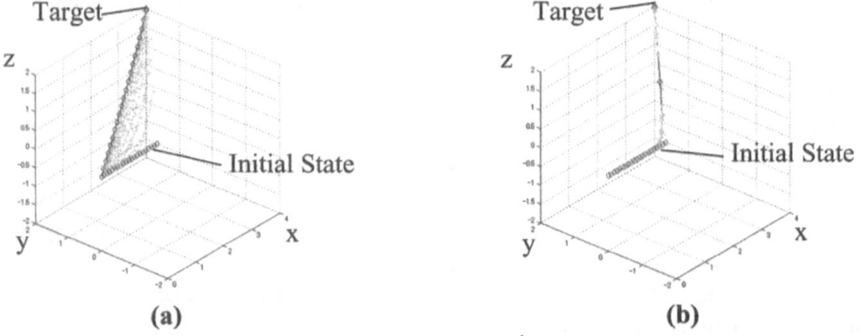

Fig. 2. Convergence to the Target (**a**) Control Based on the Heat Equation (**b**) Control Based on the Wave Equation

4 Obstacle Avoidance

4.1 Avoidance Behavior

As additional behavior which uses redundancy of the manipulator, a problem of obstacle avoidance is also considered. The obstacle is assumed as a stationary sphere. The avoidance behavior emerges when the manipulator invade the area where the danger of collision occurs. The obstacle avoidance is achieved by a simple action as follows(Fig.3):

Step 1 No.k link detect the existence of the obstacle.
Step 2 No.k link transmits the information of the obstacle to No.k-1,k+1 links.
Step 3 No.k-1,k+1 links carry out following avoidance behavior.

$$\dot{\mathbf{r}}_{k-1\mathbf{avoid}} = K_{\mathbf{avoid}} \cdot \mathbf{h}/|\mathbf{h}|, \quad \dot{\mathbf{r}}_{k+1\mathbf{avoid}} = -K_{\mathbf{avoid}} \cdot \mathbf{h}/|\mathbf{h}| \tag{11}$$

Here, $K_{\mathbf{avoid}}$ is a constant for avoidance behavior, and $\mathbf{h}$ is a vector connecting the obstacle with nearest point of the link.

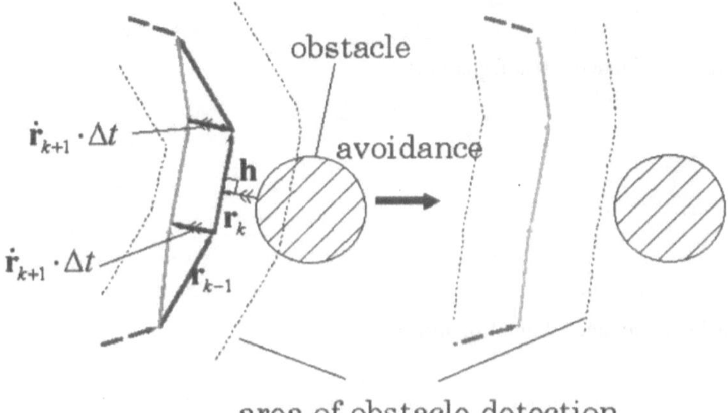

Fig. 3. Behavior of Obstacle Avoidance

4.2 Numerical Results of Obstacle Avoidance

Parameters, initial condition, and constants a_i, c are the same in Fig.2(a),(b). The position of target is $\mathbf{x}_{\text{target}} = (8,0,0)^T$. The information about obstacles and constants for avoidance behavior is shown in Table 3. Figures 4 and 5 show numerical results of the control based on heat equation and wave equation, respectively. Both control laws can realize obstacle avoidance and convergence to the target. In the case of the control based on wave equation, it is shown that there are large differences among the link vectors though obstacle avoidance can be carried out only by a few links near the end-effector. On the other hand, in the case of the control based on heat equation, the link vectors are almost even, and standard deviation of the link length is 0.0094. From these results, it is confirmed that proposed control based on heat equation is very useful for the averaging link's shapes.

Table 3. Constants for Obstacles

obstacle1	obstacle 2	radius of obstacle
$(3,0.1,0.3)^T$	$(5,0,-0.5)^T$	0.5

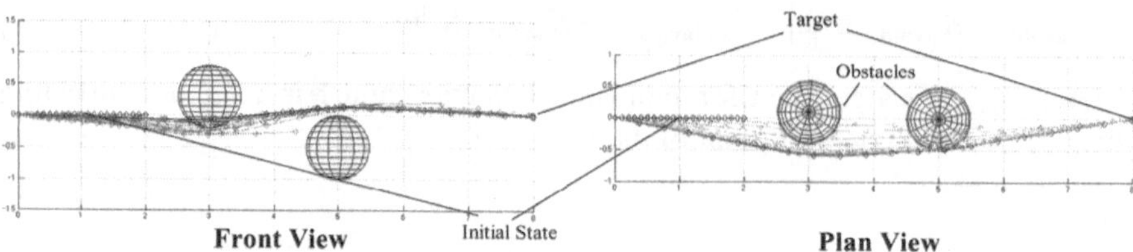

Fig. 4. Obstacle Avoidance (Heat Equation)

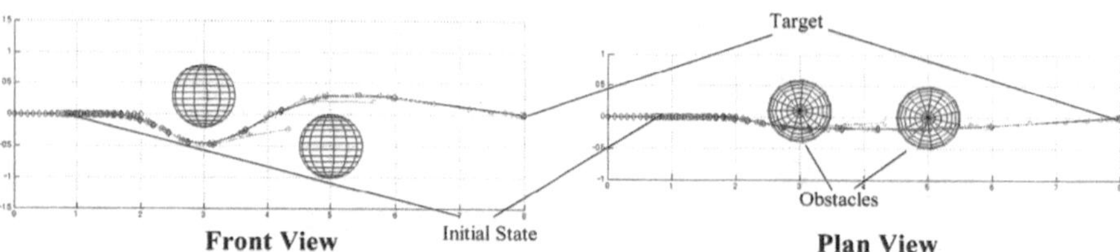

Fig. 5. Obstacle Avoidance (Wave Equation)

5 Grasping Behavior

5.1 The Control Law for Grasping

In previous researches, the parallel control of winding grasp used tactile information has already been proposed [6]. Though tactile information is treated as a digital value in the control, we reconstruct the control by treating the information as a continuous value and introducing heat equation to the control. In other words, we attempt the grasping by averaging tactile sensor value of each link. Tactile sensor value is assumed as follows.

$$m_i(t) = \begin{cases} 0 & (if \;\; p_i \geq p_{recog}) \\ K_{tact}(p_{recog} - p_i) & (else) \end{cases} \tag{12}$$

Here, p_i is distance from No. i joint to a object which we want to grasp, and p_{recog} is distance within which a object can be detected by the tactile sensor. In this case, averaging inputs for sensor value m_i are derived by the analogy of heat equation.

$$\begin{cases} \dot{m}_{1\mathbf{heat}} = a_1(m_2(t) - m_1(t)) \\ \dot{m}_{i\mathbf{heat}} = a_i(m_{i+1}(t) - m_i(t)) - a_{i-1}(m_i(t) - m_{i-1}(t)) \quad (i = 2, \cdots, n-1) \\ \dot{m}_{n\mathbf{heat}} = \qquad - a_{n-1}(m_n(t) - m_{n-1}(t)) \end{cases} \tag{13}$$

The averaging inputs $\dot{m}_i$ are transformed to the control inputs of link vectors according to following equation.

$$\dot{\mathbf{r}}_i(t) = \dot{m}_i(t)\mathbf{e}_i / |\mathbf{e}_i|, \quad \dot{\mathbf{r}}_{i+1}(t) = -\dot{m}_i(t)\mathbf{e}_i / |\mathbf{e}_i| \tag{14}$$

Here, $\mathbf{e}_i$ is a vector connecting the object and No.i joint. By Eq.(12)-(14), the control law for grasping are defined.

5.2 Numerical Results of Grasping

Initial conditions and information of the object are shown in Table 4. An example of numerical results is shown in Fig.6. Though realized behavior is near to pressing rather than grasping, the manipulator can deform its shape keeping a constant distance from the object. As further extension, grasping along the geodesic would be possible if additional control inputs such as Eq.(9),(11).

Table 4. Parameters of Simulations

constants	time step : Δt	0.1 (s)
	number of links : n	10
initial conditions	link length	0.1
	joint angles : $\theta_{i1}, \theta_{i2}, \theta_{i3}$	0 (rad)
object	position	$(0.5, 0.45, 0)^T$
	radius	0.35
	p_{recog}	1

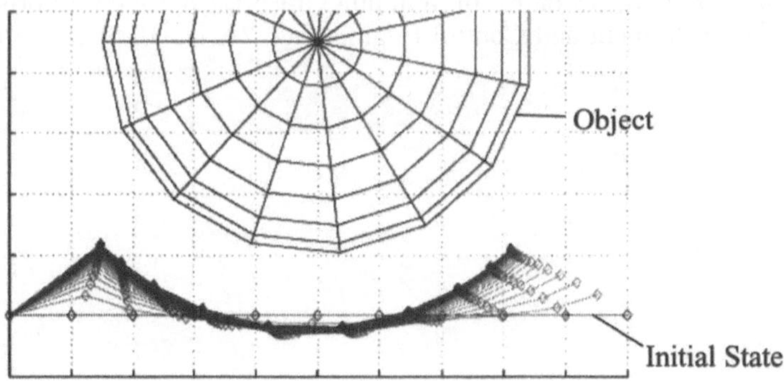

Fig. 6. Grasping Behavior

6 Conclusions

From the viewpoint of analogy of the heat and wave equations, the decentralized control methods of redundant manipulator are discussed. The capabilities of proposed methods are demonstrated about following three subjects; position control, obstacle avoidance and grasping behavior. From these results, it is concluded that the proposed control based on the heat equation is effective for the averaging physical properties of each link. The strength of averaging can be changed by the constant which are equivalent to thermal diffusivity. On the other hand, the proposed control based on the wave equation can propagate physical properties of each link from one tip link to another tip link. Furthermore, because these control methods can realize parallel processing, high speed processing is expected.

References

1. Chirikjian, G.S. (1992) Theory and Application of Hyper Redundant Robotic Manipulators, Ph.D. thesis, California Institute of Technology
2. Paljug, E., Ohm, T., Hayati, S. (1995) The JPL Serpentine Robot: A 12 DOF System for Inspection, IEEE, Robotics and Automation Society, Nagoya, Japan.
3. Salerno, R.J., Reinholtz, C.F., Dhande, S.G. (1991) Kinematics of Long-Chain Variable Geometry Truss Manipulators, Advances in Robot Kinematics, Springer-Verlag Wien, New York, 179-187
4. Ishimura, K., Natori, M.C. (2000) Deployment Behavior of Modularized Structures, ASME Int'l Mech. Eng. Cong. Exop., Adaptive Structures and Material Systems Symposium, Orland
5. Ishimura, K., Natori, M.C. and Higuchi, K. (1999) AIAA-99-1535, An Autonomous Distributed Control of Free-Floating Variable Geometry Trusses, ASIAA/ASME/AHS Adaptive Structure Forum, St. Louis, **4**, 2636-2643
6. Wada, M., Kuba, Y. (1984) Control of Winding Grasp Used Tactile Information. Trans. Soc. Instrument and Control Engineering, **20,** No.10, 973-975 (in Japanese)

Cooperating Cleaning Robots

Markus Jäger

Corporate Technology
Information & Communications
Siemens AG
81730 Munich, Germany
markus.jaeger@mchp.siemens.de

Abstract. If multiple robots are used to cooperatively clean a large room, e.g. an airport, a variety of problems have to be solved. The area which has to be cleaned must be partitioned among the robots, each robot must completely cover the area it is responsible for and collisions among the robots must be avoided.

In this paper we describe the problems in detail and provide some solutions. The methods described in the paper work completely decentralized. They do not require a centralized component or global synchronization. There is also no need for a global communication network. We demonstrate that it is sufficient for the robots to communicate in a local neighborhood.

We also introduce an agent based control architecture for mobile robots. The architecture allows us to integrate the solutions to all the problems into one system. The basic idea is that each robot is controlled by a set of agents. The agents can cooperate with each other or with agents of other robots.

1 Introduction

The predicted potential for growth in demand for service robots for the future is enormous [1]. Furthermore, specialized robots for various cleaning tasks will play an essential part. Among tasks like facade and vehicle cleaning, floor cleaning will become of essential importance [2].

Cleaning of large rooms, like airports or train stations, however, can not be done effectively by a single robot. For such tasks it is necessary to use a fleet of cooperating robots. When using multiple robots, however, a variety of new problems have to be solved. The area which has to be cleaned must be partitioned among the robots [3], each robot must completely cover the area it is responsible for [4] and collisions among the robots must be avoided [5,6].

If only a single robot is used, the robot also has to completely cover the area it is responsible for (in that case the whole area). The problem of automatically determining a path for complete coverage is, however, mostly omitted by teaching the robot a certain fixed path for each environment it is used in [2]. If multiple robots, which dynamically partition the area, are used, it is not possible to teach each robot a certain fixed path. It is therefore necessary to determine the path automatically. Although it is omitted in most cases, there are automatic methods to determine a path for complete coverage for one robot [4,7]. Since these methods can also be used in a multi robot case, we do not further consider this problem.

To partition the area among the robots, a dynamic and decentralized method is used. The area is divided into polygons, which are allocated by the robots. After a robot has allocated a certain polygon, it is responsible for cleaning the polygon. There is no global synchronization or global communication network required.

Collisions among the robots are avoided by coordinating the independently planned trajectories of the robots. Whenever the distance between two robots drops below a certain value, they exchange information about their planned trajectories and determine whether they are in danger of a collision. If a possible collision is detected, they monitor their movements and, if necessary, insert idle times between certain segments of their trajectories in order to avoid the collision. The collision avoidance also works completely in a distributed manner and does not require any global synchronization or a global communication network.

To integrate all the solutions in one system, an agent based control architecture for mobile robots is used. The basic idea is that each robot is controlled by a set of agents. The agents can cooperate with each other or with agents of other robots. The agent architecture, furthermore, provides means for user interaction so that the user can, for example, start and stop the whole system.

In the next section the core component of the system, the agent architecture, is introduced. Section three and four give a short description of the area partitioning and collision avoidance algorithms. The last two sections provide some experimental results and give a summary.

2 Agent Based Architecture

The operation of a robot in a real world scenario is a very complex task. A lot of different components have to work together to ensure that the task is solved in a robust and reliable way. The integration of all these components into one working system has been a constant research area in the past. It is necessary to integrate coordination, communication, and control structures. Nowadays the use of agent technology becomes more and more popular.

We have developed an agent based control architecture for mobile robots, called ABCHOR (Agent-Based Control architecture for Heterogeneous Open Robot communities). Our idea was to develop a flexible system which can easily be reused and which simplifies the integration of new components.

In the remainder of this section an overview of possible control structures for mobile robots is given, the ABCHOR architecture is described, and a sample architecture for the cooperating cleaning robots is introduced.

2.1 Related Work

Early robotic system were mainly based on Artificial Intelligence (AI) methods. They use a sense-plan-act cycle. They perceive the environment with their sensors, generate a model of the environment, determine a plan (based on the model), and execute the plan, by using some actuators. Nilsson [8] describes a typical system of that time. The main problem of these systems, at that time, was that the cycle time was too slow (due to the high computational complexity of the AI algorithms), which means that the world was changing faster than the robot could adapt its model.

Brooks [9] introduced the next class of systems, the reactive systems. He completely omits the modelling and planning phase. His idea is to wire the sensors directly with the actuators, via a few simple rules. These systems fit perfectly in a dynamic environment, but it is hard to represent a complex task by a few simple rules.

The fact that neither planning nor reactive systems were suitable gave birth to the concept of hybrid systems. These hybrid systems combine the reactive and planning approaches in a layered control structure, where the lower level works fast and reactively and the higher level plans and follows some complex goals. Gat [10] describes a sample hybrid system.

Recently, it became more and more popular to develop architectures for multi robot systems. This means that classical architectures have to be extended by adding a cooperation component, to handle the interactions of the single robots. There have been various approaches to extend planning, reactive and hybrid systems. Müller [11] seems to have the most promising approach. He suggests to add an additional cooperation layer to the hybrid systems. The cooperation layer of a robot cooperates with the cooperation layers of the other robots and generates goals for the planning layer.

2.2 ABCHOR Architecture

The idea of ABCHOR is to use multiple agents to control a robot, where each agent is responsible for a certain task. There could, for example, be a NavigationAgent who is responsible for all navigation tasks, so that the other agents can easily move the robot by simply telling the NavigationAgent where to go. There could also be a GripperAgent, which controls a gripper, or an ObjectTrackingAgent, which uses a camera to track certain objects. Each robot could also have a GoalAgent, which uses the other agents to fulfill some high level goals given to the robot. Figure 1 shows a few robots with some sample agents.

The partitioning of the robot control into multiple agents supports the separation of domain dependent and domain independent components. A NavigationAgent or a GripperAgent can, for example, be used the same way in many different applications. Specific domain dependent code would probably be placed into a GoalAgent.

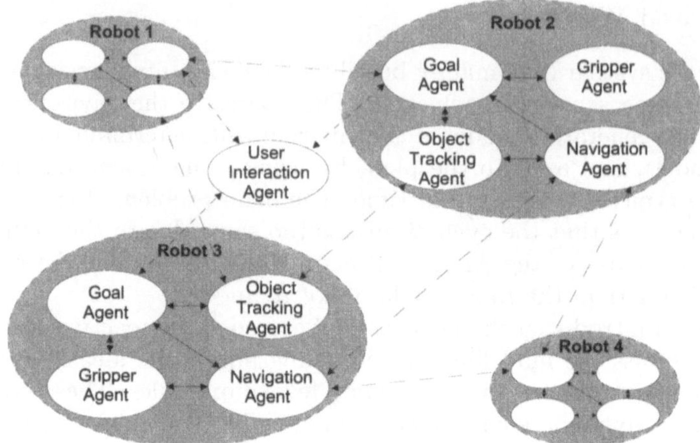

Fig.1. Sample control scenario based on ABCHOR.

The individual goals of a certain robot are achieved by its agents which cooperate with each other (solid arrows in Figure 1). Common goals of multiple robots are reached by individual agents of the robots cooperating with each other (dashed arrows in Figure 1). It is also possible that some agents cooperate with another agent, which does not belong to a certain robot. This could, for example, be a UserInteractionAgent, where the UserInteractionAgent provides an interface for the user of the robotic system.

If domain independent agents cooperate with each other, their cooperation mechanism is also domain independent. It can therefore be used in many different applications. Some examples for domain independent cooperation are multiple NavigationAgents avoiding collisions, multiple NavigationAgents building a map of the environment, two GripperAgents handing over an object, and multiple ObjectTrackingAgents cooperatively tracking an object to obtain better tracking results [12]. The cooperation can even be performed without the GoalAgents knowing it.

The use of multiple cooperating agents makes ABCHOR a very flexible system and supports heterogeneous systems in a natural way. Each robot can have a different set of agents. It also allows the design of robust and fault tolerant systems. Robots can simply have multiple agents for the same task, where one agent takes over the task of another agent, if the other agent should fail.

ABCHOR is designed as an open system. The implementation is based on FIPA (Foundation For Intelligent Physical Agents) standards [13]. ABCHOR agents can therefore cooperate with non ABCHOR agents, if the other agents also follow FIPA standards. The current implementation of ABCHOR uses JADE (Java Agent DEvelopment Framework) [14], an agent platform that is based on FIPA standards. JADE provides support for agent creation and maintenance, a communication backbone, and some standard interaction protocols. Since JADE also supports mobility of agents, the agents in the system are not bound to a specific robot, except for agents that are using hardware, like the NavigationAgent.

2.3 Cooperating Cleaning Robots

Figure 2 shows the architecture of the cooperating cleaning robots. Each robot comprises two agents, a GoalAgent and a NavigationAgent. The NavigationAgent is responsible for all navigation tasks. It is capable of navigating the robot to a certain position. The high level control of the robot is implemented in the GoalAgent. The GoalAgent tells the NavigationAgent to which position should the robot be navigated to and the NavigationAgent informs the GoalAgent when the robot has reached the position.

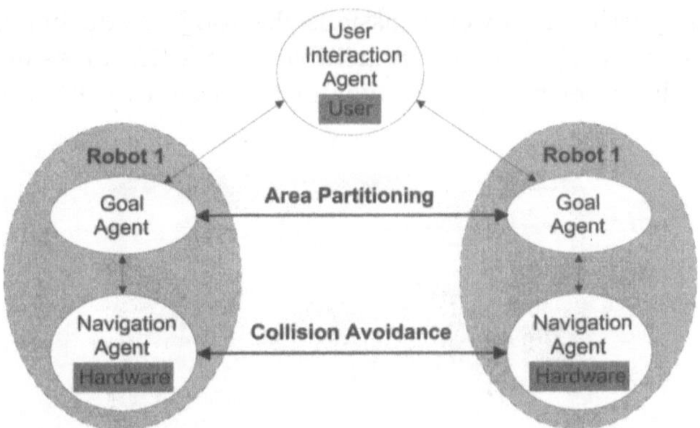

Fig.2. Architecture of the cooperating cleaning robots.

To UserInteractionAgent provides an interface for the user of the system, so that a human operator can, for example, start and stop the system. The UserInteractionAgent can run on one of the robots. It is, however, also possible to have the UserInteractionAgent running on another device, such as a PDA or a mobile phone (JADE is currently ported to these platforms).

The GoalAgents of all robots are cooperating with each other to partition the area to be cleaned. The cooperation methods are described in more detail in Section 3. Each GoalAgent determines a path for complete coverage of the partition it is responsible for, and uses the NavigationAgent to move the robot along the path.

The NavigationAgents of all the robots are cooperating with each other to avoid collisions among them. The methods they use are explained in more detail in Section 4.

The implementation of an agent is based on Jess (Java Expert System Shell) [15], a rule based expert system. The high level control of an agent is therefore controlled by a set of rules. The agents communicate with each other by exchanging facts, which are stored in the knowledge bases of their expert systems.

3 Area Partitioning

The partition of a certain area among multiple robots can be done either statically [3] or dynamically. A static approach assigns each robot a certain subarea at the beginning. Each robot is then only responsible for its assigned subarea. The main disadvantage of a static assignment is that the whole system can not be adapted dynamically to a new situation. If, for example, a robot breaks down, the other robots can not take over its work, or if a robot is slower than assumed, the other robots are not able to help him.

A dynamic partition, however, assigns the subareas during runtime. It is therefore possible to react to unpredictable events. If, for example, a robot breaks down, the other robots can dynamically take over his work.

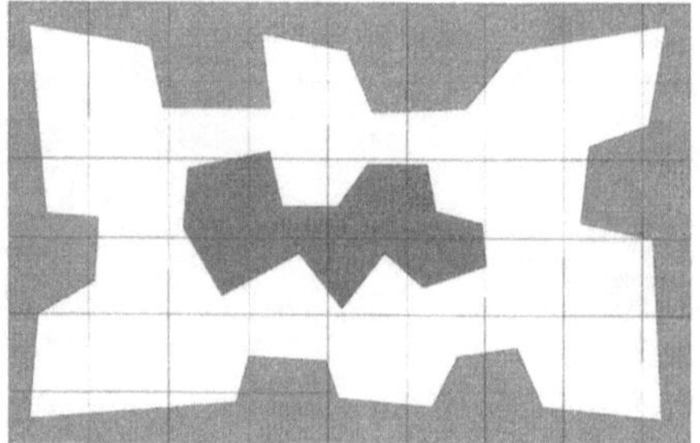

Fig.3. Division of an area into polygons.

We use a completely dynamic approach. The main idea is to divide the room into polygons, as shown in Figure 3. The robots then allocate and clean the area covered by these polygons. To allocate a polygon means that a robot intends to clean the area covered by the polygon and announces this to the other robots. In Figure 4, three robots are processing a large room. The dark polygons are cleaned regions, the light polygons are allocated regions, and the white region is not yet allocated by any robot.

The developed allocation strategy allows the robots to partition the area in a reasonable way, although they are not able to communicate with each other all the time. The radius of the circles around the robots (Figure 4) is half of the communication radius of the robots, i. e. the robots are able to communicate with each other, if their circles intersect.

Jäger and Nebel [17] give more information on the area partitioning strategy and some simulation results.

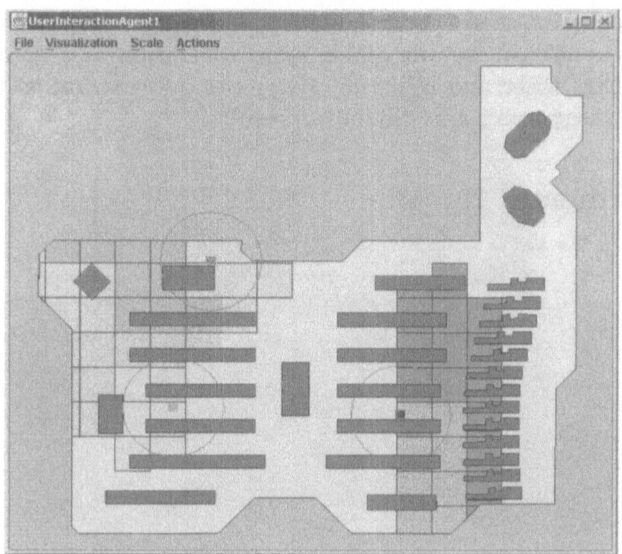

Fig.4. Three robots processing a large supermarket in Bussum, Netherlands.

4 Collision Avoidance

Whenever multiple mobile robots share the same workspace, the potential for collisions among them must be taken into account. This can be done by using a centralized component to plan collision free trajectories of all the robots simultaneously [16] or by planning the trajectories of all the robots independently and using a centralized component to coordinate these trajectories, so that no collision is possible [6].

Centralized approaches, however, have the disadvantage of high computational cost, lack flexibility, and require a global communication network. We therefore omit centralized components and, in contrast to centralized approaches, achieve global coordination by distributed algorithms and assume only local communication between pairs of physically close robots. This permits the use of less demanding communication frameworks and allows easier and more adaptive coordination between the robots.

Whenever the distance between two robots drops below a certain value they exchange information about their planned trajectories and determine whether they are in danger of a collision. If a possible collision is detected, they monitor their movements and, if necessary, insert idle times between certain segments of their trajectories in order to avoid the collision.

Deadlocks among two or more robots occur if a number of robots block each other in a way such that none of them is able to continue along its trajectory without causing a collision. These deadlocks are reliably detected. After a deadlock is detected, the trajectory planners of each of the involved robots are asked to plan an alternative trajectory until the deadlock is resolved.

We use a combination of three fully distributed algorithms to reliably solve the task. They do not use any global synchronization and do not interfere with each other. Jäger and Nebel [5] give more information on the collision avoidance strategy and some simulation results.

5 Experimental Results

Fig.5. The robots R, G and B.

All the methods have been fully implemented and extensively tested in simulation. The methods are also implemented on our three test robots R, G, and B (Figure 5). The robots R, G, and B are Pioneer robots, equipped with a Sick laser range finder, a gyroscope, and an industrial PC. We have furthermore implemented the methods on an autonomous cleaning robot (Hefter, ST82R, Figure 6). All the robots use SINAS^{TM} (Siemens Navigation System) [2] for navigation tasks.

Using these robots, we carried out various experiments in real environments. The results of the experiments fully confirm the successful results that we previously obtained by simulations. Jäger and Nebel [5,17] give a detailed discussion of the simulation results.

6 Summary

In this paper we presented ABCHOR, an agent based robot control architecture which we use to coordinate a fleet of cleaning robots. The idea of

Fig.6. Autonomous cleaning robot Hefter ST82R.

ABCHOR is to use multiple agents to control a robot, where each agent is responsible for a certain task. Agents of the same robot cooperate with each other to achieve the robots goals and agents of different robots cooperate with each other to achieve common goals of multiple robots.

We furthermore described two methods to partition an area among multiple robots and to avoid collisions among multiple robots. Both methods work completely decentralized and do not need any global synchronization. There is no need for a global communication network. It is sufficient if the robots are able to communicate regularly with each other.

The area partitioning works completely dynamic, i.e. the subareas are determined and assigned during runtime. The collision avoidance uses a combination of three fully distributed algorithms, which reliably perform the task.

References

1. R. D. Schraft and G. Schmierer, Serviceroboter: Produkte, Szenarien, Visionen, Springer, 1998
2. H. Endres, W. Feiten, and G. Lawitzky, Field Test of a Navigation System: Autonomous Cleaning in Supermarkets, *Int. Conf. on Robotics and Automation (ICRA)*, pp. 1779–1781, 1998
3. H. Bast and S. Hert, The Area Partitioning Problem, *Proc. of the 12th Canadian Conf. on Computational Geometry*, pp. 163-171, 2000
4. C. Hofner and G. Schmidt, Path Planning And Guidance Techniques For An Autonomous Mobile Cleaning Robot, *Int. Conf. on Intelligent Robots and Systems (IROS)*, pp. 610-617, 1994

5. M. Jäger and B. Nebel, Decentralized Collision Avoidance, Deadlock Detection, and Deadlock Resolution for Multiple Mobile Robots, *Int. Conf. on Intelligent Robots and Systems (IROS)*, pp. 1213-1219, 2001
6. M. Bennewitz and W. Burgard, Coordinating the Motions of Multiple Mobile Robots Using a Probabilistic Model, *8th Int. Symposium on Intelligent Robotic Systems (SIRS)*, 2000
7. E. U. Acar, Y. Zhang, H. Choset, M. Schervish, A. G. Costa, R. Melamud, D. C. Lean, and A. Graveline, Path Planning for Robotic Demining and Development of a Test Platform, *Proc. of the 3rd Int. Conf. on Field and Service Robotics (FSR)*, pp. 161-168, 2001
8. Nils J. Nilsson, Shakey the Robot, *SRI AI Center Technical Note 323*, April 1984
9. Rodney A. Brooks, A Robust Layered Control System for a Mobile Robot, *IEEE Journal of Robotics and Automation*, RA-2, No. 1, p. 14, 1986
10. E. Gat, On Three-Layer Architectures,*In AI and Mobile Robots*, AAAI Press, 1998
11. J. P. Müller, The Design of Intelligent Agents, A Layered Approach, Springer-Verlag Berlin, 1996
12. M. Dietl, J.-S. Gutmann, and B. Nebel, Cooperative Sensing in Dynamic Environments, *Int. Conf. on Intelligent Robots and Systems (IROS)*, 2001
13. FIPA (Foundation For Intelligent Physical Agents), http://www.fipa.org, 2002
14. JADE (Java Agent DEvelopment Framework), http://sharon.cselt.it/projects /jade, 2002
15. E. Friedman-Hill, Jess (Java Expert System Shell), http://herzberg.ca.sandia. gov/jess, 2002
16. J. Barraquand, B. Langlois, and J.-C. Latombe, Numerical Potential Field Techniques for Robot Path Planning, *IEEE Trans. on System, Man, and Cybernetics*, vol. 22(2), pp. 224-241, 1992
17. M. Jäger and B. Nebel, Dynamic Decentralized Area Partitioning for Cooperating Cleaning Robots, *Int. Conf. on Robotics and Automation (ICRA)*, to appear, 2002

Movement Model of Multiple Mobile Robots Based on Servo System

Mihoko Niitsuma[1], Hiroshi Hashimoto[1], Yukio Kimura[1], and Shintaro Ishijima[2]

[1] Tokyo University of Technology, Katakura 1404-1, Hachioji,Tokyo 192-0982, Japan
niituma@hiha.mech.teu.ac.jp, hasimoto@cc.teu.ac.jp, kimura@cc.teu.ac.jp

[2] Tokyo Metropolitan Institute of Technology, Asahigaoka 6-6, Hino, Tokyo 191-0065 , Japan, ishijima@ec.tmit.ac.jp

Abstract. This paper proposes the movement model of multiple mobile robots. This model is based on the servo system to advance to the destination while receiving inertia force. Furthermore, the model has Coulomb force to show the avoidance action. It is shown that the simulation result of robots progressing to one exit.

1 Introduction

Recently, the movement of multiple mobile robots has been studied [1]−[5]. In these studies, it is important to consider the collision avoidance and the decision of trajectory, and investigate the behavior of the whole group. The literatures discussed those in use of several - a dozen robots, and the dynamics of the robot in many cases is zero memory type, i.e., its inertia be very small. On the other hand, studies focused on the behavior of a large group, i.e., a crowd consisting of a dozen - hundreds robots with certain inertia are hardly found.

In this paper, the movement model of a mobile robot is proposed in order to investigate behaviors of the whole crowd consisting of over a dozen robots, when they simultaneously advance to the destination. This model contains its own inertia and functions that the decision of trajectory to a destination and the collision avoidance. Considering crowd and wall on the path, the collision avoidance frequently occurs and gives an influence on the robot movement, so its trajectory has to be sequentially corrected every unit time. From those, we realize the decision of trajectory and the motion by the inertia as the servo system. The collision avoidance, be regarded as input disturbance on the servo system, is carried out by adopting a quasi Coulomb force. The collision avoidance is not predictable and also deterministic, thus the sequential correction of the robot movement to the destination has to be required. So, the sequential correction is attained by modifying the reference point of the servo system every unit time, which be arbitrary point on the line segment connecting the present position of the robot and the destination. Sequence of the reference points forms the trajectory of robots.

Furthermore, we consider the robot movement in a corner. This is realized by imitating a movement of a human in a crowd, because a human in a crowd turns a corner well smoothly and reasonably while avoiding collision.

How to decide the reference point and designing parameters of the servo system and Coulomb force are described in section 2. In section 3, simulation results using the proposed model for two passages are discussed.

2 Model of Mobile Robot

2.1 Design of the Servo System

Let us define the dynamical characteristics of the robot i on x - y plane as following

$$m_i \ddot{\mathbf{p}}_i + \boldsymbol{\mu}_i \dot{\mathbf{p}}_i = \mathbf{u}_i \tag{1}$$

where $\mathbf{p}_i = [x_i, y_i]$ is the position of the robot i, m_i is the one's mass, $\boldsymbol{\mu}_i = [\mu_{xi}, \mu_{yi}]$ is the coefficient of viscosity and $\mathbf{u}_i$ is the driven force of the robot. The vector $\boldsymbol{\mu}_i$ consist of x-y elements defined as designing parameters individually, but the mass m_i is the same to x-y elements so the m_i is represented as a scalar in eq.(1). To design the servo system, eq.(1) is translated into the state equations such as

$$\begin{bmatrix} \ddot{\mathbf{p}}_i \\ \dot{\mathbf{p}}_i \end{bmatrix} = \begin{bmatrix} \frac{-\boldsymbol{\mu}_i}{m_i}\mathbf{I} & \mathbf{0} \\ \mathbf{0} & \mathbf{0} \end{bmatrix} \begin{bmatrix} \dot{\mathbf{p}}_i \\ \mathbf{p}_i \end{bmatrix} + \begin{bmatrix} \frac{1}{m_i} \\ 0 \end{bmatrix} \mathbf{u}_i \tag{2}$$

$$\mathbf{q} = \begin{bmatrix} \mathbf{0} & \mathbf{I} \end{bmatrix} \begin{bmatrix} \dot{\mathbf{p}}_i \\ \mathbf{p}_i \end{bmatrix} \tag{3}$$

The servo system is designed so that the robot is able to follow the ramp function [6](see Fig.1).

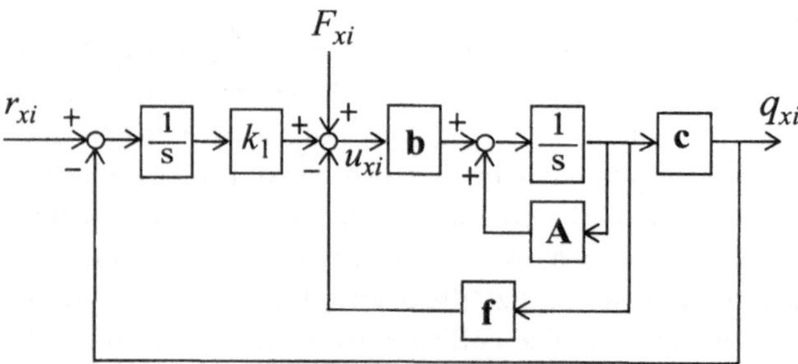

Fig. 1. Servo system composed of x-axis

In this figure, the vector $\mathbf{f} = [f_1, f_2]$ and the scalar k_1 are designing parameters to define movement characteristics of the robot, $\mathbf{r}_i = [r_{xi}, r_{yi}]$ is the reference point and $\mathbf{F}_i = [F_{xi}, F_{yi}]$ is the input disturbance. The matrices in Fig.1 are defined as

$$\mathbf{A} = \begin{bmatrix} \frac{-\mu_{xi}}{m_i} & 0 \\ 0 & 0 \end{bmatrix}, \mathbf{b} = \begin{bmatrix} \frac{1}{m_i} \\ 0 \end{bmatrix}, \mathbf{c} = \begin{bmatrix} 0 & 1 \end{bmatrix}$$

2.2 Decision Method of Trajectory

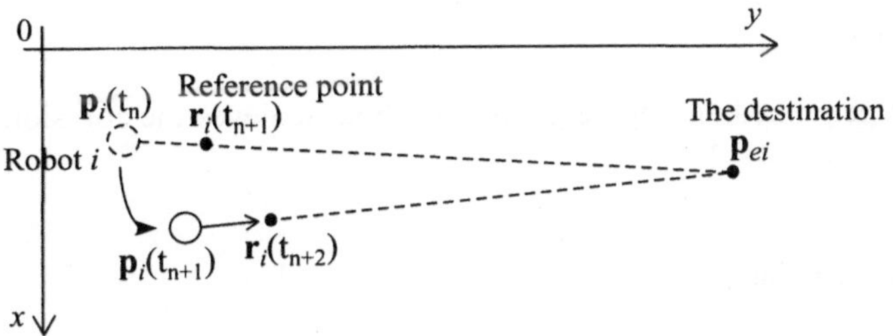

Fig. 2. Sequential correction of the reference point

It is noted that the arrival time when the robot i reaches the destination is not specified, since the collision avoidance or the interference are only considered. The reference point $\mathbf{r}_i$ is advanced toward the destination $\mathbf{p}_{ei} = [x_{ei}, y_{ei}]$ by the length $|\mathbf{l}_i|(\mathbf{l}_i = [l_{xi}, l_{yi}])$ at every Δt. For an example, let us formulate the reference point $\mathbf{r}_i$ when the robot i advances to the positive direction of y-axis (see Fig.2).

$$r_{yi} = |\mathbf{l}_i| + y_i \tag{4}$$

$$r_{xi} = \frac{x_i}{y_i - y_{ei}}(r_{yi} - y_{ei}) + \frac{x_{ei}}{y_{ei} - y_i}(r_{yi} - y_i) \tag{5}$$

In the environment with a corner (see Fig.3), the reference point $\mathbf{r}_i$ is also obtained by eqs.(4),(5). Fig.3 shows the strategy of the trajectory decision, i.e., the destination $\mathbf{p}_{ei}$ is changed according to the condition in eq.(6).

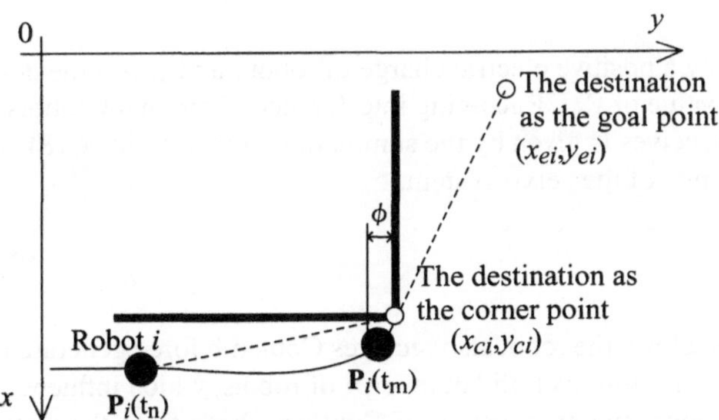

Fig. 3. Decision method of trajectory at corner

$$\mathbf{p}_{ei} = \begin{cases} (x_{ci}, y_{ci}) & \text{if } \ y_i \leq (y_{ci} - \phi) \\ (x_{ei}, y_{ei}) & \text{if } \ y_i > (y_{ci} - \phi) \end{cases} \tag{6}$$

where ϕ is the designing parameter to define the timing for the robot to change the destination into the goal point from the corner point.

2.3 Collision Avoidance

A quasi Coulomb force generated among neighboring robots achieves the collision avoidance (see Fig.4).

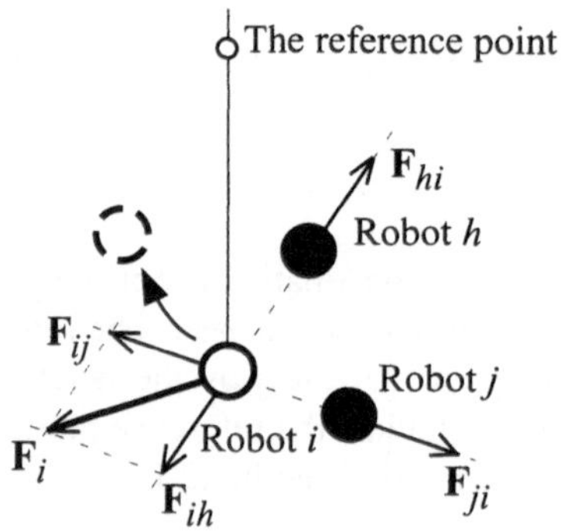

Fig. 4. Illustration of Coulomb force

To generate Coulomb force as repulsion force for the robot, a positive electric charge is given to each robot. Coulomb force, which the robot i receives from the neighboring robot j, is obtained according to Coulomb's law.

$$\mathbf{F}_{ij} = \eta \frac{Q_i Q_j (\mathbf{p}_i - \mathbf{p}_j)}{|\mathbf{p}_i - \mathbf{p}_j|^3} \tag{7}$$

where Q_i, Q_j are respectively a positive electric charge of robot i and j, η is the designing parameter to adjust value of $\mathbf{F}_{ij}$. Receiving interferences from many robots, the force which the robot i receives is given by the summation of $\mathbf{F}_{ij}$. $\mathbf{F}_i$ in eq.(8) is defined as the input disturbance of the servo system.

$$\mathbf{F}_i = \sum_j \mathbf{F}_{ij} \tag{8}$$

Approaching the wall too close, the robot also receives Coulomb force generated between the robot and the wall. Moreover, the number n of robots, which influence the robot i, is defined as proportionality constant of Coulomb force from the wall so that the effect of this Coulomb force is not lost even if many robots exist around the robot i. However, it introduces into a formula as $n + 1$, in consideration of the case of $n = 0$. For an example, the case in which the robot i approaches to the wall of the y-axis is shown in Fig.5. The robot i receives Coulomb force F_{xi} from a wall according to the condition in eq.(9).

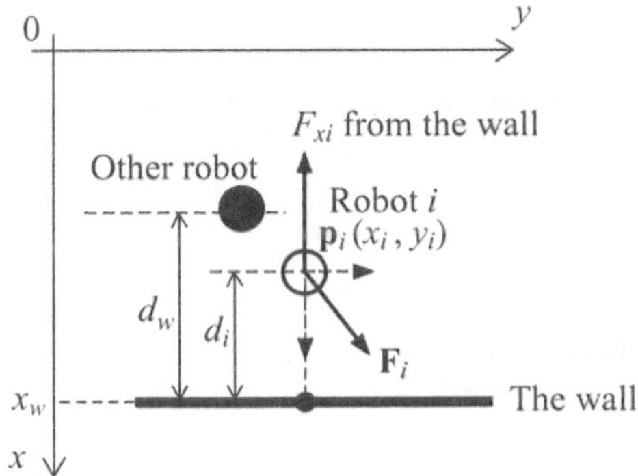

Fig. 5. Coulomb force from the wall

$$F_{xi} = \begin{cases} (n+1) \times \eta' \dfrac{Q_i{}^2 (x_i - x_w)}{|x_i - x_w|^3} & \text{if } d_i \leq d_w \\ 0 & \text{if } d_i > d_w \end{cases} \tag{9}$$

where d_i is the distance between the robot and the wall, d_w is the distance from the wall in which the robot receives the interference , x_i and x_w are respectively x coordinate of the robot i and the wall, η' is the designing parameter to adjust the magnitude of F_{xi}. The case that the robot i approaches to the wall of the x-axis is coped by the same method, too. F_{xi} is added to the x element in eq.(8).

In the environment with a corner, it is found that a human does not receive the repulsion force at a corner from the result of observation for some real data, since some human who passes very near the corner exist. Thus, the robot does not receive Coulomb force in the defined region R with the corner (see Fig.6).

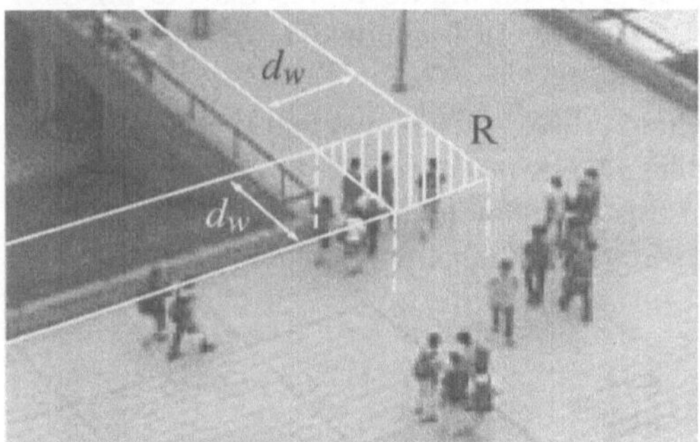

Fig. 6. The defined region R with the corner

3 Examples of Simulation

It is verified that effectiveness of the proposed model by two simulations.

- Simulation1

Behavior that fifteen robots simultaneously advance to the same exit in the straight passage is simulated.

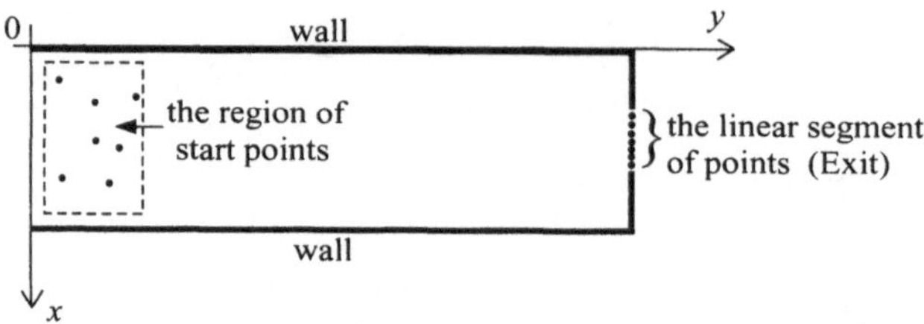

Fig. 7. Setting the passage in Simulation1

Fig.7 shows the passage used in Simulation1. The thick line in this figure represents a wall. The start point of each robot is given an arbitrary point in the start region shown with the dotted line. The destination of each robot is given an arbitrary point on the linear segment regarded as the exit.

To decide the referenece point $\mathbf{r}_i$ and to define the parameter $|\mathbf{l}_i|$, the movement direction of the robot have to be considered. In addition to the movement direction, $|\mathbf{l}_i|$ is defined by providing the velocity 1.1m/s of the robot, i.e., $|\mathbf{l}_i| = 1.65$. $\mathbf{r}_i$ is obtained by applying eq.(4) to the y-axis and eq.(5) to the x-axis, since the movement direction of robots is the positive direction of y-axis in this environment.

- Result of Simulation1

Fig.8 shows the Simulation1 result. Points in this figure represents the position which each robot passed, and is displayed each 10 seconds. The line connecting each point is the movement route of each robot.

Investigating the behavior of the whole crowd, it is shown that the distance between robots is narrow and robots is dense state at $t = 20$. At $t = 30$, the density of robots varies to a sparseness. From this phenomenon, it is found that multiple mobile robots advance to the exit while varying its density.

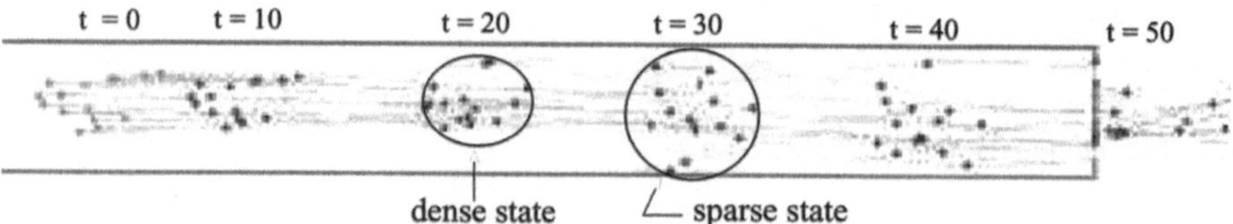

Fig. 8. Result of Simulation1

- Simulation2

Behaviors that four robots placed in every three start regions, the sum total of robots is twelve, advance to two adjacent exits is simulated. Behavior of the model in the passage with a corner is evaluated by comparing with walking behaviors of crowd humans.

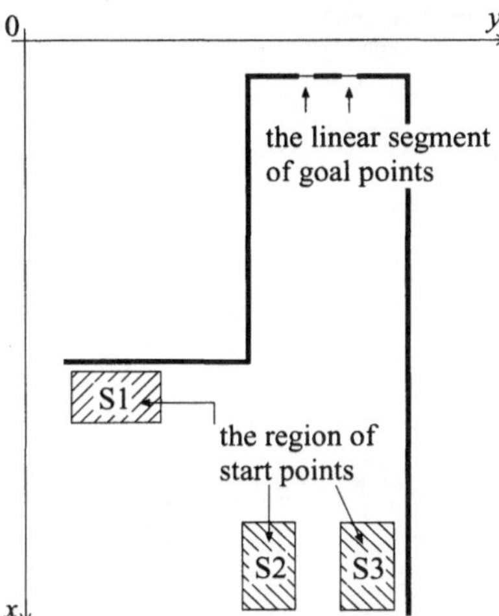

Fig. 9. Setting of the passage in Simulation2

Fig.9 shows the passage used in Simulation2. The thick line in this figure represents a wall. The start point of each robot is given an arbitrary point in the start regions called S1, S2, and S3. The destination of each robot is given an arbitrary point on the linear segment regarded as exit. In Simulation2, the reference point $\mathbf{r}_i$ and the value of $|\mathbf{l}_i|$ are set by using the same method as Simulation1 and they are explained in each start region. In the start region S1 until the robot turns at the corner, $\mathbf{r}_i$ is given by applying eq.(4) to the y-axis and applying eq.(5) to the x-axis, the advancing length $|\mathbf{l}_i| = 1.65$. After the robot passes the corner, $\mathbf{r}_i$ is given by applying eq.(4) to the x-axis and applying eq.(5) to the y-axis and $|\mathbf{l}_i| = -165$. In the start region S2 or S3, $\mathbf{r}_i$ is given by applying eq.(4) to the x-axis and applying eq.(5) to the y-axis and $|\mathbf{l}_i| = -165$.

- Result of Simulation2

Fig.10 shows the real data of crowd walking used for comparison corresponding to robots placed in S1. For comparing the movement route of the crowd robots with crowd humans' it, the region in which about 90 percent of humans pass is defined as the region of human walking route. Behaviors of remains 10 percent of humans, which do not walk in this region, are interpreted as random behaviors. The region of human walking route is set up by the same strategy also in S2 and S3.

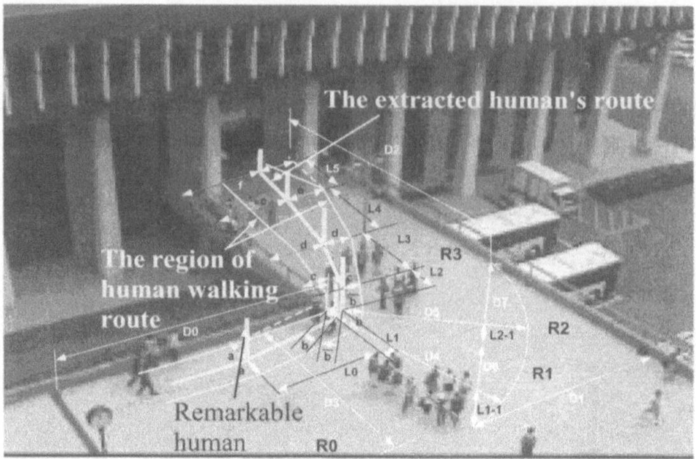

Fig. 10. The real data used for comparison

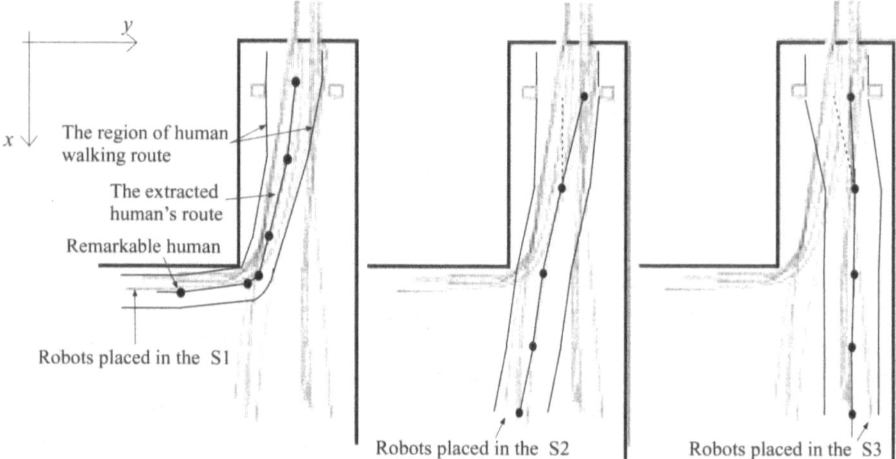

Fig. 11. Results of Simulation2

Results of Simulation2 with the real data corresponding each start region are shown in Fig.11. The left figure in Fig.11 is the result of robots placed in S1. Movement behaviors of humans extracted from the real data are classified into two patterns. The first behavior is that a human horizontally advances to the wall until passing a corner, and changes the direction of the progress at the corner. The other behavior is that a human advances to the corner point and passes near the corner. From the results, it is found that behaviors of the model like real human's ones is imitated. The center figure and the right figure in Fig.11 show each result of robots placed in S2, S3. When designing parameters **f** and k_1 are set without the consideration of movement direction of the robot, it makes the robot overshoot in the direction of y-axis and its route does not overlap the region of the human walking route. From this, designing parameters $\mathbf{f}$ and k_1 have to be set with considering the movement direction of the robot regardless of the initial position. Furthermore, several movement behaviors of robots given interference are shown. The first behavior is that a robot swerves off the perpendicular direction of trajectory. The second behavior is that a robot is pushed toward the movement direction. The third behavior is that a robot stops at its position. Giving each robot different characteristics of an electric charge, its mass and other designing parameters leads to this phenomenon. Moreover, all robots reached the destination by the sequential correction of trajectory even if a disturbance influences on the model movement.

4 Discussion and Conclusion

The proposed model contains the collision avoidance function and the decision method of trajectory. In this model, the collision free is guaranteed, because the infinite Coulomb force is generated between a robot and an obstacle when a robot infinitely approaches an obstacle. From this, Coulomb force may cause the divergence, which is defined as the case that a robot swerves from the passage or does not reach to the destination, of the whole system. Therefore, designing parameters (e.g. $\mathbf{f}, k_1, Q_i$) have to be decided according to the environment and the number of mobile robots so that the whole system does not diverge, and they are obtained through adhoc simulations. Though regrettable, mathematical representation of the stable state to the whole system is not made.

Advantages of the proposed model over a potential method [1] are deadlock free and a decrease of calculation load. The reason for deadlock free is that position of the model is controlled by the servo system, namely the model continues moving to the destination by a cumulative error as far as it does not diverge. Since calculation load to simulate this model is less than a potential method its, this model is more suitable for the case in which many robots exist around each robot than a potential method.

In this paper, guidelines for movement of multiple mobile robots in consideration of inertia and for the design method of the robot trajectory are obtained.

References

1. T. Arai, J. Ota. (1993) Motion Planning of Multiple Mobile Robots Using Virtual Impedance. Journal of the Robotics Society of Japan.Vol.11, No.7, pp.1039-1046
2. D. Kurabayashi, J. Ota, T. Arai, E. Yoshida. (1998) Motion Planning of Multiple Mobile Robots for Cooperative Sweeping. Journal of the Robotics Society of Japan.Vol.16, No.2, pp.181-188
3. C. R. Kube, H. Zhang. (1994) Stagnation Recovery Behaviours for Collective Robotics. IROS94. pp.1893-1890
4. H. Yamaguchi, G. Beni. (1996) Distributed Autonomous Formation Control of Mobile Robot Groups by Swarm-Based Pattern Generation. Distributed Autonomous Robotic Systems 2, pp.141-155

5. H. Yokoi, T. Mizuno, M. Takita, Y. Kakazu. (1996) Amoeba Like Grouping Behavior for Autonomous Robots Using Vibrating Potential Field (Obstacle Avoidance on Uneven Road). Distributed Autonomous Robotic Systems 2, pp.209-220
6. M. Hosoi, S. Ishijima, A. Kojima. (1996) On the Modeling and Simulation of the Walking Behavior. Memories of Tokyo Metropolitan Institute of Technology. No.10 pp.11-15

On-Line Collision Avoidance of Mobile Robot for Dynamically Moving Obstacles

Nobuhiro Ushimi, Motoji Yamamoto, Masanori Shimada, and Akira Mohri

Kyushu University, 6–10–1 Hakozaki, Higashi–ku, Fukuoka 812-8581, Japan
E-mail nobu@mech.kyushu-u.ac.jp

Abstract. This paper proposes a realistic on-line navigation method of mobile robot in dynamical environment where multiple obstacles (many people) are always changing their velocities and the robot does not know the velocities in advance. Considering characteristics of actual sensor system for mobile robot, a method to estimate the velocity of moving obstacles is presented. The estimated velocity and measured distance from the nearest obstacle are used to plan a velocity of mobile robot based on a new idea of Collision Possibility Cone. Then an on-line navigation method is proposed by using Collision Possibility Cone and feasible velocity space of mobile robot. By considering the dynamical constraints of mobile robot and moving obstacles, the mobile robot using the navigation method can move to it's destination without collision in the existence of multiple moving obstacles, even for the case that the moving obstacles change their velocities between sensor cycle of mobile robot. Simulational examples of on-line navigation show an effectiveness of the new navigation method. Finally a developed mobile robot to demonstrate the usefulness of the proposed method is shown.

Keywords: Mobile robot, On-line navigation, Dynamically moving obstacle, Collision Possibility Cone(CPC), Collision distance index

1 Introduction

Navigation of mobile robot to travel for given destination autonomously in various environment is considered as one of the most important capability, thus many researchers have studied about the navigation problem. Particularly, the study on on-line sensor-based motion planning or control is recently active, and some algorithms are proposed[1][2][3]. Most of these studies treat the case of multiple static obstacles, and mainly focus theoretical global convergence of the proposed algorithms. For more applications of mobile robots to real world, the robots are expected to work well in everyday space such that people are walking around. This means the robot is expected to surely reach desired point even in the dynamical environment where many people (obstacles) are moving. This problem is considered as an on-line sensor-based navigation problem among multiple moving obstacles. For a situation of such environment, a railway station's concourse is imagined. In the concourse, many people are walking to their destinations.

Considering such dynamical environment, some on-line motion planners are presented. Tsoularis[4] gives a path from start point to goal point neglecting moving obstacle's paths, then plans velocity pattern of the robot along the given path to avoid moving obstacle by changing the velocity. Another important approach for moving obstacles is the idea of time-state space which adds time axis to normal configuration space. By this method, the motion planning problem to avoid moving obstacles results in a "static" path planning problem in the augmented configuration space added one more degree of freedom[5].

These methods, however, take much time for planning, thus are not appropriate for on-line use. Fiorini et al. propose an on-line method based on the idea of Velocity Obstacle[6]. Moving obstacles are mapped into a two-dimensional "velocity space". Then velocity of mobile robot is directly planned using Velocity Obstacle in velocity space. Most studies of on-line motion planning for multiple moving obstacles including the study assume that the velocity and position of each moving obstacle can be measured using robot's sensor systems. In the situation that many people (obstacles) are crowdedly moving around the robot, it is, however, very difficult to distinguish each obstacle using sonar sensor or laser range finder, because the sensor systems have limited space resolution and the recognition of each object is essentially difficult. This means difficulties of accurate measuring of distance and velocity of each moving obstacles with reference to mobile robot. Furthermore, most on-line navigation methods assume that velocity of moving obstacle is constant during sensor cycle. It is not appropriate for the dynamical environment such as multiple moving obstacles where the obstacles may change their velocities at any time.

In this paper, an on-line navigation problem is discussed, where multiple obstacles are moving around and their paths, velocities and sizes are not given in advance. The paper also deals with the problem of velocity changes during the sensor cycle. For example, the proposed on-line navigation is applied to a situation that many people are walking as moving obstacles, like the railway station's concourse. The available information is assumed to be only distance information to obstacles at every sensor cycle considering the use of sonar sensor. Our navigation method basically selects the best velocity of mobile robot in the meaning of time-optimal considering the worst case in the view point of danger for collision with moving obstacles at every sensor cycle.

The paper first describes basic assumptions of mobile robot, moving obstacles and their workspace. Then an estimation method of approaching velocity of moving obstacle and an idea of Collision Possibility Cone are presented. Using the estimated velocity and the idea, an on-line navigation method is proposed. Some simulations are demonstrated to show effectiveness of the proposed method. Finally a developed mobile robot to demonstrate the usefulness of the proposed method is shown.

2 On-line collision avoidance for moving obstacles

Before design of a motion planner, some assumptions on mobile robot and moving obstacles are described here. Furthermore, some sufficient performance of mobile robot for collision avoidance are described. Then, an estimation method of relative approaching velocity of obstacle is presented. These assumptions and velocity estimation are needed for the later discussions on an on-line navigation problem.

2.1 Assumptions

- Mobile robot can produce every direction of velocity (omni-directional vehicle).

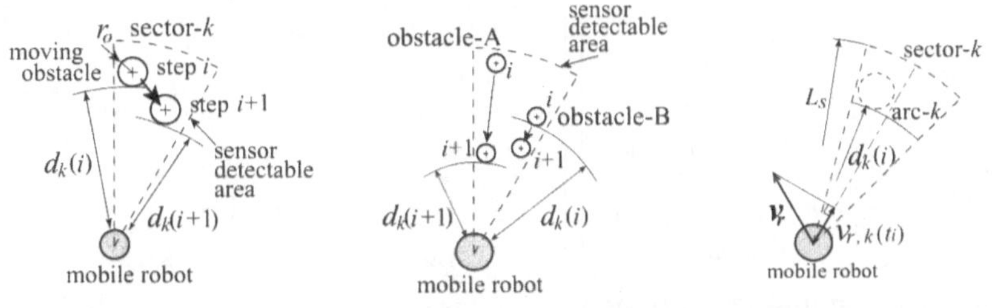

(a) Sensor Detectable area (sector-k)

(b) Multiple moving obstacles in the same area

(c) Velocity $v_{r,k}(ti)$ to the direction of sector-k

Fig. 1. Approaching velocity to obstacle

- The mobile robot and moving obstacles move on a flat floor.
- The robot does not communicate with obstacles.
- The robot and obstacles are assumed to be approximated by cylinders.
- The mobile robot has n_r ultrasonic sensors. Each sensor detects distance from the mobile robot to moving obstacle every sensor cycle T_s, if the obstacles are in the sensor detectable area which is approximated with a sector. The maximum range of the area is denoted by L_s. Note that the sensor can not distinguish individual moving obstacles, nor distinguish moving obstacles and static obstacles.
- Moving obstacles possibly change their velocities v_o and accelerations a_o at anytime within maximum values of $\pm v_{o\,\max}$ and $\pm a_{o\,\max}$.
- Maximum values $v_{o\,\max}$ and $a_{o\,\max}$ of moving obstacles are smaller than mobile robot's maximum velocity $v_{r\,\max}$ and acceleration $a_{r\,\max}$.

2.2 Dynamical constraints of mobile robot and moving obstacles

In previous assumptions, the moving obstacles are permitted to change their velocities during sensor cycle. To prevent the collision with the moving obstacles, the mobile robot should have sufficient performance ($v_{r\,\max}$, $a_{r\,\max}$) for collision avoidance. This subsection describes about necessary conditions based on dynamical property of mobile robot and moving obstacles.

- Maximum velocity

$$v_{r\ \max} > v_{o\ \max} \tag{1}$$

- Sensor cycle

$$T_s < \frac{L_s}{v_{r\ \max} + v_{o\ \max}} \tag{2}$$

- Sensor range

$$L_s > \frac{1}{2} a_{o\ \max} {T_s}^2 \tag{3}$$

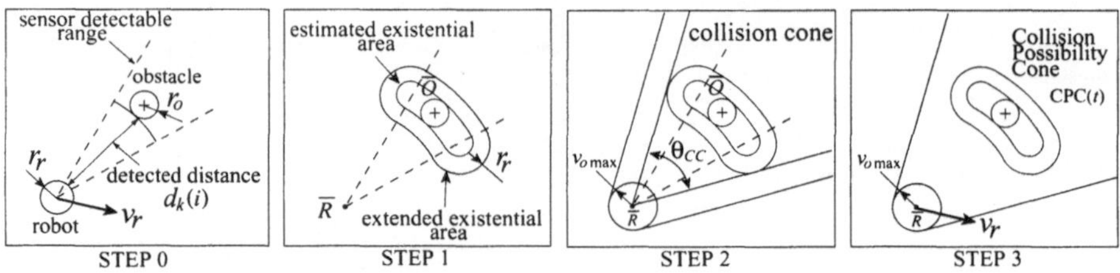

Fig. 2. Collision Possibility Cone (CPC)

2.3 Estimation of approaching velocity for obstacle

This paper discusses a realistic on-line navigation method based on the distance information by ultrasonic sensors, considering that they are most commonly used. If the robot can use the velocity information of moving obstacles in addition to the distance information, the robot is expected to avoid the obstacles more efficiently. Because the robot can estimate the trajectories (positions) of moving obstacles until the next sensor cycle by the velocity information. This subsection, thus describes about velocity estimation for moving obstacles using the distance information by ultrasonic sensor.

The ultrasonic sensor basically detects only distance $d_k(i)$ from mobile robot to a nearest moving obstacle inside the sensor detectable area (sector-k) every sensor cycle T_s(See Fig.1(a)). Where i shows a sensor information of time t_i. When the distance data by an ultrasonic sensor changes from $d_k(i)$ to $d_k(i+1)$ during sensor cycle T_s, the relative approaching velocity of moving obstacle in the direction of sector-k may be written

$$v_{ap,k}(t_{i+1}) = \frac{d_k(i+1) - d_k(i)}{T_s} \tag{4}$$

This is, however, incorrect for the case that multiple moving obstacles are in a same sensor area, because the sonar sensor can not distinguish the individual obstacles as shown in Fig.1(b). In the figure, two distances $d_k(i)$ and $d_k(i+1)$ are the result of two different obstacles. Note that actual approaching velocity for obstacle-A is larger than the estimation by (4).

To cope with the problem, the estimation of obstacle's approaching velocity is modified by

$$v_{ap,k}(t_i) = v_{r,k}(t_i) + v_{o\ \max} \ (\text{if} \ \ d_k(i) < L_s) \tag{5}$$

where $v_{r,k}(t_i)$ denotes projection of velocity $\boldsymbol{v}_r$ to the direction of sector-k as shown in Fig.1(c). This estimation basically assumes most dangerous case of velocity and moving direction of moving obstacles for the mobile robot.

3 Collision Possibility Cone

In this section, an idea of Collision Possibility Cone (CPC) is introduced to guarantee of collision free with moving obstacles using distance information and estimated approaching velocity from sonar sensor. The region of Collision Possibility Cone is used to decide a velocity of mobile robot every sensor cycle.

In this section, the mobile robot (r) and the moving obstacles (o) are described by circle (see STEP 0 in Fig.2). The mobile robot's radius and velocity are denoted with r_r, v_r. On the other hand, the radius of moving obstacles in the environment is denoted by r_o (if radius is unknown, it is defined with zero). The maximum velocity of the moving obstacles is denoted by $v_{o\,\max}$. The procedure to construct the Collision Possibility Cone is described as follows (see Fig.2).

STEP 0 Detect distance $d_k(i)$ from mobile robot to moving obstacle if a moving obstacle enters into sensor detectable area.
STEP 1 Calculate the possible existing area of the moving obstacle. Enlarge the area with r_r, then denote the extended area $\bar{O}$. The mobile robot is then represented by point robot $\bar{R}$.
STEP 2 Make two tangential lines to the extended area $\bar{O}$ from the point robot $\bar{R}$. The resultant cone region surrounded by the two lines is Collision Cone(CC). Where θ_{CC} is an extended angle from the original sensor detectable area.
STEP 3 Enlarge Collision Cone with $v_{o\,\max}$, then the resultant cone region is Collision Possibility Cone (CPC(t)).

Collision Possibility Cone indicates admissible collision free velocity of mobile robot. If the end point of the velocity vector of mobile robot is located in the region of Collision Possibility Cone, the mobile robot may collide with a moving obstacle in the future. Otherwise, if the end point of the velocity vector located out the region of Collision Possibility Cone, the mobile robot will never collide with moving obstacle. Note that collision free is guaranteed at least until next sensor step for the latter case.

In this section, the approaching velocity of moving obstacle is estimated by considering worst case, based on maximum velocities of moving obstacles. Thus, the velocity selection using Collision Possibility Cone presented here tends to an over conservative one. And the selection may leads to inefficient avoidance motion of robot. Especially for the case that the moving obstacles are far from mobile robot, it is not necessarily to use the conservative velocity of mobile robot. This problem is discussed again in the later section.

4 On-line motion planning problem

This section formulates an on-line motion planning problem where the mobile robot is navigated from a start point to a destination point in the existence of multiple moving obstacles. Then, a navigation strategy based on the idea of Collision Possibility Cone is proposed.

4.1 Formulation of on-line motion planning

In this study, the motion planning is not only path planning level but also velocity planning level for more realistic application and for more performance of resultant motion of mobile robot. To deal with velocity level planning problem, a dynamics of mobile robot should be considered. The dynamics of the mobile robot is generally described by the following equation

$$\ddot{\boldsymbol{x}}_r(t) = \boldsymbol{f}(\boldsymbol{x}_r, \boldsymbol{v}_r, \boldsymbol{u}) \quad (6)$$
$$\boldsymbol{u} \in U, \quad |\boldsymbol{v}_r| < v_{r\,\max} \quad (7)$$

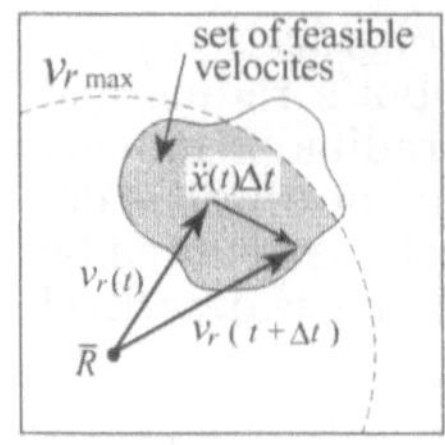

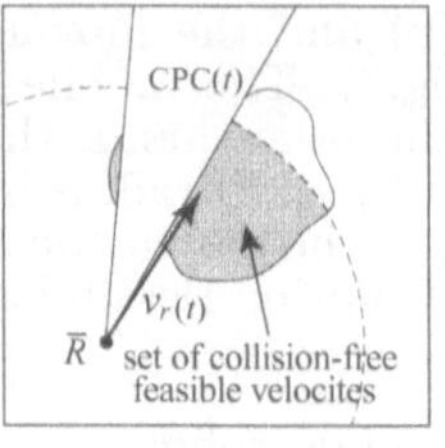

Fig. 3. Set of feasible velocities and collision-free feasible velocities

where $\boldsymbol{x}_r$ is position vector of robot, $\boldsymbol{u}$ is actuator input vector, U is admissible input set, and $v_{r\,\max}$ is maximum velocity of mobile robot. The on-line motion planning problem is to generate velocity for minimizing traveling time from an initial point to goal point avoiding moving obstacles. Where the moving obstacles change their velocities (direction and magnitude). The dynamics of moving obstacle is assumed to be unknown considering general application. However, the velocity and acceleration of moving obstacles are assumed to be limited as described in section 2.1.

4.2 On-line velocity planning

When the mobile robot moves with velocity $\boldsymbol{v}_r(t)$, the feasible velocity at $t + \Delta t$ is described with

$$\begin{aligned} \boldsymbol{v}_r(t+\Delta t) &= \boldsymbol{v}_r(t) + \ddot{\boldsymbol{x}}_r(t)\Delta t \\ &= \boldsymbol{v}_r(t) + \boldsymbol{f}(\boldsymbol{x}_r, \boldsymbol{v}_r, \boldsymbol{u})\Delta t \end{aligned} \tag{8}$$

by dynamics of mobile robot and input constraints of mobile robot (see Fig.3(left)). Where $\boldsymbol{u}$ and $\boldsymbol{v}_r(t)$ are constrained by (7). By removing the region of Collision Possibility Cone CPC(t) from the feasible velocity set, collision-free feasible velocity set is obtained as shown in Fig.3(right).

To select a velocity of mobile robot in the collision-free feasible velocity set, the following cost function J is defined by considering the travel time of mobile robot.

$$J(\boldsymbol{v}_r) = \left[\boldsymbol{v}_r(t) - \boldsymbol{v}_d(t)\right]^T \left[\boldsymbol{v}_r(t) - \boldsymbol{v}_d(t)\right] \tag{9}$$

where $\boldsymbol{v}_d(t)$ is maximum velocity of mobile robot in amplitude, and its direction is to goal position as

$$\boldsymbol{v}_d(t) = \left(\frac{x_f - x(t)}{A} v_{r\,\max}, \; \frac{y_f - y(t)}{A} v_{r\,\max} \right) \tag{10}$$

$$A = \sqrt{(x_f - x(t))^2 + (y_f - y(t))^2}$$

where $(x(t),\ y(t))$ is mobile robot's current position and $(x_f,\ y_f)$ is goal position. the mobile robot's motion is planned by selecting velocity which minimizes the cost (9) in the collision-free feasible velocity set.

To reduce calculation cost, a descritized velocity space of the collision free feasible velocity set is used. In the simulations of this paper, the velocity is selected in the grid points of the velocity space.

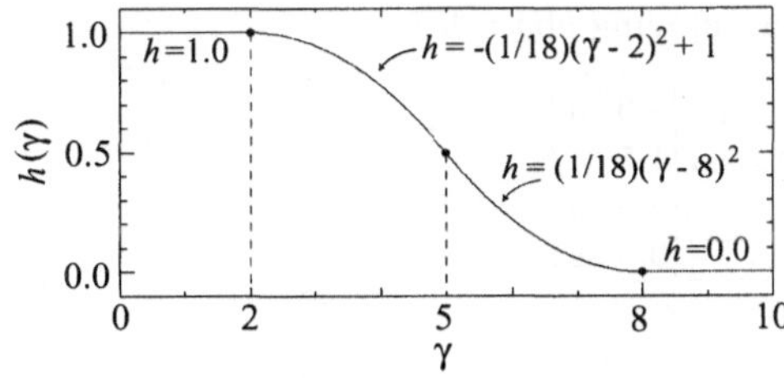

Fig. 4. Weighting function $h(\gamma)$

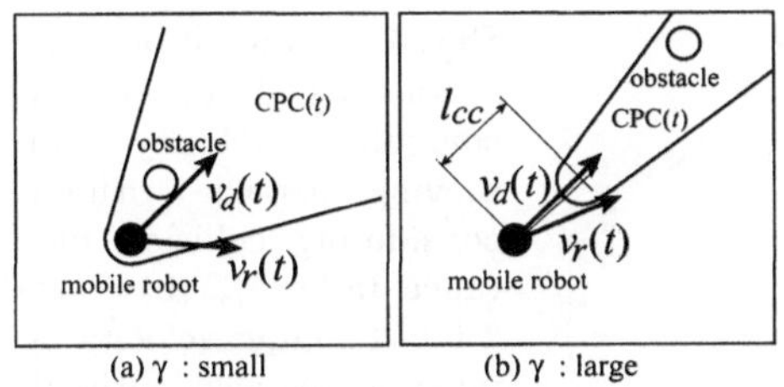

Fig. 5. Advancement of avoidance efficiency by the collision distance index γ

4.3 Collision distance index

This subsection discusses the problem of over conservative property of the proposed on-line navigation method. As described in section 3, there is no danger of collision with obstacle, if the mobile robot current position is far from moving obstacle, even if the selected velocity by the proposed method is in Collision Possibility Cone. To prevent the conservative property, the following "collision distance index" is introduced. It is defined by the detected distance $d_k(i)$ from robot to obstacle's wall and the relative approaching velocity of obstacle $v_{ap,k}(t_i)$ which is calculated by (5). The index is used for changing Collision Possibility Cone. The collision distance index is defined by

$$\gamma = \frac{d_k(i)}{v_{ap,k}(t_i)\Delta t} \tag{11}$$

Then using the index, the extended angle θ_{CC} of Collision Cone which means danger of collision in the meaning of velocity is modified by

$$\tilde{\theta}_{CC} = h(\gamma)\theta_{CC} \tag{12}$$

And Collision Cone transfers a distance l_{CC} to the direction of sensor detectable area(see Fig.5(b)). l_{CC} is calculated by

$$l_{CC} = l_k(1 - h(\gamma)) \tag{13}$$

where l_k is a positive variable, and the function $h(\gamma)$ is selected such that the value of the function is 1 when γ is small ($\gamma \leq 2$), and becomes asymptotically 0 in accordance with increasing γ as shown in Fig.4. Collision Possibility Cone is constructed after the modification of angle θ_{CC} and the transfer of distance l_{CC}. The index γ means minimum step of sensor cycle from the mobile robot's current position to collision point with moving obstacle, assuming the fixed velocities of robot and obstacle over the sensor cycle. In the function shown in Fig.4 as an example of $h(\gamma)$, the extended angle θ_{CC} and the transfer

Table 1. Parameters on mobile robot and moving obstacles

maximum velocity of mobile robot $v_{r\,\max}$	2.0 m/s
maximum velocity of moving obstacles $v_{o\,\max}$	1.5 m/s
sensor detectable range L_s	6.0 m
sensor cycle T_s	0.3 sec

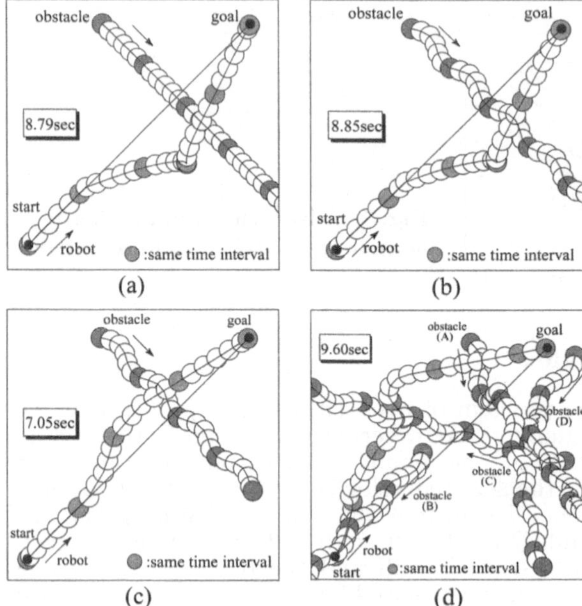

Fig. 6. Simulation results, (a) Path 1 for constant velocity of moving obstacle without considering collision distance index γ, (b) Path 2 for changing velocity of moving obstacle without considering collision distance index γ, (c) Path 3 for changing velocity of moving obstacle considering collision distance index γ, and (d) Path 4 for the effectiveness of proposed navigation method in the dynamical environment with four moving obstacles

distance l_{CC} are unchanged for $\gamma \leq 2$, and velocity of moving obstacle is neglected for $\gamma \geq 8$ where the obstacle is far from the mobile robot.

An effectiveness by using the collision distance index γ is shown in Fig.5. When the mobile robot is near from moving obstacle(γ is small), Collision Possibility Cone(CPC(t)) selects the safe velocity of mobile robot for the purpose of certainly avoiding the moving obstacles considering the worst case in view of danger for collision (Fig.5(a)). However, if the mobile robot current position is far from moving obstacle(γ is large), Collision Possibility Cone(CPC(t)) selects the avoidance efficient velocity of mobile robot by considering the collision distance index γ (Fig.5(b)).

5 Simulation of on-line navigation

This section shows simulational examples to confirm the effectiveness of the proposed on-line navigation method using the idea of Collision Possibility Cone and collision distance index. The constraints on mobile robot and moving obstacles are chosen by considering actual hardware performance as shown in Table 1. In the simulation, the size of test field is 12 m $\times$ 12 m, and the radius of mobile robot and obstacles are $r_r = 0.4$ m and $r_o = 0.4$ m.

The mobile robot has 12 ultrasonic sensors with ring shape. In the following example, the robot does not know the velocities and paths of obstacles in advance, but only knows the distance from the nearest obstacle by its sensor system every sensor cycle within sensor range.

Simulation results with one moving obstacle are shown in Fig.6(a)~(c). Circles in the figures represent positions of robot and obstacle at same time interval. Each traveling time to goal point is shown in shaded box. Path 1 (Fig.6(a)) shows the resultant path in the case of a fixed obstacle's velocity $v_o = 1.3$ m/s. The obstacle moves along a straight line on the path. The mobile robot detects the moving obstacle for the left side of moving direction, and tries to avoid from considerably large distance to the moving obstacle by taking right course. The resultant path (path 1) looks an inefficient one. The robot in this simulation does not consider the collision distance index. In path 2 (Fig.6(b)), the velocity of moving obstacle is always changing, and the path of obstacle is a meandering one, but robots does not know the path in advance. By estimating the maximum velocity of obstacle in the idea of Collision Possibility Cone, the mobile robot could arrive at goal point without collision, even for the case that the velocity of obstacle changes between sensor cycle. In path 3 (Fig.6(c)), the collision distance index γ is considered, whereas it is not considered in the simulation of path 2. This example shows that the consideration of the collision distance index γ results in more efficient path.

To confirm the effectiveness of the proposed navigation method for more realistic and complicated case such as many people walking around the robot, an example of four moving obstacles is simulated in Fig.6(d). All moving obstacles always change their velocities. Also for this example, the mobile robot successfully reaches its destination without collision.

6 Development of Experimental Mobile Robot

To confirm an effectiveness of simulation results, a mobile robot is developed. The developed robot is shown in Fig.7. The developed robot is an omni-directional vehicle with three omni-wheels, and can move any direction.

The developed robot has an ultrasonic sensor system on top of the mobile robot to detect the moving obstacles. This ultrasonic sensor is a fast measurable system, and the sensor system can detect the distance information of omni-direction at 50 msec. And sensor detectable range L_s is 6.0 m. In the future, an effectiveness of proposed navigation method is verified by using developed mobile robot, and given by a presentation.

7 Conclusions

In this paper, an on-line navigation problem for dynamical environment with multiple moving obstacles is discussed. The proposed navigation method is basically constructed using distance information from obstacles by realistic sensor system. The main points and results of the study are summarized as follows.

(1) By considering the characteristics of actual ultrasonic sensors, an estimation method of velocity for moving obstacles is presented.

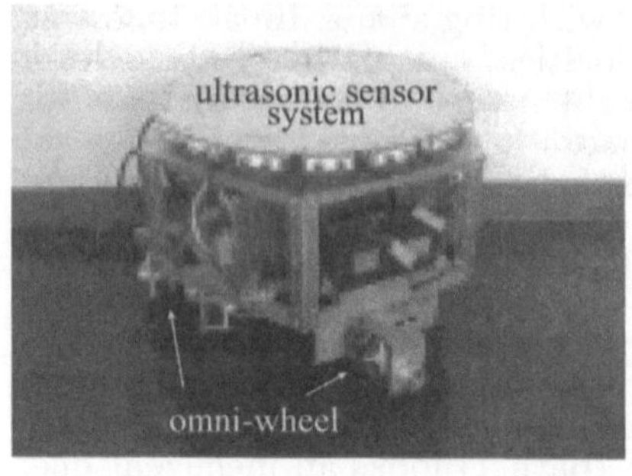

Fig. 7. Developed mobile robot with fast measurable ultrasonic sensor system

(2) For the on-line collision avoidance of moving obstacles, the idea of Collision Possibility Cone is presented. The idea makes most use of estimated velocity and distance information of moving obstacles.
(3) Using the idea of Collision Possibility Cone, new on-line motion planner is proposed. The method is based on a selection of admissible velocities in the restricted two dimensional velocity space. This enables on-line navigation of mobile robot in the dynamical environment.
(4) For an over conservative property of the proposed navigation method, the collision distance index is introduced. Generally, the use of the index leads to more efficient path of mobile robot.
(5) By showing simulational examples of on-line navigation problem for one and four moving obstacles, an effectiveness of the proposed method is presented.
(6) To confirm an effectiveness of simulation results, a mobile robot is developed. The developed robot has a fast measurable ultrasonic sensor on top to detect the moving obstacles.

References

1. V. Lumelsky and A. Stepanov. (1986) Dynamic path planning for a mobile automaton with limited information on the environment. IEEE Transactions on Automatic Control, Vol.AC-31, No.11, 1058–1063.
2. V. Lumelsky and T. Skewis. (1990) Incorporating Range Sensing in the Robot Navigation Function. IEEE Transactions on Systems, Man, and Cybernetics, Vol.20, No.5, 1058–1069.
3. I. Kamon, E. Rivlin and E. Rimon. (1996) A New Range-Sensor Based Globally Convergent Navigation Algorithm for Mobile Robots. Proceedings of the IEEE International Conference on Robotics and Automation, 429–435.
4. A. Tsoularis and C. Kambhampati. (1999) Avoiding Moving Obstacle by Deviation from a Mobile Robot's Nominal Path. The International Journal of Robotics Research, Vol.18, No.5, 454–465.
5. T. Tsubouchi, T. Naniwa and S. Arimoto. (1996) Planning and Navigation by a mobile Robot in the Presence of Multiple Moving Obstacles and Their Velocities. Journal of Robotics and Mechatronics, Vol.8, No.1, 58–66.
6. P. Fiorini and Z.Shiller. (1998) Motion Planning in Dynamic Environments Using Velocity Obstacles. The International Journal of Robotics Research, Vol.17, No.7, 760–772.

Distributed Multi Robot Reactive Navigation

Alessandro Scalzo, Antonio Sgorbissa, Renato Zaccaria

DIST – Università di Genova, Via all'Opera Pia 13, 16145 Genova, Italy
scalzo@dist.unige.it, sgorbiss@dist.unige.it, zaccaria@dist.unige.it

Abstract. In many applications, Multi-Robot Systems are able to perform tasks which cannot be accomplished by a single robot. However, when neither a central planner is available nor explicit communication is allowed, each agent must rely on some sort of local navigation strategy for computing its trajectory, thus increasing the probability of getting trapped into deadlock situations. The navigation problem for multiple agents become even harder whenever they have an arbitrary shape, since problems arise related to the encumbrance and the orientation of each agent. In this paper we extend to MRS the Micronavigation algorithm, a reactive approach to navigation which has been successfully employed for the navigation of a single agent in a two-dimensional Work Space. The result is a fully distributed algorithm which can be employed for the navigation of an arbitrary number of autonomous agents with arbitrary shapes.

Keywords: Autonomous Navigation, Potential Fields, Reactive Planning.

1 Introduction

In many industrial, service [1] and military [2] robotics applications, Multi Robot Systems allow to decompose complex tasks into subtasks: multiple robots, when coordinating themselves either through explicit or implicit communication [3], are able to perform more quickly and robustly [4], or even to carry out tasks which cannot be accomplished by a single robot [5]. However, they often require complex coordination strategies: if such strategies are missing, the complexity of the problem is very likely to increase instead of decreasing. Consider for example the navigation problem for a group of robots that do not receive navigation hints by a central planner, nor communicate with each other. In such a case, each robot has only an approximate knowledge of the environment: even if it knows the location of walls and furniture, it does not know the location nor the intentions of other robots. As a consequence, each robot must rely on some sort of local navigation strategy for computing its path to the goal, thus increasing the probability of getting trapped into deadlocks. Such kind of deadlocks have been deeply studied in literature: consider for example the case in which two robots meet in the middle of a very *narrow* corridor while heading towards opposite directions (i.e., it is not possible for the two robots to move side by side). Notice that, if we are adopting Artificial Potentials [6] or similar Force Field models [7], this situation corresponds to a local minimum. If a centralized planner is present, such problem can be solved: in [8] the trajectory of every robot is computed by finding a joint trajectory on the corresponding 6-dimensional Configuration Space (notice that planning in the C-Space becomes intractable as the number of dimensions increases). If no central planner is present but some sort of explicit communication is allowed, strategies for negotiating a path can be employed. However, if neither of the two solutions are feasible, the deadlock is very likely to last indefinitely. The navigation problem become even harder when agents have an arbitrary shape: for example, a robot which looks like an elongated object cannot be considered as a material point when finding its path through a *narrow* passage. In [8] the agent's trajectory is computed in the 3-dimensional configuration space by augmenting the obstacles' dimensions, in order to take into account the agent' shape and encumbrance. However, since this requires heavy computations, the approach is unfeasible if we want the agent to generate its trajectory in real time, on the basis of the obstacles it perceives while moving in the world. In this paper, we present a novel theoretical approach to this problem, in order to solve the multi-robot navigation problem defined as follows:

- Agents (holonomic robots with arbitrary shapes) have different goals.
- Agents have local perceptions and do not build global representations.
- Agents do not distinguish between fixed obstacles and other agents.
- Agents do not explicitly communicate.
- Agents use the same algorithm to achieve their own goals.

The approach is original in the sense that it is able to solve very complex navigation problems for a set of agents with arbitrary shapes with only local information: i.e., the information which can be easily retrieved by means of simple proximity sensing devices such as ultrasonic sensors or laser range finders. Section II describes the Micronavigation algorithm for a single robot; Section III describes a potential function which consider the presence of many agents with arbitrary shapes, and reactively computes local translations and rotations to avoid collisions. In Section IV experimental results are shown.

2 The navigation algorithm

Artificial Potential Fields [6] and similar models [7] often represent a good solution for dealing with totally or partially unknown dynamic environments in which decisions about the robot trajectory must be taken in real-time. We briefly remind the basic principle, by considering the agent A_1 as a material point moving on a two-dimensional Work Space. Some definitions are required:

- A_0 corresponds to the union shape of the fixed obstacles in the Work Space
- the Work Space is described by a fixed Cartesian coordinate system (O,X,Y); the motion of A_1 is described by the Lagrangian coordinates X_I,Y_I

- the couple X_{IS},Y_{IS} defines the start position of A_1 and the couple X_{IT},Y_{IT} defines its target position. X_{IS},Y_{IS} and X_{IT},Y_{IT} vary for every specific instance of the navigation problem.

In each configuration X_I,Y_I the model assumes that the target exerts an attractive force on A_1 and the obstacles (i.e. their union shape A_0) exert a repulsive force on A_1. Thus, the law of motion of A_0 can be seen as a descent of the gradient of the potential field U, which is given by the sum of a repulsive field U_{rep} generated by obstacles and an attractive field generated by the target U_{att}. In particular,

$$U_{rep}(X_I,Y_I)=\begin{cases}\frac{1}{2}\left(\frac{1}{\rho(X_I,Y_I)}-\frac{1}{\rho_0}\right)^2 & \rho<\rho_0\\ 0 & \rho\geq\rho_0\end{cases} \tag{1}$$

where $\rho(X_I,Y_I)$ is the minimum euclidean distance between A_1 and A_0. U_{att} is usually given a conic or paraboloid shape with its minimum in the target. Finally, U can be derived in order to find the overall force $\mathbf{F}=\mathbf{F}_{att}+\mathbf{F}_{rep}$ (a two-dimensional vector) which is exerted on the robot; such force is taken, in every moment, as the most promising direction for the robot to move to; thus, the model relies only on local information in order to compute the law of motion of the robot, since only obstacles closer than ρ_0 are considered. Notice that it is not necessary to know the shapes of obstacles in order to compute **F**, since only the minimum Euclidean distance between A_1 and A_0 is considered; in a realistic application, such distance can be easily computed by means of proximity sensors, such as ultrasonic sensors or laser rangefinders. However, it has been widely demonstrated that APF models suffer of unavoidable drawbacks [9]. In particular, since the law of motion of the robot is basically determined by descending the gradient of the potential field generated by the goal and the obstacles, it is very likely for the robot to get trapped into a local minimum. The *Micronavigation* algorithm [10] faces this problem by computing a different navigation function on the basis of the local values of the APF. That is, instead of performing a descent of the gradient, the law of motion of the robot is determined by the following function,

$$\mathbf{V}(t)=\left[W_g\cdot\mathbf{g}(X_I,Y_I)+W_t\cdot\mathbf{t}(X_I,Y_I)+W_f\cdot\mathbf{f}(X_I,Y_I)\right]\cdot V_{ref} \tag{2}$$

where

- **g** is a unitary vector along the direction $(X_{IT},Y_{IT})-(X_I,Y_I)$
- **f** is a unitary vector directed as $\mathbf{F}_{rep}$.
- **t** is a unitary vector tangential to the equipotential line passing through X_I,Y_I
- V_{ref} is a reference value for the speed.

The law of motion of the robot (i.e. the speed vector $\mathbf{V}(t)$) is thus given by the sum of three vectors **g**, **f**, and **t**, weighted on the basis of the current value $U_{rep}(X_I,Y_I)$ of the repulsive potential field (the attractive field U_{att} is no more considered). In particular, the weights W_g, W_t and W_f are set as stated by the following formulas,

$$W_g=\begin{cases}1-\frac{U_{rep}(X_I,Y_I)}{U_L} & U_{rep}(X_I,Y_I)\leq U_L\\ 0 & U_{rep}(X_I,Y_I)>U_L\end{cases} \tag{3}$$

$$W_t = \begin{cases} \frac{U_{rep}(X_I, Y_I)}{U_L} & U_{rep}(X_I, Y_I) \leq U_L \\ 1 & U_{rep}(X_I, Y_I) > U_L \end{cases} \tag{4}$$

$$W_f = \frac{1}{1 + e^{\left[U_L - U_{rep}(X_I, Y_I)\right] / \left[10^{\log_{10}(U_L) - k}\right]}} \tag{5}$$

where U_L identifies the equipotential line which the robot has to follow when avoiding obstacles. The basic agent's behaviour can be briefly described as follows: 1) when the agent is navigating far from obstacles, it heads toward the goal 2) when the agent approaches an obstacle, it follows its contour (on an equipotential line) until it can safely leave the obstacle and head towards the goal (fig. 1a). Notice that the two weights W_g e W_t are the most important in determining the law of motion of the robot, since W_f becomes relevant only when the robot is very close to obstacles. When the robot is far from obstacles ($U_{rep}(X_I,Y_I) \cong 0$) it heads towards the goal ($W_g \cong 1$, $W_t \cong 0$, $W_f \cong 0$). When approaching an obstacle ($U_{rep}(X_I,Y_I) > 0$), the robot starts following its contour ($W_g > 0$, $W_t > 0$, $W_f \cong 0$). Finally ($U_{rep}(X_I,Y_I) \cong U_L$), its trajectory approximately lies on the equipotential line defined by U_L, since $W_t \cong 1$, and $W_g \cong W_f$, but **g** and **f** have almost opposite directions, thus compensating with each other. In fig. 1 when the scalar product **gf** increases and finally reaches a positive value, the robot leaves the equipotential line.

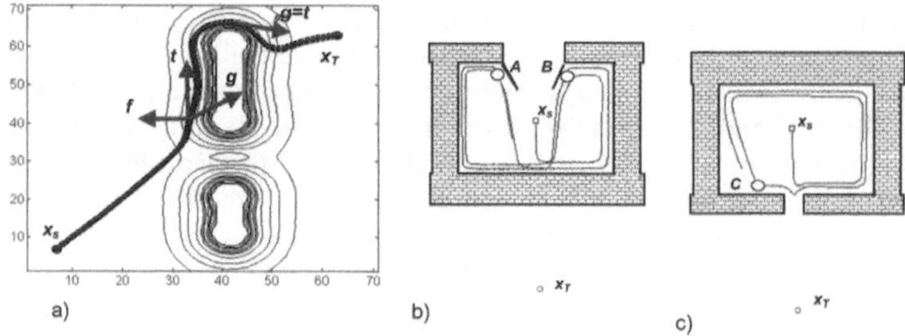

Fig. 1. a) The trajectory of the robot in a simulated environment; b) the robot cannot exit the room since it stops following the wall in points A and B c) the robot cannot exit the room because of a saddle in the potential field corresponding to the door;

In the real implementation, the rules for leaving the contour of the obstacle are more complex: in particular, the robot is forced to follow the obstacle until **g**≅**t** (we compute the APF in order to guarantee that equipotential lines are C^1 closed lines: this guarantees that the condition **g**≅**t** is satisfied at some point). Finally, notice that we have to choose between two opposite vectors **t** in X_I, Y_I. We consider the one that maximise the product **tV**(t-1): since **V**(t-1) corresponds to the current speed of the robot, this avoid oscillations and rough turns in the agent trajectory.

The law of motion described in equations 2 to 5 prevents the robot from being trapped into local minima (equipotential lines are closed lines with no drains or sources); however, other deadlocks are possible. The first deadlock situation is depicted in fig. 1b. Since in position A and B the robot is free to head towards the goal (**g**≡**t**), it never exits the room. The second deadlock situation is depicted in fig. 1c. Depending on the value of U_L, the presence of saddles in the equipotential field can prevent the robot from exploring some areas of the environment. In fact, U_L is normally assigned a low value to keep the robot far from obstacles for safety reasons; however, in some cases, a path to the goal can be only found by following an equipotential line with a higher value (i.e. passing through a narrow door). In both cases, memorising landmark positions in order to detect possible deadlocks could, at least in line of principle, solve the problem. In the fig. 1b, if the robot recognises a loop it can decide not to leave the contour of the obstacle when in point A or B; in fig. 1b, the value of U_L can be raised to explore unvisited areas. However, in a realistic scenario, it is very difficult for the robot to recognise a previously visited location without a very precise localisation system. Thus, in order to make the system more robust despite big errors in localisation, *Micronavigation* switches between two different behaviours (*following the contour of an obstacle/heading toward the goal*) on the basis of the evolution of some internal status variables and a very rough estimate of the direction of the goal, without the need for position landmarks. Describing these rules is beyond the scope of this paper: for further details refer to [10]. The algorithm has been extensively tested on simulated and real robots; even if its ability to lead the robot to its final destination has not been formally demonstrated, the behaviour of the system is very intuitive in all simple cases and it proves to be very effective even in more complex ones.

3 The potential function

When we extend the basic Micronavigation principles for the navigation of a set of agents which cannot be considered as material points, difficulties arise related to the encumbrance of each agent and the need for considering both translations and rotations. In particular, we are looking for a new definition of the distance ρ (which is the basis for computing the potential field U_{rep}) for a system which is composed of an arbitrary number of static and moving objects with an arbitrary shape (i.e. obstacles and autonomous agents). In order to achieve this, we perform computations in the C-Space; however, the final results will show that each agent requires only local information to plan its movement, thus allowing a fully distributed implementation of the algorithm. Finally, it will be shown that the overall force exerted on each agent results in having both translational and rotational components, thus allowing to deal with encumbrance and shape related problems.

Consider a set of N rigid bodies A_k $(k = 1..N)$ moving in a two-dimensional plane described by a fixed Cartesian coordinate system (O,X,Y). If we introduce the rigid body rest frame (O_k,x_k,y_k), the motion of A_k can be described by three Lagrangian coordinates X_k, Y_k, ϑ_k, where X_k, Y_k denote the components (in the fixed frame) of the vector position C_k of the origin and ϑ_k is the angle between the axes x_k and X. We describe the boundary of A_k in the rest frame by means of its parametric representation $\mathbf{x}_k=\mathbf{r}_k(\xi_k)$, where ξ_k varies in a suitable real interval. Thus the equation of the boundary of A_k with respect to the fixed frame can be described as

$$\mathbf{X}_k(X_k, Y_k, \vartheta_k; \xi_k) = \mathbf{C}_k + \mathbf{r}_k(\vartheta_k; \xi_k) \tag{6}$$

The distance vector $\mathbf{d}_{ij}$ between an arbitrary point of the boundary of A_i and an arbitrary point of the boundary of A_j is thus given by the relation $\mathbf{d}_{ij} = \mathrm{X}_i(\xi_i) - \mathrm{X}_j(\xi_j)$ which takes the explicit form

$$\mathbf{d}_{ij}(X_i, Y_i, \vartheta_i, X_j, Y_j, \vartheta_j; \xi_i, \xi_j) = (\mathbf{C}_i + \mathbf{r}_i(\vartheta_i, \xi_i)) - (\mathbf{C}_j + \mathbf{r}_j(\vartheta_j, \xi_j)) \tag{7}$$

We now introduce the $3N$ Lagrangian coordinates component algebraic vector $\mathbf{q}$:

$$\mathbf{q} = \left(q^1, \ldots, q^{3N}\right)^T = (X_1, Y_1, \vartheta_1, \ldots, X_N, Y_N, \vartheta_N)^T \tag{8}$$

that is able to represent the entire system configuration in the C-Space. Given a configuration vector $\mathbf{q}$, we can define ρ as the distance between $\mathbf{q}$ and the closest collision configuration CO in the C-Space (by considering both the collisions between agents and obstacles and the collisions among agents).

$$CO = \{\mathbf{p} \in \text{C-Space} : \exists i, j \in 0..N, i < j : A_i(\mathbf{p}) \cap A_i(\mathbf{p}) \neq 0\} \tag{9}$$

where A_0 corresponds to the union shape of the fixed obstacles in the Work Space and $A_1 \ldots A_N$ correspond to the N agents. Notice that CO is the union of all the sets of configurations CO_{ij} in which only two objects collide (i.e. a couple of agents or a couple agent-obstacle).

$$CO_{ij} = \{\mathbf{p} \in \text{C-Space} : A_i(\mathbf{p}) \cap A_i(\mathbf{p}) \neq 0\} \quad i, j \in 0..N, i < j \tag{10}$$

$$CO = \bigcup_{i,j \in 0..N, i<j} CO_{ij} \tag{11}$$

In fact, the configurations in which there is a contemporary collision among three or more objects are implicitly considered, since they are a subset of the configurations in which only two agents collide. Thus, we could define ρ as

$$\rho = \min_{ij}(\rho_{ij}) \quad \text{where} \quad \rho_{ij} = \left\| \mathbf{d}_{ij}(\hat{\xi}_i, \hat{\xi}_j) \right\| \quad \text{and} \quad \hat{\xi}_i, \hat{\xi}_j = \arg\min_{\xi_i, \xi_j} \left\| \mathbf{d}_{ij} \right\| \tag{12}$$

and consequently compute U_{rep} and the repulsive force $\mathbf{F}_{rep}$ (which is now a $3N$ dimensional vector in the C-space). The indexes i and j indicate the couple of objects (agent or obstacle) having the most critical configuration in the sense of the *navigation problem.* Notice that, although the calculus of the repulsive force exerted on $\mathbf{q}$ by the closest collision configuration is in most cases unmanageable in the C-Space, it results very simple in the Workspace, as it is shown in equation 12: in fact, it is possible to consider only the couple of objects in the most critical configuration to avoid any collision. However, to obtain a safer and smoother behaviour, we want to express ρ in a different way, in order to consider the most critical configuration *for each set CO_{ij}*. That is, for each agent, we want to take into account the minimum distance from any other object within a given distance. Finally, we wish the definition of ρ to respect the superimposition principle for the potential field U_{rep} and, consequently, for the overall repulsive force $\mathbf{F}_{\text{rep}}$.

$$U_{rep}(\rho) = \sum_{ij} U_{rep}(\rho_{ij}) \quad i, j \in 0..N, i < j \tag{13}$$

By defining the potential function as in equation 1, we obtain (when $\rho < \rho_0$)

$$\frac{1}{2}\left(\frac{1}{\rho}-\frac{1}{\rho_0}\right)^2=\sum_{ij}\frac{1}{2}\left(\frac{1}{\rho_{ij}}-\frac{1}{\rho_0}\right)\Rightarrow\rho=\left(1/\rho_0+\sqrt{\sum_{ij}\left(1/\rho_{ij}-1/\rho_0\right)^2}\right)^{-1}\quad \begin{matrix} i,j\in 0..N \\ i<j \end{matrix} \tag{14}$$

It is possible to demonstrate that $\rho \leq \min_{ij}(\rho_{ij})$; moreover, as a consequence of the superimposition principle, there will be components in $\mathbf{F}_{\mathrm{rep}}$ for all the couples of objects in mutual dangerous configuration. The overall force $\mathbf{F}_{\mathrm{rep}}$ exerted on the configuration $\mathbf{q}$ is computed as the anti-gradient of U_{rep}:

$$\mathbf{F}_{\mathrm{rep}}=\sum_{ij}\mathbf{F}_{ij}=-\sum_{ij}\left(\frac{\partial U\left(\rho_{ij}(\mathbf{q})\right)}{\partial \mathbf{q}}\right)^T=\sum_{ij}\frac{\rho_0-\rho_{ij}}{\rho_0\rho_{ij}^{\;3}}\frac{\partial \rho_{ij}}{\partial \mathbf{q}} \tag{15}$$

To compute the repulsive force $\mathbf{F}_{ij}$ that acts on two generic objects A_i and A_j, we have to calculate the derivative of ρ_{ij} with respect to $\mathbf{q}$:

$$\rho_{ij}=\sqrt{\mathbf{d}_{ij}\left(\hat{\xi}_i,\hat{\xi}_j\right)\mathbf{d}_{ij}\left(\hat{\xi}_i,\hat{\xi}_j\right)}\Rightarrow\frac{\partial \rho_{ij}}{\partial \mathbf{q}}=\frac{1}{\rho_{ij}}\mathbf{d}_{ij}\frac{\partial \mathbf{d}_{ij}}{\partial \mathbf{q}} \tag{16}$$

If we assume that the boundaries of A_i and A_j are C^1 functions (this is not restrictive, because in a real case each object's boundary can be approximated by such a function with arbitrary precision), then the shortest distance between the two objects corresponds to the case in which $\mathbf{d}_{ij}\cdot\mathbf{d}_{ij}$ is minimal. Since $\mathbf{d}_{ij}$ is a function of ξ_i and ξ_j, this happens when

$$\frac{\partial}{\partial \xi_i}\left(\mathbf{d}_{ij}\cdot\mathbf{d}_{ij}\right)=0 \qquad \frac{\partial}{\partial \xi_j}\left(\mathbf{d}_{ij}\cdot\mathbf{d}_{ij}\right)=0 \tag{17}$$

by choosing the points of absolute minimum. Since

$$\frac{\partial \mathbf{d}_{ij}}{\partial \xi_i}=\frac{\partial \mathbf{r}_i}{\partial \xi_i} \qquad \frac{\partial \mathbf{d}_{ij}}{\partial \xi_j}=\frac{\partial \mathbf{r}_j}{\partial \xi_j} \tag{18}$$

the previous relation implies the conditions

$$\mathbf{d}_{ij}\cdot\left.\frac{\partial \mathbf{r}_i}{\partial \xi_i}\right|_{\hat{\xi}_i,\hat{\xi}_j}=0 \qquad \mathbf{d}_{ij}\cdot\left.\frac{\partial \mathbf{r}_j}{\partial \xi_j}\right|_{\hat{\xi}_i,\hat{\xi}_j}=0 \tag{19}$$

Such a relation have a straightforward geometric interpretation: the shortest distance is achieved when the distance is orthogonal to both the tangents to the boundaries of A_i and A_j. In the following we assume that all the relevant quantities are evaluated in correspondence of such values of $\mathbf{d}_{ij}$. The repulsive force F_{ij} is thus a formal column vector whose $3N$ components F_{ij}^{α}, $\alpha=1,\ldots,3N$ are defined as

$$F_{ij}^{\alpha}=\frac{\rho_0-\rho_{ij}}{\rho_0\rho_{ij}^{\;3}}\mathbf{d}_{ij}\frac{\partial \mathbf{d}_{ij}}{\partial q^{\alpha}}=\frac{\rho_0-\rho_{ij}}{\rho_0\rho_{ij}^{\;3}}\mathbf{d}_{ij}\frac{\partial\left[\left(\mathbf{X}_i+\mathbf{r}_i\right)-\left(\mathbf{X}_j+\mathbf{r}_j\right)\right]}{\partial q^{\alpha}} \tag{20}$$

After omitting straightforward calculation, $\mathbf{F}_{ij}$ is explicitly evaluated as in eq. 21. The three components of $\mathbf{F}_{ij}$ in places i acts on A_i configuration, while those in places j acts on A_j configuration. As we can see, even if our reasoning is based on the configuration space, the calculus of the repulsive force can be decoupled in a distributed algorithm, in which each agent requires only knowledge about the presence of obstacles and other agents within the distance ρ_0.

$$\mathbf{F}_{ij} = \frac{\rho_0 - \rho_{ij}}{\rho_0 \rho_{ij}^{\,3}} \left[\underset{1}{\begin{bmatrix} 0 \\ 0 \\ 0 \end{bmatrix}}^T \cdots \underset{i}{\begin{bmatrix} d_{ij}^X \\ d_{ij}^Y \\ r_i^X d_{ij}^Y - r_i^Y d_{ij}^X \end{bmatrix}}^T \cdots \underset{j}{\begin{bmatrix} -d_{ij}^X \\ -d_{ij}^Y \\ -\left(r_j^X d_{ij}^Y - r_j^Y d_{ij}^X\right) \end{bmatrix}}^T \cdots \underset{N}{\begin{bmatrix} 0 \\ 0 \\ 0 \end{bmatrix}}^T \right]^T \tag{21}$$

Once again, notice that it is not necessary to know the shapes of obstacles in order to compute the repulsive force, since only the minimum Euclidean distance between agents is considered; such distance can be easily computed by means of proximity sensors. The third component in each of the column vectors is exerted on the ϑ_k (k=1..N) coordinates, thus introducing a rotational effect which changes agents' orientation when required to avoid obstacles. We can express the resultant force $\mathbf{F}_{\text{rep}}$ as:

$$\mathbf{F}_{\text{rep}} = \left[\mathbf{f}_1^{\,T} \quad \mathbf{f}_2^{\,T} \quad \cdots \quad \mathbf{f}_N^{\,T}\right]^T = \sum_{i,j=0..N,i<j} \mathbf{F}_{ij} \tag{22}$$

where the generic $\mathbf{f}_k$ is the three component vector corresponding to the force exerted on A_k's configuration (X_k, Y_k, ϑ_k). In the following, since each agent A_k can autonomously compute the force $\mathbf{f}_k$, we will refer exclusively to the behaviour of A_k , by supposing that all the agents are behaving in a similar way. Once each agent has computed the repulsive force $\mathbf{f}_k$, it needs computing a $\mathbf{t}_k$.vector normal to $\mathbf{f}_k$ (as stated in eq. 2) However, since we are considering also the agent's orientation, the solutions of the equipotential equation

$$U(X_k, Y_k, \vartheta_k) = const \tag{23}$$

now describe an equipotential surface instead of a line. Thus, when computing $\mathbf{t}_k$, the agent has to choose between an infinite number of tangents to the equipotential surface in $\mathbf{q}$. By imposing that the rotational component of the vector $\mathbf{t}_k$ must be null (this is an heuristic, which gives $\mathbf{f}_k$ the whole responsibility of rotating the agent, while $\mathbf{t}_k$ is responsible only for translation) the agent has, once again, to choose between two opposite tangent vectors. To choose between these two, the agent refers to the usual Micronavigation rules. The computation of $\mathbf{g}_k$, W_g, W_t, and W_f is straightforward.

4 Experimental results

We performed many tests in a simulated environment. All the agents are considered as totally autonomous, holonomic vehicles; no communication is allowed between agents. Since every agent is completely autonomous and has only local knowledge about the environment, it generates its own trajectory (as stated in eq. 2) independently from all the other agents. Moreover, agents do not distinguish between objects and other agents. In fig. 2, goal configurations are closed areas marked with a dashed line; agents are depicted as black rectangles (A, B, and C), whilst obstacle are depicted as grey rectangles (in the current version of the simulator only rectangular shapes are allowed). Notice that A is the first agent to reach its own goal (fig. 2c). As a consequence, it prevents other agents to reach their own goals, since it blocks the access to the corridor. This situation corresponds to a local minimum in the overall potential field. However, as we said before, local minima are not a problem with the Micronavigation algorithm, since agents follow equipotential lines instead of descending the gradient. In fig. 2d (and subsequent ones), notice that agent B perturbs the equilibrium state of A (which has reached the goal). A perceives B as an obstacle which is coming closer and closer: the potential U_{rep} in A's configuration increases, and the two weights W_g and W_f increase as well, thus forcing A to start following an equipotential line and to move away from the goal. Notice that C gets benefit of this situation to enter the corridor (fig. 2e-f), while B -which perceives A as an obstacle- moves away while attempting to avoid it. In fig. 2g, both A and C have reached their goals. Since B has not reached its goal yet, it increases the value of U_L, thus starting to move on equipotential lines with a higher potential value (i.e. closer to obstacles). In fig. 2i, B perturbs again the equilibrium state of A: but this time, since B is allowed to move closer to obstacles, it manages to push A away and to enter the corridor, finally reaching its goal configuration (fig. 2k). In fig. 2l, every agent has reached its goal.

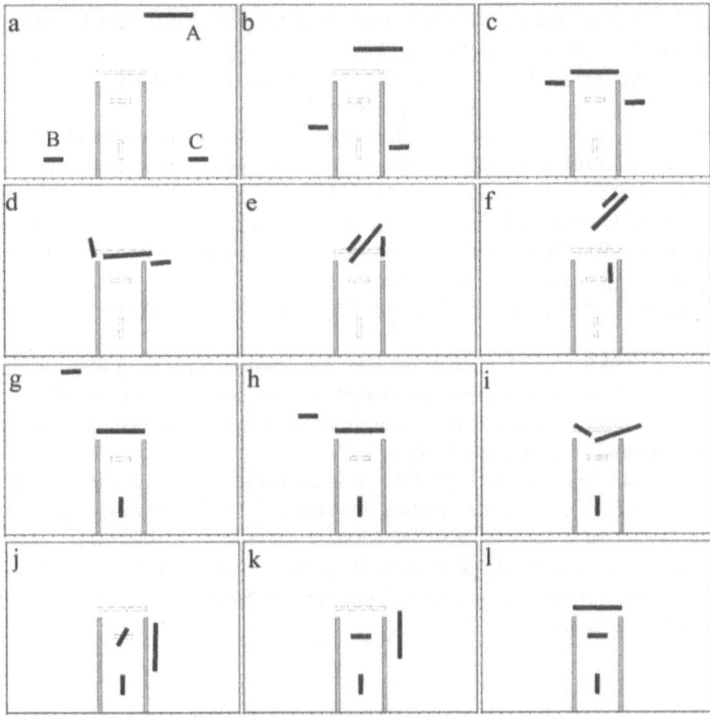

Fig. 2. Snapshots of a simulated experiment.

Conclusions

In this paper we presented a distributed algorithm that allows to reactively solve the navigation problem for a set of *N* autonomous agents. The algorithm is a natural extension of Micronavigation, it has a computational complexity $O(N^2)$ and shows to perform well even in complex cases. As a future work, we aim at considering non-holonomic vehicles and to extend the same principles to different domains (assemblage problems, manipulator motion planning, etc.).

Acknowledgements

Authors wish to thank Professor Franco Bampi and Professor Clara Zordan for their fundamental contribution.

References

1. Evans J. M. (1995) "HelpMate: a Service Robot Success Story," *ServiceRobot: An International Journal*, 1(1), 19-21.
2. Blitch, J. (1999) "Tactical Mobile Robots for Complex Urban Environment," *Mobile Robots XIV*, Boston, MA, 116-128.
3. Balch, T., and Arkin, R.C. (1994) "Communication in Reactive Multiagent Robotic Systems," *Autonomous Robots*, 1(1), 27-52.
4. Parker, L.E. (1999) "Cooperative Robotics for Multi-Target Observation," *Intelligent Automation and Soft Computing*, special issue on Robotics Research at Oak Ridge National Laboratory, 5 (1), 5-19.
5. Guibas, L.J., Latombe, J.C., LaValle, and S.M., Lin, D. (1999) "A Visibility-Based Pursuit-Evasion Problem," *International Journal of Computational Geometry and Applications*, 9, 471-494.
6. Latombe, J.C. (1991) *Robot Motion Planning*, Kluwer Academic Publisher, MA.
7. Kathib, O., 1986, "Real-Time Obstacle Avoidance for Manipulators and Mobile Robots," International Journal of Robotics Research, 5(1), 90-98.
8. Arkin, R.C. (1989) "Motor Schema-Based Mobile Robot Navigation", International Journal of Robotics Research, 8(4), 92-112.
9. Koren, Y. and Borenstein, J. (1991) "Potential Field Methods and Their Inherent Limitations for Mobile Robot Navigation," *Proceedings of the IEEE International Conference on Robotics and Automation*, 1398-1404
10. Piaggio, M., Sgorbissa, A., Zaccaria, R. (2000) "Micronavigation," *From Animal To Animats 6 - Proc. Sixth Int. Conf. on the Simulation of Adaptive Behavior*, MIT Press, 2000

Chapter 8
Distributed Sensing and Mapping

Omnidirectional Distributed Vision for Multi-Robot Mapping

Emanuele Menegatti[1] and Enrico Pagello[1 2]

[1] **Intelligent Autonomous Systems Laboratory**
Department of Informatics and Electronics
University of Padua, Italy
[2] Institute LADSEB of CNR Padua, Italy

Abstract. In this paper, we propose our ideas for realizing a Distributed Vision System for building a map of a large environment with a team of mobile robots equipped only with vision sensors. The vision sensors used in this work are catadioptric omnidirectional vision systems. The mirrors of these catadioptric systems have profiles expressly designed for the robot's body and for the robot's task. We report preliminary experiments on the interactions between heterogeneous vision systems both mounted on the same robot and mounted on different robots.

1 Introduction

The exploration and the map building problems attracted the attention of robotics researchers since the beginning. These problems have been widely studied in the past, but the attention has been mainly focused on single robot exploration. Lately, the attention of some researchers moved to multi-robot systems able to explore an unknown environment in a parallel fashion. This technique is appealing not only because permits to reduce the time needed to cover the whole environment, but also because using a multi-robot system, the redundancy of the observation and of the observers permits to increase the accuracy of the map built by the robots. As an example, refer to the work of Burgard, Thrun *et al.* [13] [2] that are porting on multi-robot systems the expertises and techniques successfully developed on single robot systems like Minerva [14] and Rhino [1]. The sensor used by the robots in these works is a laser range finder. Every robot builds a local map of the environment with the data collected by its own range finder and it uses this local map to pre-process successive scans that are sent to an "*external central mapper*" that create a global map of the environment. The global map is not created from the local maps of the different robot, but directly from their preprocessed readings. The cited works focus on the issue of a coordination strategy to maximize the information gain and then to minimize the exploration time.

In this paper, we want to report our efforts to build a distributed vision system intended for exploration and mapping using mobile robots. In this work we do not focus on the coordination of the robots in order to maximize the information gain or the coverage, we are more interested on the processes of fusion and matching of the visual information gathered from the different robots. We apply a very simple coordination strategy that we called **misrobot**, i.e. the robots choose the exploring direction that increases the distance from the other robots present in their field of view. The exploration is carried on without an explicit strategy, every robot moves according to simple reactive control laws, like wall-following or corridor following. The only sensors used by these robots are the on-board vision systems. The robot team is composed of heterogeneous robots fitted with heterogeneous vision sensor, like standard perspective cameras and catadioptric omnidirectional vision system.

The mapping strategy we use is the same developed for a single robot in the Caboto Project [11]. A single robot equipped with an omnidirectional vision sensor build a map using the Spatial Semantic Hierarchy proposed by Kuipers [6]. The robots then fuse the local maps to generate a global map. In the future we want to be able to use the redundancy of the observation, i.e. the fact that two robots observe the same objects, in order to improve the accuracy of the map.

In Section 2, we describe the mapping strategy used by the single robots and the catadioptric omnidirectional sensors used by the robots. In Section 3, we introduce the idea of Vision Agent (i.e. a vision system with communication capabilities that is embodied in the robot but for some aspects transcends the single robot) and we report the two stream of research we are following to develop the prototypes of Distributed Vision Systems. In Section 4, we will present our final target: a community of Vision Agents that realises a Distributed Vision System able to explore and realise a map of an unknown environment.

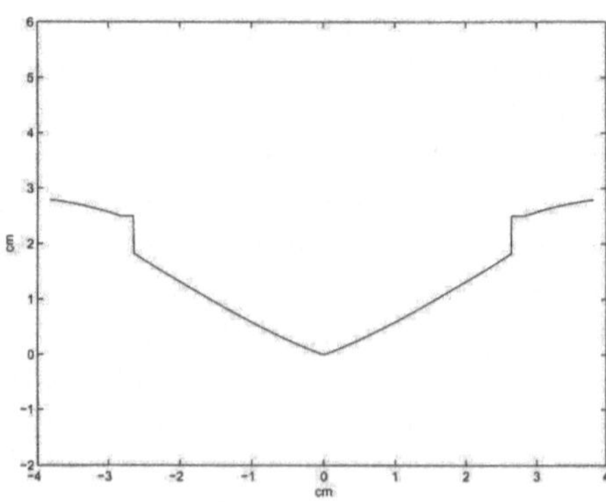

Fig. 1. (a) The robot with its omnidirectional sensor. (b) The actual mirror profile of the new omnidirectional mirror designed for this application.

2 The Caboto project

In the *Caboto Project* we proposed a new approach to the map building task: the fusion of a well understood method, the Spatial Semantic Hierarchy (**SSH**) [5] [6] with a relatively new sensor, an omnidirectional vision system [16]. The Caboto Project was the first work that created a link between the Spatial Semantic Hierarchy and the omnidirectional images [12]. We proposed a set of topological events, that it is possible to identify in the sequence of omnidirectional images acquired while the robot moves. These topological events can be used to pose a discrete set of places that will be the nodes of the topological map. We designed a mirror profile that permits to simply identify topological events while the robots moves [11]. In the next section we will discuss the omnidirectional sensor used, but first let us see the basic structure of the Spatial Semantic Hierarchy (**SSH**).

2.1 Spatial Semantic Hierarchy

The **SSH** (Spatial Semantic Hierarchy) is a "method for robot exploration and map building" [5] inspired to the knowledge of large-scale spaces of humans. The SSH is made up of several interacting layers that can be implemented separately. Let us see briefly what each layer is about:

- The *Sensory Level* is the interface with the agent's sensory system. It extracts the useful environmental clues from the continuous flux of information it receives from the robot's sensors.
- The *Control Level* describes the world in terms of continuous actions called *control laws.* A control law is a function which relates the sensory input with the motor output. A control law is retained until a transition of state is detected. A transition of state can be detected with a function called a **distinctiveness measure**.
- The *Causal Level* abstracts a discrete model of the environment from the continuous world. In other words, it is at the causal level that a discrete set of experiences (actions and perceptions) is extracted from the continuous world. These will be the nodes of the topological map. The discrete model of the environment is composed of **views** , **actions** and the causal relations between them. A **view** is defined as the sensor's reading at a **distinct** place. A **distinct place** is a place where a transition of state is detected. An **action** is defined as the application of a sequence of control laws.
- The *Topological Level* represents an environment with geographical features in the world, such as places, paths and regions connected or contained one in the other.
- The *Metrical Level* augments the topological representation of the environment by including metrical properties such as distance, direction, shape, etc. This may be useful, but is seldom essential.

In the following, we will briefly present the omnidirectional sensor used for these experiments and we will report why an omnidirectional sensor is a good sensor for building a topological map within the Spatial Semantic Hierarchy frame[1].

2.2 The Omnidirectional Sensor

The robot we used in this work is depicted in Figure 1(a). Its omnidirectional sensor is composed of a perspective camera pointed upwards at the vertex of a multi-part omnidirectional mirror.

The mirror used is a multi-part mirror, where each segment is designed to view a specific region of space[2]. The design of this mirror was inspired by the work of Marchese and Sorrenti [7]. To understand the mirror profile consider the rough sketch in Fig. 2. The inner part of the mirror (Part A in Fig. 2) is designed to view objects from 60 cm around the robot up to six meters, without displaying the body of the robot . This part produce the main part of the image. The middle ring (Part B) permits to view very distant objects and can be used for a better planning of the exploration movements, using the ideas about the catastrophe theory exposed in [15]. The external ring (Part C) displays at higher resolution (compared to the resolutions attainable in the other two sections) the area closer to the robot. This is useful for the design of control laws like *corridor following* and *wall following.* The actual mirror profile is displayed in Figure 1 (b) .

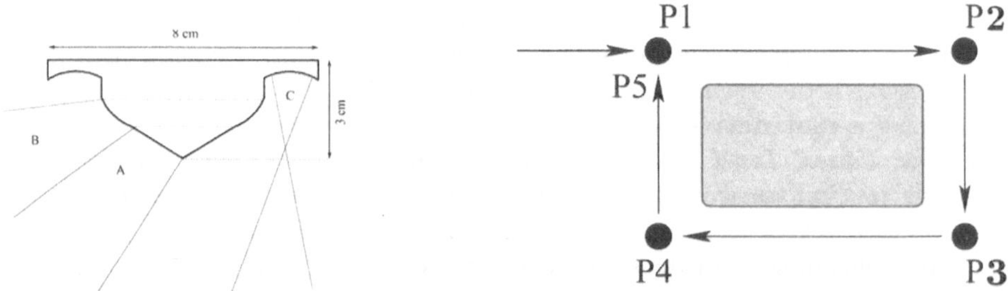

Fig. 2. (a)A rough sketch of the mirror profile where the curvatures of the different sections are exaggerated for sake of clearness. (b) The "exploring around the block" problem. The problem of recognizing the same place under different state labels.

[1] For a deeper discussion of these topics refer to [11].

[2] The shape of the mirror is designed in order to maximize the resolution in the regions of interest. For details on the procedure we used to design the custom profile of the mirror, please refer to [9]

2.3 Omnidirectional vision and the Spatial Semantic Hierarchy.

As we demonstrated in [12], the omnidirectional images can be strictly correlated with the **views**[3] introduced in the causal level of the SSH. A **view** is the sensor reading at a **distinct** place, the omnidirectional image is a global reading of the surrounding at a certain place. The association of **views** with omnidirectional images simplifies the interpretation of data and then the construction of the map.

As an example consider the *exploring around the block* problem, i.e. the problem of recognizing the same place under different state labels, see Figure 2 (b). The rotational invariance of the omnidirectional **views** permits a direct solution. The robot moves around the block following the arrows. When it reaches Place 5 from Place 4, it is very difficult to recognize it as the previously visited Place 1, unless it is equipped with an omnidirectional camera and it makes use of the rotational invariance. Using the SSH terminology, it is easy to spot whether the current **view** is the same which has been experienced before and therefore to consider this view not as a different **place** but as the same **place** reached from a different direction.

In the next sections, we will outline how this mapping technique can be scaled to a team of heterogeneous robots performing a mapping task with their vision systems as only sensors.

3 The Vision Agents

First of all, let us precise a terminology issue: in the following we will prefer the term *Vision Agent* instead of "vision system". The term Vision Agent emphasizes that the vision system can interact with the other vision systems to create an intelligent distributed system[4] and it is not just one of the several sensors of a single robot. At the moment, we are working on two research streams: interaction of two VAs mounted on the same robot, coordination of several VAs mounted on different robots [10].

3.1 Heterogeneous VAs on the same robot

First we realised a Cooperative behavior between two heterogeneous vision agents embodied in the same robot. We want to create a Cooperative Vision System using an omnidirectional and a perspective vision system mounted on the same robot, see Fig.3 (a) for a close view of the two vision sensors.

With these two vision systems we intend to equip the robot with something similar to the animals vision system, replicating the relationship between the peripheral vision and the foveal vision. The peripheral vision gives

[3] In the following, the bold font is used to indicate we are using the SSH meaning of the words.

[4] Like in the definition of Ishiguro [4] (one of the most active researcher in distributed vision).

a general, and less accurate, information on what is going on around the observer. The foveal vision determines the focus of attention and provides more accurate information on a narrow field of view. In a similar way, the omnidirectional camera, mounted on the top of the robot, offers a complete view of the surroundings at low resolution and the perspective camera, mounted in the front, offers a more accurate view of objects in front of the robot. Inspired by our previous researches presented in [3], we realised a focus of attention via the exchange of messages between the two vision agents. The omnidirectional vision agent monitors the surroundings of the robot to detect the occurrence of particular events. When one of these events is detected, the Omnidirectional Vision Agent (OVA) send a message to the Perspective Vision Agent (PVA). If the PVA is not already focused on a task, it will move the robot in order to put the event in the field of view of the perspective camera.

At the moment, solved some technical problems, we are carrying on tests and experiments to validate this approach. These experiments will provide a deeper understanding about the cooperation of heterogeneous vision agents.

Fig. 3. (a) A close view of the vision system of Nelson. On the left, the perspective camera. In the middle, pointed up-ward the omnidirectional camera. (b) A close view of two of our robots. Note the different vision systems.

3.2 VAs on different robots

The second stream of research is the creation of a Cooperative Distributed Vision System for a multi-robot team. The practical implementation we decided to tackle is the realization of the Cooperative Object Tracking Protocol proposed by Matsuyama. In the experiments presented in [8], Matsuyama used active cameras mounted on a special tripod. The active cameras were pan-tilt-zoom cameras modified in order to have a fix view point. This allowed the use of a simple vision algorithm, not very different from the case of static cameras. We want to introduce mobile robots and robot vision algorithms in such a system and realise a cooperative distributed tracking of an object within a team of autonomous mobile robots, as more deeply described in [10].

Matsuyama's work is centered on the concept of *agency*. In the definition of Matsuyama, an **agency** is *the group of the VAs that sees the objects to be tracked and keeps an history of the tracking.* This group is neither fixed nor static. In fact, a VA exits the agency, if it is not able to see the tracked object anymore and a new VA can joint the agency as soon as the tracked object comes in its field of view.

Let us sketch how the agency works, using an example draw from our application field: the RoboCup domain. Suppose to have a team of robots in the field of play. Each robot is fitted with a Vision Agent. None of the Vision Agents is seeing the ball. In such a situation no agency exists. As soon as a Vision Agent see the ball, it creates the agency sending a broadcast message to inform the other Vision Agents the agency has been created and it is the master of the agency. After this message a second message follows, telling the other Vision Agents the estimated position of the ball. The other Vision Agents maneuver the robots in order to see the ball. Once a Vision Agent has the ball in its field of view, it asks permission to joint the agency and send to the master its estimation of the ball position. If this information is compatible with the information of the master, i.e. if the new Vision Agent has seen the *correct ball*, it is allowed to joint the agency. Also this experiment is in progress and it is a preliminary testbed for the realization of the more complex system described in the next section.

4 The Distributed Vision System

Our main target is the realization of a Distributed Vision System mounted on a team of mobile robots able to explore and map an unknown environment.

The only sensors mounted on the fleet of robots are vision systems. Some robots mounts only omnidirectional vision sensors, while other mounts omnidirectional and perspective vision sensors. The vision sensors are part of the previously introduced Vision Agents that in addition to the image processing capabilities have also some communication capabilities, i.e. they can send and receive messages over the network.

The basic idea is that every robot perform a local map of the environment while it moves through it. When two of the robots sees each other, they perform a matching on the overlapping regions of the two local maps and if this matching is successful they merge their maps, so every robot has a bigger map composed of the two local maps.

The first experiment we are settling is the cooperative exploration and mapping of a simple environment with three robots mounting omnidirectional vision agents. Consider Fig. 4 (a), the robots start from a common location. Every robot moves using one of the designed control laws (wall-following, corridor-following) and moves in a direction chosen in order to increase its distance from the other robots[5]. While the robots move they build a local map of the surrounding using the vertical edges of the environment as described in Section 2. The local map of every robot is represented in Fig. 4 (b) by the lines beside the objects' boundary. The robots are able to detect the teammates and to recognize them thanks to a markers placed on top of the robots. If a robot spots in its field of view one or more teammates, it creates a link with them and they start a matching process on the local features they see in the omnidirectional image. The matching/merging process is repeated every time the robot spots a new vertical edge in the environment, in order not to merge maps that contains non-new information again and again. To simplify the matching process we decided to use the colours in the image. We built the maze with boxes with a different colour for every side, so if the colours and the dimensions of the boxes match, it is possible to create a correspondence between the two local maps built by each robot and merge them in a single map. At the end of the merging process, every robot has the single map.

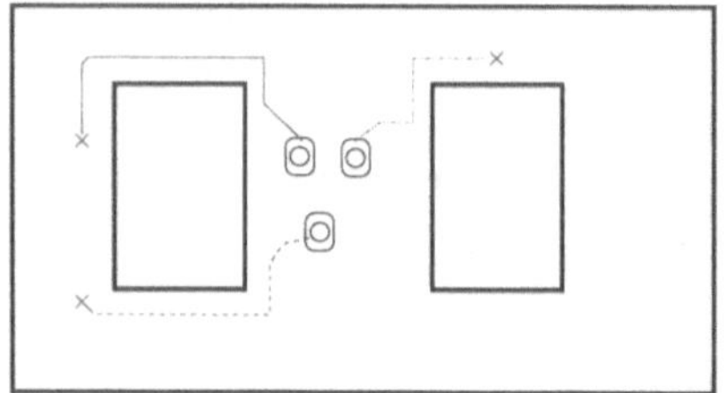

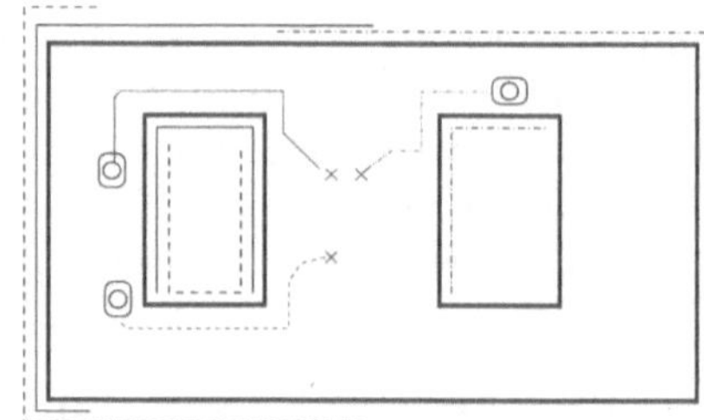

Fig. 4. (a) The robots at the starting point with the path each robot will follow. (b) The robots during the exploration and for each robot the local map it built. Note the two robots on the left, they can see each other, so they will perform a matching on the features they can see and merge their local maps.

This project is at a preliminary stage. At the moment we did not implemented a system to automatically stop the exploration when the map is complete and we are not concerned on the fact that every robot has a complete map of the environment. In the end, every robot will have a map that contains only the portion of the environment it traversed and the portions explored by teammates it met on its way.

5 Future works

The next step will be to improve the accuracy of the map using the possibility that two robots are observing the same objects from different points of view.

[5] As we stated in the Introduction, in this work we are not focusing on coordination issues and we use the simple *misrobot* coordination we discussed in Section 1.

In other words we want to exploit the redundancy of the observers and of the observations.

Another issue we want to investigate is the exploitation of the heterogeneity of the Vision Agents fitted on the robots. Our idea is to use the multi-robot system to perform a search and inspect task. The task will be to locate an object in the maze and observe it with high resolution vision sensor: the perspective camera. Therefore, if the object is discovered by one of the robots equipped only with the omnidirectional sensor, this robot should be able to request the intervention of the robot equipped with the perspective camera.This means that the robot must be able to store the path it traveled and to communicate it to the other robots. Therefore, a compact representation of the map and of the traveled path is required. If the two robots already merged their maps, the problem is quite trivial. If this is not the case, the called robot should first return to the starting position using its own map and then reach the calling robot using the map build by the latter. The starting location is the bridge between the two maps, in fact this is represented in both maps because both robots started from this common position.

6 Conclusions and Acknowledgments

In this paper we presented the research stream we are following to realise a map of an unknown environment with a Distributed Vision System using only omnidirectional vision sensors. The approach we used is to port the work done on the single robot within the Caboto Project on to a multi-robot system. The mapping strategy chosen for Caboto allows this porting without major modifications. The interaction between the Vision Agents is under study with the two experiments presented in the text. The exploitation of omnidirectional mirrors with a profile tailored for the particular application showed to be effective on the field.

At the time of writing experiments are running on such a systems providing theoretical and practical insight.

We wish to thanks the student of the ART-PD and Artisti Veneti Robocup teams who built the robots. This research has been partially supported by: the Italian Ministry for the Education and Research (MURST), the Italian National Council of Research (CNR) and by the Parallel Computing Project of the Italian Energy Agency (ENEA).

References

1. W. Burgard, A. Cremers, D. Fox, D. Hhnel, G. Lakemeyer, D. Schulz, W. Steiner, and S. Thrun. Experiences with an interactive museum tour-guide robot. *Artificial Intelligence (AI)*, 114((1-2)), 1999.
2. W. Burgard, D. Fox, M. Moors, R. Simmons, and S. Thrun. Collaborative multi-robot exploration. In *Proceedings of the IEEE International Conference on Robotics and Automation (ICRA)*. IEEE, 2000.

3. S. Carpin, C. Ferrari, E. Pagello, and P. Patuelli. Bridging deliberation and reactivity in cooperative multi-robot systems through map focus. In M.Hannebauer, J. Wendler, and E. Pagello, editors, *Balancing Reactivity and Social Deliberation in Multi-Agent Systems,*, LNCS. Springer, 2001.
4. H. Ishiguro. Distributed vision system: A perceptual information infrastructure for robot navigation. In *Proceedings of the Int. Joint Conf. on Artificial Intelligence (IJCAI97)*, pages 36–43, 1997.
5. B. Kuipers. The spatial semantic hierarchy. *Artificial Intelligence*, 119:191–233, February 2000.
6. B. J. Kuipers and T. Levitt. Navigation and mapping in large scale space. *AI Magazine*, 9(2):25–43, 1988. Reprinted in *Advances in Spatial Reasoning, Volume 2*, Su-shing Chen (Ed.), Norwood NJ: Ablex Publishing, 1990, pages 207–251.
7. F. Marchese and D. G. Sorrenti. Omni-directional vision with a multi-part mirror. In P. Stone, T. Balch, and G. Kraetzschmar, editors, *RoboCup 2000: Robot Soccer World Cup IV*, LNCS. Springer, 2001.
8. T. Matsuyama. Cooperative distributed vision: Dynamic integration of visual perception, action, and communication. In W. Burgard, T. Christaller, and A. B. Cremers, editors, *Proc. of the Annual German Conf. on Advances in Artificial Intelligence (KI-99)*, volume 1701 of *LNAI*, pages 75–88, Berlin, Sept. 1999. Springer.
9. E. Menegatti, F. Nori, E. Pagello, C. Pellizzari, and D. Spagnoli. Designing an omnidirectional vision system for a goalkeeper robot. In A. Birk, S. Coradeschi, and S. Tadokoro, editors, *RoboCup-2001: Robot Soccer World Cup V.* Springer, 2002 (TO APPEAR).
10. E. Menegatti and E. Pagello. Cooperation between omnidirectional vision agents and perspective vision agents for mobile robots. In *Proceedings of the 7th International Conference on Intelligent Autonomous Systems (IAS-7) (TO APPEAR)*, Los Angeles, USA, March 2002.
11. E. Menegatti, E. Pagello, and M. Wright. Using omnidirectional vision sensor within the spatial semantic hierarchy. In *IEEE International Conference on Robotics and Automation (ICRA2002)(TO APPEAR)*, Washinton, USA, May 2002.
12. E. Menegatti, M. Wright, and E. Pagello. A new omnidirectional vision sensor for the spatial semantic hierarchy. In *IEEE/ASME Int. Conf. on Advanced Intelligent Mechatronics (AIM '01)*, pages 93–98, July 2001.
13. R. G. Simmons, D. Apfelbaum, W. Burgard, D. Fox, M. Moors, S. Thrun, and H. Younes. Coordination for multi-robot exploration and mapping. In *Proceedings of the AAAI National Conference on Artificial Intelligenc AAAI/IAAI*, pages 852–858, 2000.
14. S. Thrun, M. Beetz, M. Bennewitz, W. Burgard, A. Cremers, F. D. Fox, D. Haehnel, C. Rosenberg, N. Roy, J. Schulte, and D. Schulz. Probabilistic algorithms and the interactive museum tour-guide robot minerva. In *International Journal of Robotics Research*, volume Vol. 19, pages pp. 972–999, November 2000.
15. M. Wright and G. Deacon. A catastrophe theory of planar orientation. *Int. Journal of Robotics Research*, 19(6):531–565, June 2000.
16. Y. Yagi, Y. Nishizawa, and M. Yachida. Map-based navigation for a mobile robot with omnidirectional image sensor copis. *IEEE Transaction on Robotics and automation*, VOL. 11(NO. 5):pp. 634–648, October 1995.

How Many Robots? Group Size and Efficiency in Collective Search Tasks

Adam T. Hayes

Collective Robotics Group
136-93 California Institute of Technology, Pasadena CA 91125
athayes@caltech.edu
www.coro.caltech.edu/People/athayes/athayes.html

Keywords: Collective Autonomous Robotics, Distributed Exploration

Abstract. This paper presents a quantitative analysis of the tradeoffs between group size and efficiency in collective search tasks that considers both the time-sensitive nature of search completion and the system operating cost. First, the search task is defined and a performance metric is presented that can account for all of the costs associated with the task. Next, for both random and coordinated search strategies, analytical expressions are derived that can be used to predict optimal system performance bounds given a particular task description, and the performance benefit of using coordinated search is shown to be dependent on the relative values of the different cost components. Finally, an embodied computer simulation is used to support the analytical results, suggesting that the assumptions involved in their derivation are sound.

1 Introduction

Search tasks, because they submit well to parallelization, are an ideal application for multi-agent systems. Search is a well studied problem (for a review, see [1]), and there has been a significant amount of investigation into the efficiency tradeoffs between random and coordinated search strategies [2]. However, how to assess the performance of multi-agent search systems is still an open problem. Some researchers take into account only energy used [2], while others consider only the time required until completion [3] when analyzing the performance of multi-agent systems on similar search tasks. Clearly, the performance metric used must be appropriate for the task being studied, but there is reason to believe that a more complete cost metric might offer further insight into the design tradeoffs present and aid in the comparison of results across research groups.

2 Search Task Description

The search task examined in this paper can be described as follows: a group of N agents each having a sensor radius r must locate a single target contained within an enclosed 2-D arena. For simplicity consider this arena to be a square of length L, with $L \gg r$ so that the agents are likely to disperse throughout the arena before the target is found. To ensure that the agents do not begin with full coverage of the arena (thus driving the search time to 0), initial agent deployment must be within a single deployment area of radius R. It is assumed that $L \gg R$, although the deployment area may be located anywhere within the arena. Figure 1 shows a schematic of an example task layout.

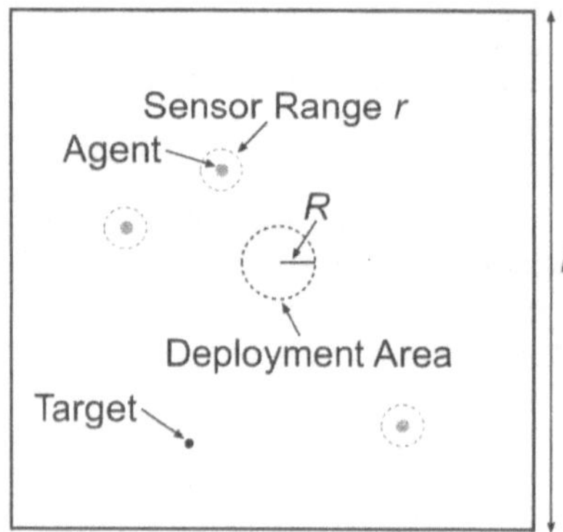

Fig. 1. Example task layout in which $N = 3$

2.1 Performance Metric

Performance on this search task can be measured in terms of T, the time elapsed before an agent detects the target, and D, the sum of the distances traveled by each of the agents. D then correlates to the amount of energy needed for system operation. There are also setup costs that need to be considered in a complete system evaluation. Since these measures are physically independent, a composite metric incorporating a task-specific weighting of these basic factors can be considered. For N agents:

$$C = \alpha T + \beta D + \gamma N \tag{1}$$

There are three basic cost components. α is taken to be the cost per unit time of not completing the task, β is the cost per unit distance of running the system, and γ is the initialization cost per agent. C represents the total cost incurred before the task is completed. By choosing specific values for α, β, and γ the appropriate relationship between time required, energy used, and initial cost can be generated for evaluating any particular application.

To simplify the analysis, if the control algorithm used maintains an average speed v across time, the total distance traveled can be approximated by the time required to complete the task:

$$D = TNv \tag{2}$$

Substituting into equation 1 above,

$$C = \alpha T + \beta TNv + \gamma N \tag{3}$$

Thus, for any given group size, the system cost can be obtained directly from the time required. Although C is the metric used in the analysis section of this paper, in order to facilitate comparison across environments, it can be normalized by the minimum completion cost to generate a unitless performance metric. The minimum cost is based on the optimum values for the given task (T_{MIN}, D_{MIN}) for a single agent with prior knowledge of the source location, as determined from the average distance between starting location and target location as well as maximum agent speed.

$$P = \frac{\alpha T_{MIN} + \beta D_{MIN} + \gamma}{C} \tag{4}$$

This form of P ensures that for any cost α, β, or γ greater than 0, the optimal system will achieve a performance of 1, and any that requires more time, distance, or agents will have a performance less than 1.

3 Deriving Performance

The stochastic nature of real systems (e.g. from sensor noise, agent movement, or deployment and target location variation), means that for each trial the cost to complete a search task is drawn from some distribution. For some applications the designer is interested in minimizing the average cost of system operation, and for other tasks the value of interest is a composite of the average cost and its variation. This work focuses on bounding the cost of a given percentage of trials, that is, determining the cost C which exceeds the cost of some fraction S of all trials in that particular environment.

Expressions for the optimal cost of random and coordinated search strategies are derived in the following sections. For clarity, a summary of the variables used is provided in Table 1.

3.1 Random Search

In a system performing random search the agents move randomly while searching for the target without any explicit attempt to partition the space amongst agents or avoid searching the same area multiple times. Given that a system has some probability g of finding the target during a time interval t, the probability S that the target is found before some time T can be expressed as a geometric series:

$$S = \sum_{t=1}^{T} g(1-g)^{t-1} \tag{5}$$

Table 1. Summary of Parameters and Variables

N	Number of agents	r	Sensor radius
L	Arena length	R	Deployment area radius
T	Time to complete task	D	Total distance to complete task
α	Cost of not finishing	β	Cost of operation
γ	Initialization cost per agent	C	Total system cost
v	Average agent velocity	P	System performance measure
S	Desired performance bound	g	Probability of system finding target
t	Time interval	k	Minimum dispersion time
η	Sensor detect probability	p	Probability of agent finding target
$x^{\star}$	Optimal value for variable x	U	Single agent trial search time

To solve for T, the series can be simplified as follows:

$$S - (1-g)S = g - g(1-g)^T \tag{6}$$

$$T = \frac{\log(1-S)}{\log(1-g)} \tag{7}$$

The above equation describes the time to complete the task based on search success probability and desired performance bounds. To be more accurate, however, a term needs to be added to account for the fact that the agents cannot begin the task with full coverage of the entire search area (because all agents start within the deployment area).

$$T = \frac{\log(1-S)}{\log(1-g)} + k \tag{8}$$

The factor k represents the time required to cover the distance between the deployment area and target point, and serves as a lower bound of the time needed to perform the task (i.e. $k = T_{MIN}$).

g can be decomposed in terms of the number of individual agents N performing the task, and the probability p of a single agent scanning the target per time period t. p can be approximated using the ratio of the area of scanned per t to the total area of the arena L^2. A sensor detect probability η, modeled here as the probability of target detection given that the target enters the sensor range, factors in as well.

$$p = \frac{rv\eta}{L^2} \tag{9}$$

Assuming that the probability of each agent succeeding is fully independent, given p and a group size of N agents, the probability g of the system locating the target during a time period t can be calculated to be:

$$g = 1 - (1-p)^N \tag{10}$$

Plugging this value into equation 8:

$$T = \frac{\log(1-S)}{N\log(1-p)} + k \tag{11}$$

Now the optimum number of robots, the optimum time, and the optimal cost for a given task can be derived. Beginning with a substitution, let:

$$U = \frac{\log(1-S)}{\log(1-p)} \tag{12}$$

U represents the length of time necessary for S percent of trials using a single agent to locate the target (after the initial dispersion period k). Substituting into equation 11:

$$T = \frac{U}{N} + k \tag{13}$$

And substituting this value into equation 3, another form of the total system cost is derived:

$$C = \frac{\alpha U}{N} + \alpha k + \beta v U + \beta N v k + \gamma N \tag{14}$$

Assuming that all the parameters in the system are fixed except N, determining the critical points leads to an expression for the optimal number of robots $N^\star$. Taking the derivative of C, setting it equal to 0, and then solving for $N^\star$:

$$\frac{\delta C}{\delta N} = -\frac{\alpha U}{N^2} + \beta v k + \gamma = 0 \tag{15}$$

$$N^\star = \sqrt{\frac{\alpha U}{\beta v k + \gamma}} \tag{16}$$

The positive root is taken because the number of agents must be positive, and the second derivative $\frac{\delta^2 C}{\delta N^2}$ is positive so $N^\star$ occurs at a minimum value of C. Plugging this value into equation 13 produces the optimal search time $T^\star$.

$$T^\star = \sqrt{\frac{U(\beta v k + \gamma)}{\alpha}} + k \tag{17}$$

Equations 14 and 16 can be combined to arrive at the optimal cost $C^\star$ for searching a particular environment using random search:

$$C^\star = \alpha k + 2\sqrt{(\beta v k + \gamma)\alpha U} + \beta v U \tag{18}$$

$C^\star$ breaks down into essentially three terms. The first, αk, represents the minimum cost of having to disperse throughout the arena before finding the target. Generally, however, because the sensor radius is assumed to be small compared to the arena size, $U \gg k$ so this term will not have a substantial influence on the overall cost. The second term represents the cost of not finishing the task accrued while performing the task (e.g., the damage done by the target before it can be located and neutralized). It will dominate when α is the dominant cost component. β and γ play a role in this term as well because they influence the optimal number of agents and thus the speed at which the task can be accomplished. The third term represents the cost of searching the required area to complete the task. It will dominate when β is the dominant cost component. It has a relatively simple form because the number of agents in the system does not influence the size of the area that must be searched. Substituting back in for U, the optimal random search cost can be specified in terms of the component costs and basic task parameters:

$$C^{\star} = \alpha k + 2\sqrt{(\beta v k + \gamma)\alpha \frac{\log(1-S)}{\log(1-\frac{rv\eta}{L^2})}} + \beta v \frac{\log(1-S)}{\log(1-\frac{rv\eta}{L^2})} \tag{19}$$

3.2 Coordinated Search

The performance of coordinated search algorithms has been well-studied [1]. In terms of the variables described in this paper, the results are as follows. Coordinated search for N agents requires breaking the search space into N equal partitions, and assigning one agent to sequentially search each. The total amount of time T_{Pass} required for each agent to make a single pass over its entire partition can be stated in terms of the arena size L, agent speed v, and sensor range r:

$$T_{Pass} = \frac{L^2}{Nrv} \tag{20}$$

Given a sensor detect probability η, the total number of passes M each robot must make can be expressed similarly to equation 7 above:

$$M = \frac{\log(1-S)}{\log(1-\eta)} \tag{21}$$

Thus the total time required for the optimal system to search the arena is as follows:

$$T = T_{Pass} M = \frac{\log(1-S)L^2}{\log(1-\eta)Nrv} + k \tag{22}$$

Where k represents the time required for the robots to move from the deployment area to their respective partitions. If U_{cor} is defined as follows:

$$U_{cor} = \frac{\log(1-S)L^2}{\log(1-\eta)rv} \tag{23}$$

Equation 13 is again reached:

$$T = \frac{U_{cor}}{N} + k \tag{24}$$

All of the optimal value derivations in the previous section now apply.

3.3 Performance Comparison

Comparing the optimal costs of different search algorithms can provide insight into the conditions under which each type might be more suitable. This can be done by looking at the ratio of the optimal cost of random search $C^{\star}_{rnd}$ to the optimal cost of coordinated search $C^{\star}_{cor}$. The choice of algorithm influences only the value U, and U_{rnd} (equation 12) and U_{cor} (equation 23) are defined above. As shown in [4], the ratio U_{rnd} to U_{cor} simplifies as follows:

$$\frac{U_{rnd}}{U_{cor}} = \frac{\frac{\log(1-S)}{\log(1-\frac{rv\eta}{L^2})}}{\frac{\log(1-S)L^2}{\log(1-\eta)rv}} \approx \frac{-\log(1-\eta)}{\eta} \tag{25}$$

The approximation holds when $\frac{rv\eta}{L^2}$ is close to 0, as is typical when the search arena is large. This equation indicates that as the sensor reliability decreases, the performance gap between random and optimal search strategies closes. However, the cost components play a role as well.

As stated in section 3.1, when α is the dominant cost component, the second term in the cost function (equation 18) will dominate, so assuming all cost components remain constant across the different algorithms,

$$\frac{C^{\star}_{rnd}}{C^{\star}_{cor}} = \sqrt{-\frac{\log(1-\eta)}{\eta}} \tag{26}$$

Likewise, when β dominates, the third term in the cost function is the most important, so

$$\frac{C^{\star}_{rnd}}{C^{\star}_{cor}} = \frac{-\log(1-\eta)}{\eta} \tag{27}$$

Therefore, aside from sensor detect probability, tasks for which there is considerable time pressure will be more suited to random search strategies than tasks that emphasize economy of effort. This is not an unexpected finding, but this analysis formalizes the tradeoffs involved.

Because the cost γ of building and maintaining different types of robots suitable for each algorithm is difficult to deal with abstractly, it is not considered here. However, it is worthwhile to note that robots capable of the coordinated action will likely cost more than robots suitable for random search.

4 Supporting Simulations

Formulation of the optimal search cost is straightforward, but the analysis of the random search algorithm required assumptions about the independence of the success probability over sequential time periods for a single agent as well as across agents. To verify that these assumptions are valid for this type of task, the search task was implemented in simulation and the time and distance required for groups of various sizes to succeed was recorded. To implement the random search behavior, the agents moved forward at a constant speed, making random turns (between $\frac{\pi}{4}$ and $\frac{3\pi}{4}$ rad) away from obstacles (walls and other agents) when necessary.

4.1 Webots

To maintain a close correspondence with the structure and function of real robots, Webots [5], a 3D sensor-based, kinematic simulator, originally developed for Khepera robots [6] was used to simulate the search task. This embodied simulator has previously been shown to generate data that closely matches real robot experiments [7], [8], so there is reason to believe that the results are representative of a real system.

4.2 Results

1000 simulations of the random algorithm with group sizes from 1 to 80 agents were run and the time and group distance required to complete the task were measured. The deployment area was always placed in the arena center, and the target was placed randomly throughout the arena for each trial. The dispersal time k was calculated from the arena length and the agent speed. The task and cost parameter values selected are shown in Table 2. Note here η is significantly less than one and $\alpha \gg \beta$, so the random algorithm is expected to perform similarly to the coordinated search.

Table 2. Task and Cost Parameter Values

Agent radius		.5 [m]
Sensor radius	r	.5 [m]
Arena length	L	100 [m]
Deployment area Radius	R	10 [m]
Average agent velocity	v	2.9 [m/s]
Minimum dispersion time	k	17 [s]
Desired performance bound	S	.95
Sensor detect probability	η	.5
Cost of not finishing	α	10 [$/s]
Cost of operation	β	.1 [$/m]
Initialization cost	γ	82 [$/agent]

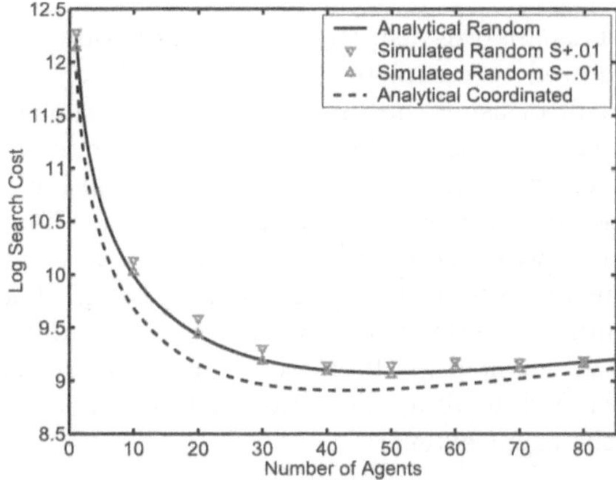

Fig. 2. Simulated and analytical results for this search task. For the simulated data the lower triangles are above $S - .01$ of the cost data and the upper triangles exceed $S + .01$ of the cost data. Good agreement between the simulated and analytical results indicates the random search model assumptions are sound

Figure 2 shows the results of calculating the costs in this system analytically compared to the costs derived experimentally. There is good quantitative agreement between the analytical and simulated results for the random algorithm, suggesting that for this task the assumptions of independence hold and the analytical results are valid. Also, it is worthwhile to note that the optimal group size for both algorithms is well above 1 (so the interest in multiple agents completing this task is warranted), and the optimal cost of the random algorithm is fairly close to that of the optimal system. This suggests that if the increased cost of adding coordination and fault tolerance into the optimal system is significant, the random system (which has fault tolerance built in because all of the agents perform the same actions) may be the most efficient.

5 Conclusion

This paper presented a quantitative analysis of the tradeoffs between group size and efficiency in collective search tasks that considers both the time-sensitive nature of search completion and the system operating cost. First, the search task was defined and a performance metric was presented that can account for all of the costs associated with the task. Note that computation of the cost parameters may not be simple, but estimates are feasible. Also, while the costs used in this paper were linear functions of the task metrics, any differentiable functional form can be used in this framework. Next, for both random and coordinated search strategies, analytical expressions were derived that can be used to predict optimal system performance bounds given a particular task description. This analysis also allowed the prediction of the optimal number of agents required to complete a task most efficiently. In addition, the performance benefit of using coordinated search was shown to be dependent on the relative values of the different cost components, with coordinated search being less favored when the cost of not completing the task significantly outweighs the cost of operating the search system. Finally, an embodied computer simulation was used to support the analytical results, suggesting that the assumptions involved in their derivation are sound. These assumptions, which include minimal interference between agents and uniform coverage of the given arena, will not hold in all environments, but they will be approximately correct for many difficult applications where the area to be searched is much larger than the agent extent.

6 Acknowledgments

I would like to thank Sarah Farivar for her mathematical expertise and her help reviewing this manuscript. This work is supported in part by the Center for Neuromorphic Systems Engineering as part of the NSF ERC Program under grant EEC-9402726. Support has also been received from the ONR under grant N00014-98-1-0821 and from the ARO under MURI grant DAAG55-98-1-0266. Adam Hayes is supported by a NSF Graduate Research Fellowship.

References

1. S. Benkowki, M. Monticino, and J. Wiesinger. A survey of the search theory literature. *Naval Research Logisitics*, 38(4):469–494, 1991.
2. D. W. Gage. Randomized search strategies with imperfect sensors. In *Proceedings of SPIE Mobile Robots VIII*, volume 2058, pages 270–279, Boston, September 1993.
3. T. Balch and R. C. Arkin. Behavior-based formation control for multi-robot teams. *IEEE Robotics and Automation*, 14(6):926–939, December 1998.
4. D. W. Gage. Many-robots mcm search systems. In A. Bottoms, J. Eagle, and H. Bayless, editors, *Proceedings of Autonomous Vehicles in Mine Countermeasures Symposium*, pages 9.55–9.63, Monterey, April 1995.
5. O. Michel. Webots: Symbiosis between virtual and real mobile robots. In *Proceedings of the First International Conference on Virtual Worlds, VW'98*, pages 254–263, Paris, France, July 1998. Springer Verlag.
6. F. Mondada, E. Franzi, and P. Ienne. Mobile robot miniaturization: A tool for investigation in control algorithms. In T. Yoshikawa and F. Miyazaki, editors, *Proc. of the Third International Symposium on Experimental Robotics ISER-93*, pages 501–513, Kyoto, Japan, 1993. Springer Verlag.
7. A. Martinoli, A. J. Ijspeert, and F. Mondada. Understanding collective aggregation mechanisms: From probabilistic modelling to experiments with real robots. *Robotic and Autonomous Systems*, 29:51–63, 1999.
8. A. T. Hayes, A. Martinoli, and R. M. Goodman. Swarm robotic odor localization. In *Proc. of the IEEE Conf. on Intelligent Robots and Systems*, pages 1073–1078, Wailea, HI, October 2001. IEEE Press.

Mobile Sensor Network Deployment using Potential Fields: A Distributed, Scalable Solution to the Area Coverage Problem

Andrew Howard, Maja J Matarić, and Gaurav S Sukhatme

Robotics Research Laboratories, Department of Computer Science, University of Southern California, Los Angeles, CA 90089-0781
ahoward@usc.edu, mataric@usc.edu, gaurav@usc.edu

Abstract. This paper considers the problem of deploying a mobile sensor network in an unknown environment. A mobile sensor network is composed of a distributed collection of nodes, each of which has sensing, computation, communication and locomotion capabilities. Such networks are capable of self-deployment; i.e., starting from some compact initial configuration, the nodes in the network can spread out such that the area 'covered' by the network is maximized. In this paper, we present a potential-field-based approach to deployment. The fields are constructed such that each node is repelled by both obstacles and by other nodes, thereby forcing the network to spread itself throughout the environment. The approach is both distributed and scalable.

Keywords: Distributed robotic systems, sensor networks, deployment, potential fields.

1 Introduction

This paper considers the problem of deploying a mobile sensor network in an unknown environment. A mobile sensor network is composed of a distributed collection of *nodes*, each of which has sensing, computation, communication and locomotion capabilities. It is this latter capability that distinguishes a *mobile* sensor network from its more conventional static cousins. Locomotion facilitates a number of useful network capabilities, including the ability to *self-deploy*; that is, starting from some compact initial configuration, the nodes in the network can spread out such that the area 'covered' by the network is maximized.

Our approach is motivated by the need to deploy sensor networks in environments that may be both hostile and dynamic. Consider, for example, a scenario involving a hazardous materials leak in a damaged structure. We would like our sensor network, whose nodes are equipped with chemical sensors, to rapidly deploy throughout the environment and return real-time data indicating the location and concentration of hazards. This kinds of scenario imposes two important constraints on our deployment algorithm: prior models of the environment are either incomplete, inaccurate or unavailable, and network nodes may be lost or destroyed.

In this paper, we describe a potential-field-based approach to deployment, in which nodes are treated as virtual particles, subject to virtual forces. These forces repel the nodes from each other and from obstacles, and ensure that an initial, compact configuration of nodes will quickly spread out to maximize the coverage area of the network (it should be noted that nowhere do we reason about coverage explicitly; rather, coverage is an emergent property of the algorithm). In addition to these repulsive forces, nodes are also subject to a viscous friction force. This force is used to ensure that the network will eventually reach a state of static equilibrium; i.e., all nodes will ultimately come to a complete stop. The viscous force does not, however, prevent the network from reacting to *changes* in the environment; if something is moved, the network will automatically reconfigure itself for the modified environment before return once again to a static equilibrium. Thus, nodes move only when it is necessary to do so, saving a great deal of energy.

The potential field approach described in this paper relies on only one assumption: that each node is equipped with a sensor that allows it to determine the range and bearing of both nearby nodes and obstacles (suitable sensors can be constructed using scanning laser range-finder or omni-camera). Using this information, the node can determine the virtual forces acting it, and convert this information into a control vector to be sent to its motors. No other information is required. It should be emphasized that this approach *does not* require models of the environment, localization, or communication between nodes. As a result, the algorithm is both robust and highly scalable.

In the remainder of this paper, we develop the potential field theory underlying the deployment algorithm, and demonstrate that this algorithm has the desired property that the network will converge to a state of static equilibrium. We describe a series of simulation experiments that both validate the general approach and reveal some of its emergent properties. These experiments include realistic sensor-based simulations of a network containing 100 nodes in a large, complex environment.

2 Related Work

The concept of *coverage* as a paradigm for evaluating many-robot systems was introduced by Gage [5]. Gage defines three basic types of coverage: blanket coverage, where the objective is to achieve a static arrangement of nodes that maximizes the total detection area; barrier coverage, where the objective is to minimize the probability of undetected penetration through the barrier; and sweep coverage, which is more-or-less equivalent to a moving barrier. According to this taxonomy, the deployment problem described in this paper is a blanket coverage problem.

Potential field techniques for robotic applications were first described by Khatib [10] and have since been widely used in the mobile robotics community for tasks such as local navigation and obstacle avoidance. The related concept of 'motor schemas', which utilizes the super-position of spatial vector fields to generate behavior was introduced by Arkin [1]. Both techniques have since been applied to the problem of formation control for groups of mobile robots [13,2]. The formation problem is similar, in some respects, to the deployment problem described in this paper, in that the robots will attempt to maintain a formation based on local sensing and computation. A key difference, however, is that there is no requirement that the formation reach a state of static equilibrium.

The deployment problem also is also similar, in some respects, to the multi-robot exploration and mapping problem. Here, the aim is to build a global map of the environment by sequentially visiting each location with one or more robots. This problem has been considered by a number of authors [4,15,14,3] who use a variety of techniques ranging from topological matching [4] to fuzzy inference [11] and particle filters [16]. Two good examples are provided by Simmons [14] and Burgard [3], both of whom build global maps, apply heuristics to select goal locations for exploration, and use explicit communication to prevent more than one robot from heading for the same goal. This approach to *exploration* contrasts markedly with the approach to *deployment* described in this paper. As we will show in Section 4, potential field methods are able to achieve good coverage without global maps, without communication, and without explicit reasoning. Instead, area coverage is an emergent, system-level property.

Finally, we note that the problem of deployment is related to the traditional *art gallery* problem in computational geometry [12]. The art gallery problem seeks to determine, for some polygonal environment, the minimum number of cameras that can be placed such that the entire environment is observed. While there exist a number of algorithms designed to solve the art gallery problem, all of these assume that we possess good prior models of the environment.

3 Potential Fields

Potential fields are a commonly used and well understood method in mobile robotics, where they are typically applied to tasks such as local navigation and obstacle avoidance. In this paper, we apply potential fields to the deployment problem. The fields are constructed in such a way that each node is repelled by both obstacles and by other nodes, thereby forcing the network to spread itself throughout the environment.

The basic potential field method is as follows. Each node is subject to a force $\mathbf{F}$ that is the gradient of a scalar potential field U; i.e.,

$$\mathbf{F} = -\nabla U \tag{1}$$

We divide the potential field into two components: the field U_o due to obstacles, and the field U_n due to other nodes; these fields give rise to repulsive forces $\mathbf{F}_o$ and $\mathbf{F}_n$, respectively. Thus $U = U_o + U_n$ and $\mathbf{F} = \mathbf{F}_o + \mathbf{F}_n$.

Consider the potential field due to obstacles. If we imagine that each node and each obstacle carries an electric change, we can write down an expression for the resultant 'electrostatic' potential:

$$U_o = k_o \sum_i \frac{1}{r_i}. \tag{2}$$

The summation is over all obstacles that can be seen by the node, k_o is a constant describing the strength of the field, and r_i is the Euclidean distance between the node and obstacle i. Let $\mathbf{x}$ denote the position of the node and let $\mathbf{x}_i$ denote the position of obstacle i. The distance r_i is then given by $r_i = \mid \mathbf{x}_i - \mathbf{x} \mid$. Using these

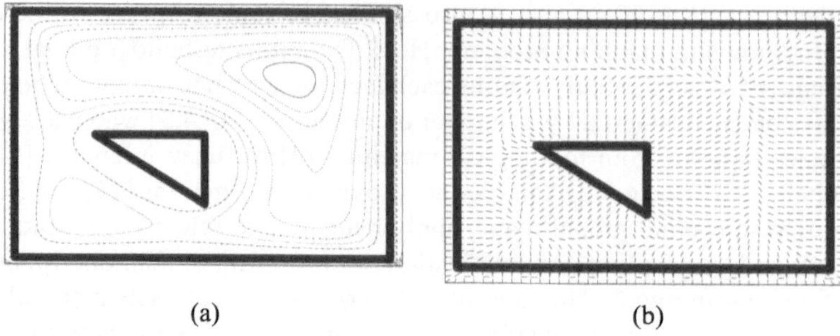

(a) (b)

Fig. 1. (a) Potential field generated by a simple environment; the contours show the lines of equal potential. (b) Force fields generated by this potential; the arrows indicate the direction (but not magnitude) of the force.

definitions, the total force F_o due to obstacles can be computed using Equation 1. We re-write the equation and expand using the chain rule, as follows:

$$\mathbf{F}_o = -\frac{\mathrm{d}U_o}{\mathrm{d}\mathbf{x}} = -\sum_i \frac{\mathrm{d}U_o}{\mathrm{d}r_i} \cdot \frac{\mathrm{d}r_i}{\mathrm{d}\mathbf{x}}. \tag{3}$$

We then insert the appropriate derivatives to obtain:

$$\mathbf{F}_o = -k_o \sum_i \frac{1}{r_i^2} \cdot \frac{\mathbf{r}_i}{r_i} \tag{4}$$

where $\mathbf{r}_i = \mathbf{x}_i - \mathbf{x}$. Note that the force is expressed entirely in terms of the relative positions $\mathbf{r}_i$ of obstacles, rather than their absolute positions $\mathbf{x}_i$. This allows us to compute the force directly from sensor data, without the need for global localization. Figure 1 shows the potential field U_o and force field $\mathbf{F}_o$ generated by a simple environment.

Consider now the potential field U_n due to other nodes. By analogy with the obstacle field, we can derive expressions for the potential U_n and force $\mathbf{F}_n$ by replacing a summation over visible obstacles with a summation over visible nodes; thus:

$$U_n = -k_n \sum_i \frac{1}{r_i} \qquad \text{and} \qquad \mathbf{F}_n = -k_n \sum_i \frac{1}{r_i^2} \cdot \frac{\mathbf{r}_i}{r_i} \tag{5}$$

where $\mathbf{r}_i$ is the relative position of node i.

3.1 The Equation of Motion and The Control Law

The trajectory of a node subject to force $\mathbf{F}$ can be computed using an appropriate *Equation of Motion*. We use an equation of the following form:

$$\ddot{\mathbf{x}} = (\mathbf{F} - \nu\dot{\mathbf{x}})/m \tag{6}$$

where $\ddot{\mathbf{x}}$ denotes the acceleration of the node and m denotes its mass. The second term on the right hand side of this equation is a *viscous friction* term in which ν is the viscosity coefficient and $\dot{\mathbf{x}}$ is the node velocity. This term is used to ensure that, in the absence of external forces, the node will eventually come to a standstill. [1] In Section 3.2 we will show that the viscous friction term also guarantees that the network *as a whole* will ultimately reach a state of static equilibrium (i.e., a state in which all nodes have stopped moving).

Our discussion up to this point has focused entirely on a *virtual* physical system, i.e., one in which the forces, accelerations, masses, etc. are entirely imaginary. This virtual physical system must, however, be mapped onto a *real* physical system made up of real nodes. Typically, these nodes will have some form of velocity controller; consequently, the mapping from virtual to real physical system is achieved by defining a *control law* that maps a virtual force onto a velocity *control vector*. In most applications using potential field techniques, such as single-robot obstacle avoidance, the control law can be entirely arbitrary; in this case, there is little to gain from preserving the correspondence between a robot's real and virtual dynamics. In our application however, it is extremely useful to retain this correspondence: if our control law is such that the nodes obey the Equation of Motion, we can be certain that the network *will* reach static equilibrium (see Section 3.2). This is not the case for any arbitrary control law.

In deriving a control law for a real node, we must be cognizant of the fact that real nodes are not 'free particles': they have both kinematic and dynamic constraints. The kinematic constraints can be largely ignored if we make the assumption that the nodes have holonomic drive mechanisms (i.e. they can move equally well in any direction). [2] The dynamic constraints, however, cannot be ignored: the node will have both a maximum velocity and a maximum acceleration, which must be captured by the control law.

Our control law can be expressed algorithmically as follows. Let $\mathbf{v}$ denote the commanded velocity at some time t, and let $\Delta\mathbf{v}$ denote the change in the commanded velocity between times t and $t + \Delta t$. The change in commanded velocity is determined using a piecewise-constant approximation to the Equation of Motion (Equation 6):

$$\Delta\mathbf{v} \longleftarrow (\mathbf{F} - \nu\mathbf{v})/m \cdot \Delta t. \tag{7}$$

The x and y components of $\Delta\mathbf{v}$ are subsequently 'clipped' such that $-a_{\max} \leq \Delta\mathbf{v} \leq a_{\max}$ where $a_{\max}$ denotes the largest allowable change in velocity. The commanded velocity $\mathbf{v}$ is determined using:

$$\mathbf{v} \longleftarrow \mathbf{v} + \Delta\mathbf{v} \tag{8}$$

[1] Strictly speaking, the node will never come to a complete stop; rather, its velocity will approach zero asymptotically.

[2] Even a standard differential drive mechanism can be treated as a holonomic platform if one is prepared to sacrifice the rotational degree of freedom. Furthermore, if one has omni-directional sensors, this sacrifice has no functional impact.

and is then clipped to the domain $-v_{\max} \leq \mathbf{v} \leq v_{\max}$ where $v_{\max}$ is maximum allowed velocity.

Using this control law, the real node dynamics will closely approximate those described by the Equation of Motion. There are, however, two regimes in which the correspondence will fail. Firstly, for small $\mathbf{v}$, the viscous friction term will tend to produce oscillation rather than asymptotic convergence to zero velocity; this kind of behavior is typical of discrete control systems and can be eliminated by introducing a velocity 'dead-band'. Secondly, large accelerations and velocities will simply be clipped, in which case the deviation from the virtual dynamics may become arbitrarily large. This deviation is significant only if it prevents the network as a whole from reaching static equilibrium, or if it significantly increases the time taken to reach this equilibrium. We assert (without proof) that the acceleration and velocity limits act like additional non-linear friction terms, and that therefore, these limits will not prevent the system from reaching static equilibrium. The limits may, however, impact on the time taken to reach equilibrium, and this impact must be determined empirically.

3.2 Static Equilibrium

One can show that the network as a whole will reach a static equilibrium (i.e. a situation in which all nodes are stationary) by considering the total energy of the system. Each node has both potential and kinetic energy: the former arises from the node's interaction with the potential field, the latter from the node's motion. The total energy of the system is determined by summing these energies for all nodes. The Equation of Motion includes a viscous friction term that has the effect of removing energy from the system; i.e., the system is said to be *dissipative* [8]. For such systems, the total energy will decrease monotonically over time, and since the potential energy of the system is bounded from below, this necessarily implies that the total kinetic energy of the system will asymptote to zero. Clearly, therefore, the network as a whole must asymptotically approach static equilibrium.

This argument rests on the assumption that the environment itself is static, and therefore does not introduce additional energy into the system or modify the space of reachable states. In a dynamic environment, however, energy may be added to or subtracted from the system whenever an object is moved by some agency other than the network itself. Furthermore, states which where previously unreachable may now become reachable, and vice-versa. As a consequence, in an environment that is continually changing, we do not expect the network to reach a state of static equilibrium. If, however, the environment is changing *periodically* or *intermittently*, the network will reach static equilibrium, but the equilibrium state may be different after each change. Consider, for example, a network placed in a closed room: the network will deploy to fill the room and then stop. If the door to the room is now opened, the network will start deploying again, spreading beyond its original confines to seek a new equilibrium state.

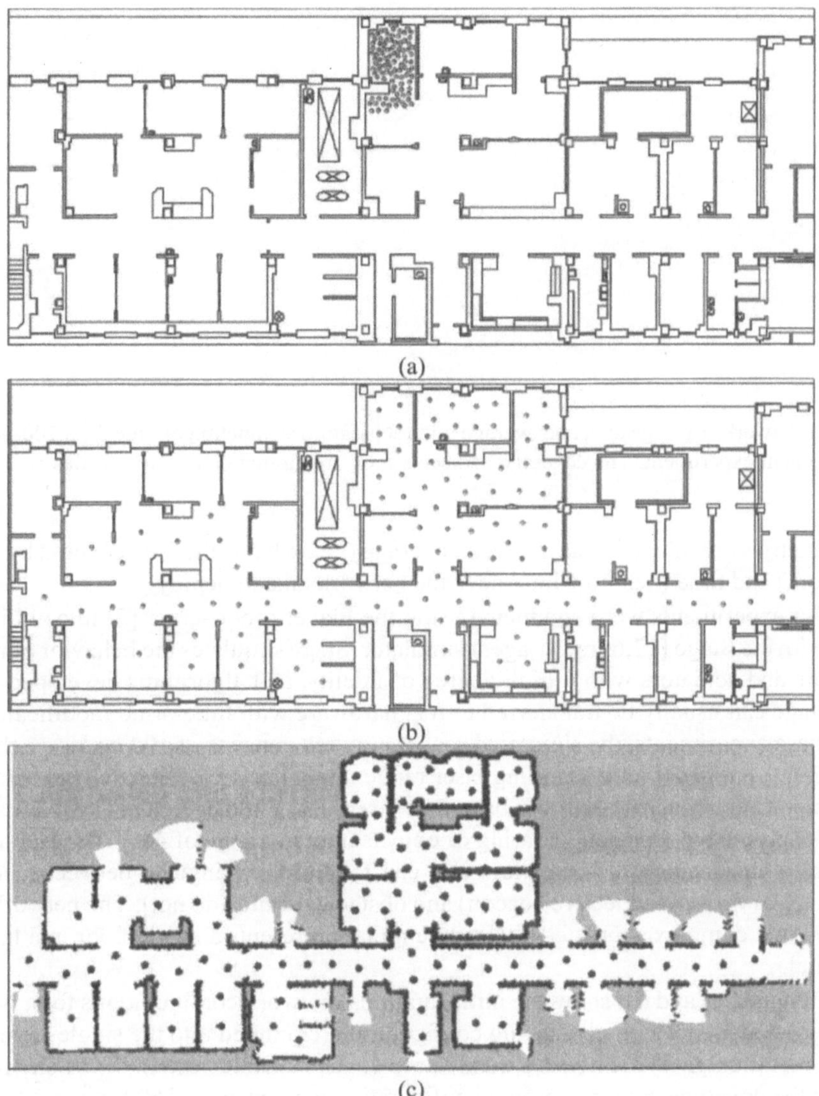

Fig. 2. A proto-typical deployment experiment for a 100-node network. (a) Initial network configuration. (b) Final configuration after 300 seconds. (c) Occupancy grid generated for the final configuration; visible space is marked in black (occupied) or white (free); unseen space is marked in gray.

4 Experiments

We have conducted a series of simulation experiments aimed at both validating and investigating the use of potential fields for the sensor network deployment problem.

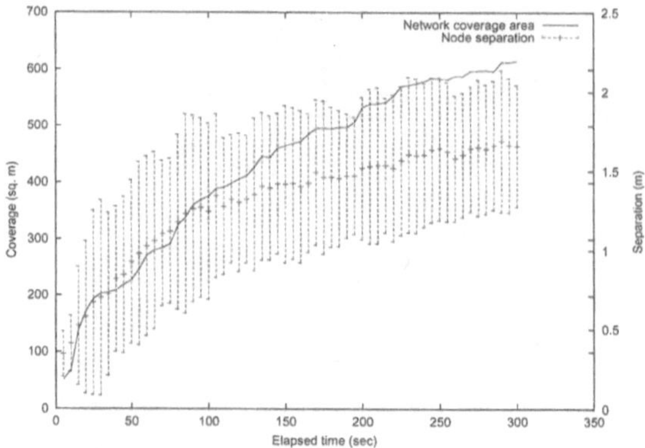

Fig. 3. Network coverage area and average node separation as function of time for a 100-node deployment experiment. The coverage and separation are plotted on different scales.

Two metrics are of particular interest: coverage (i.e., what is the area covered by the network) and time (i.e., how long does the network take to deploy).

Our experiments were conducted using the Player robot server [7] in combination with the Stage [17,6] multi-agent simulator. Stage simulates the behavior of real sensors and actuators with a high degree of fidelity, and algorithms developed using Stage can usually be transferred to real hardware with little or no modification. For these experiments, the simulated sensor network consists of 100 nodes, each of which is equipped with a scanning laser range finder, a retro-reflective beacon and an omni-directional mobile robot base. The laser has a 360 degree field-of-view and can determine the range and bearing of objects out to a range of 4m. The laser also returns some intensity information, and can therefore distinguish between nodes (which carry a retro-reflective beacon) and obstacles (which do not). The network is placed in a complex simulated environment that represents a single floor in a large hospital.

Figures 2(a) and (b) show the initial and final network configurations for a typical deployment. From their starting configuration (crammed into the single room at the top of the figure) the nodes spread out to cover a sizable portion of the environment; the coverage area in the final configuration is in excess of 500 m^2, a 10-fold improvement over the initial coverage of around 50 m^2. The temporal behavior of the network is captured Figure 3(a), which shows a plot of coverage versus deployment time. From this plot, it is apparent that the rate of coverage decreases with time, and that the total coverage was still increasing when the experiment was terminated after 300 seconds. We plan to investigate and characterize this curve more carefully in future experiments.

The deployment is quite fast. The elapsed time for this experiment is 300 seconds, in which time the nodes on the network boundary have traveled a distance of around 40 m. Since the maximum permitted velocity in this experiment is 0.5 m/s, the average velocity of the boundary nodes during deployment is just under 15% of the theoretical maximum. If we restict ourselves to the early phase of the deployment, the average velocity is higher still: over the first 180 seconds, the nodes reach 35% of the theoretical maximum velocity.

An unexpected, but appealing, feature of the deployment is the evenness of the node spacing: the average nearest-neighbor separation in the final configuration is $1.6 \pm 0.4m$. The variance is surprisingly low given the lack explicit coordination between nodes and the structural variability of the environment. As for the separation distance, it is unclear, at this point, whether the distance is related to external phenomena, such as the scale of features in the environment, or is a function of purely internal factors, such as the relative weights on the potential fields. In an open environment, we might expect the average separation to approach 4 m, corresponding to the range limit on the laser's field-of-view; it is not obvious, however, what value we should expect in a highly structured environment such as the one used in this experiments.

The network coverage produced in this experiment is of a very high quality. Figure 2(c) shows an occupancy grid generated for the final configuration: areas that can be seen by the network are shown in black (for obstacles) or white (for open space); unseen areas are shown in gray. Note that there are no gaps or breaks in the coverage. The high quality of this coverage can be attributed to the even spacing of nodes, combined with the fact that the average node separation is about half the sensor range. This effectively creates a dense, highly redundant network.

Animations of this and other experiments can be found at:

`http://robotics.usc.edu/~ahoward/movies.html`.

5 Conclusion and Further Work

The experiments described in Section 4 are far from complete. To fully characterize the approach described in this paper, we need to perform a much more extensive series of experiments, in which we vary both external factors (such as network size, environment, and initial conditions) and internal factors (such as the weights k_o and k_n, the node mass m and viscosity coefficient ν). We are currently in the process of conducting such experiments.

The experiments described in this paper are, however, quite sufficient to demonstrate that a potential field approach *can* be used to deploy mobile sensor networks. The approach has the advantage that it does not require centralized control, localization or communication, and will therefore scale to very large networks. Furthermore, as demonstrated in Section 3.2, this approach has provable convergence characteristics.

There are a number of directions in which we would like to expand this research. We are interested, for example, in how one might apply this approach to coverage problems in which *line-of-sight connectivity* is important [9]. For these problems, would like the deployment to proceed such that the network is fully connected at all times by line-of-sight relationships. In principle, this requires form of communication between nodes; in practice, however, it may be the case that connectivity, like area coverage, can emerge from a combination of purely local rules.

References

1. R. C. Arkin. Motor schema based mobile robot navigation. *International Journal of Robotics Research*, 8(4):92–112, 1989.
2. T. Balch and M. Hybinette. Behavior-based coordination of large-scale robot formations. In *Proceedings of the Fourth International Conference on Multiagent Systems (ICMAS '00)*, pages 363–364, Boston, MA, USA, July 2000.
3. W. Burgard, M. Moors, D. Fox, R. Simmons, and S. Thrun. Collaborative multi-robot exploration. In *Proc. of IEEE International Conferenceon Robotics and Automation (ICRA)*, volume 1, pages 476–81, 2000.
4. G. Dedeoglu and G. S. Sukhatme. Landmark-based matching algorithms for cooperative mapping by autonomous robots. In L. E. Parker, G. W. Bekey, and J. Barhen, editors, *Distributed Autonomous Robotics Systems*, volume 4, pages 251–260. Springer, 2000.
5. D. W. Gage. Command control for many-robot systems. In *AUVS-92, the Nineteenth Annual AUVS Technical Symposium*, pages 22–24, Hunstville Alabama, USA, June 1992. Reprinted in Unmanned Systems Magazine, Fall 1992, Volume 10, Number 4, pp 28-34.
6. B. Gerkey, R. Vaughan, and A. Howard. Player/Stage homepage. http://robotics.usc.edu/player/, September 2001.
7. B. P. Gerkey, R. T. Vaughan, K. Støy, A. Howard, G. S. Sukhatme, and M. J. Matarić. Most valuable player: A robot device server for distributed control. In *Proc. of the IEEE/RSJ Intl. Conf. on Intelligent Robots and Systems (IROS01)*, pages 1226–1231, Wailea, Hawaii, Oct. 2001.
8. H. Goldstein. *Classical mechanics*. Addison-Wesley, 1980.
9. A. Howard, M. J. Matarić, and G. S. Sukhatme. Localization for mobile robot teams: A maximum likelihood approach. Technical Report IRIS-01-407, Institute for Robotics and Intelligent Systems Technical Report, University of Sourthern California, 2001.
10. O. Khatib. Real-time obstacle avoidance for manipulators and mobile robots. *International Journal of Robotics Research*, 5(1):90–98, 1986.
11. M. López-Sánchez, F. Esteva, R. L. de Mántaras, C. Sierra, and J. Amat. Map generation by cooperative low-cost robots in structured unknown environments. *Autonomous Robots*, 5(1):53–61, 1998.
12. J. O'Rourke. *Art Gallery Theorems and Algorithms*. Oxford University Press, New York, August 1987.
13. F. E. Scheider, D. Wildermuth, and H.-L. Wolf. Motion coordination in formations of multiple mobile robots using a potential field approach. In L. E. Parker, G. W. Bekey, and J. Barhen, editors, *Distributed Autonomous Robotics Systems*, volume 4, pages 305–314. Springer, 2000.
14. R. Simmons, D. Apfelbaum, W. Burgard, D. Fox, M. Moors, S. Thrun, and H. Younes. Coordination for multi-robot exploration and mapping. In *Proc. of the Seventeenth National Conference on Artificial Intelligence (AAAI-2000)*, pages 852–858, 2000.
15. S. Thrun, W. Burgard, and D. Fox. A real-time algorithm for mobile robot mapping with applications to multi-robot and 3d mapping. In *Proceedings of the IEEE International Conference on Robotics and Automation (ICRA2000)*, volume 1, pages 321–328, 2000.
16. S. Thrun, D. Fox, W. Burgard, and F. Dellaert. Robust monte carlo localization for mobile robots. *Artificial Intelligence Journal*, 128(1–2):99–141, 2001.
17. R. T. Vaughan. Stage: a multiple robot simulator. Technical Report IRIS-00-393, Institute for Robotics and Intelligent Systems, University of Southern California, 2000.

Map Acquisition in Multi-Robot Systems based on Time Shared Scheduling

Yoshikazu ARAI[1], Hajime ASAMA[2], and Hayato KAETSU[2]

[1] Iwate Prefectural University, Sugo 152-52, Takizawa, Iwate 020-0173, Japan
[2] The Institute of Physical and Chemical Research (RIKEN), Hirosawa 2-1, Wako, Saitama 351-0198, Japan

arai@soft.iwate-pu.ac.jp, {asama, kaetsu}@cel.riken.go.jp

Abstract. In this paper, we propose a map acquisition method in multi-robot systems. Distance information for robot's surrounding environment is very useful to acquire environmental maps and plan robot's paths. However it is difficult for active sensors, such as ultrasonic sensors, to measure distances accurately in an environment in which multiple robots exist because emissions from the sensors interfere each other. An exclusive operation of emission for the sensors is one of solutions to avoid interference of the emission, when a robot measures distances for its surrounding objects in such environment. We have already proposed the time shared scheduling to make synchronization between sensors for achieving the exclusive operation using local communication. In this paper, we introduce the occupancy grid method and propose map acquisition method in multi-robot systems based on the time shared scheduling. After a concept of the time shared scheduling to make synchronization, the environmental map acquisition method based on the scheduling is described. A computational simulation is conducted to show efficiency of proposed method.

Keywords: Environmental Map, Occupancy Grid, Time Shared Scheduling, Multi-Robot System, Local Communication

1 Introduction

To recognize a working environment and plan paths of movements, distances between a robot and surrounding objects are very useful information. Ultrasonic sensors are used widely to acquire distance information [5,6]. However, in Distributed Autonomous Robotic Systems (DARS) [2,3,9,11], it is difficult for the sensors to measure distances accurately due to interference of ultrasonic waves emitted from multiple sensors at the same time. It is very important to resolve this problem for distance measurement. Kawabata et al. have proposed a method which detects a target wave from interfered ultrasonic waves [7]. In this kind of method based on such detection, special signal processings are required to calculate such as a cross correlation between interfered waves and a wave which a robot emitted. On the other hand, there is another method based on controlling not to interfere for emissions and letting a wave occupy a space momentarily. Although synchronization of timing between multiple sensors is very important, robots have to communicate each other to carry out such synchronization. As a method to exchange information between robots, global communications have been used generally [4,13]. These kinds of methods are not suitable for the above use which requires frequent communication in a short cycle because of large overheads for communication. In addition, since robots which should synchronize each other are close, it is reasonable to communicate locally only between them. For this problem, we have proposed a method for distance measurement in multi-robot systems based on the time shared scheduling by synchronizing timing for emitting ultrasonic waves with local communication between robots [1]. By applying this method, robots can measure accurate distance for surrounding objects and recognize a working environment even in multi-robot systems without complex signal processings.

In this paper, we propose a method of environmental map acquisition based on distance information in multi-robot system for individual robot. To build the environmental map, the occupancy grids method is introduced and existence of any objects for each grid in the environment is calculated stochastically. Finally, a validity of proposed method is shown through an experiment of map acquisition in a simulation environment.

2 Distance Measurement

To measure distances for surrounding objects without interference of emission, it is important to make synchronization of emitting between sensors mounted on multiple robots. In this section, a local communication system for multi-robot systems, LOCISS (LOcally Communicable Infrared Sensory System), which we have developed and the time shared scheduling using the LOCISS to make synchronization between sensors are described.

2.1 Local Communication among Multiple Robots

The LOCISS is a sensing system which utilizes infrared light as media and achieves local communication between robots by adjusting intensity of emissions [12]. By receiving a signal of the LOCISS, a robot detects surrounding objects. When an object is detected by the LOCISS, it is important to discriminate whether the object is other robot or an obstacle to know how the robot should schedule a timing of emitting for distance measurements as described later. For this purpose, an unique ID number for each robot (robot ID) is transmitted as a code which shows a source of signal as shown in Fig. 1. When the robot receives other robot's ID number, it can recognize and discriminate the other robot besides it. When the robot receives its own ID number, it can recognize an obstacle because the signal is reflected by the obstacle.

A transmitter and a receiver of sensors are configured as shown in Fig. 2. The devices of an infrared and an ultrasonic sensors are synchronized and

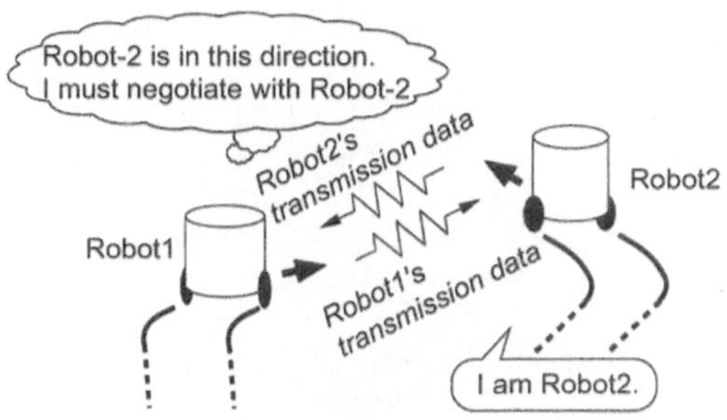

Fig. 1. Concept of discriminating surrounding robot

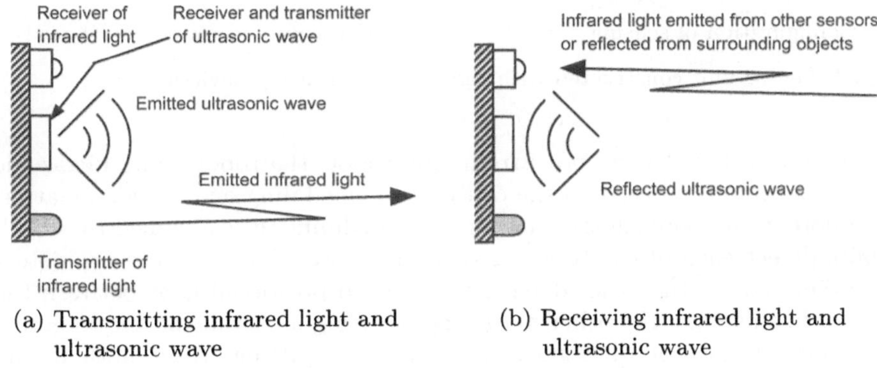

(a) Transmitting infrared light and ultrasonic wave

(b) Receiving infrared light and ultrasonic wave

Fig. 2. Structure of LOCISS and ultrasonic sensor

repeat transmitting (Fig. 2(a)) and receiving (Fig. 2(b)) in a same sampling cycle. A sensor unit has eight pairs of such devices radially, as shown in Fig. 3(a). Each pairs of infrared devices in the LOCISS has a directivity of 45 degrees to communicate with robots in all direction by eight pairs of devices. A communicable area of the device is limited to about 1 [m] according to characteristics for movement of a robot. The sensor unit mounted on the omni-directional mobile robot is shown in Fig. 3(b).

2.2 Time Shared Scheduling

Although a scheduling of timing for emitting based on time sharing scheme (time shared scheduling) expects that accurate distance measurement is accomplished, it also brings reduction of samplings. Therefore application of this scheduling should be minimized and a scheduling of timing for continuous emitting (continuous scheduling) should be usually applied. The necessity of the time shared scheduling depends on existence of surrounding robots. Therefore a robot detects the surrounding robot by using the LOCISS and switches the above two scheduling policies. The robot emits ultrasonic waves based on the timing planed on the scheduling policies.

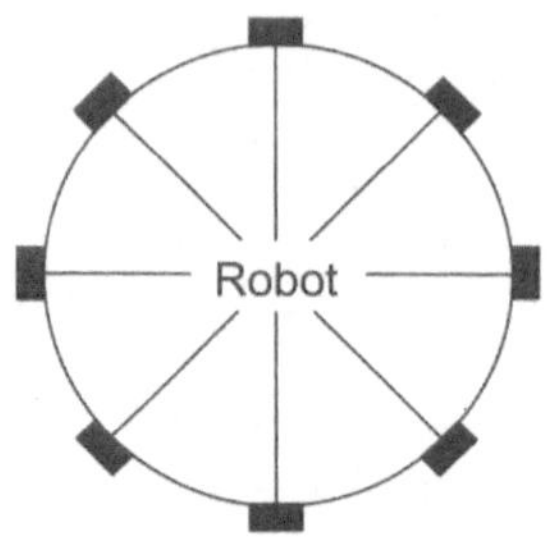

(a) Configuration of sensing devices

(b) Sensor unit mounted on robot

Fig. 3. Sensor unit constructed with eight pairs of sensing devices

While a robot detects no surrounding robot, the robot emits ultrasonic waves continuously based on the continuous scheduling on one-sided matter. Therefore, to accomplish the time shared scheduling, it is assumed that both robots detect each other. However there is a case, that one dose not detect the other though the other detects one, due to positional gaps between the robots. In this case, the other must stop emitting and wait for one to detect the other because one has no timing to stop emitting. After both robots detect each other, they alternate emitting and waiting frequently. Based on the above consideration, the following states and behaviors which a robot should assume and execute in each state are designed to accomplish the time shared scheduling.

- S_0 : No surrounding robot is detected $\rightarrow$ emitting
- S_1 : The robot detects surrounding robot, but oneself is not detected $\rightarrow$ waiting
- S_2 : The time shared scheduling is accomplished and the robot should emit in this turn $\rightarrow$ emitting
- S_3 : The time shared scheduling is accomplished and the robot should wait in this turn $\rightarrow$ waiting

In addition, the factors which lead changes between the above states are also designed as follows.

- d_0 : The robot detects surrounding robot
- d_1 : The robot is detected by surrounding robot
- d_2 : robot ID of the robot $<$ robot ID of surrounding robot

where d_2 denotes a factor to decide which robot emits firstly. These factors are recognized by analyzing own information and one received from surrounding robot through the LOCISS, that is, a robot ID and a detection flag. The

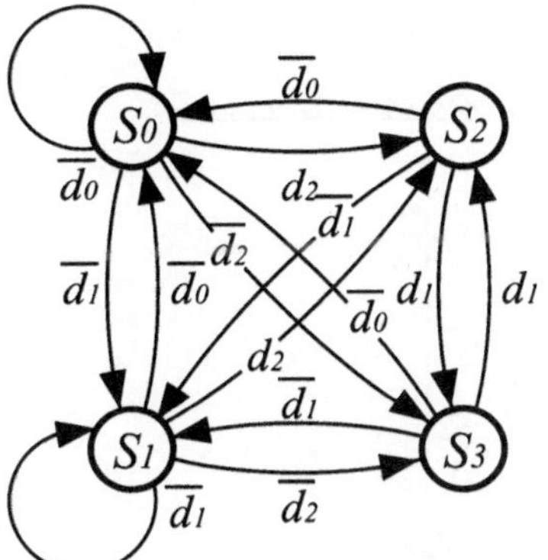

Fig. 4. Transition diagram for time shared scheduling

detection flag is set to ON when a robot receives a signal from surrounding robot, or to OFF when a robot receives no signals. The flag shows a state for detection of surrounding robot.

By catching the above factors as inputs to occur transitions of states, the time shared scheduling model is represented as a kind of automaton. The transition diagram of this automaton is shown in Fig. 4. The initial state of the diagram is state S_0. $\overline{d_n}(n = 0, 1, 2)$ means negatives of factors $d_n(n = 0, 1, 2)$, respectively. The inputs have higher priority in order of d_0, d_1, d_2 and they are evaluated until the transition of state is decided based on the diagram in the order.

3 Acquisition of Environmental Map

As a method to build an environmental map, the occupancy grids method [10][8] is introduced. In this method, an environment is divided by grids and probability of object's existence in each grid is estimated based on distance information. A position of grid i is represented as (r_i, θ_i) on the polar coordinates originated on a sensing device (Fig. 5). When a distance D between a robot and an object is measured by the sensing device, probability density of object's existence on grid i is calculated as follows.

$$p_i = \frac{\alpha(r_i)}{2\pi\delta(r_i)\sigma} e^{-\frac{\theta_i{}^2}{2\sigma^2}} e^{\frac{(D-r_i)^2}{2\delta(r_i)^2}}$$

where $\alpha(r_i)$ and $\delta(r_i)$ denote the attenuation of detection with distance and the range variance increasing with distance, respectively. σ is the emission width of sensing device. Final probability of object's existence on the grid is estimated considering not only p_i but also probability density in which no object exists on the grid. Probability of objects' existence on all of grid in sensing area of each sensing device is set as shown in Fig. 6 by estimating

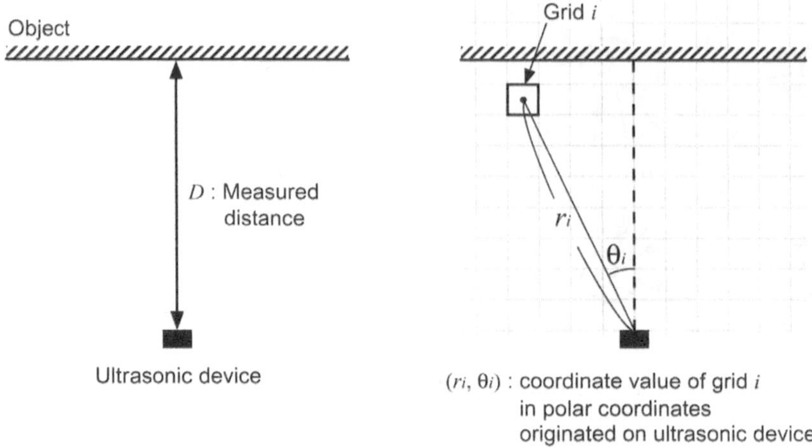

Fig. 5. Parameters for occupancy grid method on each grid

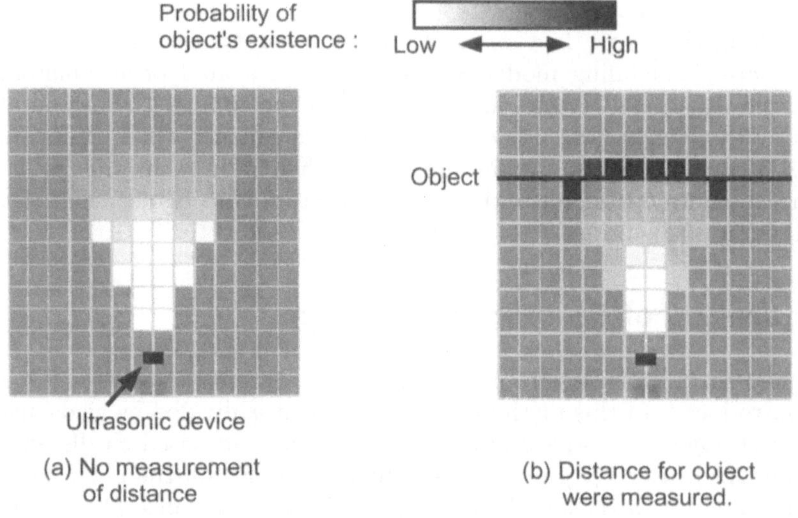

Fig. 6. Estimation of probability density of object's existence on each grid

according to the above procedure. In the case that no distance is measured, probability becomes lower on near grids and becomes unclear on far grids due to attenuation. In the other case that a distance between an object and a sensing device is measured, probability on grids located at interval of the distance from the device becomes higher.

In general, it is desired to put only static obstacles on an environmental map because moving object, such as a robot, does not always exist in a fixed point. Therefore, distance information measured for moving object should not be reflected for building of environmental maps in a dynamic environment.

Stationary object
"usualy exists there".

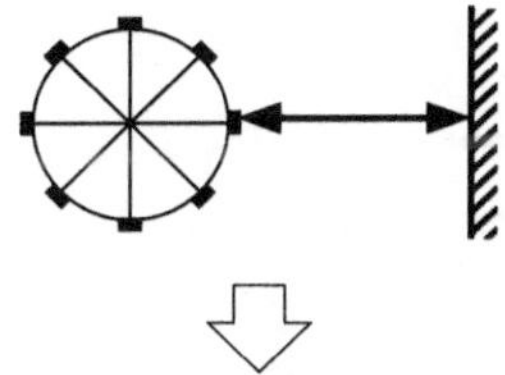

Recording into environmental map
is neccessary.

(a) Static environment

Moving object (robot)
"exists there unexpectedly".

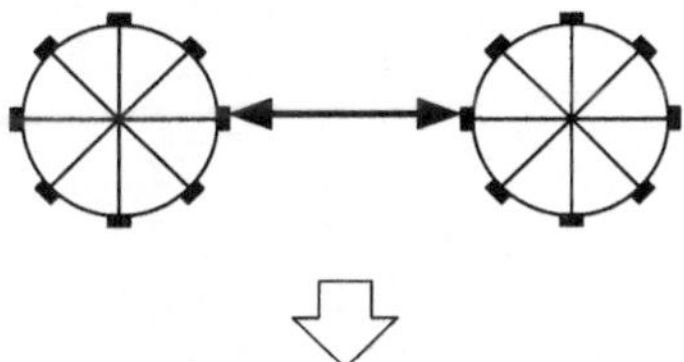

Recording into environmental map
is unneccessary.

(b) Dynamic environment

Fig. 7. Recording probability of object's existence into environmental map according to target of measurement

In this method, when a distance for an object is measured, the object is discriminated whether it is a robot or not based on information received by the LOCISS. If the object is a robot, distance information is canceled as shown in Fig. 7. Consequently, an environmental map only for static objects is acquired without influence of moving robots.

4 Simulation Experiment

To confirm validity of proposed method, an experiment of environmental map acquisition based on the time shared scheduling was conducted on a simulation environment. In this experiment, robots have omni-directional mobility and can move toward any direction without changes of its orientation. The robots mount sensor units composed of eight pairs of ultrasonic devices and infrared devices and can communicate and measure distances for all directions. A communicable area of the LOCISS is designed about 1 [m] and a sensing area of ultrasonic device is also set adjusting LOCISS's range.

An experimental environment is shown in Fig. 8. There are two robots on the passage environment with 2000 [mm] wide. Each robot moves along given path from its start to goal while measuring distances for surrounding objects. Two robots encounter each other at a center of the environment which are drawn with black line. Then they communicate each other with the LOCISS and start sensing based on the time shared scheduling.

Environmental maps which two robots acquired based on the occupancy grids method individually are shown in Fig. 9. To put up picture lower means that time has passed. Probability of objects' existence is illustrated as a gradation of grids. White one means lower probability and black one means higher probability. An initial value of probability with no estimation is set as

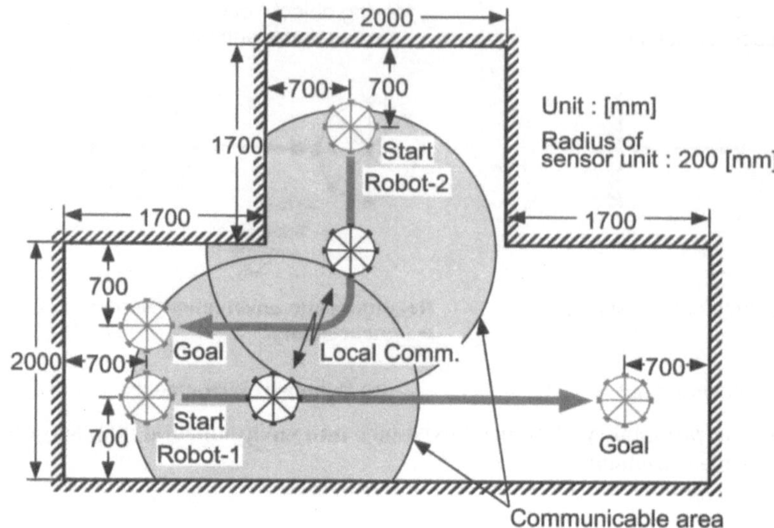

Fig. 8. Environment of experiment

gray which means unknown. Black dots illustrate trajectories of the robots at a constant time interval.

In the environmental maps Fig. 9(a), (b) which the robots acquired, it is confirmed that grids on passage area became white and girds on near objects became black. On third picture for each robot, the robots encountered and the time shared scheduling was started. As a result of discrimination of objects, areas of white grids were distorted just as if no distance was measured because distance information was canceled. Consequently, it is confirmed that multiple robots can build and acquire an environmental map only for static objects without influence from surrounding robots each other individually.

5 Conclusion

We proposed a new environmental map acquisition method based on the time shared scheduling in an environment in which multiple robots exist. The robot can measure distance information for surrounding environment without interference of emissions because they can make synchronization based on local communication. The synchronization can let an ultrasonic wave occupy a space momentarily and allow sensors to measure accurate distances even in multi-robot systems. Therefore this method can cope with situations in which more robots exist. An achievement of robot navigation based on correspondence of given environmental map and acquired one is future work.

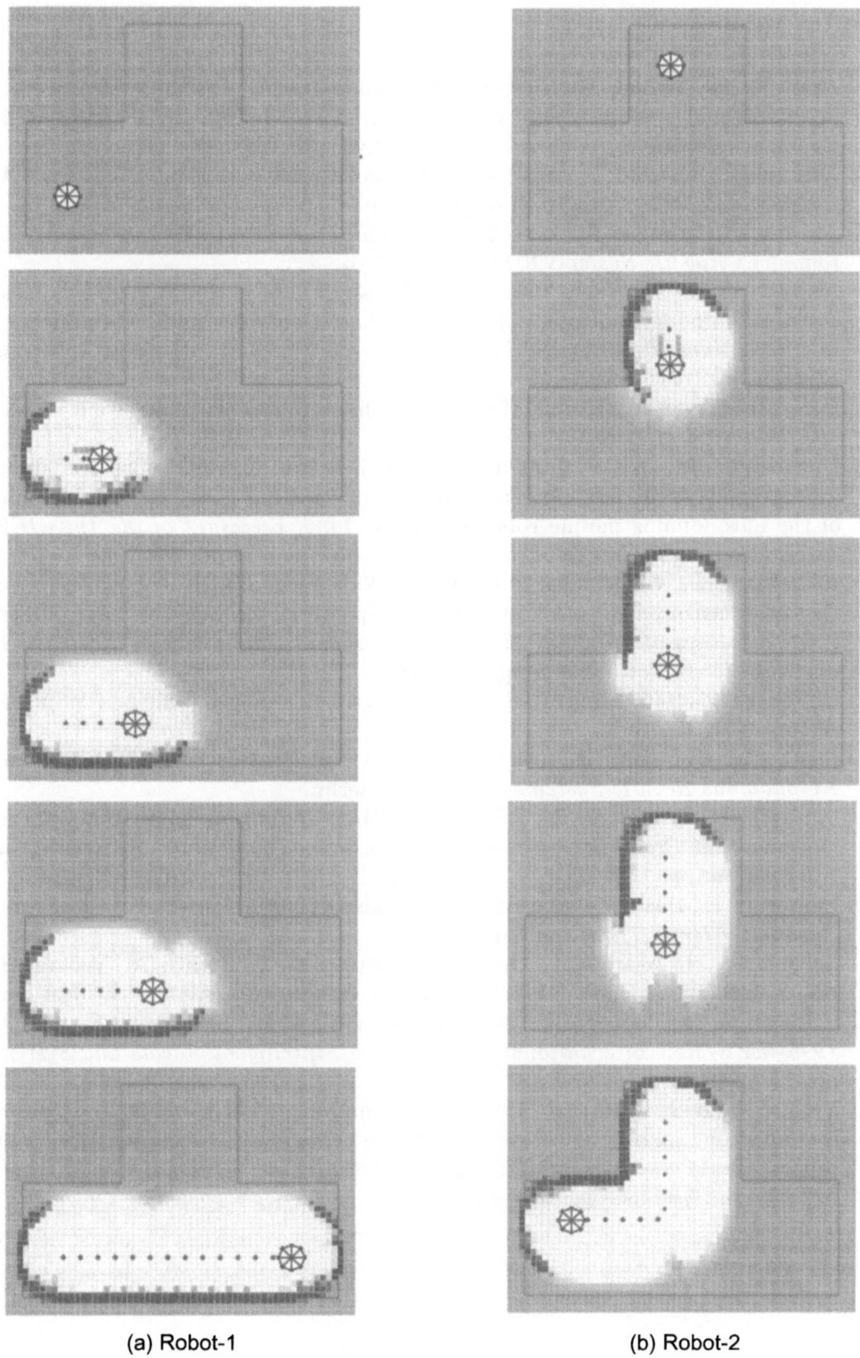

Fig. 9. Result of individual map acquisition

References

1. Arai, Y., H. Asama, H. Kaetsu and I. Endo (2000). Distance measurement in multi-robot systems based on time shared scheduling. In: *Distributed Autonomous Robotic Systems 4*. Springer Tokyo. pp. 189–198.
2. Asama, H., Fukuda, T., Arai, T. and Endo, I., Eds.) (1994). *Distributed Autonomous Robotic Systems*. Springer Tokyo.
3. Asama, H., Fukuda, T., Arai, T. and Endo, I., Eds.) (1996). *Distributed Autonomous Robotic Systems 2*. Springer Tokyo.
4. Asama, H., K. Ozaki, A. Matsumoto, Y. Ishida and I. Endo (1991). Collision avoidance among multiple mobile robots based on rules and communication. In: *Proceedings of IEEE/RSJ International Workshop on Intelligent Robots and Systems*. pp. 1215–1219.
5. Borenstein, J. and Y. Koren (1988). Obstacle avoidance with ultrasonic sensors. *IEEE Journal of Robotics and Automation* **4**(2), 213–218.
6. Buchberger, M., K.-W. Jorg and E. von Puttkamer (1993). Laserradar and sonar based world modeling and motion control for fast obstacle avoidance of the autonomous mobile robot mobot-iv. In: *Proceedings of the 1993 IEEE International Conference on Robotics and Automation*. pp. 534–540.
7. Kawabata, K., N. Nishioka, P. C. Lin, H. Nakamura and H. Kobayashi (1996). Distance measurement method under multiple ultra sonic sensors environment. In: *Proceedings of the 22nd Annual International Conference of the IEEE Industrial Electronics Society*. pp. 812–816.
8. Konolige, K. (1997). Improved occupancy grids for map building. *Autonomous Robots* **4**(4), 351–367.
9. Lueth, T., Dillmann, R., Dario, P. and Wörn, H., Eds.) (1998). *Distributed Autonomous Robotic Systems 3*. Springer Berlin.
10. Moravec, H. P. and A. E. Elfes (1985). High resolution map from wide angle sonar. In: *Proceedings of IEEE International Conference on Robotics and Automation*. pp. 116–121.
11. Parker, L. E., Bekey, G. and Barhen, J., Eds.) (2000). *Distributed Autonomous Robotic Systems 4*. Springer Tokyo.
12. Suzuki, S., H. Asama, A. Uegaki, S. Kotosaka, T. Fujita, A. Matsumoto, H. Kaetsu and I. Endo (1995). An infra-red sensory system with local communication for cooperative multiple mobile robots. In: *Proceedings of the 1995 IEEE/RSJ International Conference on Intelligent Robots and Systems*. pp. 220–225.
13. Yuta, S. and S. Premvuti (1992). Coordination of autonomous and centralized decision making to achieve cooperative behaviors between multiple mobile robots. In: *Proceedings of IEEE/RSJ International Workshop on Intelligent Robots and Systems*. pp. 1566–1574.

Principled Monitoring of Distributed Agents for Detection of Coordination Failure

Brett Browning, Gal A. Kaminka, and Manuela M. Veloso

School of Computer Science, Carnegie Mellon University, Pittsburgh PA 15213
{brettb, galk, veloso}@cs.cmu.edu

Abstract. There is a very rich variety of systems of autonomous agents, be it software or robotic agents. In particular, multi-agent systems can include agents that may be part of a team and need to coordinate their actions during their distributed task execution. This coordination requires an agent to observe, i.e., to monitor, the other agents in order to detect a possible coordination failure of the team. Several researchers have addressed the problem of monitoring for single or multiple agent systems and have contributed successful, but mainly application-specific, approaches. In this paper, we aim at contributing a unifying, domain-independent statement of the distributed multi-agent monitoring problem. We define the problem in terms of a pre-defined *desirable joint state* and an *observation-state mapping*. Given a concrete joint observation during execution, we show how an agent can detect a possible coordination failure by processing the observation-state mapping and the desirable joint state. To illustrate the generality of our formalism, one of the main contributions of the paper, we represent several previously studied examples within our formalism. We note that basic failure detection algorithms can be computationally expensive. We further contribute an efficient method for failure detection that builds upon an off-line compilation of the principled relations introduced. We show empirical results that demonstrate this effectiveness.

1 Introduction

Agents in distributed systems may need to coordinate. In the absence of allowed communication between the agents, coordination needs to be based on observations of the other agents. An agent therefore needs to monitor the other agents. Observation-based coordination (OBC) is a key challenge to the multi-agent and multi-robot systems. Increasingly, robots and synthetic agents are being deployed in multi-agent virtual environments for training [1] and entertainment [2], robotic soccer [3], hazardous cleanup tasks [4], formation-maintenance tasks [5,6], and more. Many of these applications rely on agents to coordinate with one another based on their observations of each other [7].

OBC is often a challenging process, mainly because it is computationally expensive. In general, OBC requires an agent to observe its peers and infer their state. Typically, the actual state is hidden and it is only partially revealed through observations. The agent must then decide what action it should take based on its own goals and internal state, and the observation-based inferred state of its peers. Specific examples of this process include OBC in self-interested agents [2], observation-based teamwork [8]. The inference and decision process are often considered computationally too intense for resource-limited robots. Indeed, there have been many investigations of ways to generate robust and predictable globally-coordinated behavior using agent-local control rules that shortcut the inference and decision processes, (e.g. [6,5]). However, these approaches often require painstakingly hand-crafted control rules and have been applied mostly in spatial coordination tasks (see Section 2 for an in-depth exploration of OBC in the literature).

We focus in this paper on an important component of OBC: *observation based monitoring*, which is used when agents use observations to verify that a group of observed agents does not suffer from a coordination failure. We thus limit ourselves to determining the existence of a coordination failure, and do not consider the actions that the agent may take to respond to it (if detected).

We introduce a novel formalism for describing observation-based monitoring of *simultaneous activities*. This formalism can be used to describe a significant class of coordination processes reported in the literature and commonly found in real-world multi-agent systems. We use this formalism to describe on-line coordination, and explore the complexity of this task. We then show how an off-line compilation process emerges as a result from this description, essentially transferring the on-line run-time complexity of the task off-line, allowing for quick execution. To evaluate our work empirically, we examine several examples of OBC in the literature. We present simulation results demonstrating the significant run-time computational resources saved by the off-line computation process. This compilation process takes a first step towards unifying the two themes of observation-based coordination.

This paper is organized as follows: Section 2 presents an overview of previous investigations into the OBC process. Section 3 presents a formalism describing OBC. Section 4 how OBC can be compiled into reactive rules. Section 5 presents an evaluation, and Section 6 concludes.

2 Observation-Based Coordination

Observation-based coordination (OBC) begins by each agent observing its fellow agents. Observations can be as simple as relative distance, or location, or as complex as intended goal, or plan. Based on these observations, each agent decides on what actions it will perform so as to achieve its own goals. Note that we are using the phrase coordination here in its broadest sense: We consider agents to be coordinating when the actions of each agent are dependent on the state of the other agents. Thus, both collaborative and adversarial settings are included in this definition.

Work on OBC has traditionally followed two parallel themes. The first such approach, which we call *state-based coordination* (SBC), emphasizes coordination at the level of internal states of agents, e.g., at the level of their selected plans, goals, or behaviors. Thus, SBC requires that the (unobservable) internal state of observed agents be known to the coordinating agent. The other approach, which we call *reactive coordination* (RC), shortcuts this process by directly mapping from observations of others to subsets of actions to be executed. It is distinct from SBC in that it does not explicitly consider the internal state of the observed agents.

SBC uses explicit knowledge of the internal state of the observed agents. Unfortunately, this knowledge is typically inaccessible. The agent must therefore use whatever observations it has of the other agents to *infer* the state of the other agents from the observations. Based on this inferred state, the coordinating agent selects its own internal state and actions. A key benefit of state-based coordination is that the designer can specify such selections in a *coordinating policy*. A coordinating policy is often easier to understand and design since it relates the plans and goals of others to those of the coordinating agent at a convenient level of abstraction.

Unfortunately, SBC often requires substantial computational resources. Inferring the state of the other agents based on observations (a process sometimes known as plan-recognition [2], behavior recognition, or agent modeling) is often expensive because, in general, the same observation can be interpreted as evidence for multiple internal states. A common approach relies on probabilistic networks for plan recognition, e.g., [2]. Washington[9] relies on POMDPs in coordination. Both probabilistic networks and POMDPs are intractable in the general case. Washington shows that under certain conditions, the coordination problem using POMDPs can be polynomial, however these conditions require the coordinating agent to not affect the behavior of the observed agents–a significant restriction in collaborative and adversarial settings. Tambe developed the non-probabilistic polynomial-time $RESC_{team}$ algorithm for reasoning about adversaries hierarchical behaviors [10]. This algorithm gains its computational advantage by always adapting and committing to a single interpretation of the opponents' actions.

Independently from investigations of SBC (mostly in software agent settings), researchers in the multi-robot community have focused their efforts on approaches appropriate for resource-limited hardware. The key ideas in these is to shortcut the inference and decision making process of SBC by introducing *reactive coordination* (RC) behaviors that tie specific observations of other agents with actions by the coordinating agent. For instance, Mataric demonstrated that many spatial group behaviors can be achieved by combinations of relatively simple agent-local rules, that directly tie spatial observations (e.g., distance and angle to other robot) with actions that should be taken [6]. Balch investigated methods for reliable execution of group tasks, such as foraging and formation maintenance, using hierarchical reactive behaviors that emphasized coordination by reliance on simple relative observations of other agents [5]. Unfortunately, while RC does indeed offer significant execution-time advantages for resource-constrained settings, it is difficult to design and maintain. Coordinated group behavior emerges in RC out of the interaction of reactive behaviors, but these interactions are difficult to predict.

This paper takes a step towards unifying these themes, by showing how an important class of coordination relationships can be specified in a more understandable manner (i.e., at the internal state level), but still executed quickly by resource-constrained agents. We focus on an important component in the OBC task: The observation-based monitoring that allows agents to detect coordination failures in their peers, thus allowing them to decide on an action if a failure is detected. The idea is to compile the expensive on-line failure detection process at design time into a set of fast-to-execute reactive rules. These rules can then be used at run-time to detect failures directly from observations without having to infer the joint agent state.

3 A Formalism for Describing Observation-Based Coordination

This section defines OBC formally and uses a running example from a coordination process from the literature to illustrate its meaning.

3.1 Examples from the literature

These examples are taken from the ModSAF domain, a high-fidelity virtual environment for military training that allows thousands of agents (synthetic and human) to interact in battlefield scenarios [1]. Several coordination scenarios for synthetic helicopter pilots in this domain have been described in [11,8]. These rise from the scenario described below.

> A team of 3–6 helicopters take off from their base and fly in formation (with the *fly-flight-plan* (f) behavior) until they reach an area marked by a visible landmark. Upon arrival, they select the *wait-at-point* (w) behavior, in which they split into two subteams. The scouting subteam moves towards the estimated enemy location to identify its exact position (l). The attacker subteam stays behind (b). When the scouting team identifies the enemy position, it radios for the attackers to join it (all team-members select the *join-scout* (c) behavior) and waits for it to arrive. Once joined, they engage the enemy together (the *engage* (e) behavior). Each helicopter *masks* (m) by hiding behind trees or hills and *unmasks* (u) to come out of hiding to *shoot* (s) at the enemy. It then masks again before moving to a new location to shoot. When the enemy is destroyed, the entire team joins together and flies in formation (f) back to base. At any point during the mission, pilots may be ordered to *halt* (h) their activities and await further instructions.

The designers of these synthetic pilots use hierarchic behaviors to implement portions of the tasks. That is, higher-level behaviors control the instantiation and sequential activation of lower-level behaviors for each pilot. To coordinate the activities between pilots, the designers sought to make sure that specific, selected team-level tasks are jointly selected by different agents. We extract several general cases of coordination:

Agreement. Two or more agents have to simultaneously and synchronously select a single plan for joint execution by the entire team [11,8]. In the scenario above, agreement was to be achieved on the team-level behaviors, e.g., f, j, w, and h.

Simultaneous role selection coordination. Two or more agents have to select from a pool of behaviors, such that when one agent has selected a specific behavior another agent has selected the appropriate corresponding behavior. In the scenario above, such coordination takes place when the scouting subteam members select l while the attackers select b.

Sequential coordination. Two or more agents select states in a particular coordinated sequence. For instance, when engaging the enemy, the helicopters may want to fire in particular sequence such that all helicopters go through $m \to u \to s \to m \to \ldots$ in the same order. When one helicopter is executing a particular step in this sequence, its teammate executes the next step in the sequence.

The problem is to verify that the individual selection of team-level behaviors is coordinated [11]. Under the limited communication range and reliability conditions in this domain, as well as security restrictions imposed on the pilots, an OBC scheme was required [8]. The key to this approach is for each pilot agent to observe its teammates using radar. Based on their velocity and altitude, the agent would infer what behaviors are being executed (a behavior-recognition process), and detect failures in coordination. For most cases, the same observation can be interpreted as being indicative of different behaviors. Hence, there can be multiple interpretations for the observation. The agent can then use an optimistic or pessimistic monitoring policy to select between these interpretations, thereby guaranteeing sound or complete detection quality (see [8]) for details).

We will now show how these examples can be formalized. We start by defining the basic elements of OBC before continuing to the failure detection (FD) policies and algorithms. To illustrate the formalism, we use the Simultaneous Role Selection coordination as described above.

3.2 Definitions

Consider a team of n agents. Each of the agents has several internal states that are in general unobservable to any other agent but itself. We define the state space to be identical for each agent. In practice agents may use only a subset of the full state space (i.e., different agents have different state spaces). The state space is then:

$$S = \{s_1, \ldots., s_m\} \tag{1}$$

In our working example, we define the state space to be $\{b, c, l, f, w, m, u, s, h\}$

As an agent's state is typically not directly observable, each agent must engage in behavior recognition. An agent's state must be inferred by observing its behavior. We formalize observations as discrete members of a set

$$O = \{o_1, \ldots o_k\} \tag{2}$$

Observations map to states in some predefined manner. We define the observation-state mapping, which we call the recognition function, for agent j to be:

$$\begin{gathered} M_j : O \to S' \in S = \{(o_1, S'_{j,1}), \ldots (o_k, S'_{j,k})\} \\ \text{where } o_1, .., o_k \in O \text{ and } S'_{j,1}, .., S'_{j,k} \subseteq S, \text{ for } j = 1..n \end{gathered} \tag{3}$$

In our working example, observations and recognition function may be:

$$O = \{(o_1 : \mathit{speed} < 3), (o_2 : (\mathit{altitude} > 3) \wedge (\mathit{speed} > 3))\} \tag{4}$$

$$M_1 = M_2 = \ldots = M = \{\langle o_1, b\rangle, \langle o_1, l\rangle, \langle o_2, b\rangle\} \tag{5}$$

Intuitively, the agent slows down when waiting (b) and moves forward at altitude when locating the enemy's position (l). It may also slow down in l, making observation o_1 ambiguous.

Observations on the team of agents form a joint observation space defined as:

$$JO = O \times ... \times O = \{\langle o_{i_1}, ..., o_{i_n}\rangle \,|o_{i_j} \in O, \; \forall i_j = 1..k, j = 1..n\} \tag{6}$$

We will use the following notation for an element of the joint observation set.

$$jo_{i_1,...,i_n} = \langle o_{i_1}, ..., o_{i_n}\rangle \in JO \tag{7}$$

Each element of the joint observation set, via the recognition function, maps to one or more elements in a joint state space, formed by the cross products of the state spaces.

$$JS = S \times ... \times S = \{\langle s_{i_1}, .., s_{i_n}\rangle \,|s_{i_j} \in S, \; \forall i_j \in \{1..m\}, j = 1..n\} \tag{8}$$

We will use the following notation for an element of the joint states set.

$$js_{i_1,...,i_n} = \langle s_{i_1}, ..., s_{i_n}\rangle \in JS \tag{9}$$

For a given joint observation, $jo_{i_1,...,i_n}$, we call the set of possible joint states the team could be in, the observed joint states OJS or hypotheses set. We can construct OJS as the cross-product of the mapped sub-sets from M_j for each agent $j = 1..n$. That is for the given joint observation we form:

$$\begin{aligned} OJS_{i_1,...,i_n} &= M_1(o_{i_1}) \times ... \times M_n(o_{i_n}) \\ &= S'_{1,i_1} \times S'_{2,i_2} \times ... \times S'_{n,i_n} \end{aligned} \tag{10}$$

In our example, given a joint observation $\langle o_1, o_1, o_2\rangle$, the observing agent may conclude that the OJS for the observed agent is: $\{\langle b,b,l\rangle, \langle b,l,l\rangle, \langle l,b,l\rangle, \langle l,l,l\rangle\}$. In order to detect failures in coordination, the hypotheses set is compared to the set of *desired* joint states (DJS) as specified by the designer. DJS, a subset of JS, is defined as:

$$DJS = \{js_{i_1,...,i_n} | js_{i_1,...,i_n} \in JS \wedge js_{i_1,...,i_n} \text{desired}\} \subseteq JS \tag{11}$$

In principle, if $DJS \cap OJS = \emptyset$ then the observed agents' joint state is definitely not desired and a clearly recognizable coordination failure has occured. In contrast, if $DJS \cap OJS = OJS$ then no failure occured. In general, however, DJS and OJS only partially overlap meaning some joint states are acceptable and some are not. Therefore, we must classify failures via some policy in these ambiguous cases.

We define two failure detection policies: π_{PFD} a pessimistic policy and π_{OFD} an optomistic policy. π_{PFD} takes any possibility of a failure to mean a failure occurred, while π_{OFD} reports failures only when a clear failure occurred. Mathematically, π_{PFD} reports failures when *any* element of OJS is not an elements of DJS. In contrast, π_{OFD} reports failures only when *no* element of OJS is an element of

DJS. It can be shown that π_{OFS} will never report a false-positive, while π_{PFS} will never report a false-negative. Thus:

$$\pi_{PFD}\left(jo_{i_1,\dots,i_n}\right) = \begin{cases} \text{failure} & OJS_{i_1,\dots,i_n} \not\subseteq DJS \\ \text{okay} & \text{otherwise} \end{cases} \tag{12}$$

$$\pi_{OFD}\left(jo_{i_1,\dots,i_n}\right) = \begin{cases} \text{failure} & OJS_{i_1,\dots,i_n} \cup DJS = \emptyset \\ \text{okay} & \text{otherwise} \end{cases} \tag{13}$$

3.3 Examples

To better clarify the formalism, let us consider a simple three agent system with three states s_1, s_2, s_3 and two observations o_1, o_2. Let us define the recognition functions and DJS as:

$$M_j = M = \{\langle o_1, s_1\rangle, \langle o_1, s_2\rangle, \langle o_2, s_3\rangle\} \text{ for } j = 1, 2, 3 \tag{14}$$

$$DJS = \{\langle s_1, s_1, s_1\rangle, \langle s_2, s_2, s_2\rangle, \langle s_3, s_3, s_3\rangle\} \tag{15}$$

Let us consider the three joint observations $\langle o_1, o_1, o_1\rangle, \langle o_1, o_1, o_2\rangle$ and $\langle o_2, o_2, o_2\rangle$. Thus we have:

$$jo_{1,1,1} \to M(o_1) \times M(o_1) \times M(o_1) = \{s_1, s_2\} \times \{s_1, s_2\} \times \{s_1, s_2\} \tag{16}$$

$$= \{\langle s_1, s_1, s_1\rangle, \langle s_1, s_1, s_2\rangle, \dots, \langle s_2, s_2, s_2\rangle\} \tag{17}$$

$$jo_{1,1,2} \to \{\langle s_1, s_1, s_3\rangle, \langle s_1, s_2, s_3\rangle, \langle s_2, s_1, s_3\rangle, \langle s_2, s_2, s_3\rangle\} \tag{18}$$

$$jo_{2,2,2} \to \{\langle s_3, s_3, s_3\rangle\} \tag{19}$$

The first observation contains some elements that are also in DJS and so passes the optomistic FD policy but fails the pessimistic one. The second observation contains no elements from DJS and thus fails both policies and is a clear failure. The final observation is a strict subset of DJS and therefore is a success using either policy.

As a final example, let us return to our earlier helicopter example. Let us assume that the first agent is a scout and the other two agents are attackers. We have $DJS = \{\langle l, b, b\rangle\}$. Given a joint observation $\langle o_1, o_1, o_2\rangle$, OJS is $\{\langle b, b, l\rangle, \langle b, l, l\rangle, \langle l, b, l\rangle, \langle l, l, l\rangle\}$. The π_{PFD} policy would report a failure as $OJS \not\subseteq DJS$. π_{OFD} would also report a failure as $OJS \cap DJS = \emptyset$. In contrast, for $\langle o2, o1, o1\rangle$, π_{PFD} reports a failure while π_{OFS} does not. In general, the set of joint observations that are failures according to π_{OFD} are a subset of those classified as failures by π_{PFD}.

4 Compiling Reactive Monitoring Rules

Performing the failure detection (FD) policy evaluation on-line becomes intractable in many problems, particularly where computational resources are at a premium. The core computational cost in the policy evaluation occurs in the OJS calculations.

The calculation for OJS, see 10, are exponential in the number of agents. Off-line compilation of reactive rules offers one method for addressing this issue.

To compile reactive rules, we want to generate from our apriori knowledge a mapping from JO to the two-element space, $\{fail, pass\}$, representing the output of the policy evaluation. Our approach divides JO into two subsets, one containing joint observations that map to failures FJO, and the other containing joint observations that map to successes SJO. Thus we define FJO and SJO for failure detection policy π_{FD} as:

$$FJO = \{jo_{i_1,\ldots,i_n} | jo_{i_1,\ldots,i_n} \in JO \wedge \pi_{FD}(jo_{i_1,\ldots,i_n}) = \text{failure}\} \subseteq JO \quad (20)$$

$$SJO = \{jo_{i_1,\ldots,i_n} | jo_{i_1,\ldots,i_n} \in JO \wedge \pi_{FD}(jo_{i_1,\ldots,i_n}) = \text{success}\} \subseteq JO \quad (21)$$

FJO and SJO partition JO such such that $FJO, SJO \subseteq JO$ and $FJO \cup SJO = JO$. For the ensuing discussion we will use the non-failure set SJO with the optimistic π_{OFD} policy. A similar approach could be taken for the π_{PDF} policy but will not be discussed here.

It is readily apparent that building SJO by testing each individual element of OJS with the policy function, π_{OFD}, will take an exponential amount of time. Clearly, a better approach is required. Our approach operates by recognizing that we only need to find those joint observations that have any corresponding joint states in DJS. Thus, for each joint state in DJS we generate the set of observation tuples that could possibly map to that joint state. For a given element of a joint state that is in DJS the $GENO$ function, short for GENerate Observations, will perform this tasks as:

$$GENO(j, js_{i_1,\ldots,i_n}) = \{o_v | s_{i_j} \in \bigcup_{\alpha=1..N} M_\alpha(o_v)\} \ where\ js_{i_1,..i_j..,i_n} \in DJS \quad (22)$$

Generating the set of joint observations that could possibly map to the given joint state is a cross product operation as:

$$GENJO(js_{i_1,..,i_n}) = GENO(1, js_{i_1,..,i_n}) \times .. \times GENO(n, js_{i_1,..,i_n}) \quad (23)$$

The complete compilation operates over all the elements in DJS with the results combined with the union operator to form SJO as:

$$SJO(DJS) = \bigcup_{js_{i_1,\ldots,i_n} \in DJS} GENJO(js_{i_1,\ldots,i_n}) \quad (24)$$

5 Evaluation

We implemented a system that accepts monitoring examples written in the presented formalism. The system performs monitoring when given observation tuples using both uncompiled and compiled mechanisms. To evaluate the performance of compilation, we translated all three examples described in section 3 using the formalism and compared the mean execution times for 100 randomly selected observation tuples. Trials were performed with between 1 and 20 agents using the optomistic failure detection policy.

Figure 1 shows the average running times for the compiled and non-compiled versions. The X axis shows the number of agents in each configuration. The Y axis measure monitoring time in seconds. Figure 1-a shows that the run-time curve as the number of agents increase is non-linear in all three examples, indeed clearly exponential in the case of the agreement coordination example (in which the observations lead to very ambiguous interpretations).

The compiled monitoring results (Figure 1-b) follow very similar trends to those of the non-compiled versions, however their running times are *orders of magnitude* smaller (note the difference in range on the Y axis between the two figures). While the curves for the compiled "Simultaneous Role" and "Sequential" coordination monitoring seem almost linear, the curve for compiled agreement monitoring again grows exponentially, though much slower than its non-compiled version. As described in Section 4, the compiled rules may still require exponential time in the worst case.

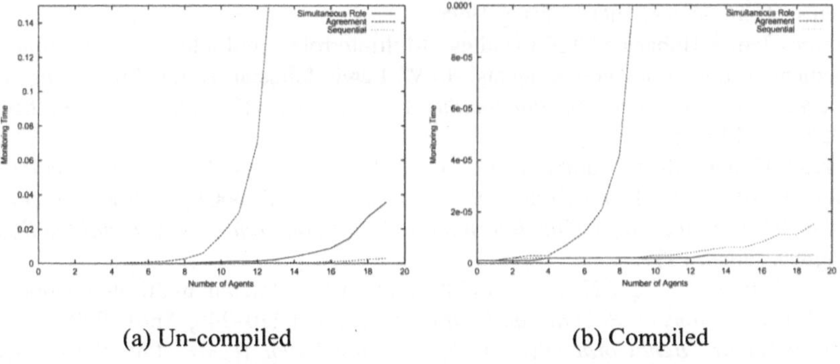

(a) Un-compiled (b) Compiled

Fig. 1. Average running times for the non-compiled and compiled monitoring. *Note the Y-axis range is significantly different in Figures a and b.*

6 Conclusions and Future Work

This paper introduces a novel formalism for describing observation-based monitoring for coordination failure, an important component in observation-based coordination. The formalism allows application-independent investigation of the monitoring process, and facilitates analytical treatment of monitoring. We provide an example of such analytical treatment, a process that compiles the exponential run-time process of monitoring into a set of reactive rules that can be executed quickly. We evaluate our work empirically on varying-scale examples of observation-based monitoring taken from the literature, and show that the compilation process results in orders-of-magnitude faster run-time. Future directions for our work include in-depth study of the compilation process and the effects of ambiguous observations on monitoring run-time.

Acknowledgements This research was sponsored by the United States Air Force and the United States Army under Cooperative Agreements Nos F30602-00-2-0549, F30602-98-2-0135 and DABT63-99-1-0013. The views and conclusions contained in this document are those of the authors and should not be interpreted as necessarily representing the official policies or endorsements, either expressed or implied, of the Defense Advanced Research Projects Agency (DARPA), the Air Force, the Navy, or the US Government.

We thank Guy Lebanon and Yuval Kaminka for useful discussions.

References

1. Milind Tambe, W. Lewis Johnson, Randy Jones, Frank Koss, John E. Laird, Paul S. Rosenbloom, and Karl Schwamb. Intelligent agents for interactive simulation environments. *AI Magazine*, 16(1), Spring 1995.
2. Marcus James Huber and Tedd Hadley. Multiple roles, multiple teams, dynamic environment: Autonomous netrek agents. In W. Lewis Johnson, editor, *Proceedings of the International Conference on Autonomous Agents*, pages 332–339, Marina del Rey, CA, 1997. ACM Press.
3. Hiroaki Kitano, Milind Tambe, Peter Stone, Manuela Veloso, Silvia Coradeschi, E. Osawa, H. Matsubara, Itsuki Noda, and M. Asada. The RoboCup synthetic agent challenge '97. In *Proceedings of the International Joint Conference on Artificial Intelligence*, Nagoya, Japan, 1997.
4. Lynne E. Parker. ALLIANCE: An architecture for fault tolerant multirobot cooperation. *IEEE Transactions on Robotics and Automation*, 14(2):220–240, April 1998.
5. Tucker Balch. *Behavioral Diversity in Learning Robot Teams*. PhD thesis, Georgia Institute of Technology, 1998.
6. Maja J. Mataric. *Interaction and Intelligent Behavior*. PhD thesis, Massachusetts Institute of Technology, 1994.
7. Yasuo Kuniyoshi, Sebastien Rougeaux, Makoto Ishii, Nobuyuki Kita, Shigeyuki Sakane, and Masayoshi Kakikura. Cooperation by observation—the framework and the basic task patterns. In *the IEEE International Conference on Robotics and Automation*, pages 767–773, San-Diego, CA, May 1994. IEEE Computer Society Press.
8. Gal A. Kaminka and Milind Tambe. Robust multi-agent teams via socially-attentive monitoring. *Journal of Artificial Intelligence Research*, 12:105–147, 2000.
9. Richard Washington. Markov tracking for agent coordination. In *Proceedings of the International Conference on Autonomous Agents*, pages 70–77, Minneapolis/St. Paul, MN, 1998. ACM Press.
10. Milind Tambe. Tracking dynamic team activity. In *Proceedings of the National Conference on Artificial Intelligence (AAAI)*, August 1996.
11. Milind Tambe. Towards flexible teamwork. *Journal of Artificial Intelligence Research*, 7:83–124, 1997.

Chapter 9
Multi-Agent and Group Systems

Cooperative Behavior of Interacting Simple Robots in a Clockface Arranged Foraging Field

Ken Sugawara[1] and Masaki Sano[2]

[1] PRESTO,JST. Graduate School of Information Systems, Univ. of Electro-Communications, Chofu Tokyo 182-8585, JAPAN
[2] Depertment of Physics, Univ. of Tokyo, Hongo, Tokyo 113-0033 ,JAPAN

Abstract. Emergence of swarm intelligence is investigated through the foraging task. As the robots assumed in this paper is very simple in order to discuss their behavior analytically, there are a few parameters to characterize their behavior such as interaction duration and interaction range. The behavior of the group is investigated on computer simulation and mathematical model. In this paper, we discuss the definition of swarm intelligence at first, and shows the emergence of swarm intelligence in a clockface arranged foraging field.

Keywords collective behavior, foraging, interaction

1 Introduction

The research of multi-robot system is active in these days, and many researchers have been studied the behavior of multi-robot system [1–3]. One of the good example of multi-robot system is a Robo-Cup, which is a soccer tournament by autonomous robots[4].

The most important characteristic of multi-robot system is "they can work cooperatively." It means that they have a possibility to achieve the task which single robot cannot do. Most researchers have focused on qualitative aspect of multi-robot system, but we consider the number effect is also important characteristics of multi-robot system. In this paper, we discuss the efficiency of the system quantitatively focusing on the relation between the number of robots and their performance by foraging task.

2 What is Swarm Intelligence ?

The behavior emerging from interacting elements is called "swarm intelligence"[5]. There is not strict definition of swarm intelligence, but Beni claimed the meaning of swarm intelligence from the viewpoint of quantitative aspect[6]. Let's assume a situation that N autonomous elements act asynchronously in an environment. Beni's claim is that swarm intelligence emerges from interacting among N elements only when N exceeds a critical number N_C. It means that an assembly of independent elements does not exhibit any useful work.

Let's denote $W(N)$ as the work achieved by N interacting elements and $W_0(N)$ as the work achieved by N independent elements. Then, $W(N)$ is described by a function with a critical number N_C, and $W_0(N)$ is described as zero for all N. The definition of swarm intelligence by Beni is schematically represented in Figure 1.

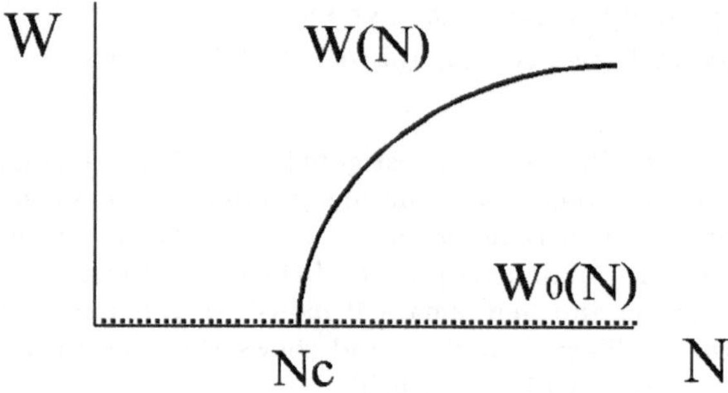

Fig. 1. Schematic of swarm intelligence defined by Beni.

In this paper, we report the emergence of swarm intelligence defined by Beni applying foraging task to the simple multi-robot system.

3 Foraging Behavior of Multi-robot

3.1 What is foraging behavior?

Remarkable characteristic of multi-robot system is that the robots work co-operatively. In our research, we choose foraging behavior as a task (Figure 2). Forage is the act of searching or hunting for food. This task is suitable and popular task for multi robot system[7–9]. It is composed of parallel searching and cooperative transportation, and it is easy to extend another kind of task such as mutual observation or structure formation. The most popular phenomenon of cooperative foraging behavior by simple interaction is observed in social insects such as ants and bees[10,11].

3.2 Basic behavior of each robot

Cooperative foraging behavior can be described by the combination of the following basic behaviors: Wandering, Broadcasting, Attracted, Homing, Staying and Avoiding.
<Wandering>
To move straight at constant velocity.

$$\boldsymbol{v} = v \cdot \boldsymbol{s_i}, \tag{1}$$

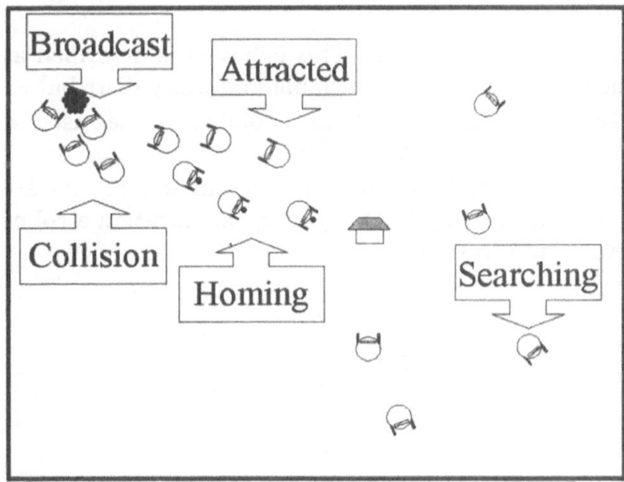

Fig. 2. Schematic of foraging task.

where v,$\boldsymbol{s_i}$ are a constant and heading unit vector of robot i respectively ($\boldsymbol{s_i} = (\cos\theta_i, \sin\theta_i)$).

<Broadcasting>

To radiate isotropic signal light. The condition is described as

$$l_i = \begin{cases} 1 \text{ (radiating)} \\ 0 \text{ (otherwise)} \end{cases} \tag{2}$$

<Attracted>

To move toward the signal source. Angular velocity obeys the following equation.

$$\frac{d\theta_i}{dt} = \tanh(c \cdot \sum_{j \neq i}^{n}((\boldsymbol{s_i} \times \frac{\boldsymbol{r_{ij}}}{||\boldsymbol{r_{ij}}||})_z \cdot \frac{l_j}{||\boldsymbol{r_{ij}}||^2})) \cdot \Omega, \tag{3}$$

where $\boldsymbol{r_{ij}}$ is a vector directed from robot j to i, c and Ω are constants, and $()_z$ implies z component of the vector.

<Homing>

To move toward home. Angular velocity obeys the following equation.

$$\frac{d\theta_i}{dt} = \tanh(c \cdot (\boldsymbol{s_i} \times \frac{\boldsymbol{r_{ih}}}{||\boldsymbol{r_{ih}}||})_z) \cdot \Omega, \tag{4}$$

where $\boldsymbol{r_{ih}}$ is a vector directed from home to robot i.

<Staying> To stay there.

$$\boldsymbol{v} = 0. \tag{5}$$

<Avoiding>
To avoid collision with other robots and walls. This behavior is described as follows: $\theta_i \leftarrow \theta_i + \Delta\theta$, where $\Delta\theta$ is a random variable uniformly distributed within $(0, -2\pi)$. It takes a constant time $\frac{\Delta\theta}{\Omega}$ for avoiding. Values used in computer simulation are as follows: $v = 10, \Omega = \frac{2}{3}\pi$.

Figure 3 is a schematic of robots behavior. As you can see, robots are interacting while a robot is broadcasting. In this paper, the duration a robot broadcasts is called "interaction duration."

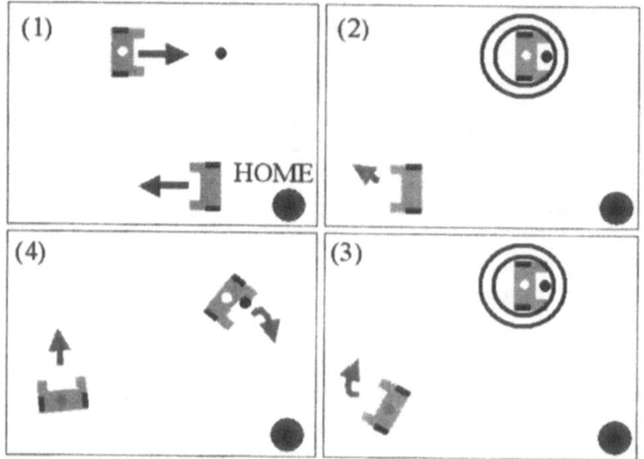

Fig. 3. A schematic of interaction. While a robot is broadcasting, others react to this signal.

3.3 Simulation Field

We chose a spatially discrete model to accelerate computation. We assume a circular field with home located at the center. The field is partitioned into 800×800 cells. Each cell can have three states: "empty", "robot", or "puck." The position and direction of each robot are calculated as continuous variables. A cell which includes a robot is occupied and has the state "robot," implying that the size of the robot is one cell.

4 Foraging Behavior in a Localized Distribution Field

In this section, we show the foraging behavior in the field where all foods are locally placed. Behavior of robots is analyzed by computer simulation and mathematical model.

4.1 Computer Simulation

As mentioned above, movement of each robot and their interaction is very simple, We can easily simulate their behavior by computer simulation. Figure 4(a) shows the result of robot simulation. In this figure, horizontal axis is the number of robots and vertical axis is The number of collected food per a unit time. A parameter in this figure is the interaction duration. As you can see, the performance of the group greatly depends on the number of robots and there is a critical number for their performance.

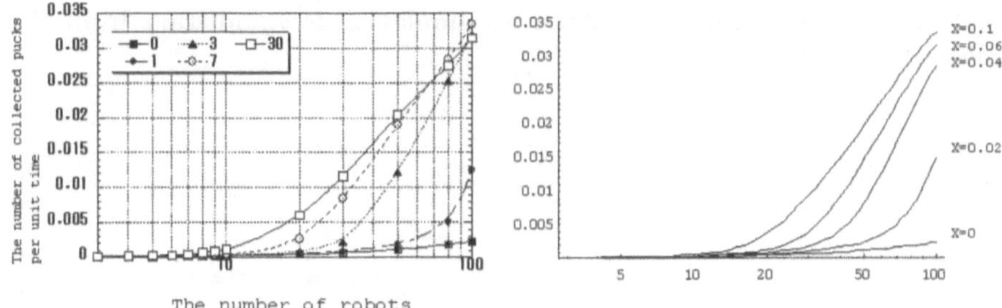

Fig. 4. (a) The result of robot simulation. Parameter is an interaction duration. (b) Static feature of the equations. α=0.0002, b=0.01, d=200, v=10, γ=0.02, τ=100, c=0.05.

4.2 Mathematical Analysis of the system

The behavior of robots can be described by state transition diagram and we can get the following equations[12].

$$\begin{aligned}\dot{S} &= -\alpha S + \tfrac{1}{\tau}R - l(x)\cdot S\cdot B + bM\\ \dot{B} &= -\tfrac{1}{x+1}B + \alpha S + \gamma\cdot\tfrac{1}{1+cA}A\\ \dot{R} &= \tfrac{1}{x+1}B - \tfrac{1}{\tau}R\\ \dot{M} &= -\tfrac{v}{d}M + l(x)\cdot S\cdot B - bM\\ \dot{A} &= \tfrac{v}{d}M - \tfrac{\gamma}{1+cA}A,\end{aligned} \qquad (6)$$

where S: the number of robots in searching mode, B: the number of robots in broadcasting mode, R: the number of robots in returning mode, M: the number of robots that move to signal source, A: the number of robots that avoid collision.

Meaning of parameters in the equation is as follows: α: probability to find a puck independently, τ: time to return home, x: interaction duration, d: average distance between interacting robots, v: velocity of robot, γ: probability to find a puck by following other robots. b: probability to lose the direction to signal source. $l(x)$ describes a probability to turn to the broadcasted signal source.

Figure 4(b) shows static feature of the equations. It is clear the equations describe their behavior well.

5 Foraging in Clockface Arranged Field

In this section, we show the behavior of the robots in the field that some localized foods are arranged like clockface. We assume 6 localized foods accumulations around the home. Each accumulation is equidistant from the home and has 400 pucks (Figure 5(a)).

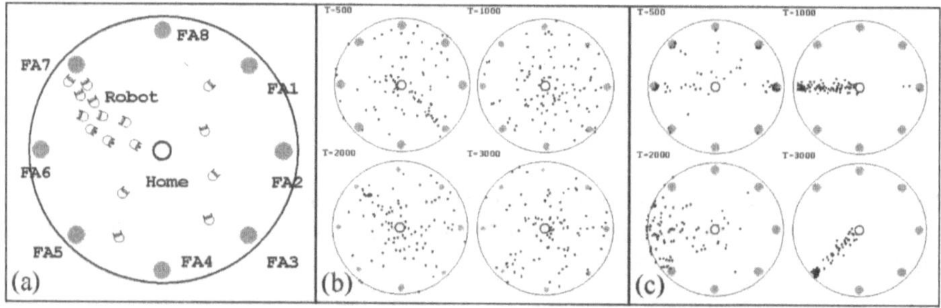

Fig. 5. (a) Schematic of foraging behavior in clockface arranged field. Each accumulation contains 400 foods. (b) In case of weak interaction. (c) In case of strong interaction.

Strength of the interaction between each robot can be controlled by "interaction duration" and "interaction range."

The behavior of the group depends on their interaction. When the interaction is weak (that is, short interaction duration or short interaction range), they work in parallel (Figure 5(b)). It means the decreasing rate of each food accumulation is almost same. On the other hand, when the interaction is strong, we can observe they work in series (Figure 5(c)). It means that all robots flock around an accumulation intensively and when the accumulation is eaten out, they flock around other accumulations. Figure 6 shows the time evolution of the number of the remained foods in each condition.

Paying attention to Figure 5(a) and Figure 6, we notice that their behavior is rotational. Each accumulation is processed counter-clockwise. In this figure, the process is carried out clockwise, but we confirmed the probability counter-clockwise process occurs and clockwise process occurs is same.

As described before, each robot does just carrying by simple mechanism, but globally ordered phenomenon emerges. In addition, the probability of

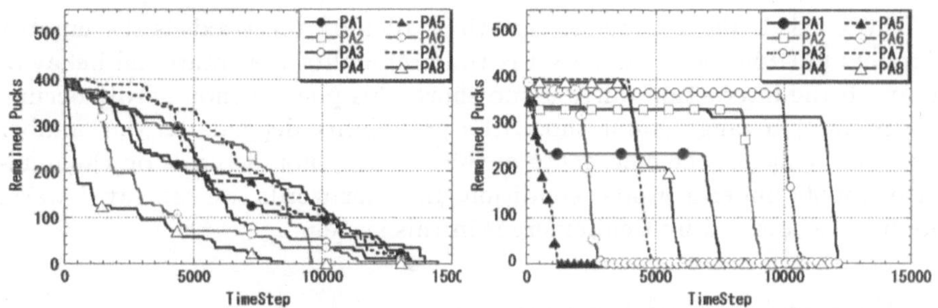

Fig. 6. Time evolution of the number of the remained pucks. In case of weak interaction (left) and of strong interaction(right).

the emergence of such a rotational phenomenon depends on the number of robots, the interaction range and interaction duration. Figure 7 shows the probability this phenomenon occurs.

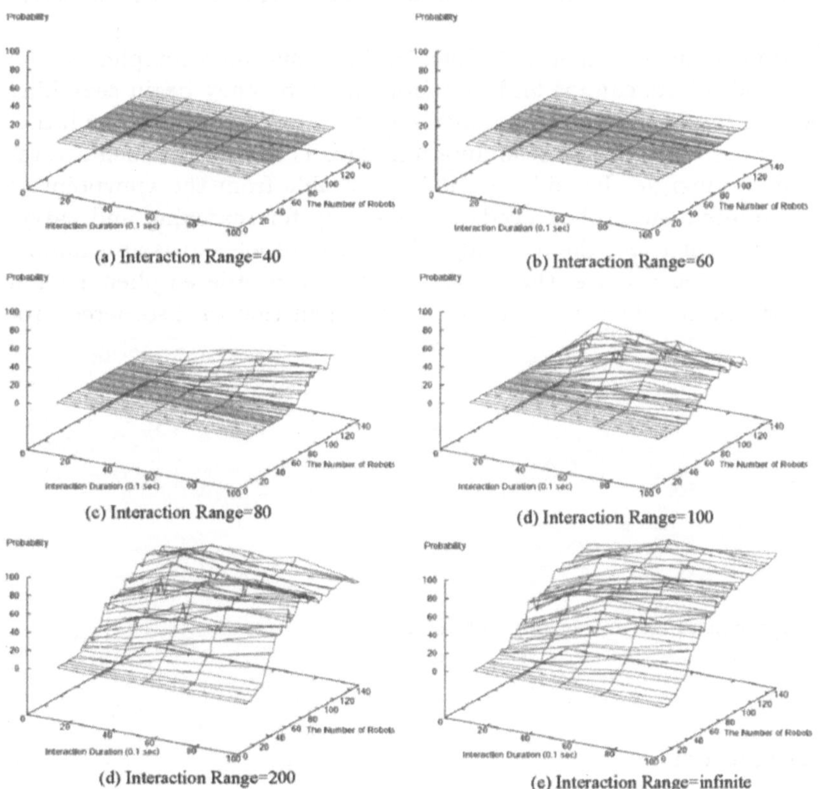

Fig. 7. Probability the rotational behavior occurs.

The x-axis is the interaction duration (10-100), the y-axis is the number of robots (10-140), and the z-axis is the probability the rotational behavior occurs. If the interaction range is too short, this phenomenon hardly occurs, but interaction range is not short, the probability depends on the number of robots. It means the emergence of this phenomenon depends on the number of robots under a proper condition. In other words, we can say "swarm intelligence" defined by Beni emerges in this system.

6 Discussion and Conclusion

In this paper, we introduced simple interacting robots and observed their behavior mainly focusing on the quantitative aspect. In spite of simple interaction, the group shows ordered behavior under a proper condition. It means clockwise (or counter-clockwise) processing emerges in clockface arranged field, and the probability this phenomenon occurs depends on the number of robots and the strength of interaction. It is generally called the collective behavior of ant's foraging is a good example of "swarm intelligence." In this paper, we showed another example of swarm intelligence in their collective behavior.

Why this phenomenon emerges? When an accumulation is completely processed, crowded robots cannot find any food there. So they begin searching other foods. In this situation, probability they find the nearest accumulation is high compared with the far accumulation. This is the reason such a rotational behavior emerges. In addition, it is profitable from the viewpoint of costs. By assuming each robot needs some energy for searching and carrying foods, we can measure the dissipated energy. Figure 8 shows an example of the result. As you can see, the energy consumed in ordered phenomenon (clockwise or counter-clockwise process) is less than that of disordered process.

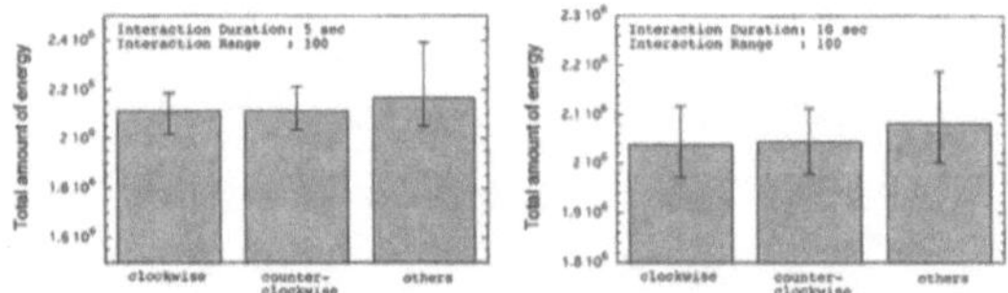

Fig. 8. Energy consumed in foraging behavior.

Acknowledgement

We acknowledge Prof. I. Yoshihara, Prof. K. Abe and Prof. T. Watanabe for helpful comments. This work was partially supported by a Japanese Grandin-Aid for Encouragement of Young Scientists from the Ministry of Education, Science and Culture (No.12750399).

References

1. Y.U.Cao, A.S.Fukunaga and A.B.Kahng, "Cooperative Mobile Robotics:Antecedents and Directions," *Autonomous Robots*, 4, (1997) pp.7-27.
2. D. Kurabayashi, "Toward Realization of Collective Intelligence and Emergent Robotics", *Proc. of 1999 IEEE Int. Conf. on Systems, Man, and Cybernetics*, (1999) pp.748-752.
3. L.E.Parker, G.Bekey and J.Barhen Eds. *Distributed Autonomous Robotic System 4*, Springer-Verlag, (2000).
4. http://www.robocup.org/
5. E.Bonabeau, M.Dorigo and G.Theraulaz, *Swarm Intelligence - From Natural to Artificial Systems*, Oxford University Press, (1999).
6. G. Beni and J. Wang, "Distributed Robotic Systems and Swarm Intelligence," *Proc. of IEEE Int. Conf. Rob. Autom.*, April, (1991), pp.1914.
7. T. Balch and R.C. Arkin, "Communication in Reactive Multiagent Robotic Systems," *Autonomous Robots*, 1, (1994), pp. 27.
8. R. Beckers, O.E. Holland and J.L. Deneubourg, "From Local Actions To Global Tasks: Stigmergy and Collective Robotics," *Artificial Life IV*, MIT Press, (1994), pp. 181-189
9. M.J.B. Krieger, J.B. Billeter and L. Keller, "Ant-like task allocation and recruitment in cooperative robots," *Nature*, 406, (2000) pp.992 - 995.
10. B. Hölldobler and E. O. Wilson, *JOURNEY TO THE ANTS*, Harvard University Press, (1994).
11. T. D. Seeley, *THE WISDOM OF THE HIVE*, Harvard University Press, (1995).
12. K. Sugawara, M. Sano, I. Yoshihara, K. Abe and T. Watanabe, "Foraging Behavior of Multi-robot System and Emergence of Swarm Intelligence," *Proc. of 1999 IEEE Int. Conf. on Systems, Man, and Cybernetics*, (1999) pp.257-262.

Graduated Spatial Pattern formation of Micro Robot Group

Toshio Fukuda[1], Yusuke Ikemoto[1], Fumihito Arai[1] and Toshimitsu Higashi[2]

[1] Nagoya University, Japan,
[2] Murata Machinery, LTD., Japan

Abstract. This study proposes how to make a spatial pattern of a micro robot group in distributed autonomous systems. The autonomous formation of spatial pattern is important for the advancement of cooperative robotic systems. Although some people are also studying about spatial pattern formation by the swarm of robots, their way of pattern formation is mainly that robot can get global information such as the positions of all robots. However, micro robot is not usually able to get such information because micro robot's scale effect restricts their functions. Therefore, we propose the method of spatial complicated pattern without getting global information. Then we propose a dynamical spatial pattern formation using CA theory, and examine the group behavior with actual hardware, micro robot "MARS„. Finally, we propose inverse design of a more complicated polygon pattern formation from a circle state using Turing Instability theory.

Keywords: graduated spatial pattern formation, micro robot experiment by hardware, homogeneous system, CA, Turing Instability

1 Introduction

It is desirable for a robot to reach dangerous places so humans do not need to go there. For example, in the big facilities such as a nuclear power plant, various parts like pipes need to be safely. In these situations, each robot is designed not only inspection but also secondary works that mean transportation in fragile article and repairs. If we set a robotic system on such tasks, a robot should be required to make its size as small as we can. However, it cannot avoid a scale effect. It means that there is a trade off relationship between a scale value and function of a robot.

Some researches have presented methodologies of cooperation of swarm intelligence to solve these subjects[1][2]. For example, a swarm of robots can transfer a big object that one robot cannot transfer and a swarm of robots can infiltrate into narrow places by keeping line formation. Thus, it is important for robots to make and change spatial formations by swarm intelligence.

 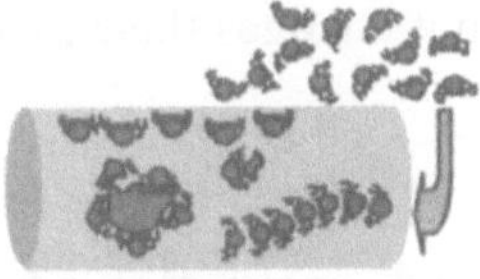

Fig. 1 Image of micro robot works

Some researchers have proposed about a spatial pattern formation by using Distributed Autonomous Robotic Systems. A general overview about their study allows agents to acquire global information[3][4]. However, as the communication range of micro robots is very small, it is difficult for the robots to acquire information of its position and the other robot's positions. Therefore, we assume the following simulations and experiments.

- A robot can communicates with other robots, which are in local domain of its position. The robot can also sense obstacle or robots in local domain.
- Since each robot is required to be changeable when a robot is broke down in working area, target system is homogeneous, in which each robot is composed of same hardware and is programmed with same code.
- Each robot cannot acquire global information. Therefore, they do not know their position and direction.
- All robots can move with two stepping motors.

In this paper, we propose "graduated spatial pattern formation", which is the method of autonomous pattern formation by using a swarm of robots without acquiring global information. Then we indicate the experimental results by hardware.

2 Related work

Some researches of pattern formation of robots have proposed based on real hardware. Murata [5][6] achieves two or three dimensional pattern formation and conversion to plural pattern by real modular robot. Yoshida[7] also studies about self-pattern formation of micro robots by using shape-memory alloy. His approach is applied robot's actuator and pattern formation without referring agent's autonomy.

Although the main issue of their paper also refer to pattern formation by hardware, the way of pattern formation is that the state of robot's pattern is given by observer demonstratively and each agent's procedures of conversion to some pattern is prompted by observer in sequence. In this paper, we propose the algorithm that a swarm of robot can form into some gradual patterns autonomously. Then the observer send only signal of to start formation to robots in a swarm. Each robot changes their function and move to goal pattern by self-organizing function.

3 Outline of graduated pattern formation algorithms

Think that we make a swarm of robots form spatial complex pattern from condition that every robot's position changes abruptly. Each robot recognizes their position with global coordinates and acquires only local information. Although each robot has a communicating function with each other, we assume that communication domain is effective within narrow range. In this case, it is difficult for them to acquire their global or local coordinates.

Ideal False Robot

Fig. 2 Examples of goal

Assigned to vertex

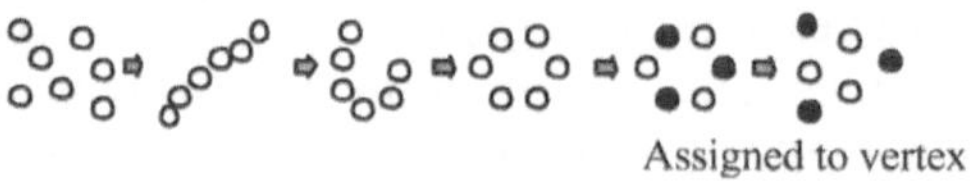

Fig. 3 The way of graduated pattern formation

The spatial pattern formation can be expressed by relative position among robots. Therefore, it is difficult to transform to the goal condition from random positions that each robot does not have global or local coordinates. When a swarm of robots tries to make certain pattern with only local information, the following failure can occur (Fig. 2). As shown in this figure, the goal formation is not always validity even if each robot's position is located on the correct position.

In the case of a global pattern formation by robots, generally, the robots need to not only design an interaction but also acquire the global information. In this paper, we use the method of graduated pattern formation like the process shown in Fig. 3. It is important to make an adjustment between each robot's locality and directions when forming a polygon pattern. At first, robots make a line pattern. Secondly, robots make a circle pattern, which guarantee the condition to make a polygon pattern. Thirdly, robots assign vertex and edge by themselves. Finally, robots move to each position.

4 Hardware

This section is about hardware. We call the following robot (Fig. 4) MARS (Micro Autonomous Robotic Systems). Each MARS robot has one CPU, two stepping motors, eight optical sensors, infrared light communication device, and secondary battery in a small body.

Each MARS robot can communicate with other MARS robots located 70 [mm] away from itself on an average and host computer by infrared light communication device. However, the communication range is not a circle completely because of the directivity of LED. Optical sensors enable MARS robot to recognize obstacles. MARS robot has two stepping motors and it can move to front back left right and turn. The size of MARS robot is Diameter of 30 [mm] and height of 35 [mm].

Fig. 4 MARS (Micro Autonomous Robotic Systems)

Fig. 4 shows a MARS and comparing MARS to a packet of cigarettes. Optical sensors locate the side of MARS robot's body at even regular intervals. By using sensors, a MARS robot can acquire information about some obstacle or other robots in eight directions even though it cannot distinguish obstacle from other robots because catoptrical devices apply to the sensors. Each robot can ascertain a distance to the seven-grade steps by changing voltage supply to optical device. A robot can acquire existence of obstacle located 35 [mm] away from it.

5 Line pattern formation

A robot joining a group one after another can form a line pattern. The pattern formation begins from one robot. The robot, which is the top position of the line, confirms its position, and searches the other robots. Other robots stand still at the same position or move at random and avoid other robots. If the top robot recognizes another robots with obstacle sensor, the top robot sends a message that is a request for changing a top to another robot. The robot, which was received the message, changes a direction to look forward and become a new top robot. They continue a series of behavior to make a line pattern. Then a top robot tells the new top robot the sum of the line, and they stop moving at their decision time.

Fig. 5 Control sequence of making a line Pattern

The result of the experiment with hardware shows in Fig. 6. In Fig. 6, there are three robots. After they perform a series of behavior; a line pattern formation is generated. Sometimes the mission is impossible when a communication error occurs. For example, in the consequence of broadcasted communications of the robots, only one robot does not usually takes the lead when plural robots send a request to change a part. Then the communication range needs to have adjusted.

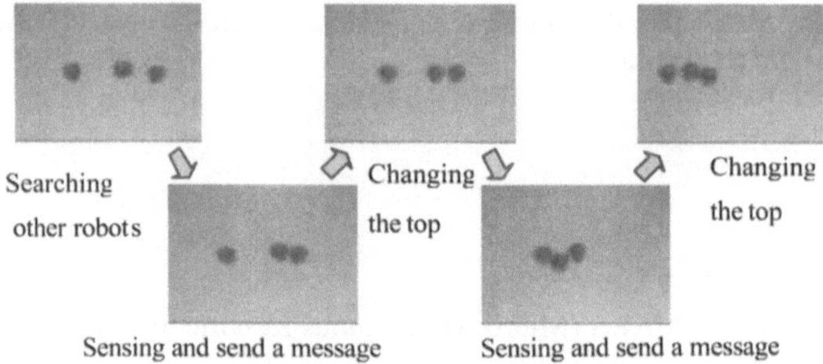

Fig. 6 Experimental result of line pattern formation by hardware

6 Circle pattern formation

The algorithm of circle pattern formation is very simple. A robot following the top robot forms a circle pattern as its own mechanism. Trailing robots keep the distance to the top robot. If one robot does not detect any obstacle during a sampling time, the trailing robot changes into a top robot. Therefore, another robot can become a new top if the top robot has trouble or parts from swarm. Then the top robot moves in an arc. If the top robot finds the end robot of line, the mission regards to finish. At the end of the formation, the top robot of the above situation sends information to other robots to let them know the mission finished. The information diffuses spatially and all robots can recognize that they finished making a circle pattern. It is not necessary to acquire a number of robots. As the curvature of the arc becomes small gradually, the top robot can reach the end of line absolutely.

Fig. 7 shows the result of an experiment by using seven robots. We can success to make a circle pattern as Fig. 7.

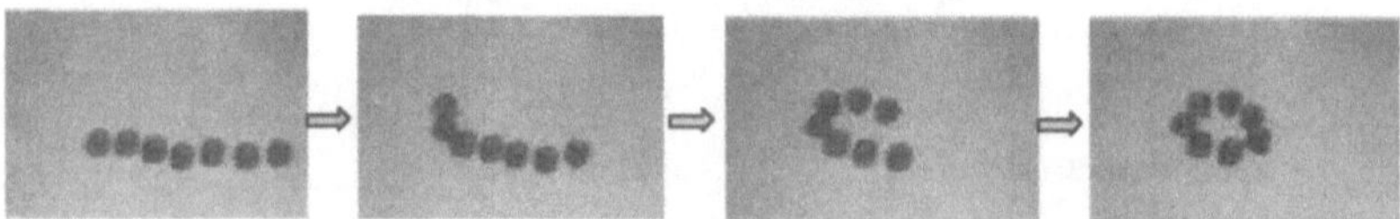

Fig. 7 Experimental result of circle pattern formation by hardware

7 Polygon pattern formation by using Cellular Automata theory

In this section, we express the elementary experiment that robots form a polygon using local communication with hardware.

When a swarm of robots tries to make polygon pattern, robots must recognize their position whether a side or a vertex. How can a robot decide its function whether it is a vertex or not without order of supervisor? To solve this problem, Cellular Automata theory (CA) is available. CA is also available for simulations of fluid mechanics or stock and so on. The algorithm is that individuals separated to finite element decide their condition or next action by using a certain definite rule. In this way, a certain global pattern appears when a system is describable with a bottom-up approach and agents update their condition given an initial state.

As the above, we consider assumptions as follows.

- A system consists of homogeneous elements.
- Each element can take ***k*** kinds of states.
- The state of the following element is decided by only the individual rule of the state of the elements, which adjoins the present state.
- The generated global pattern is decided by the initial state of an element and applied rule.

A simple simulation for a hardware experiment was performed based on these considerations. Here, the state of each element is set into white and black, and the case where the interaction of only neighboring elements is taken into consideration. Therefore, location rule of each element in a line is expressed the following eight kinds as shown in Fig. 8. Next, the updating rule over each state is decided as shown in Fig. 9.

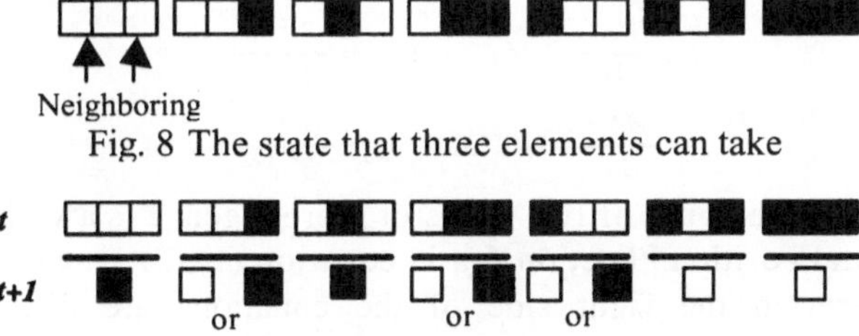

Fig. 8 The state that three elements can take

Fig. 9 Rule of updating

The simulation was performed by six elements using these rules. The simulation result converges to two kinds of patterns, as shown in Fig. 10, where the element of a left end and a right end adjoins.

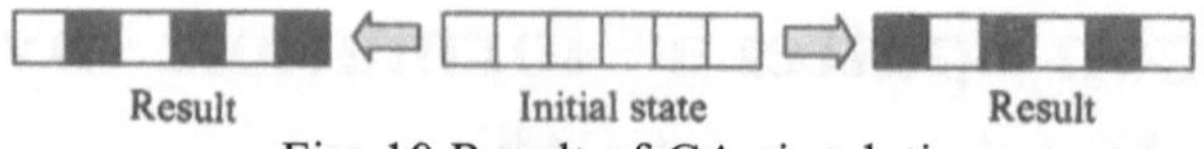

Fig. 10 Result of CA simulation

Fig. 10 shows a simulation result to make a triangle with six robots from assigning black and white to a vertex and side respectively. As the above simulation, the robot assigned to a vertex and a side by moving toward the outward ness of the center of a circle and inner respectively.

Based on the result of the above simulation, we succeeded in making a triangle from a circle by six robots shown in Fig. 11, where the transfer distance of a robot is designed a priori. Fig. 11 shows not only Successful result but also unsuccessful result. When communication is not working well between robots, an experiment often ends in failure.

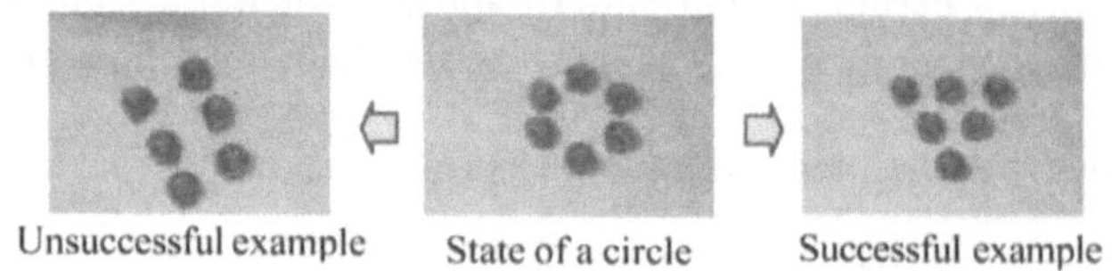

Fig. 11 Experimental result of polygon pattern formation by

8 Complex pattern formation algorithm

In order to make a polygon pattern from circle pattern by distributed theory, it is required to assign the vertex of a Polygon. In this section, in order to make more complicated polygon pattern, we propose how to assign vertex by using Turing instability theory. Alan Turing explains that pattern formation in a living body is produced by reaction and diffusion of two kinds of morphogen, which are activator and inhibiter. Activator is substance prompting functional changes and inhibitor is substance controlling functional changes. This model is expressed by the following two general equations.

$$\frac{\partial X}{\partial t} = f(X, Y) + \frac{\partial^2 X}{\partial x^2} \quad (1)$$

$$\frac{\partial Y}{\partial t} = g(X, Y) + \frac{\partial^2 Y}{\partial x^2} \quad (2)$$

where ***X***, ***Y*** are the amount of two kinds of morphogen, ***t*** is a variable of a time and ***x*** is a spatial position.Function *f*, *g* in equation (1),(2) are reaction term and the second term in the right side of the equations are diffusion term of morphogen. Now if the reaction is expressed linear function, equation (1), and (2) can be converted into equation (3) and (4).

$$\frac{\partial u}{\partial t} = au - bv + D_u \frac{\partial^2 u}{\partial x^2} \quad (3)$$

$$\frac{\partial v}{\partial t} = cu - dv + D_v \frac{\partial^2 v}{\partial x^2} \quad (4)$$

where $\boldsymbol{u}$, $\boldsymbol{v}$ are activator and inhibitor respectively, $\boldsymbol{a}$, $\boldsymbol{b}$, $\boldsymbol{c}$, $\boldsymbol{d}$ are parameters to control reaction, $\boldsymbol{D_u}$, $\boldsymbol{D_v}$ are diffusion coefficient of activator and inhibitor respectively. All parameter are positive. The heterogeneous spatial pattern appears if the parameter sets are designed suitably (Fig. 12). This phenomenon is generally known as Turing Instability. The series of this phenomenon are explained by the following dynamics.

Activator $\boldsymbol{u}$ tries to increase itself and inhibitor $\boldsymbol{v}$ tries to reduce the value of $\boldsymbol{u}$. $\boldsymbol{v}$ tries to reduce itself and $\boldsymbol{u}$ try to reduce the value of $\boldsymbol{v}$. When diffusion coefficient, $\boldsymbol{D_u}$, $\boldsymbol{D_v}$ are related to $\boldsymbol{D_u} < \boldsymbol{D_v}$, the value of $\boldsymbol{u}$ diffuse spatially faster than the value of $\boldsymbol{v}$. Although the couple of equation (3), (4) is singularity and unstable at origin point, if swing is added to $\boldsymbol{u}$ and $\boldsymbol{v}$, like some noises, the value of $\boldsymbol{u}$ and $\boldsymbol{v}$ increase. Both of $\boldsymbol{u}$ and $\boldsymbol{v}$ are usually inhibited by $\boldsymbol{v}$. However the condition $\boldsymbol{D_u} < \boldsymbol{D_v}$ make the value of $\boldsymbol{v}$ diffuse faster than $\boldsymbol{u}$. So the value of $\boldsymbol{v}$ doesn't increase so much, and the peak of $\boldsymbol{u}$ isn't lost.(Fig. 12)

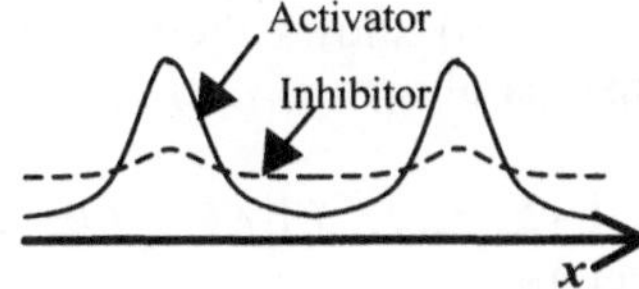

Fig. 12 Turing Instability

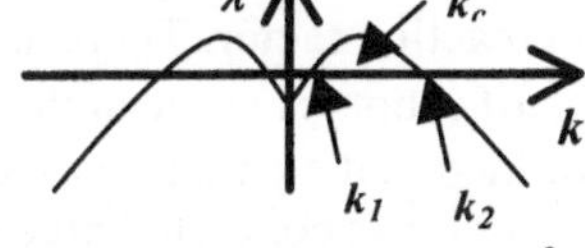

Fig. 13 Relation between $\boldsymbol{\lambda}$ and $\boldsymbol{k}$

Now parameters of equation (3), (4) must be designed to generate the peak of activator at the desired position. If $\boldsymbol{u \sim u_k exp(\lambda\ t+ikx)}$, $\boldsymbol{v \sim v_k exp(\lambda\ t+ikx)}$ are substituted for equation (3), (4) with Fourier analysis, the equation (5) is obtained. Fig. 13 shows the relations between $\boldsymbol{\lambda}$ and $\boldsymbol{k}$. When $\boldsymbol{K_c}$ is between $\boldsymbol{K_1}$ and $\boldsymbol{K_2}$ in an equation (5), a certain frequency ingredient will grow to time, and a global pattern will be generated.

$$?^2 - \{(a - D_x k^2) - (d + D_y k^2)\}? + bc - (a - D_x k^2)(d + D_y k^2) = 0 \qquad (5)$$

When a pattern is not to vibrate or divergent, it is necessary to make $\boldsymbol{\lambda}$ and $\boldsymbol{k}$ a solution of real number. In order to satisfy that stability, the following conditions of equation (6), (7) are needed.

$$aD_y - dD_x > \sqrt{(aD_y + dD_x)^2 - 4bcD_xD_y} \Rightarrow bc - ad > 0 \qquad (6)$$

$$(aD_y + dD_x)^2 - 4bcD_xD_y > 0 \qquad (7)$$

The equation (8) is what the value of $\boldsymbol{k_1}$, $\boldsymbol{k_2}$ was solved concretely.

$$k_1 = \sqrt{\frac{aD_y - dD_x - \sqrt{(aD_y + dD_x)^2 - 4bcD_xD_y}}{2D_xD_y}} \quad k_2 = \sqrt{\frac{aD_y - dD_x + \sqrt{(aD_y + dD_x)^2 - 4bcD_xD_y}}{2D_xD_y}} \qquad (8)$$

It is considerable that reverse designing is to design the value of a parameter using these condition equations. At first we decide a $\boldsymbol{k_c}$ which is the value of a number of wave we hope desire. Secondly we decide the value of parameter $\boldsymbol{a}$ and diffusion coefficient $\boldsymbol{D_u}$, $\boldsymbol{D_v}$. Then $\boldsymbol{k_1}$ and $\boldsymbol{k_2}$ are the value satisfied the condition $\boldsymbol{k_1<k_c<k_2}$. Thirdly parameter $\boldsymbol{b}$, $\boldsymbol{c}$, $\boldsymbol{d}$ can be solved using a set of equation (8). If the parameters set solved satisfy condition equation (6), (7), The second process is repeated. The set of parameters is decided by these searches. Based on the above consideration, the preliminary experiment was conducted with a computer. The couple of equation (3) and (4) are converted into difference equations for implementation to robots.

$$\frac{\partial u_i}{\partial t} = au_i - bv_i + (N/2p)^2 D_x(u_{i+1} - 2u_i + u_{i-1}) \tag{9}$$

$$\frac{\partial v_i}{\partial t} = cu_i - dv_i + (N/2p)^2 D_y(v_{i+1} - 2v_i + v_{i-1}) \tag{10}$$

Where Na is a number of robots, $\boldsymbol{u_i}$ and $\boldsymbol{v_i}$ are activator and inhibitor of the robot i respectively. The couple of equation (9) (10) shows their dynamics of diffusion and reaction factor. There are relationships of $\boldsymbol{u_1 = u_{i+1}}$, $\boldsymbol{v_1 = v_{i+1}}$ because the formation of robots is circle at the time.

It is determined that a parameter was set to (i) $\boldsymbol{k_c = 2.0}$, (ii) $\boldsymbol{k_c = 3.0}$, and (iii) $\boldsymbol{k_c = 4.0}$. Table1 shows concrete value of parameters.

Table 1 Simulations condition

k_c	a	b	c	d	D_u	D_v	k_1	k_2
(i) *2.0*	*0.5*	*0.651*	*0.651*	*0.840*	*0.01*	*0.02*	*2.05*	*1.95*
(ii) *3.0*	*0.5*	*0.580*	*0.580*	*0.640*	*0.01*	*0.02*	*3.05*	*2.95*
(iii) *3.0*	*0.75*	*0.834*	*0.834*	*0.836*	*0.01*	*0.02*	*4.05*	*3.95*

Fig. 14 shows the results of computer simulation in $\boldsymbol{N=24}$ which is a number of robots with parameter sets of Table1 (i), (ii) and (iii) in Fig. 14 are results gotten after enough time. Fig. 14 shows that the peak of a wave in (i), (ii) and (iii) are 2, 3, and 4 pieces respectively.

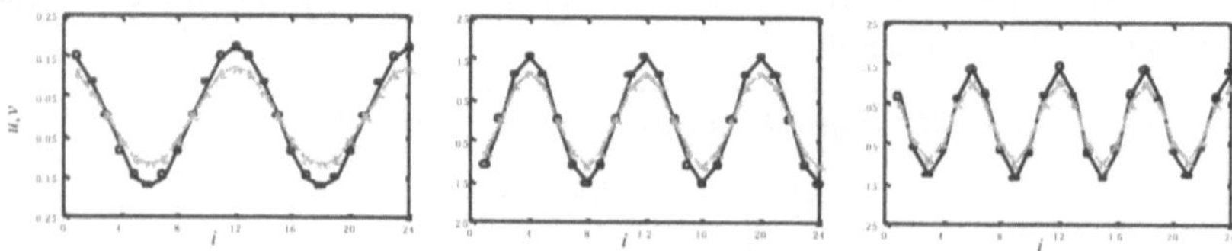

Fig. 14 Simulation Results

When the robot that located in the peak of wave is assigned to a vertex, arbitrary polygons can be generated from a circle pattern. Then overlapping plural waves can generate complicated spatial pattern.

9 Conclusion and future work

In this paper, we have proposed the method of a spatial pattern by micro robots without getting global information. Then amount of success by hardware has achieved. As concerns, line pattern formation, the experiment successes by using three robots. In this experiment, the line pattern by many robots will be generated if the swam of robot can meet other robot which move at random or stop. About circle pattern formation, the experiment has indicated feasibility no matter what the number of robots is extended. However, there are some requirements for further improvement in that way of circle pattern formation, because robots must make a polygon pattern as a gradual spatial pattern formation. Concerning polygon pattern formation, we use a simple method of CA and experiment by hardware as a robot can send only eight bits data. If Turing Instability theory will be used by the experiment with hardware, communication function is required to get advanced.

Turing Instability theory is available for more complicated pattern formation. Since the inverse design is seemed to be important from an engineering position, we will apply Turing Instability theory to a swarm of robots on spatial pattern formation.

References

1. Toshio Fukuda and Tsuyoshi Ueyama, Cellular Robotics and Micro Robotic System, World Scientific, World Science in Robotics and Automated Systems, vol.10, 1994.
2. F.E.Schneider, D.Wildermuth, and H.-L. Wolf, Motion coordination in formations of multiple mobile robots using a potential field approach, Proceeding of Distributed Autonomous Robotic Systems, Pages 307-314, 2000
3. Yoshiki Shimomura, Keisuke Sasae and Yasuo Otuki, Pattern Formation Technology based on Distributed Autonomous Control, Journal of the Robotics Society of Japan, Vol.19 No.3, Pages 316-319. 2001, in Japanese.
4. Hiroshi Yamaguchi, Adaptive Formation Control for Distributed Autonomous Mobile Robot Groups, Proceeding of IEEE Robotics and Automation Society. The Robotics and Automation, Pages2300-2305, 1997.
5. Haruhisa Kurokawa, Kohji Tomita, Eiichi Yoshida, Satoshi Murata, and Shigeru Kokaji, Motion Simulation of a Modular Robotic System, IEEE Int. Conf. Industrial Electronics, Control and Instrumentation (IECON2000), CD-ROM.
6. Satoshi Murata, Eiichi Yoshida, Kohji Tomita, Haruhisa Kurokawa, Akiya Kamimura and Shigeru Kokaji, Hardware Design of Modular Robotic System, IEEE/RSJ Int. Conf. Intelligent Robots and Systems (IROS2000), CD-ROM.
7. Eiichi Yoshida, Satoshi Murata, Shigeru Kokaji, Kohji Tomita and Haruhisa Kurokawa, Micro Self-Reconfigurable Robotic System using Shape Memory Alloy, Distributed Autonomous Robotic Systems 4, Lynne E. Parker, George Bekey, and Jacob Barhen, eds., , pp.145-154, 2000 (Knoxvill, USA).

Heterogeneous Agents Cooperating: Robots, Field Devices, and Process Automation Integrated with Agent Technology

Pekka Appelqvist[1], Ilkka Seilonen[2], Mika Vainio[1], Aarne Halme[1], Kari Koskinen[2]

Helsinki University of Technology
[1] Automation Technology Laboratory
[2] Information and Computer Systems in Automation
P.O. Box 5400, FIN-02015 HUT, Finland

Email: pekka.appelqvist@hut.fi

Abstract. In this paper a novel application area for heterogeneous multi-agent cooperation is studied. A functional specification of an agent-based system as an extension to a process automation environment is described. In the proposed concept the agent technology is used as a method to interconnect multiple mobile robots, mechatronics devices, various software modules, and other intelligent objects. The approach is expected to promote flexibility, modularity, fault tolerance, as well as distributed autonomy and problem solving capacity into process automation systems.

Key Words. Heterogeneous agents, Multi-robot system, Process automation, Integration, Cooperation.

1 Introduction

The application-oriented multi-agent concept presented in this paper is addressed to the field of process automation. By using agent technology, mobile robots, mechatronics field devices, software modules of the process automation, and other intelligent objects can be integrated. Traditionally, such concepts as the agent software technology or multi-robot systems have had very little to do with the process control applications and automation systems. However, in the long run, things can be seen in a different perspective.

In the development of the agent technology, e-business and telecommunications have been in the forefront of the practical applications [1]. Typical tasks for agents in these systems have included, e.g., coordination of negotiations, information brokering, or handling of autonomous recovery in faulty situations. The nature of these agents is purely software based. Industrial applications for the agent technology have been introduced in such areas as production planning, supply-chain management, process automation, and robotics; as an example, see [2], [3], and [4]. An essential motivation for the application of agent technology has been the possibility to utilize the distributed coordination mechanisms of agents [5]. Problems that can naturally be modeled as agent systems have been the primary applications. These problems may typically be decomposed into fairly independent sub-problems and solved autonomously.

When it comes to the multi-robot systems, terms like *robot colony*, *distributed autonomous robotic systems*, *cellular robotics*, *collective robotics* or *robot society* are used to describe systems, where multiple agents are working together towards a common goal. Although, the word "robot" is normally emphasized in this context, these systems are also illustrative examples of physical autonomous agents. If a large number of homogenous robot members are applied, the level of redundancy becomes very high. The benefits and drawbacks of homogenous multi-robot systems compared to heterogeneous systems have also been studied, see, e.g., [6] and [7].

The scope of practical applications for these systems has been enormously wide, from micro-scale manufacturing to extra-terrestrial planetary missions. Another application for a distributed multi-robot system has previously been suggested by the authors: robots targeted to distributed autonomous perception and task execution in the internal, three-dimensional on-line monitoring of various liquid flow-through processes, introduced originally in [8]. The prototype robots, SUBMARs, are controlled by the nature-inspired Robot Society control architecture developed in the same context.

The development of the Robot Society approach and SUBMAR robots have been previously reported also in the DARS forum: see [9] for the early development stages of collective behaviors and navigation methods of the robots; and [10], where the control architecture of the Robot Society system is described in detail along with the results from cooperative task execution. Recently, the effect of limited communication for the performance has been analyzed in details [11].

As one possible answer to the expected future needs of a process automation system (PAS), a synthesis and further development of the previously mentioned issues has been proposed. This concept integrates SUBMAR type of mobile robotic instrumentation, intelligent field devices, and process automation software. Decentralized heterogeneous agent-based system is capable of handling such tasks as distributed sensor data fusion, planning, and problem solving. The resulting main objective for the multi-agent concept in process automation applications is to gain increased accuracy in control, improved flexibility in reconfiguration situations, and autonomous recovery in faulty situations. Eventually, the utilization of agent technology can be seen as an opportunity to better handle the complexity and lifespan of a modern process automation system by raising the level of abstraction.

2 The concept of agent-augmented process automation

Applications of agent technology in the field of process automation have not been so numerous. One reason for this has obviously been the real-time requirements of process automation applications that current agent technology hardly can fulfil. Another reason is the characteristics of process control tasks with complex interrelationship between various controlled variables.

The basic idea in the concept of an agent-augmented process automation system is to complement an existing PAS with agent technology rather than replacing it. The agent system forms an additional *agent layer* on top of the PAS. The purpose of this layer is to supervise the underlying PAS and re-configure it when needed. This means that the agents do not normally interfere with the closed loop real-time control domain. With the existing technology, the agent layer is typically run in a Java-based agent platform. The agent-augmented PAS conforms to a layered architecture as illustrated in Fig. 1. The agent layer can receive data directly from process instrumentation or via a PAS.

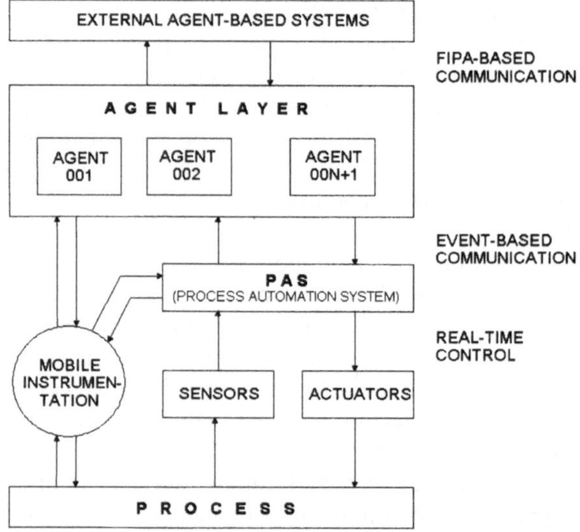

Fig. 1. Architecture of the agent-augmented process automation system.

The communication between the agents complies with the proposed FIPA standard [12], [13]. This standard defines languages and protocols for inter-agent messaging. Nowadays, an increasing number of the implemented agent platforms support FIPA specifications. In the future, FIPA-compatibility may well develop into a de facto standard for industrial agent systems. This would greatly promote the openness of totally different kind of sub-systems, and therefore facilitate the integration of separately designed agents.

The agent layer forms a FIPA interface to the underlying PAS. This enables the integration of the PAS to external agent-based information systems, e.g., e-business, production planning management, or remote monitoring and diagnostics applications.

3 Heterogeneous agents for process automation

The agent concept for process automation does not define which device, object, or logical entity could exist as an agent in the system. However, at least the following categories can be recognized: pure software agents, mechatronics agents, robot agents, and composite (unit) process agents.

3.1 Software agents

The software agents have the functionality to reconfigure the underlying PAS in a distributed and collaborative fashion, e.g., in various fault situations, or when the control logic of the application needs to be altered. This means that the software agents have selected negotiation methods and decision rules for each type of reconfiguration event.

The precise internal architecture of the software agents depends on the agent platform used in the implementation. In the test case presented in this paper, this is a Java-based open-source agent platform called FIPA-OS [14]. FIPA-OS is an application domain independent platform providing agent services, e.g., message brokering and agent registration, as well as building blocks for the implementation of the agents. Therefore, the software agents presented here can be regarded as application domain specific extensions to FIPA-OS agents. The resulting internal architecture of the agent with examples of some process control modules is illustrated in Fig. 2.

Interface modules consist of an agent communication mechanisms and data transfer mechanisms to non-agent systems, particularly the underlying PAS. The agent communication mechanism is provided by the FIPA-OS, while the PAS interface is typically an OPC-client (OPC = OLE for Process Control).

Model modules contain models of the controlled process, PAS, and agent society itself. These models are needed for all process control agents regardless of the application.

Application modules contain the application specific logic that is defined with cooperative negotiation mechanisms and decision rules. While the negotiation mechanisms are defined in the FIPA standard, e.g., contract net and auctions, the decision rules are partially specific to the controlled process.

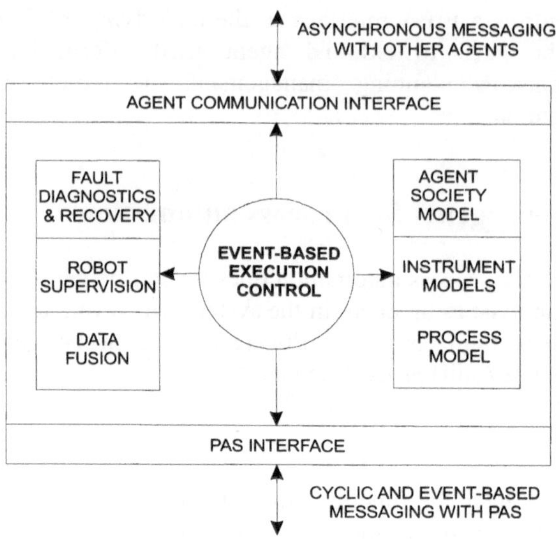

Fig. 2. An example of the internal architecture of the FIPA compatible process control agent.

The execution model of the agents is provided by the underlying FIPA-OS platform. The agents have *tasks* that define how and when the agent participates in negotiations and when applies its decision rules. The execution model is essentially event-based concurrent processing of the agent's multiple tasks. This kind of execution model is quite different from the typically cyclic execution of the control logic in a PAS.

3.2 Mechatronics agents

In some cases, it may be motivated that even a single intelligent device, sensor, or actuator, operates as an independent agent. Basically, mechatronics agents have the same inner structure than the software agents, although, the interface module is inseparably connected to the physical I/O-functions of the agent. An intelligent field device, e.g., such as Foundation Fieldbus connected Neles control valves used in the demonstration process (see Section 4.1), could be configured to the system as stand-alone agents. The functionality is in the device itself, while the intelligent agent features can physically be mapped elsewhere into the agent platform.

3.3 Robot agents

Autonomous mobile robots can naturally be outlined as agents. In Fig. 3., the structure of a mobile sensor/actuator robot is illustrated. By combining features from dataloggers, in-pipe robotic vessels, underwater robots, and mobile multi-robot systems, the SUBMAR robot has been developed [15].

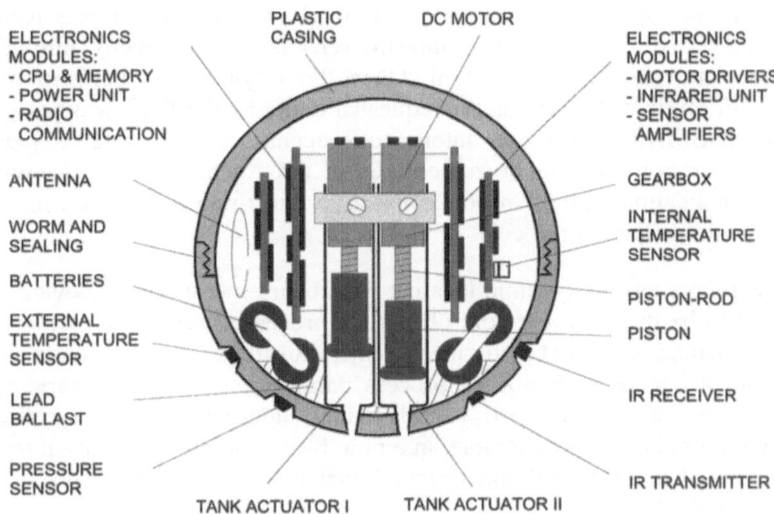

Fig. 3. Cross-section of a SUBMAR robot. The diameter for the robot is 11 cm.

SUBMAR robots are intended as mobile instrumentation for submerged use in a liquid process environment. The robots control their vertical motion only, otherwise they move passively along the process flow. However, due to the active navigation and positioning systems, the robots are capable to perform 3D on-line measurement tasks with their onboard sensors. The obvious limitations caused by the simplified design of the robots will be complemented by multiplicity and cooperation. Functioning of the individual robots and their cooperation is defined by the Robot Society control structure. The SUBMAR Society is shown in Fig. 4.

Fig. 4. Situated and embodied agents, the SUBMAR Robot Society. Most of the robots are shown as cover opened; the electronics unit and battery pack are visible.

When the robots operate in an autonomous mode, a single robot gets feedback from its actions directly from the onboard sensors, and indirectly through the mutual robot-to-robot communication. Once the robot agents perform on-line measurement routines the actions are requested either by the PAS or other agents. Although SUBMAR robots are autonomous agents as such, due to practical reasons, each robot can also be represented in the Agent layer augmenting the PAS. Communication with the robots is based on short-range radio network and a special Ethernet-inspired protocol.

SUBMAR robots enable dynamic locating of onboard measurement devices. This is expected to be useful in the control of such processes where the measurement needs may change or current static measurement locations cannot provide enough information of spatially distributed process parameters. Complementary mobile instrumentation can be used to reveal hidden non-laminar flow profiles or concentration gradients. For example, in a pulp bleaching process, certain residual chemicals could be measured and watched over to indicate that the reaction has reached the end.

3.4 Unit process agents

As mentioned earlier, all reasonable combinations are possible to formulate an agent. An agent can be composed of several different types of objects. As is further described in Chapter 4., a unit process is a natural approach. A certain unit process agent is responsible for all the instruments which belong to the corresponding physical space or area.

4 Test case

In the preceding research projects the task for the SUBMAR Robot Society was the mapping of the laboratory test environment, along with the searching and hunting down of the emulated biomass growth locations found from the environment [15]. For the testing of heterogeneous multi-agent concept, the same environment is used.

4.1 Demonstration environment

To study the technological feasibility of the FIPA based agent-approach, a practical implementation into a laboratory test environment will be accomplished. A fully transparent test environment is not a model of any existing industrial process, but consists of different types and shapes of typical process parts, see Fig. 5. The total volume of 700 liters is filled with fresh water. In order to imitate process flow, water is circulated with a jet-flow pump. Other instrumentation includes, e.g., several temperature and pressure sensors. Valves controlling hot and cold water inputs can be used to generate plug flows and temperature gradients to be monitored and controlled by the PAS. The forming, monitoring, and controlling of certain temperature distributions is one of the defined task scenarios which will be carefully examined and analyzed with the test environment.

Fig. 5. Layout and instrumentation for the laboratory test process environment.

To manage the resources in the process, the natural division into sub-areas can be made on the unit process basis. The unit processes are responsible for the instrumentation existing in their areas. Furthermore, for more finely split zones, division can be made to form partial processes depending on the needs of an application. This is illustrated in Fig. 6., where Unit Process I is divided into three horizontal zones to control the vertical temperature distribution more precisely.

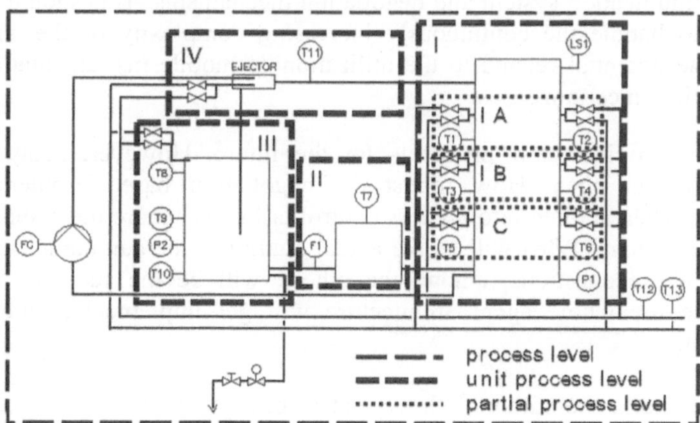

Fig. 6. Hiérarchical division of the process into unit processes and sub-ordinate partial processes IA, IB, and IC.

4.2 Functioning of the agents

In general, the role of the agent layer is to react to such events which require changes in the control logic or parameterization of the underlying PAS. The events originate both from the tasks requested from the PAS and changes in the PAS itself. The functioning of the agents can be illustrated with a simple application scenario related to autonomous fault detection and recovery. For example, the agent IB (in Fig. 6.) might observe a malfunction in its control valves and infers that it can't fulfill its control objectives any more. It initiates a negotiation with agents that can possibly help or which are influenced by the emerged problems. In this example, these agents would be the other temperature controlling agents IA, IC, and the agent responsible of the flow circulation pump. The agent IB requests other agents to change their control settings in order to help the situation. Based on the answers, the agent IB decides if the new setting is acceptable, or should it commence the process shutdown.

In other words, the responsibility of the agent layer is to maintain stable mapping between the requested control tasks and the equipment of the PAS. In order to agree with all the active tasks, the agents *negotiate contracts*. The contracts define the responsibilities the agents take for the process control operations.

5 Conclusions and future work

In this paper a novel approach has been suggested to integrate multiple mobile robots, mechatronics devices, various software modules, and other intelligent objects related to the distributed process automation environment with the agent technology. The heterogeneous agent approach is expected to provide increased system flexibility and fault tolerance to process automation applications through a mutual communication system and negotiation mechanisms. This should also give assistance to handle the continuously increasing complexity of the automation systems. The potential related to the utilization of mobile instrumentation seems also interesting in certain cases.

The concept provides new prospects for the future. However, many practical questions remain open. How robust the negotiation based problem solving methods are when applied into process control tasks? What are the response times of the agent system? What will be the actual quality of the autonomously iterated contracts? These questions, among the others, will be the subjects of further studies once the whole agent architecture will get implemented into the test environment.

As a final remark, the authors believe that in order to promote the spreading of multi-robot installations towards practical applications, the interoperability characteristics of the robot agents to interact with heterogeneous (non-robot) agents should be carefully developed.

References

[1] A.L.G. Hayzelden, R. Bourne (Eds). (2001) *Agent Technology for Communication Infrastructures*, 1st edition, John Wiley & Sons, 316 p.

[2] D.H. van Parunak. (1999) Industrial and practical applications of DAI, in *Multiagent systems* (ed. G. Weiss), Cambridge, MA, USA, pp. 377-421.

[3] N.R. Jennings and M.J. Wooldridge. (1998) Applications of Intelligent Agents, in *Agent Technology: Foundations, Applications, and Markets* (eds. N.R. Jennings and M. Wooldridge), pp. 3-28.

[4] D. Cockburn and N.R. Jennings. (1996) ARCHON: A Distributed Artificial Intelligence System for Industrial Applications, in *Foundations of Distributed Artificial Intelligence* (eds. G.M.P. O'Hare and N.R. Jennings) Wiley, pp. 319-344.

[5] J. Ferber. (1999*) Multi-Agent Systems: An Introduction to Distributed Artificial Intelligence*, Addison-Wesley Pub Co, 528 p.

[6] L.E. Parker. (1994) Heterogeneous Multi-Robot Cooperation, *Ph.D.Thesis*, Massachusetts Institute of Technology, Artificial Intelligence Laboratory, MA, MIT-AI-TR 1465, USA.

[7] T. Balch. (1998) Behavioral Diversity in Learning Robot Teams, *Ph.D. Thesis*, College of Computing, Georgia Institute of Technology, USA.

[8] A. Halme, P. Jakubik, T. Schönberg, M. Vainio. (1993) The concept of robot society and its utilization, in *Proceedings of the IEEE/Tsukuba International Workshop on Advanced Robotics*, pp. 29-35.

[9] M. Vainio, A. Halme, P. Appelqvist, P. Kähkönen, P. Jakubik, T. Schönberg, Y. Wang. (1996) An Application Concept of an Underwater Robot Society, in *Distributed Autonomous Robotic Systems 2 (DARS'96)*, Asama H., Fukuda T., Arai T., Endo I., (Eds.), Springer-Verlag, Tokyo, Japan, pp. 103-114.

[10] M. Vainio, P. Appelqvist, A. Halme. (2000) Control architecture for an underwater robot society, in *Distributed Autonomous Robotic Systems 4 (DARS 2000)*, Parker L.E., Bekey G., Barhen J., (Eds.), Springer-Verlag, Tokyo, Japan, pp. 15-24.

[11] P. Appelqvist, M. Vainio, A. Halme. (2001) The Effect of a Limited Communication within an Underwater Robot Society, in *Proceedings of the IASTED International Conference on Robotics and Applications (RA'2001)*, ACTA Press, USA, pp. 174-179.

[12] (2001) http://www.fipa.org.

[13] Y. Labrou, T. Finin, Y. Peng. (1999) Agent communication languages: The current landscape, in *IEEE Intelligent Systems*, Volume 14, Issue 2, March-April, pp. 45-52.

[14] (2001) http://fipa-os.sourceforge.net.

[15] P. Appelqvist. (2000) Mechatronics Design of a Robot Society - A Case Study of Minimalist Underwater Robots for Distributed Perception and Task Execution, *Doctoral Thesis*, Automation Technology Laboratory, Helsinki University of Technology, Finland.

SUBMAR® is the registered trademark of ISS Ltd.

Chapter 10
Multi-Robot Motion Planning

Heuristic Approach of a Distributed Autonomous Guidance System to Adapt to the Dynamic Environment

Daisuke Kurabayashi[1] and Hajime Asama[2]

[1] Tokyo Institute of Technology, 2-12-1 Ookayama, Meguro-ku, Tokyo, 152-8550 Japan
[2] The Institute of Physical and Chemical Research, 2-1 Hirosawa, Wako-shi, Saitama, 351-0198 Japan

Abstract. In this paper, we built a device and algorithm for distributed autonomous guidance system (DAGS). In this system, autonomous mobile robots construct guidance information structure by means of an Intelligent Data Carrier (IDC). We make models of dynamic environments, and investigate the behaviors of autonomous robots guided by DAGS. We create an algorithm that estimates the validity of knowledge in an IDC and allows the IDC to renew the knowledge autonomously. We verify effectiveness of the proposed algorithm by means of simulations.

1 Introduction

Currently, researchers are trying to create a robotic system that can function in any general environment. Realization of autonomous task execution would be especially advantageous in hazardous environments. Most robotic systems require a model environment in order to execute tasks effectively. Autonomous robotic systems should create models of the environment by themselves. However, such tasks are not easy for current autonomous robots because they have only limited ability to sense and thus survey the environment. In such cases, the method of knowledge acquisition and sharing becomes very important. It is very advantageous that autonomous robots can build and maintain guidance information about their environment automatically.

Let us regard the social insects. Ants make use of their pheromone trails to search and obtain their foods. Simulation studies by Drogoul and Feber[1] have suggested that the pheromone trails are very effective for the completion of iterative transportation tasks to be performed by autonomous agents. The authors have proposed a device and algorithm to apply this kind of data storage and communication system for robotic systems for the sharing of knowledge and for cooperation. The device is named as "intelligent data carrier (IDC)[2]-[4]" that is small computer including local communication ability via radio waves. By utilizing the device as intelligent artificial pheromone, the authors have established distributed autonomous guidance system (DAGS) in which autonomous robots construct information structure by themselves[5][6].

In this paper, we create a model of a dynamic environment in which the destinations are frequently changed. We investigate the behaviors of autonomous mobile robots that navigate in dynamic environments by means of the intelligent data carrier system, and propose an algorithm that allows each IDC to revise its own knowledge.

This paper is organized as follows: In section 2, we introduce the intelligent data carrier, and the basic algorithm of DAGS. In section 3, we make models of a dynamic environment, and in section 4 propose an algorithm to adapt the IDC system to the dynamic environment. Section 5 provides a summary of the paper.

2 Scope of DAGS

2.1 IDC System

The authors have developed an Intelligent Data Carrier (IDC) in order to provide local communication links and local information management functions. By reading information from and writing it into the IDCs, robots can use them as media for inter-robotic communication.

The IDC system consists of portable information storage (tags) and read-write devices carried by the robots (Fig. 1(a)(b)). Size of a tag is about 100×65×25[mm]. Tags are usually refereed to as an "IDC". A tag has its own CPU, memory, and battery. A user can download and execute original programs into the tags. By placing the IDCs in specific locations in a particular environment, robots can allocate functions to act as agents for information storage and management (Fig. 1(c)). The communication range is up to 3.0 [m].

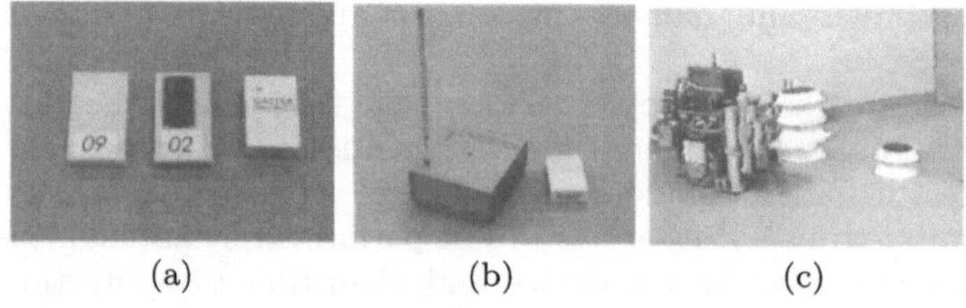

(a) (b) (c)

Fig. 1. The prototype of the IDC system: (a) Tags, (b) A reader/writer device, (c) Handling system for autonomous robots.

2.2 Problem settlement

In this paper, we consider iterative transportation tasks (e.g. [7]). A robot has to carry objects to given destinations. We posit an environment in which several destinations are located. When a robot arrives at a desired destination, it receives instructions regarding the next destination and then continues with the task at hand.

We assume that robots do not have maps, because fixed maps, made by humans, may decrease the flexibility of autonomous robotic systems. We consider adaptability to unknown environments as crucial. We assume that a robot consists of the following characteristics.

- A robot does not have a map and does not estimate its global position.
- A robot can sense junctions, and destinations.
- A robot can remember the last visited destination and count, in steps, the duration of running time.

A robot cannot understand its global position, but it can understand its position in relation to its immediate context.

In this example, we assume that the transportation task is as follows:

- A robot is given only the ID number of a destination. When a robot arrives at a destination, it receives another destination ID at random.
- The work area is a maze-like environment that consists of square cells and walls (Fig. 2). We locate an IDC at each junction.
- A robot can move to neighboring cells, but only at one step at a time.
- We do not consider cases of collisions among robots.

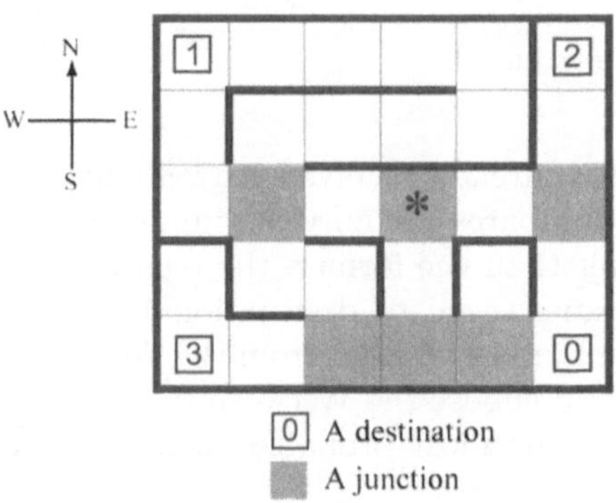

Fig. 2. An example of a work area: Four destinations are located in a maze-like environment.

2.3 Basic algorithm of DAGS

In this section, we denote basic algorithm of DAGS. The algorithm builds knowledge for guidance autonomously according to experiences of robots. Upon reaching a junction, a robot should select the most feasible branch that allows the shortest path to its current destination. We set an IDC at each junction to facilitate robots' decisions. Robots can store and share their fragments of experience by means of the IDCs.

As we assumed, a robot does not know its global position. Thus we have proposed an algorithm to locate where branches connect at junctions leading to expected destinations. This proceeds according to the last visited destination and entered branch. For example, when we see that a robot, which has started from destination 1, enters a junction through a southern branch, we can expect the southern branch may lead us to destination 1.

We describe the data structure of an IDC in Fig. fig:idcdata. This example illustrates a possible status of an IDC located at the junction marked "*" in Fig. 2. At the initial state, no valid data is recorded. When a robot enters a communication area of an IDC at a junction, it reports to the IDC the branch of entry, the ID number of the most recently visited destination, and the running step measured from the destination.

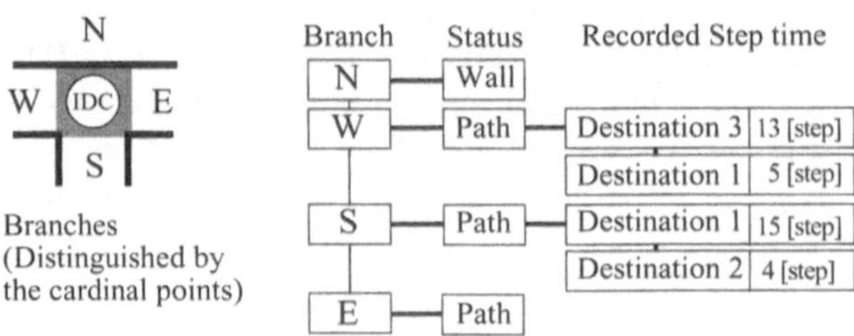

Fig. 3. Data structure in an IDC

When the IDC has already received data about the same destination in the same branch, it compares the current running step with a former one. If the new one is shorter than the former, the record is renewed.

When a robot wants to go to destination 1, it should choose a branch involving the fewest numbers of steps from the destination. In the example of Fig. 3, the robot should choose the W (western) branch. We implemented an algorithm that selects the most probable path. The steps of this algorithm are as follows.

(0) When a robot can not communicate with an IDC at a junction, it chooses a branch at random.

(1) A robot whose destination is j comes to a junction which has m branches. We describe the recorded steps from destination j in branch $i \in m$ as t_{ij}.

We can find four data, $t_{W1} = 8$,$t_{W2} = 19$,$t_{S1} = 15$,$t_{S3} = 21$ in the example shown in Fig. 3

(2) When the IDC has no record regarding destination j, the robot chooses a branch at random.

(3) When the IDC has one or more branches which contain records regarding destination j, it sets $s_{ij} = t_{ij}^{-1}$. If direction i has no data about destination j, set $s_{ij} = 0$.

(4) Calculate probability p_{ij} to choose branch i by equation (1). Then it selects to proceed via branch i according to the probabilities (This procedure is often called "roulette selection").

Note that a robot has the ability to choose a branch at random. We set P_{min} which denotes a fixed probability that a robot chooses a branch at random. The fewer steps to j involving branch i, probability $p_i j$ becomes the larger. A robot does not choose simply "a branch involving shortest step", because DAGS system has small but certain possibility to got stuck on non-optimal guidance sequence. By the algorithm, a destination is certainly achieved [5].

$$p_{ij} = (1 - p_{min}) \frac{s_{ij}}{\sum_j^m s_{ij}} + \frac{p_{min}}{m} \tag{1}$$

2.4 DAGS in static environment

Here we show simulation results by DAGS in static environment. We evaluated the number of achieved destinations by counting a constant number of steps. We assume an environment like the one given in Fig. 2 and performed simulations with or without utilizing IDCs at the junctions. Each robot worked for 1000 steps. We set $P_{min} = 0.01$.

Figure 4 shows a comparison of results achieved by a single robot and by ten robots. We can see that the proposed algorithm using IDCs achieved about 600% more destinations than did the algorithm without IDCs. We did not provide the robot any knowledge about the environment in advance of performing simulations. Additionally, other robots can share the knowledge stored in IDCs by just the same algorithm. In the case of ten robots, the average number of achieved destinations increases about 10% more than that of a single robot. Without the use of the IDCs, the number of results achieved by a single robot and the average of the results achieved by ten robots was approximately the same. Those results suggest that the proposed algorithm and IDC system realized implicit cooperation among autonomous robots without explicit communication and without a priori knowledge.

3 DAGS in Dynamic Environments

3.1 Model of Dynamic Environment

We consider changing destinations in an environment. Figure 5 shows destination 2 having moved to a new place. In this case, IDCs that have previous

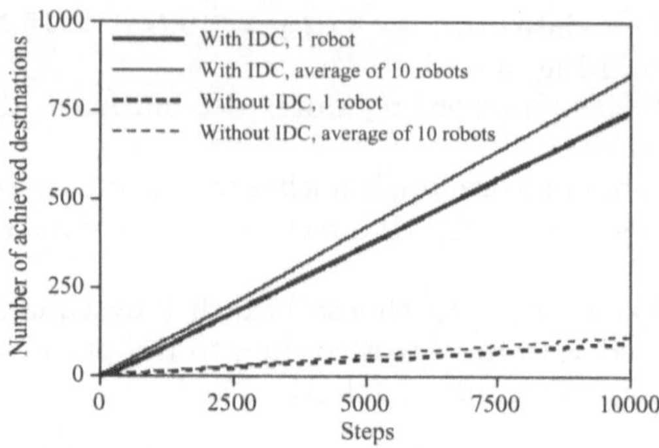

Fig. 4. Comparison of task execution

knowledge will lead robots in incorrect directions, thus lowering the effectiveness of transportation. In some case, robots that do not make use of knowledge in an IDC perform better than robots misled by IDCs. We have to realize an algorithm that allows each IDC to evaluate and renew its knowledge by itself.

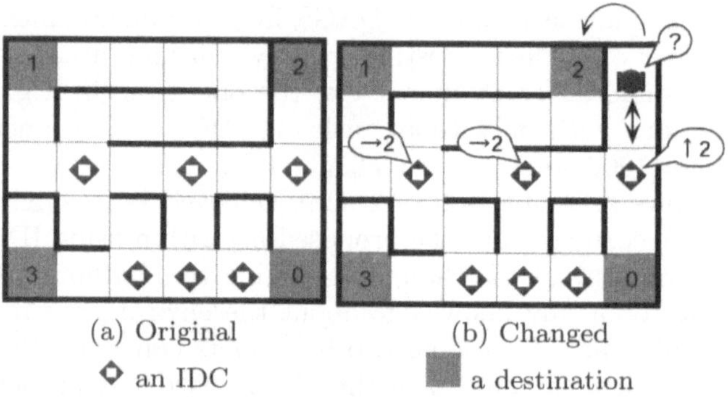

Fig. 5. An example of incompatible information.

3.2 Behaviors in Dynamic Environment

Before building a new algorithm, let us investigate the behavior of robots utilizing DAGS in a dynamic environment. Much like the consciousness of a pheromone, an IDC is provided a very simple algorithm in which the IDC's knowledge is erased at particular constant intervals.

We performed simulations in which the intervals are 10, 100, 500, 1000, 2000, 5000, and 10000. Note that IDCs are not synchronized. To create the dynamic environment, we exchange locations of destinations as $1 \rightarrow 2$, $2 \rightarrow 3$, $\cdots$ every constant step. We create four cases in which the constant step is 500, 1000, 2000 and ∞ (static environment). Figure 6 shows the number of achieved destination in simulations for 10000 steps. Figure 6(a) and (b) denote the simulation results of a single robot and ten robots, respectively.

In static environments, the longer the interval, the more effectively the robotic system works. Curves have peaks in dynamic environments (constant steps=1000 or 2000). The environment may demand special reset interval value to ensure the greatest effectiveness, which depends on both the condition of the environment and the number of robots. However, each IDC cannot know the most effective interval because it works individually. Thus, we need an algorithm in order to estimate timing to reset knowledge in an IDC.

When the destination locations are changed every 500 steps, the performance level of the robotic system becomes quite low. This may suggest that such an environment is too dynamic to allow autonomous robots to construct the knowledge necessary for navigation.

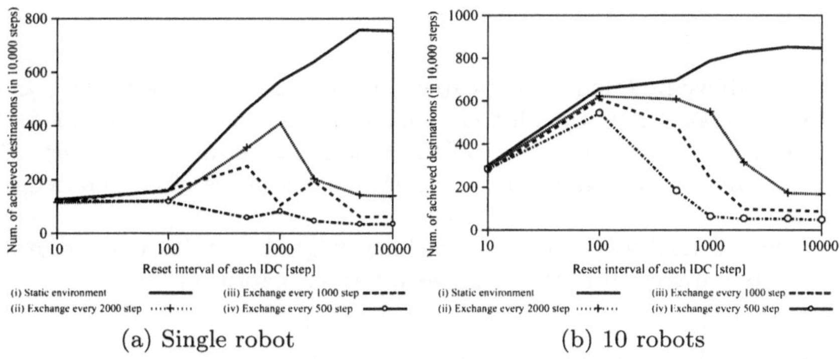

(a) Single robot (b) 10 robots

Fig. 6. Achieved destinations in dynamic environments.

4 Adaptation to Dynamic Environments

4.1 Autonomous Data Renewal

Because an IDC works individually in DAGS, it has to evaluate the validity of its knowledge by itself. An IDC can obtain only reports of steps from autonomous robots. It cannot contain global knowledge regarding an environment. Thus, an IDC calculates probability in order to erase its current knowledge according to the number of steps reported by robots that denotes steps from the arrival of a robot at a previous destination. An IDC decides whether delete its knowledge or not based on probability.

(1) A robot r started from previous destination i and proceeds to new destination j. After t_r steps from destination i, it meets IDC k and reports the steps. When the IDC k has knowledge about both i and j, we call them d_i and d_j, respectively. When the IDC k does not have both, skip (2).

(2) We can expect that the robot r can reach IDC k by d_i steps, and it will arrive at j after d_j steps when there are no changes in the environment. So we compare t_r with d_i and d_j to calculate probability p_{del} to erase the IDC's knowledge. We apply logistic function to calculate (equation 2).

$$p_{del} = K \frac{e^{m(t_r-(d_i+d_j+d_j^2))}}{1+e^{m(t_r-(d_i+d_j+d_j^2))}} \tag{2}$$

The range of this function is (0 K). The function derives $K/2$ when $t_r = d_i + d_j + d_j^2$. K and m are constants. We set $K = 0.5$. Equation 3 determines m, which results in $p_{del} = 0.01$ when $t_r = d_i + d_j$.

$$m = \frac{\log 99}{d_j^2} \tag{3}$$

By the above functions, the greater t_r a robot reports, the higher p_{del} an IDC calculates. Additionally, the smaller d_j knowledge is contained, the higher p_{del} because the center of the function is set at $d_i + d_j + d_j^2$. This means that the nearer the destination j is, the more often knowledge is deleted.

(3) When the IDC does not erase its knowledge, it obtains data from and suggests a direction to the robot as a former algorithm.

4.2 Simulations in dynamic environments

We use simulations to verify the effectiveness of the proposed algorithm. In the simulations, we set $K = 0.5$ and $p_{min} = 0.01$. As the dynamic environment, we exchange locations of destinations as $1 \rightarrow 2$, $2 \rightarrow 3$, $\cdots$ every constant step, as in section 4.2. We create four cases in which the constant steps are 500, 1000, 2000, and ∞ (static environment). Figure 7 shows the number of achieved destinations in simulations involving 10000 steps. We compare the results of the proposed algorithm with those of simple "metabolism" shown in section 4.2. We used both single robot and ten robots in the simulations.

In the case of a single robot in a static environment, the proposed algorithm achieved about 90% of its destinations as compared with the best value of the constant reset interval (described section 4.2). The proposed algorithm lost about 10% of its destinations as the overhead. In the case of ten robots, the proposed algorithm achieved about 80% of its destinations as compared with the best value of the constant reset interval. The proposed algorithm lost about 20% of its destinations as the overhead.

In the case of a single robot in a dynamic environment (constant steps=1000, 2000), the proposed algorithm achieved 69% and 108% of its destinations, respectively, as compared with the best results of the simple periodic resetting method. In the case of ten robots, the proposed algorithm achieved 91% and 98% of its destinations, respectively, as compared with the best results of the simple metabolism method. These results show that the proposed algorithm realizes autonomous adaptation to both dynamic environments and the number of robots, even though individual IDCs cannot know the global state of an environment.

When the locations of destinations are changed every 500 steps, the performance by a single robot system becomes quite low even if we apply the proposed algorithm. This may suggest that the environment is too dynamic to construct knowledge to allow the navigation of single autonomous robots. However, in the case of ten robots, the proposed algorithm achieved about 512% of its destinations as compared with the single robot system. This result demonstrates that the knowledge sharing IDC system works quite effectively to accelerate cooperation among individual robots.

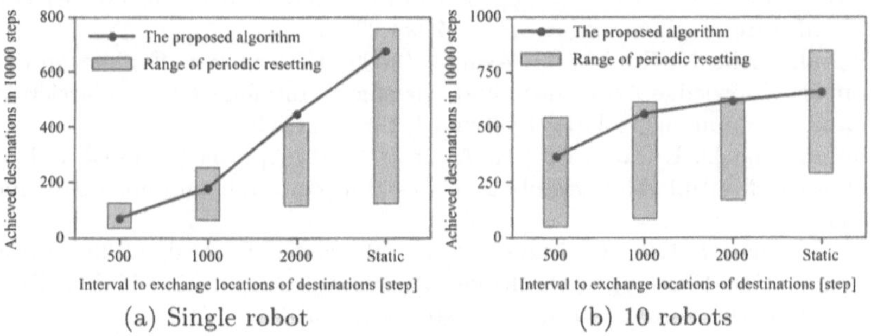

Fig. 7. Comparison of the number of achieved destinations.

5 Conclusion

In this paper, we proposed a device and an algorithm intended to enhance the efficiency of autonomous robots by means of autonomous knowledge sharing and acquisition. In conclusion, we accomplish the following:

We introduced distributed autonomous guidance system (DAGS) utilizing a device that enables local communication and information management, and we refer to this device as an "Intelligent Data Carrier (IDC)".

We investigated the behaviors of autonomous robots that utilize DAGS in dynamic environments, and proposed an algorithm to adapt the IDC system to a dynamic environment.

When robots break down, we need only to replace them with new ones. The new robots can perform as effectively as the original robots. This is because the knowledge is derived from the environment, and this knowledge can be easily shared among the robots. We conclude that the proposed system successfully realized a robust autonomous robotic system.

References

1. A. Drogoul, J. Feber. (1992) From Tom Thumb to the Dockers: Some Experiments with Foraging Robots, From Animals to Animats 2, 451-459.
2. T. Fujii, H. Asama, T. Numers, T. Fujita, H. Kaetsu, I. Endo. (1996) Co-evolution of a Multiple Autonomous Robot System and its Working Environment via Intelligent Local Information Storage, Robotics and Autonomous Systems, Elsevier, 19, 1-13.
3. D. Kurabayshi, H. Asama. (2000) Knowledge Sharing and Cooperation of Autonomous Robots by Intelligent Data Carrier System, Proceedings of IEEE Int. Conf. on Robotics and Automation, 464-469.
4. D. Kurabayashi, H. Asama, K. Noda, I. Endo, H. Hashimoto. (2001) Information Assistance in Rescue using Intelligent Data Carriers, Proceedings of IEEE/RSJ Int. Conf. Intel. Robots and Systems, 2294-2299.
5. D. Kurabayashi, K. Konishi, H. Asama. (2000) Performance Evaluation of Autonomous Knowledge Acquisition and Sharing by Intelligent Data Carriers, Distributed Autonomous Robotic Systems 4, Springer, 69-78.
6. I. Paromtchik, D. Kurabayashi, H. Asama. (2001) Approach Global and Local Guidance of Multiple Mobile Robots, Proceedings of Intel. Autonomous Vehicles, 350-355.
7. Y. Yoshimura, J. Ota, K. Inoue, D. Kurabayashi, T. Arai. (1996) Iterative Transportation Planning of Multiple Objects by Cooperative Mobile Robots, Distributed Autonomous Robotics Systems 2, Springer, 171-182.

Spreading Out: A Local Approach to Multi-robot Coverage

Maxim A. Batalin and Gaurav S. Sukhatme

Robotic Embedded Systems Laboratory, Robotics Research Laboratory
Department of Computer Science, University of Southern California,
Los Angeles CA, 90089-0781
{maxim, gaurav}@robotics.usc.edu

Abstract. The problem of coverage without *a priori* global information about the environment is a key element of the general exploration problem. Applications vary from exploration of the Mars surface to the urban search and rescue (USAR) domain, where neither a map, nor a Global Positioning System (GPS) are available. We propose two algorithms for solving the 2D coverage problem using multiple mobile robots. The basic premise of both algorithms is that *local dispersion* is a natural way to achieve global coverage. Thus, both algorithms are based on local, mutually dispersive interaction between robots when they are within sensing range of each other. Simulations show that the proposed algorithms solve the problem to within 5-7% of the (manually generated) optimal solutions. We show that the nature of the interaction needed between robots is very simple; indeed anonymous interaction slightly outperforms a more complicated local technique based on ephemeral identification.

Keywords: Coverage, distributed, mobile robots, sensor network

1 Introduction

We address the problem of deploying a mobile sensor network into an environment with the task of maximizing sensor coverage of the environment. We restrict ourselves to the case where every node in the network is a mobile robot. We describe two algorithms, which perform the coverage task successfully using only local sensing and local interaction between robots. The fundamental constraint that we impose on the problem is that the system does not have global information (either a map or access to global positioning information). Our algorithms also do not build maps or acquire global positioning information in the process of exploration. We are motivated by a number of applications ranging from Mars surface exploration to urban search and rescue (USAR) scenarios, both examples of situations where the environment is unknown *a priori* and global positioning is unavailable. For example, in a USAR application, we envisage a scenario where the team of robots would be "thrown" into the catastrophic site and activated. The system would automatically spatially distribute itself to maximize its sensor coverage. The resulting sensor network could be used by rescue workers to find humans, as a communications backbone etc. In this paper we are concerned with the problem of deploying a sensor network and its spatial self-organization, which results in a high degree of sensor coverage. Specifically, we address the planar coverage problem of a bounded area using robots equipped with laser range finders and cameras. Each robot is equipped with two 180° field of view planar laser range finders positioned back-to-back (equivalent to a 2D omnidirectional laser range finder), color camera and vision beacons. All robots have wireless communication.

Our premise is that in order to achieve good coverage as a team, robots must 'spread out' over the environment, i.e. if robots are too close to each other, their coverage areas overlap resulting in poor overall coverage. This premise is loosely inspired by the diffusive motion of fluid particles. In our system, robots not only perform obstacle avoidance, but are mutually repelled by each other within the range of their sensors. The first approach, which we call *Informative*, is based on the idea of assigning local identities to robots when they are within sensor range of each other. This approach relies on ephemeral identification where temporary local identities are assigned and mutual relative location information is exchanged between the interacting robots, allowing them to spread out in a coordinated manner. The second approach, called *Molecular*, is simpler than the first. Robots do not perform any directed communication, and no local identification is made. Instead each robot selects a direction 'away' from all its immediate sensed neighbors and moves in that direction without communicating with its neighbors. Both these approaches do depend on the ability of a robot to distinguish another robot from other objects in its environment. A third approach (termed *Basic*), in which there is no inter-robot interaction (other than obstacle avoidance) is also presented and compared with the two proposed techniques. In this approach robots make no distinction between robots and other objects in the environment. In all three approaches the motion of every robot is guided by its perceived coverage area. The major difference is that the *Informative* and *Molecular* techniques address interaction between the robots, whereas the *Basic* technique is based only on individual coverage maximization. Simulations show that the *Informative* and *Molecular* techniques solve the problem to within 5-7% of the (manually generated) optimal solutions and significantly outperform the *Basic* technique. We show that the nature of the interaction needed between robots is very simple; indeed anonymous interaction (*Molecular*) slightly outperforms ephemeral identification (*Informative*).

2 Related Work

Exploration and map building by a single robot in an unknown environment has been studied by several authors [6,7,8]. The frontier-based approach, described in detail in [6,7], concerns itself with incrementally constructing a global occupancy map of the environment. The map is analyzed to locate the 'frontiers' between the free and unknown space. Exploration proceeds in the direction of the closest frontier. The multi-robot version of the same problem was addressed in [3, 9, 10].

In [9] an incremental approach for deploying a mobile sensor network was introduced with the assumption that every robot in the network is equipped with an 'ideal' localization sensor. Even though there are inherent similarities between [6,7], and [9], the approaches differ fundamentally in that [9] uses live sensor data whereas [6,7] use stored data. [5] discusses the problem of deployment of distributed sensors (robots) in the wireless ad hoc network domain. In their setup, the communication ranges between the robots are assumed to be limited and the environment is assumed to be big enough so that the network connectivity cannot be maintained. A random-walk algorithm is used to disperse the robot network into the environment to support communication. The two techniques proposed in this paper differ from the above mentioned approaches in a number of ways. We use neither a map, nor localization in a shared frame of reference. The proposed techniques are adaptive (as opposed to [9]). Despite the similarity of the idea of dispersion, our techniques differ from [5], since every robot performs local visibility maximization rather than a random walk.

3 Architecture

Both techniques proposed in this paper are behavior-based [1] and have the same architecture. Laser, Vision and Position are the sensors being used. Position is a virtual sensor that includes odometry and compass. Arbitration is used for

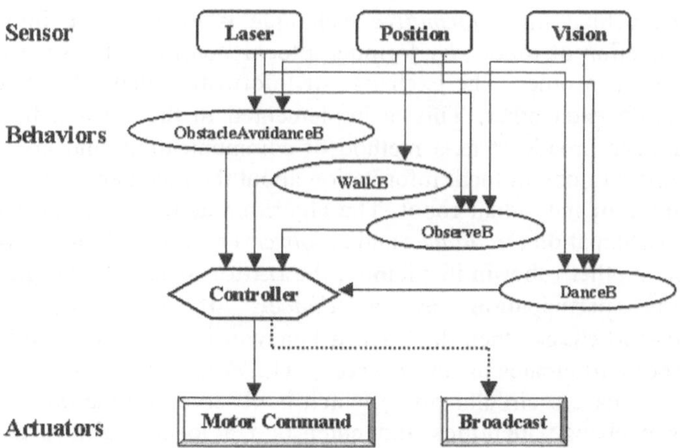

Fig. 1. System Architecture.

behavior coordination. Priorities are assigned to every behavior *a priori*. As shown in Figure 1, there are four behaviors in the system: `Obstacle Avoidance`, `Walk`, `Observe`, and `Dance`. In addition to priority, every behavior has an activation level, based on sensor information, which decides whether the behavior should be in an active or passive state (1 or 0). Each behavior computes the product of its activation level and corresponding priority and sends the result to the Controller, which picks the maximum value, and assigns the corresponding behavior to command the Motor Controller for the next cycle. Note that the only difference between the *Informative*, *Molecular* and *Basic* techniques is in the implementation of the `Dance` behavior (or the lack of it, in the *Basic* approach).

The `Observe` behavior chooses the most 'promising' direction for exploration. `Observe` is triggered if the visibility area has decreased compared to the visibility area of the previous cycle. `Observe` consists of two algorithms, which determine 'promising' direction for motion depending on a timer. If the timer is below a threshold, the direction that maximizes the frontal visibility of the robot is found. Otherwise `Observe` causes the robot to move in a circle, to explore for a better vantage point. `Observe` results in a suggested direction of motion, which locally increases the sensor coverage of the robot. The timeout mechanism is used to avoid local minima. `Obstacle Avoidance` causes robots to steer away from each other and other objects in the environment. `Walk` causes a robot to move forward in the direction it is currently facing. Thus the *Basic* approach is a greedy algorithm where each robot tries to find the best direction to move to improve its coverage while avoiding obstacles.

4 Informative Technique

The idea behind the *Informative* technique is to exchange information for better coordination of robots, by forming a local *'coalition'* between robots when they are near each other. The exchange of information depends on robots being able to identify each other. This is implemented in the `Dance` behavior. This behavior utilizes broadcast as a method of communication, and laser and vision sensors in order to obtain local information about the members of a coalition from the perspective of individual robot. The algorithm assumes two possible variants for robot's participation – *a dancer* and *an observer.* If the robot identifies a vision beacon atop another robot in its vicinity, the `Dance` behavior is triggered and the robot starts participation as a *dancer*. If the robot receives a '*DanceRequestMessage*', then the `Dance` behavior is triggered as well, but in this case the robot participates as an *observer*. The *dancer* robot performs a stylized motion (in our case a circular orbit) which is observed by the *observer* robot(s). The robots exchange identities and enter into a local coalition to decide the subsequent motion. Based on exchanging relative position and bearing, each robot computes the sensor coverage of the local coalition with the goal of selection a direction of motion such that the total coverage increases as a result of the interaction. The exact details of the local geometry of interaction and detailed algorithm are omitted here. The reader is referred to [11] for details. In this technique the vision system is used to detect robots, and the laser is used to compute local coverage.

5 Molecular Technique

The *Molecular* technique does not use direct communication for coordination and relies only on vision. Thus, the range of view is significantly less then that of the omnidirectional laser in the previous technique. As before, if the vision system detects a vision beacon atop another robot, the `Dance` behavior is triggered. The *dancer* in this implementation is only concerned with identifying a direction of motion for itself. No stylized motion is performed, nor is the *dancer* identified by other robots. The *dancer* simply selects a direction of motion for itself, which is diametrically opposite to the average angle subtended by all its neighbors in its visual field. It is thus 'repelled' away from its neighbors. Unlike the *Informative* technique no local coverage analysis is performed in this technique during the interaction.

6 Experiments and Results

We experimentally tested the three techniques in simulation using a planar environment. The general setup of the experiments is the same – three trials for each team size (3, 5, 7, and 9 robots). Figure 3 shows the simulation environment. The simulation engine used in our experiments is Player/Stage, developed at the USC Robotics Research Lab and described in detail in [2,4]. A trial terminates either when a pre-specified time threshold is exceeded or if the locations of the robots have not changed for a certain amount of time.

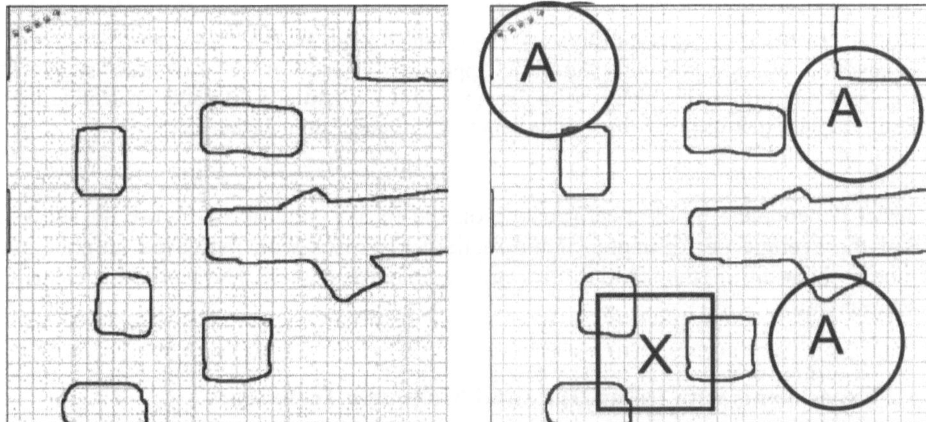

Fig. 3. (left) The simulation environment. **(right)** Areas marked with a circle and letter **'A'** represent good vantage points where the sensor coverage of a robot is high. The square, marked with letter **'X'**, shows a poor visibility area. In order to get to any of the high visibility areas, however, the robots need to move past the local minimum areas ('passages')

6.1 Experiments with the Basic Approach

The first sequence of experiments was to test the *Basic* approach. A series of 12 experiments with varying initial conditions were conducted. Figure 4b presents the results of the experiments. The first column represents the experiments with all robots starting from the top-left corner of the environment shown in Figure 3. The second column represents the experiments with robots initially spread out randomly throughout the environment. The third column represents the results of the experiments with robots initially positioned in the areas of maximum visibility (areas **'A'** in Figure 3 (right)). The results show a

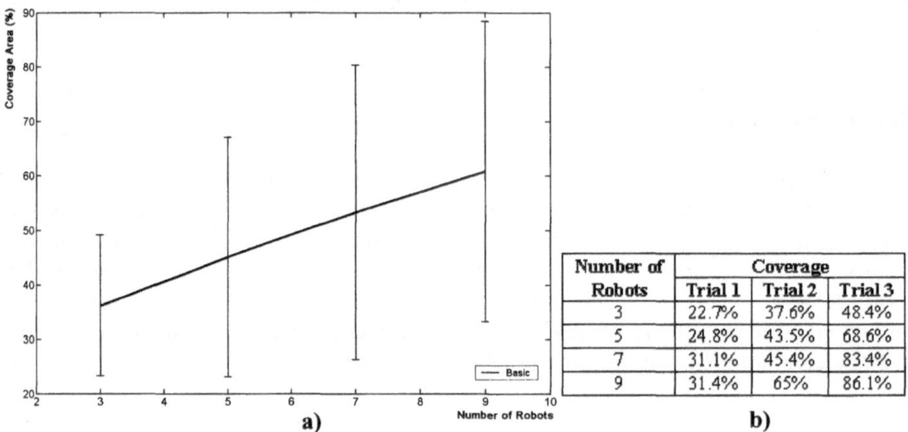

Number of Robots	Coverage		
	Trial 1	Trial 2	Trial 3
3	22.7%	37.6%	48.4%
5	24.8%	43.5%	68.6%
7	31.1%	45.4%	83.4%
9	31.4%	65%	86.1%

Fig. 4. The Basic Technique results. **a)** The resulted graph and the "error bars" show the standard deviation functions for the Basic Approach. The size of the "error bars" suggests the variability of possible results with changing initial conditions. **b)** Table presenting the results of the experiments in three trials using 3, 5, 7, and 9 robots team sizes.

direct dependence of the performance on the initial conditions. Figure 4a shows that the average performance of the method increases with increased number of robots. In addition, the standard deviations increase as well. The method is clearly impractical.

6.2 Experiments with Informative and Molecular Techniques

Trials were performed with the *Informative* and *Molecular* approaches with the same group sizes (3, 5, 7, and 9). The two techniques differ in the amount of information they use for the coverage task. The *Informative* approach utilizes the on-board sensors to the fullest, but pays the price of speed and convergence time (the whole system has to stop in order to scan for the *dancer*). At the same time, the *Molecular* approach uses only vision for the indirect communication with the robots, but is fast and adaptive to changes in robot's physical parameters (demonstrates the same performance with increased speed). In addition, while considering the results of the experiments, note that the optimal configuration (derived manually by the experimenter) required 9 robots for approximately 99% coverage. Figure 5a shows the respective graph of the two techniques and their corresponding standard deviations. The deviations are very small, which suggests that the results of the experiments are independent of the initial conditions. In order to check the validity of the distributions of the results and the confidence intervals for these distributions, a T-test was performed. The T-test computes the significance values and confidence intervals taking into account the data from *Informative* and *Molecular* techniques. Figure 5c presents the table of T-test results at the 95% significance level. The significance values and confidence

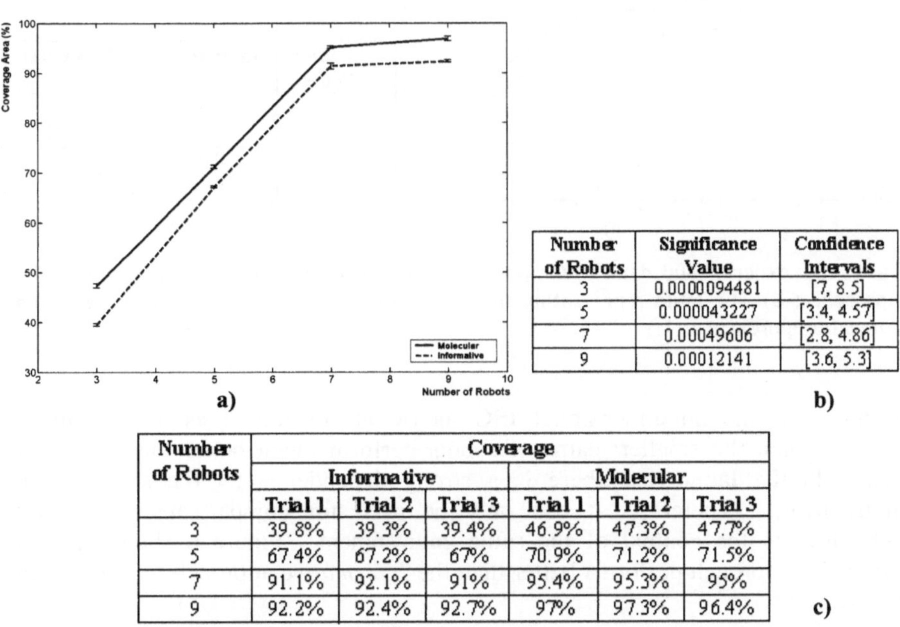

Number of Robots	Significance Value	Confidence Intervals
3	0.0000094481	[7, 8.5]
5	0.000043227	[3.4, 4.57]
7	0.00049606	[2.8, 4.86]
9	0.00012141	[3.6, 5.3]

Number of Robots	Coverage					
	Informative			Molecular		
	Trial 1	Trial 2	Trial 3	Trial 1	Trial 2	Trial 3
3	39.8%	39.3%	39.4%	46.9%	47.3%	47.7%
5	67.4%	67.2%	67%	70.9%	71.2%	71.5%
7	91.1%	92.1%	91%	95.4%	95.3%	95%
9	92.2%	92.4%	92.7%	97%	97.3%	96.4%

Fig. 5. Experimental data from experiments with *Informative* and *Molecular* approaches. **a)** The graph of the experimental results; **b)** The table summarizing the results of the T-test. **c)** The table of the experimental results conducted in three trials with teams of 3, 5, 7, and 9 robots;

intervals are small, suggesting that the *Molecular* technique performed statistically better than the *Informative* approach.

Instead of measuring coverage as a percentage of the total area, an equally valid metric is to measure the resulting coverage as a fraction of the best possible outcome. This can be done by nondimensionalizing the coverage by computing the ratio of the total coverage area to the sum of coverage areas in 'ideal' conditions (without obstacles and sensory reading overlapping). We term this the *Independent Sensor Characteristic* (ISC) metric. This metric is especially important in the context of the coverage problem, since the meaning of the sensor coverage may vary and the characteristics of different sensors may vary as well. Figure 6a presents the performance graph using this metric and corresponding values. Note that if we consider the metric of the performance to be the coverage area (Figure 5), then the two techniques improve with increased number of robots.

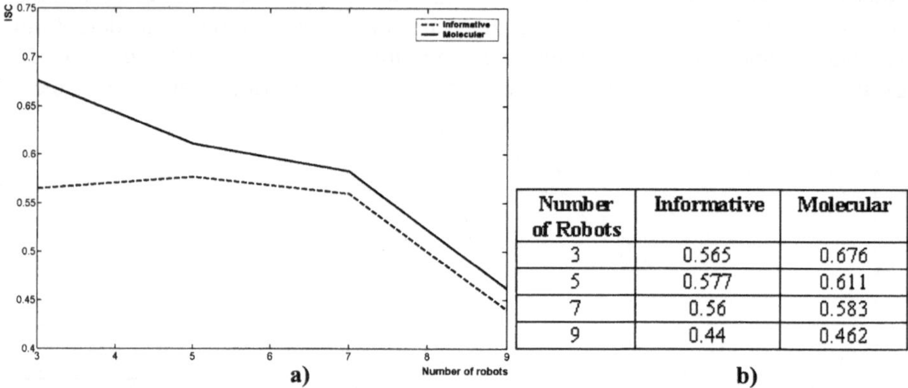

Number of Robots	Informative	Molecular
3	0.565	0.676
5	0.577	0.611
7	0.56	0.583
9	0.44	0.462

Fig. 6. ISC experimental data. **a)** Graph of the ISC metric for *Informative* and *Molecular* approaches; **b)** The table representing the values of the ISC metric for *Informative* and *Molecular* approaches.

If, however, the nondimensional ISC metric is considered as a performance measure, then the smaller teams of robots perform much better than the larger ones. In the latter case there is a strong dependence on the shape of the environment. Therefore, it can only be used for comparisons of the two techniques in the context of the same environment. Figure 6 shows that the *Molecular* approach performs better than the *Informative* approach in terms of the ISC as well.

7 Discussion

Despite the differences between the two techniques presented, their results are quantitatively similar. Both approaches outperform the *Basic* approach and both rapidly saturate to nearly complete coverage. Even though the *Molecular* approach is simpler, it slightly outperforms the *Informative* approach. We hypothesize that this is due to the additional overhead of pausing in the *Informative* approach for coalition formation. It is not clear that ephemeral identification actually helps in such cases though this warrants further investigation. What is clearly obvious is that the ability of robots to tell each other apart from obstacles is critical, both the Molecular and the Informative approaches use this and significantly outperform the Basic approach. The Molecular approach performed better in terms of the ISC metric as well, despite the fact that ISC decreases with increased number of robots.

One other question that we wanted to address is the question of the steady state. The question of a steady state in the problem of coverage and in the context of the two techniques presented in this paper arises naturally. How one would determine a steady state? Does steady state necessarily mean static state? Presently, the system 'times-out' for us to make an evaluation of its performance. On the other hand, a number of other applications would require the solution to have some kind of patrolling behavior, which, in turn would signify a patrolling steady state (a limit cycle rather than a limit point). The problem with defining a patrolling steady state, however, is that it is inherently difficult to compute when this state is achieved and provide guarantees that it will not diverge.

During the course of conducting the experiments in both the *Informative* and *Molecular* approaches, the patrolling behavior seemed to control the system. Imagine, for a moment, that we could view the environment in such a form where every point would be colored with respect to the visibility area possible to cover from it. The resulted picture may reveal the *tracked* nature of the environment. That is, the environment has high visibility tracks or ridges (like a Voronoi diagram). It would be interesting to reformulate the problem of coverage in terms of finding the *tracks* of maximum coverage. The problem of patrolling in a steady state could be answered in a formal way by identifying the limit cycles with the tracks. In our future work, we plan to approach the coverage problem from this point of view.

8 Future Work

An interesting metric that is omitted from the discussion in this paper is the time it takes the techniques to converge. In this paper we abstract the discussion from the metric of time, providing a number of other interesting metrics, like ISC, for example. In future work, however, we plan to extend our work with time-oriented metrics. The time metric is important, because real life applications often require fast response rather than optimality. We also plan to study the adaptability of the mobile sensor network in our future work with particular emphasis on the dynamic addition and removal of robots to/from the network. The results of the present experiments suggest that the *Informative* approach performed successfully but could not outperform the *Molecular* approach in spite of access to more information. In future work we plan to modify the 'dancing' stage so that it will not require stopping for *dancer* identification. On physical robots, this can be accomplished by attaching a bulb that would flash in case of the `Dance` behavior, for example, in order to attract the attention of *observers*. Thus, the performance may be improved significantly (especially in terms of the time metric). In addition, we plan to investigate coverage algorithms in which robots deploy a static sensor network into the environment to improve coverage. We also plan to address the general problem of designing algorithms for both coverage and exploration.

Acknowledgments

This work is supported in part by DARPA grant DABT63-99-1-0015 and NSF grants ANI-9979457 and ANI-0082498. We thank Andrew Howard for valuable suggestions.

References

[1] M. J. Mataric, Behavior-Based Control: Examples from Navigation, Learning, and Group Behavior, *Journal of Experimental and Theoretical Artificial Intelligence,* special issue on Software Architectures for Physical Agents, 9(2-3), H. Hexmoor, I. Horswill, and D. Kortenkamp, eds., 1997, 323-336

[2] B. P. Gerkey, R.T. Vaughan, K. Stoy, A. Howard, G.S. Sukhatme, and M.J. Mataric. Most valuable player: A robot device server for distributed control. *Proc. Of IEEE/RSJ Intl. Conf. On Intelligent Robots and Systems* (IROS), Wailea, Hawaii, Oct. 2001

[3] I. M. Rekleitis, G. Dudek, and E.E. Milios. Graph-based exploration using multiple robots. In L. E. Parker, G.W. Bekey, and J. Barhen, editors, *DARS*, vol. 4, pp. 241-250, Springer, 2000.

[4] R.T. Vaughan. Stage: a multiple robot simulator. Technical Report IRIS-00-393, Institute for Robotics and Intelligent Systems, University of Southern California, 2000

[5] A. F. Winfield. Distributed sensing and data collection via broken ad hoc wireless connected networks of mobile robots. In L.E. Parker, G.W. Bekey, and J. Barhen, editors, *DARS*, vol. 4, pp. 273-282. Springer, 2000.

[6] B. Yamauchi. Frontier-based approach for autonomous exploration. In *Proceedings of the IEEE International Symposium on Computational Intelligence, Robotics and Automation*, pages 146-151, 1997

[7] B. Yamauchi, A. Schultz, and W. Adams. Mobile robot exploration and map-building with continuous localization. In *Proceedings of the 1998 IEEE/RSJ International Conference on Robotics and Automation*, volume 4, pages 3175-3720, 1998

[8] A. Zelinsky. A mobile robot exploration algorithm. *IEEE Transactions on Robotics and Automation*, 8(2): 707-717, 1992

[9] A. Howard, M. J. Mataric and G. S. Sukhatme, "An Incremental Self-Deployment Algorithm for Mobile Sensor Networks", *Submitted to Autonomous Robots Special Issue on Intelligent Embedded Systems*, 2001

[10] W. Burgard, D. Fox, M. Moors, R. Simmons, and S. Thrun (2000) Collaborative Multirobot Exploration, Proc IEEE ICRA 2000

[11] M. Batalin and G. S. Sukhatme. Computing Dispersion via Local Interaction Geometry in a Robot Group. Institute for Robotics and Intelligent Systems Technical Report IRIS-01-408, 2001

Real-Time Cooperative Exploration by Reaction-Diffusion Equation on a Graph

Chomchana Trevai[1], Keisuke Ichikawa[1], Yusuke Fukazawa[1], Hideo Yuasa[1], Jun Ota[1], Tamio Arai[1], and Hajime Asama[2]

[1] University of Tokyo, 7-3-1 Hongo, Bunkyo-ku, Tokyo, 113-8656, JAPAN.
[2] RIKEN:The Institute of Physical and Chemical Research, 2-1 Hirosawa, Wako-shi, Saitama, 351-0198, JAPAN.

Abstract. In this paper we present a path planning method for the environment exploration using the reaction-diffusion equation on a graph. We autonomously arrange sub-goals for robots to go along and explore the environment. By using reaction-diffusion on a graph for sub-goals arrangement, we can deal with the emergence of unknown objects or obstacles in the dynamic environment. This method also can explore the environment in real-time and make sure that robots completely explore all the area in the environment. The results given in this paper demonstrate that our method successfully solves exploration problems both in simulations and in real robot experiments.

1 Introduction

The problem of exploring the environment belongs to the fundamental problems in mobile robotics. Nowadays the real-time environment exploration by robots is in high demand to be applied in places such as construction sites and office environments. As the periodic maintenance patrol or security patrol have to be done, many approaches have been done for the environment exploration problem. There is the technique for exploration in unknown structure environment by moving each robot to the closest frontier [1] which is the closest unknown area around the robot.And there is the method to coordinate robots while they explore the environment by probabilistic approach [2,3]. In this paper we consider the problem of effective exploration of known structure environment by the multiple mobile robots. The definition of known structure in this paper is the environment that only its shape information can be obtained.However there are the unknown obstacles and objects inside the environment. The key problem for the environment exploration is to explore the environment completely and effectively so that the robots can get all the information of the environment in real-time. This problem is already NP-hard in the graph-like environments. Especially in dynamic environment with unknown obstacles, the area covering problem can be described as one of the hard computational geometry problems [7,8].With the mobile robots, the complete recognition of the dynamic environment is the real challenging goal. There are also the method using grid for as a spatial representation, however the accuracy of the exploration is depending on the detail of the grids.In our method we used sub-goals and the reaction-diffusion equation on a graph [6] to deal with the dynamic environment exploration. By using reaction-diffusion equation on a graph to coordinate the sub-goals for the robots to move along, our method has a tolerance for the emergence of unknown obstacles. For an exploration problem in the changing dynamic environment, the arrangement of sub-goals can flexibly adapt to the environment. Then our method does not have to re-compute all the sub-goals arrangement from the first state, therefore the robots can re-plan the path effectively.

2 Problem Settlement

In this research, we target the problem as the exploration of the known structure office environment. The environment composed of the unknown obstacles(Fig.1a). The robots at first can get only the shape information of the environment. But there are markers to be attached on the objects to give the robots the information about each object. The robot has a camera to read the marker when the marker is in the visible area(*Sensing Area*)(Fig.1b).

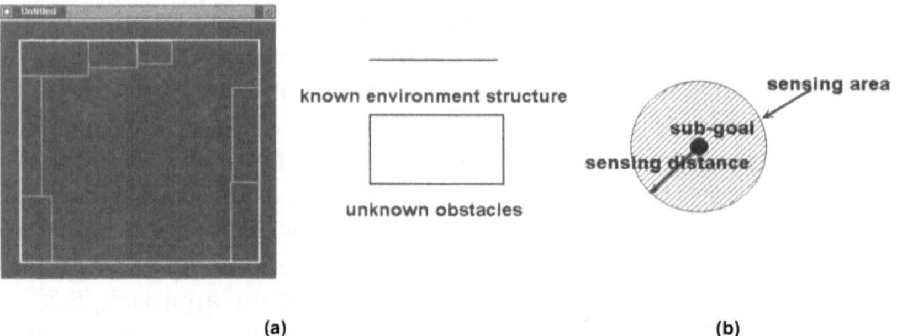

Fig. 1. (**a**)Environment with unknown obstacle(**b**)Structure of sub-goal and sensing area

In other exploration problems when the environment's structure is unknown, the first step for the robots to move in the environment is limited to random-wandering or wall-following. Then it is hard to plan the effective exploration path for the robots. But in the real-world, as the environment is the man-made, the structure information of an environment is accessible. Then in our research, we give the robots only the knowledge about the structure of the environment at first.

3 Approach to Multiple Mobile Robots Cooperative Exploration

In our method, we define a sensing area around a sub-goal as a circle with the radius of the sensing distance(Fig.1b). The environment exploring problem can be defined as coordinate sub-goals to cover the environment by sensing area of each sub-goal. However to cover the area by the circles is known as very hard mathematical problem, even in the static area [7,8]. In dealing with the dynamic environment as in our research, the covering problem can hardly be solved.In our approach, we use reaction-diffusion equation on a graph to arrange sub-goals and coordinate them into the same distance. Then we confirm the covering of the environment by detecting the overlapping of the sensing area. By using of the reaction-diffusion equation on a graph, we can arrange sub-goals with the status information of a few neighboring sub-goals. However the changing of each sub-goal's status can be transmitted to all the environment.

4 Sub-Goal positioning

In this section we will show how to use the reaction-diffusion equation on a graph to reduce the computing time to coordinate the sub-goals in the environment and to deal with the dynamic environment and the emergence of unknown obstacles. The number of the sub-goals is in proportion to the size of the environment.

4.1 Potential Designing for Reaction-Diffusion Equation on a Graph

To coordinate sub-goals by reaction-diffusion equation on a graph, at first we have to give a scalar function $f(u)$ to define the status of each sub-goal. In this method we need to arrange sub-goals into the same distance(r_i). Then we design the scalar function in proportion to the sensing distance (r_{sen}) of the robot.

$$f(u) = \sum_{e=1}^{6} |r_e - r_i| \tag{1}$$

While $|r_e - r_i|$ is the distance from the neighboring sub-goal e, The neighboring sub-goals are defined as six of the nearest sub-goals. To arrange sub-goals into the target status, the reaction-diffusion equation on a graph uniformly minimize the scalar function $f(u)$ of all of each sub-goal. The minimization of size and the unification of all sub-goals $f(u)$ is described by the reactive term W_o and the diffusive term W_1, where V is the set of sub-goals.

$$W_o(f) = \sum_{u \in V} f^2 \tag{2}$$

$$W_1(df) = \sum_{u \in V} \|\boldsymbol{df(u)}\|^2 \tag{3}$$

The potential state of the target environment $W(f)$ can be defined as the sum of 2 and 3.

$$W(f) = W_o(f) + W_1(df) \tag{4}$$

Where $\boldsymbol{df(u)}$ is the derivative vector of $f(u)$ starting from the sub-goal u to its neighboring sub-goals.

4.2 Calculate Moving Direction of Sub-Goal

The scalar function of the sub-goal to move $\boldsymbol{x}$ in distance can be defined as the scalar function $\boldsymbol{g(x)}$.

$$\boldsymbol{g(x)} = \sum_{e=1}^{6} |\boldsymbol{r_e} - \boldsymbol{x} - r_i \frac{\boldsymbol{r_e} - \boldsymbol{x}}{\|\boldsymbol{r_e} - \boldsymbol{x}\|}| \tag{5}$$

when $\boldsymbol{g(x)}$ is satisfied $\boldsymbol{f(u)} = \boldsymbol{g(0)}$. Then if we partial differentiate between $\boldsymbol{g(x)}$ and $\boldsymbol{x}$, we can obtain the vector to minimize $f(u)$. This vector is the moving direction of the sub-goal ($\boldsymbol{e}_{move}$).

$$\nabla \boldsymbol{g(0)} = -\sum_{e} (1 - \frac{\|\boldsymbol{r_e}\|}{r_i}) \frac{\boldsymbol{r_e}}{\|\boldsymbol{r_e}\|} \tag{6}$$

$$\boldsymbol{e}_{move} = \frac{\nabla \boldsymbol{g(0)}}{\|\nabla \boldsymbol{g(0)}\|} \tag{7}$$

4.3 Calculation of Moving Distance for Sub-Goal

After the moving distance of a sub-goal was obtained, we have to calculate the moving distance to move the sub-goal. The moving distance of the sub-goal can be calculated as the differential of its scalar function.

$$\frac{\partial f}{\partial t} = -\frac{\delta W(f)}{\delta f} = -2f(u) - 4\sum_{e=1}^{6}(f(u) - f(e)) \tag{8}$$

Therefore the moving distance of each sub-goal minimizes the potential of the environment, then the moving distance D is described by $\boldsymbol{\nabla g(0)}$.

$$D = \frac{\partial f}{\partial t} / \|\boldsymbol{\nabla g(0)}\| \tag{9}$$

After all we can describe the moving vector $\boldsymbol{v}$ for each sub-goal as

$$\boldsymbol{v} = D \cdot \boldsymbol{e}_{move} \tag{10}$$

As mentioned above, the behavior of sub-goals at last will uniformly minimize the potential of the environment. The behavior of sub-goals relate to only the status of the neighboring sub-goals $f(u)$, $r(e)$ and $f(e)$. Then even the unknown obstacles are detected the computing time of sub-goals behavior does not change.However the reaction to the detection of unknown obstacles can be transmitted to all the environment. The sub-goal will move and optimize its coordinate to the environment.

4.4 Border Line for Environment and Object

In this section we will describe how we handle the status of a sub-goal when it moves near the border of the environment or obstacle. If we positioned a sub-goal at distance r_i from the border line as in Fig.2a, we can not cover all the environment with the sensing area.Then we have to give the weight k for the sub-goal to imitate that there is another sub-goal at the distance $k \cdot r_{wall}$ inside the border line. Thereupon the border line position of the obstacles or the environment the sub-goals can be positioned into the optimized distance r_{wall}.

$$r_{wall} = \sqrt{r_{sen}^2 - (\frac{r_i}{2})^2} \tag{11}$$

$$k = \frac{r_i}{r_{wall}} \tag{12}$$

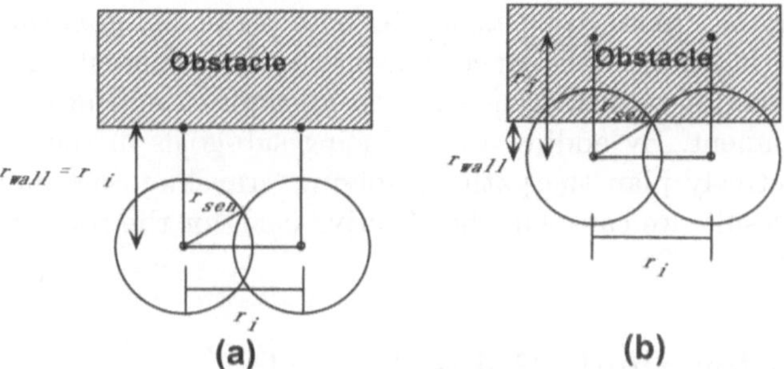

Fig. 2. Sub-goals at the border line:(**a**) $k = 1$ (**b**) Optimized r_{wall}

4.5 Covering Detection

To make sure after the path was planned for the robots to move along the sub-goals that their sensing area completely cover all the environment. We have to confirm the covering of the environment by the sensing area. The environment covered by the sensing area can be detected by checking that the circumference of the sensing area around each sub-goal was covered by its neighboring sub-goals. The covering will be detected after the changing of potential function $f(u)$ of each sub-goal reach the setting range.

4.6 Adding and Reducing Sub-Goal

Even all the obstacles in the environment have been discovered, the number of sub-goals to cover all the environment cannot be determined. Therefore we have to add or reduce the number of sub-goals to optimize the path for the robots. Regarding adding sub-goals, if there is the sub-goal that its circumference is not covered by its neighboring sub-goals. We put the new sub-goal to its circumference and minimize all the scalar function of sub-goals again.

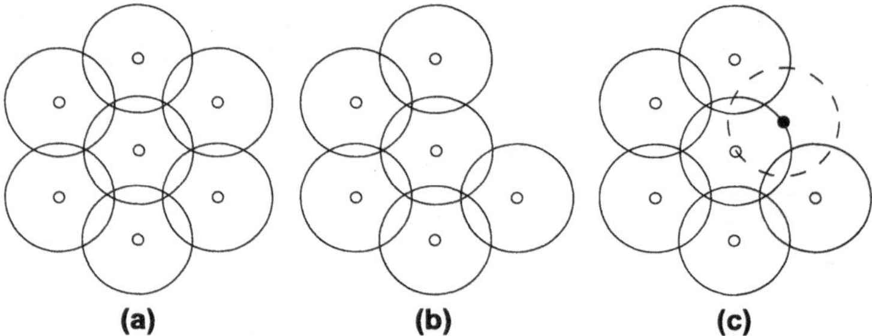

Fig. 3. Adding new sub-goal:(**a**) Covered (**b**) Not covered (**c**) Added

By the detection for the unknown obstacles, if there is the sub-goal that can not be reached by the robot, we reduce it from the environment. And again we balance the reaction-diffusion equation on a graph to optimize all sub-goals in the environment. By adding and reducing sub-goals in the environment, we can effectively plan the path by solving Traveling Salesman Problem (TSP) . As a result, we can plan the effective path for the robot to explore the environment.

5 Path Planning for Single Robot by TSP

Traveling Salesman Problem (TSP) is a deceptively simple combinatorial problem. There is a lot of algorithm to solve the problem. In this paper even for single robot to explore the environment the number of sub-goals in the problem is about ten to few hundreds. Then we choose the famous and easy to implement, Simulated Annealing Algorithm (SA Algorithm), to solve the TSP in our problem.To find the shortest path for N sub-goals, we define the distance between the sub-goals as 13 for the criterion function of the SA Algorithm. The expectation (Boltzmann's factor) of the Algorithm was described as p.

$$E = \sum_{i=1}^{N} \sqrt{(x_i - x_{i+1})^2 + (y_i - y_{i+1})^2} \tag{13}$$

$$p = exp[-(E_2 - E_1)/kT] \tag{14}$$

6 Path Planning for Multiple Mobile Robots by TSP and Sweep Algorithm

In multiple mobile robots Exploration, we used Sweep Algorithm [10] to part the sub-goals into groups for each robot to move along. We computed each sub-goal polar coordinate and split up the sub-goals on a per capita basis for each robot. The sub-goals are then sorted by increasing polar angle with respect to the depot point. The depot point in this simulation was given as the center of the environment. As the sub-goals was coordinated into the same distance, only the simple partition was needed to part the exploration path for each robot. Then we apply the SA Algorithm to each group of the sub-goals. The planned path will be presented as the experimental result in the next section.

7 Experimental Result

The described approach has been implemented and tested on both simulation experiments and the real robots in the real environments. We performed a series of simulation experiments and real robot experiments to get a quantitative assessment of our approach.

7.1 Simulation Experiment for Single Robot Exploration

The first simulation experiment in this section is designed to demonstrate the advantage of our method to coordinate sub-goals by using reaction-diffusion equation on a graph. We use the 6x5.8m^2 environment depicted in Fig.4, which is an outline of our experimental space. Fig.4a shows how the environment was covered by the sensing area.Therefore the path was planned as in Fig.4b. In Fig.5, the obstacles was detected after that the sub-goals inside the unknown obstacle was reduced. Then the path was re-planned.

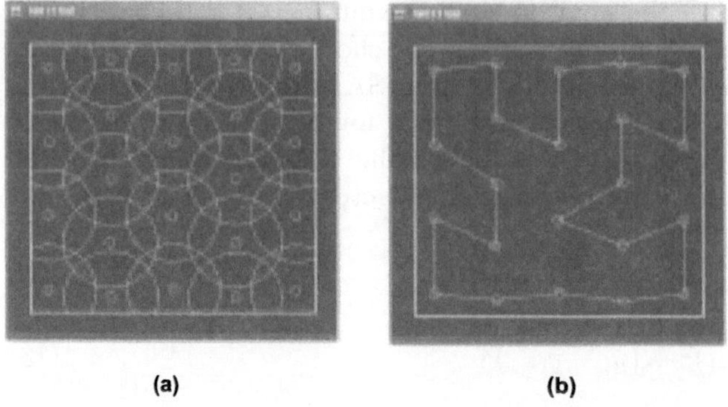

Fig. 4. (**a**)The covered environment by sensing area (**b**)The planned path

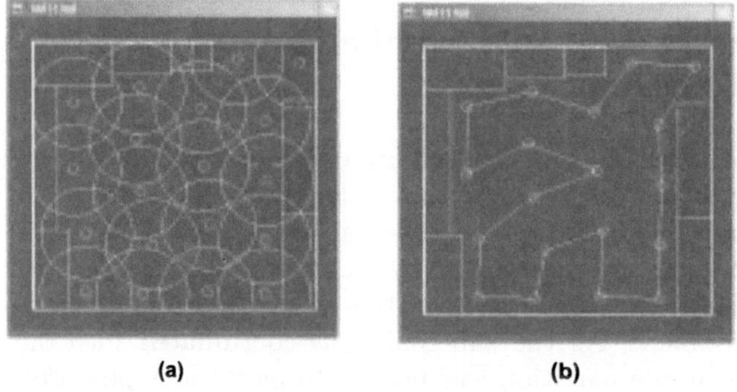

Fig. 5. (**a**)The covered environment after the obstacles was detected (**a**)The re-planned path

7.2 Single Robot in Real Robot Experiment

In the real robot experiments, we used the RIKEN's ZEN robot equipped with the camera. The vision range for the camera , due to the detection of a marker, was limited to 1m. Fig.6a shows the robot moves along and explores the environment by the coordinated path. The environment in this experiment was described referring to our experimental space. Fig.6b shows our 4x4m^2 real robot experimental space.

7.3 Simulation Experiment for Multiple Mobile Robots

The simulation experiments for the multiple mobile robots have been working on. For now we can plan the paths for the multiple mobile robots. Fig. 7 shows the paths for a robot to explore the same environment as in the single robot simulation experiment. Since we part the sub-goals in the environment before applying the SA for TSP, the time to plan the paths for the multiple mobile robots is shorter than that for the single robot. This result will prove that

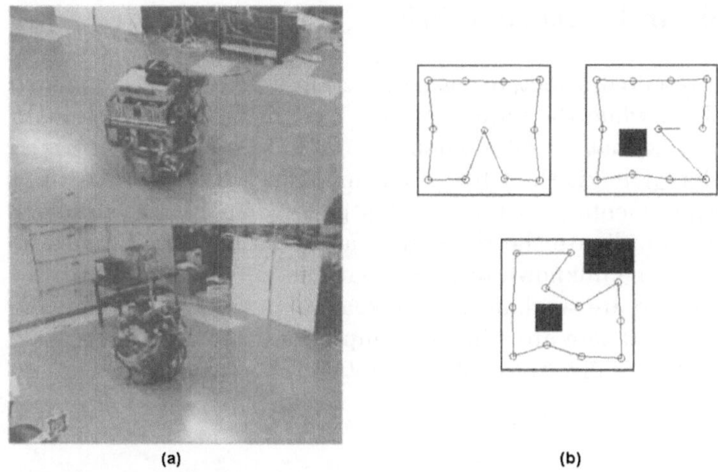

Fig. 6. (**a**)The real robot experiment (**b**)The planned path

the multiple mobile robots can explore the environment more effectively than single robot.

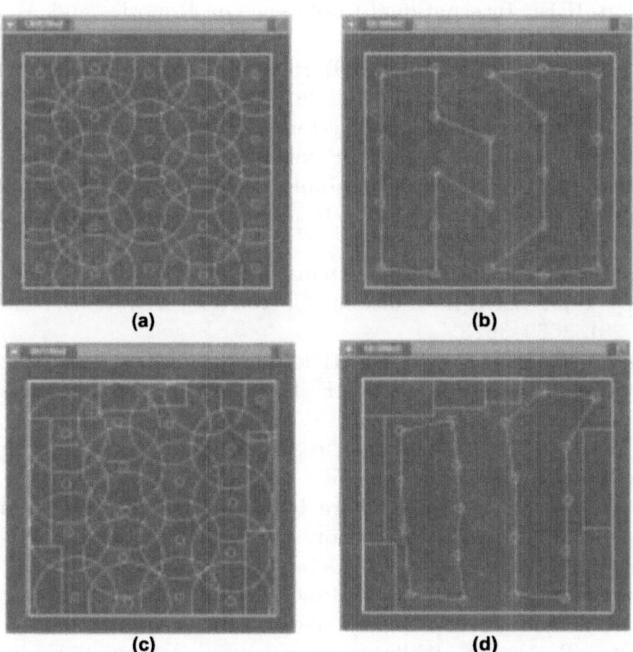

Fig. 7. The multiple mobile robot simulation:(**a**)The covered environment (**b**)The planned path (**c**)The covered environment after the obstacles was detected (**d**)The re-planned path

8 Conclusion and Future Work

In this paper we presented a method using reaction-diffusion on a graph to coordinate the robots while they are exploring the environment. The main idea of this method is that it autonomously coordinates sub-goals for the robots to move along.Our method has been implemented and tested on a real robot. The experiments presented in this paper demonstrate that our approach is able to coordinate the robot so that it can flexibly explore the environment that has the unknown obstacles within.
As the future work, we are working on the real robot experiments by the multiple mobile robots. There are still some improvements which have to be done for our method to realize the multiple mobile robots experiments.

References

1. B. Yamauchi. (1999) Decentralized coordination for multirobot exploration. Robotics and Autonomous Systems, **29**, 111–118.
2. S. Thrun, W. Burgard, D. Fox. (2000) A Real-Time Algorithm for Mobile Robot Mapping With Applications to Multi-Robot and 3D Mapping. IEEE International Conference on Robotics and Automation, 321–328.
3. W. Burgard, D. Fox, M. Moors, R. Simmons, S. Thrun. (2000) Collaborative Multi-Robot Exploration. IEEE International Conference on Robotics and Automation, 476–481.
4. N. Miyata, J. Ota, Y. Aiyama, T. Arai. (1999) Real-time Task Assignment for Cooperative Transportation by Multiple Mobile Robots. IEEE/RSJ International Conference on Intelligent Robots and Systems, 1167–1174.
5. A. Yamashita, K. Kawano, J. Ota, T. Arai, M. Fukuchi, J. Sasaki, Y. Aiyama. (1999) Planning Method for Cooperative Manipulation by Multiple Mobile Robots using Tools with Motion Errors. IEEE/RSJ International Conference on Intelligent Robots and Systems, 978–983.
6. H. Yuasa, M. Ito. (1998) Internal Observation Systems and a Theory of Reaction-Diffusion Equation on a Graph. IEEE International Conference on Systems, Man and Cybernetics(SMC'98), 3669–3673
7. E. Gonzalez, A. Suarez, C. Moreno, F. Artigue. (1996) Complementary Regions: a Surface Filling Algorithm, Proceeding of the IEEE International Conference on Robotics and Automation, 909–914.
8. J.B.M Melissen, P.C. Shuur. (2000) Covering a rectangle with six and seven circles, Discrete Applied Mathematics 99, 149–156.
9. S. Ichikawa, F. Hara. (1996) Experimental Characteristics of Multiple-Robots Behaviors in Communication Network Expansion and Object-Fetching, Distributed Autonomous Robotics Systems,**2**, 183–194.
10. The Traveling Salesman Problem. (1985) Edited by E.L. Lawler, J.K. Lenstra, A.H.G. Rinooy Kan, D.B. Shumoys, A Wiley-Interscience Publication.
11. D. Kurabayashi, J. Ota, T. Arai, S. Ichikawa, S. Koga, H. Asama, I. Endo. (1996) Cooperative Sweeping in Environments with Movable Obstacles, Distributed Autonomous Robotics System, **2**, 257–267.

Research on Cooperative Capture by Multiple Mobile Robots – A Proposition of Cooperative Capture Strategies in the Pursuit Problem –

Kazunari TAKAHASHI, Masayoshi KAKIKURA

Graduate School of Engineering, Tokyo Denki University
2-2, Nishiki-cho, Kanda, Chiyoda-ku , Tokyo 101-8457 Japan
kazu@isl.d.dendai.ac.jp, kakikura@isl.d.dendai.ac.jp

Abstract. The aim of this research is to propose behaviour planning and cooperative capture strategies for distributed autonomous robots systems. These propositions are accomplished through the examination of preceding studies on the "Pursuit Problem" taken up as an object task of multi-agent systems. In this paper, we demonstrate a deadlock evasion by making the mobile robot repeat simple behaviors. We have made some experiments on the simulation system we have built. The results of cooperative capture are reported.

Keywords. Pursuit Problems, Deadlock-free strategy, emergence

1. Introduction

The recent evolution of robots is amazing. However, we must make the intelligence and function of robots higher enough as the work range of robots increases and the contents of work become complicated. No matter how we make the intelligence and function of a single robot higher, it is clear that there is a limit to its capabilities.

A system with two or more robots is one solution for this problem [1]. Moreover, many researchers are studying the distributed autonomous robots systems in order to control two or more of these systems. The key to these systems is "cooperation".

On the other hand, the problem of cooperation by two or more systems is also a subject of research under Distributed Artificial Intelligence (DAI). Multi-agent systems, as they are called, are a method of problem solving by knowledge sharing and/or adjustment behaviors. The "Pursuit Problem" is one many researchers [2] [3] [5] have taken up as an object task of such multi-agent systems.

We are engaged in research into a solution to a cooperation problem, using multiple mobile robots are given restricted capability and only partial information about the world. The aim of this research is to propose behavior planning and cooperative capture strategies for distributed autonomous robots systems. These propositions are accomplished through the examination of preceding studies on the "Pursuit Problem". The strategies are based only on simple intelligence and function.

In this paper the Pursuit Problem dealt as DAI problems are discussed in Chapter 2. Moreover, we re-set this problem as the problem on a multiple mobile robots system. We propose behavior planning and cooperative capture strategies of this robot system in Chapter 3. The results in simulation experiment are shown in Chapter 4.

2. Pursuit Problem

2.1 Problem definition

The Pursuit Problem is that four blue agents (Predator) aim to enclose one red agent (Prey). Each agent moves vertically or horizontally on a 2-dimensional grid (Figure. 1 (a)). The purpose terminates when the red agent has been surrounded by the blue agent (Figure. 1 (b)).

Because this Pursuit Problem has many parameters (e.g., the definition of capture, the agent's action rules, the range of which predator recognition, whether or not there is communication between predators, etc.), there could be many ways to approach a solution by changing these parameters. Many researchers are evaluating their own studies by adjusting the parameters to the suitable forms.

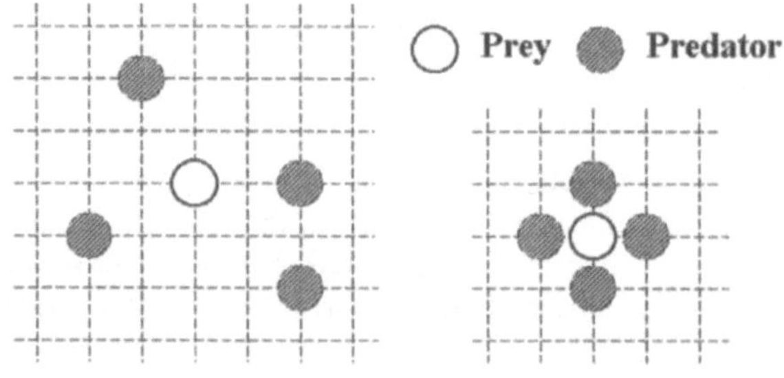

Fig. 1. (a) Pursuit Problem (b) Success State

2.2 Research example

Some heuristics on the Pursuit Problem are proposed to cope with this problem. Gasser et al. [2] studied the Pursuit Problem in order to investigate a distributed cooperation mechanism. They proposed a heuristics where a blue agent (Predator) shortens the distance from a red agent (Prey) under the state shown Figure. 2. By using this, the capture is successful in the end of the process. This state is called the "Lieb Configuration."

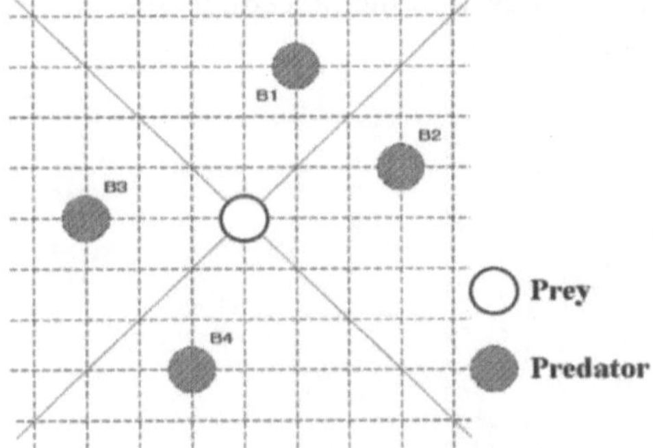

Fig. 2. Lieb Configuration

On the other hand, there is a proposal of capture by the agent based on experience, such as the learning. Haynes et al. [3] are engaged in research on multi-agent interaction, in particular, conflict resolution between agents for restricted resources. They used the Pursuit Problem to propose one solution for this problem. As the solution of this problem, the agent was made to learn the state called Case Window (Figure. 3), when an agent runs into a deadlock state. They have proposed to modify the agent behavior so that this learned state might not happen.

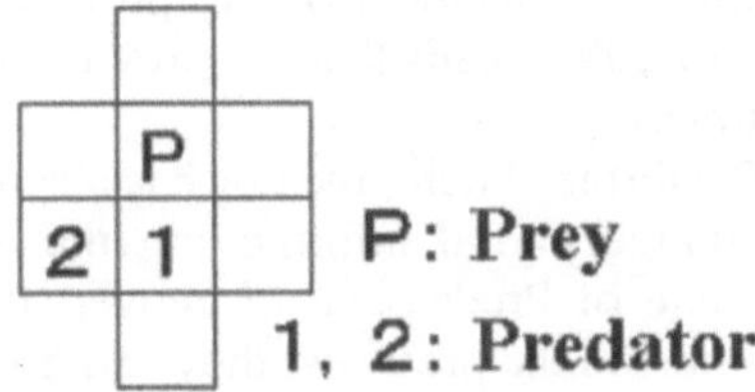

Fig. 3. Case Window for Predator1

However, these strategies have guaranteed capture only under limited conditions because they tried to realize capture by avoiding running into a deadlock state in the last stages of this problem.

2.3 Problem set up

We set up the problem dealt with this research as follows, and built the simulation system [4]. In this paper, we suppose that the mobile robots (Prey and Predator) expressed by this simulation system are called the "agents", and this problem is called the "game".

(1) Experiment environment: An agent moves within the framework of a 2-dimensional lattice. (Henceforth, we suppose that this framework is called the "cell".) This game is performed in either the infinite plane called Toroidal World

(Figure. 4), which has been used in the preceding studies, or the plane limited by the boundary of Toroidal World.

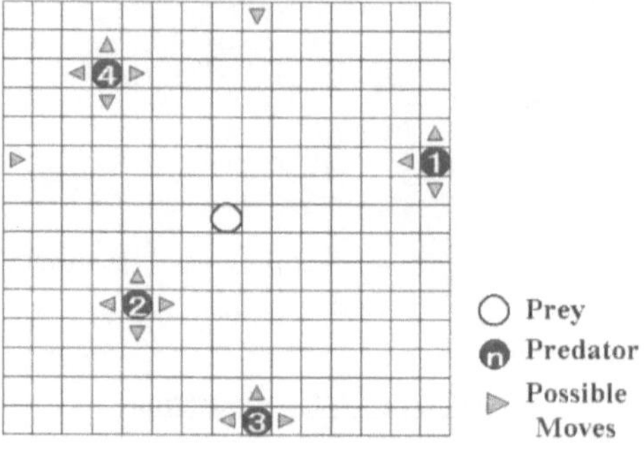

Fig. 4. 15x15 Toroidal World

(2) The rule common to each agent: One unit time is one cycle. A prey and predators choose actions by turns. Each agent moves vertically or horizontally, if not, the agent stops moving at the same place. No agents can occupy the same cell another agent exists.
(3) The action rule of Prey: A prey chooses a direction at random or chooses not to move. We can restrain the movement of the prey with arbitrary probability. (e.g., mobility restriction of 20% means that the prey is caused to stay in the same position for 20% of the time.)
(4) The action rule of Predator: Each predator has a sensor on its circumference and can recognize the limited partial information (the hatched part of Figure.5 shows the recognition range of Predator1). Moreover, they each have a camera which can recognize and pursue the prey, and they can use it for the judging a plan of operation with the partial information they have. It is necessary to negotiate (e.g., communication) between agents in order to realize the agents' cooperative action. So, we made the agents have a partial communication function.

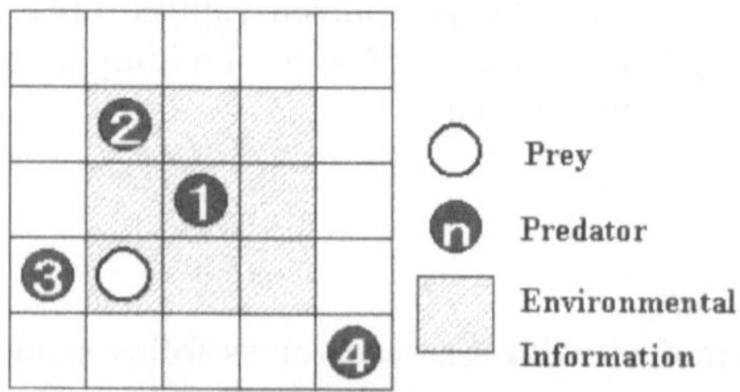

Fig. 5. Experiment Environmental Model

(5) The end of a game: A game ends when capture succeeds or when each agent's positions did not change continuously twice in a row. We take this state as failure of capture.

3. Capture Strategies

3.1 An agent's plan of operation

Generally, in object oriented programming, mobile robots (agents) can be supposed as one object. Each object can be considered as a thread, because they move independently. Therefore, we used the idea of multithread programming for cooperation of the agents in this Pursuit Problem.

Since two or more agents have the same purpose, an agent may move to the same position and "Collision" (Figure. 6 (a)) may be caused. Or, each agent may hold back and "Compromise" (Figure. 6 (b)) may be caused.

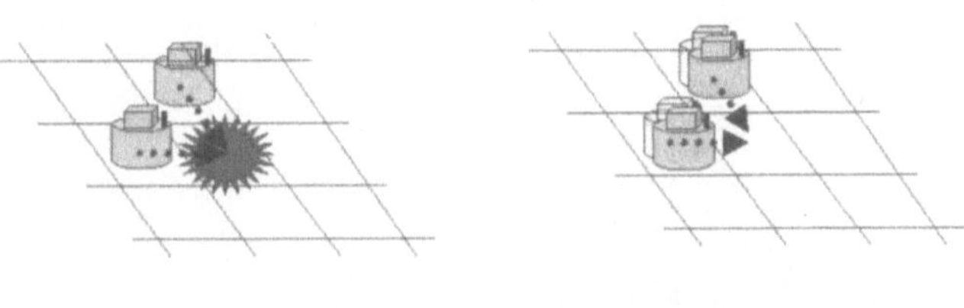

Fig. 6. (a)Collision (b)Compromise

In this research we gave the agents a priority to solve this problem. The agents which exist in the range of partial communication publicize each priority and determine their order. When the agents have the same priority, they determine in order of communicative formation. The agents set each plan of operation in order of their priority and publicize the plan of operation to other agents. The agents have solved this "Collision" and "Compromise" (Figure. 7) by acting in ways that will not interfere with another agent's plan of operation publicized in advance.

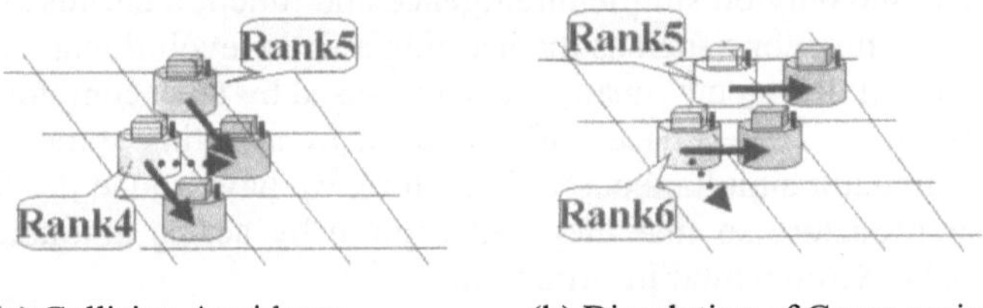

(a) Collision Avoidance (b) Dissolution of Compromise

Fig. 7. Solution upon Priority

3.2 The deadlock problem in the last game

A strategy where the predator shortens the distance from a prey is needed. We have adopted the basic strategy of greedy heuristics called the Manhattan distance (MD) algorithm [5] (A predator moves to the direction lessen the Manhattan distance from a prey).

As a result of capture experiments (two kinds of the cases are prepared; that is, one is that make action of a prey choose vertically or horizontally at random and the other is that make action of a prey stay) using only this greedy strategy (MD), we recognized nine organization structures where the agent failed in capture. Those nine kinds of patterns are shown in Figure. 8. These situations happened without regard for the behavior of the prey.

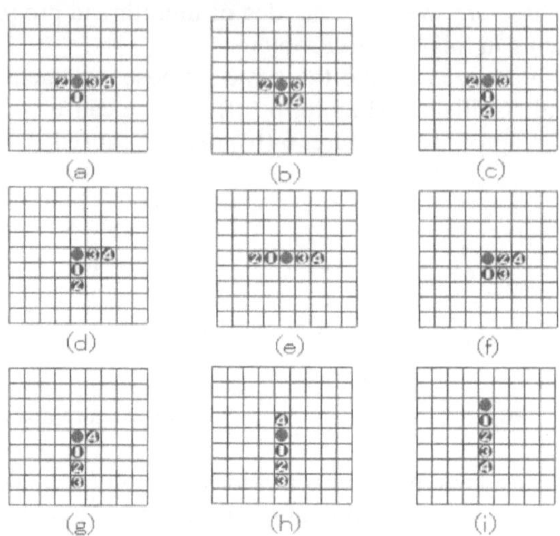

Fig. 8. Deadlock Pattern

We tried to create robust and flexible robots systems to solve such a deadlock state by using the partial environmental information the agent has. The agents must behave based only on simple intelligence and function because the agents are not given higher functions such as the learning and the evolutional technique.

As a solution of the agents' deadlock state caused by their common purpose, we have established the "emergence" [6] produced by adopting some simple behaviors for each predator simultaneously. Therefore, we have added the following two functions for predators so that they can capture by acting autonomously based only on partial environmental information.

(1) Deadlock detection: We newly introduced the deadlock flag for predators. When all predators finished moving, they simulate their next movement based on the partial environmental information they have. When there is another agent on the cell the predator is supposed to select, the predator sets up the deadlock flag and broadcasts the situation with its priority.

(2) Evasion: When all predators publicized deadlock flag, Some predators run an evasion algorithm to temporarily nullify the deadlock state by moving to some cells they can move. We have defined that the predator runs evasion under a condition where there is at least one predator and one prey in its adjoining cells.

3.3 Improvement of algorithm

Haynes et al. [3] performed an analysis and evaluation of preceding typical capture strategy algorithm, and reported the effectiveness of their improved algorithm. We also improved the basic strategy (MD). It is not as complicated as Haynes's algorithm. When there are two or more cells where the predator can move, we make a rule that it moves in the direction where there are fewer agents in the cell which they can move and the next-door cells on both sides (i.e., enclosed in dotted line in the state of Figure. 9 (a)). In the state of Figure. 9 (a), there are two choices for Predator1 (shown in (b) and (c)) to keep the minimum Manhattan distance. Predator1 will choose (c) instead of (b). The reason for this is because the number of empty cell in the left enclosed in dotted line is 1, and that of empty cell in the down enclosed in dotted line is 2.

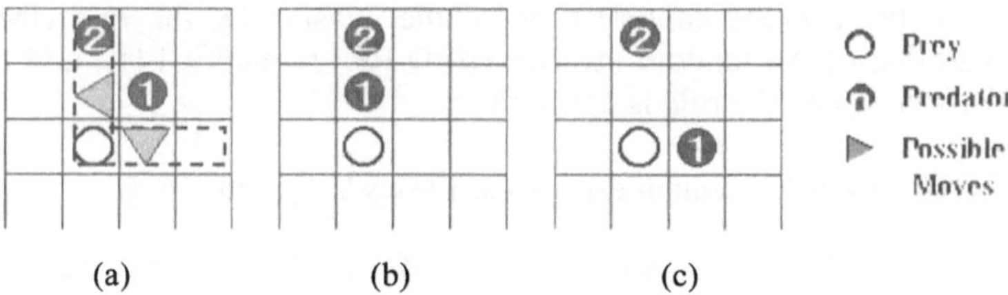

Fig. 9. Improved MD-Algorithm

Similarly, when there are two or more cells which they can avoid in running evasion algorithm, we make the predator does not chose a direction at random, but choses a direction so that each predator's action is not affected if possible. And we make a rule that the predator moves to the direction where there are fewer agents in the cell which they can move and the next-door cells on both sides. In the state of Figure. 10 (a), there are two choices for Predator1 (shown in (b) and (c)) in order to evasion. Predator1 will choose (c) instead of (b). The reason for this is because the number of empty cell in the left enclosed in dotted line is 2, and that of empty cell in the right enclosed in dotted line is 3.

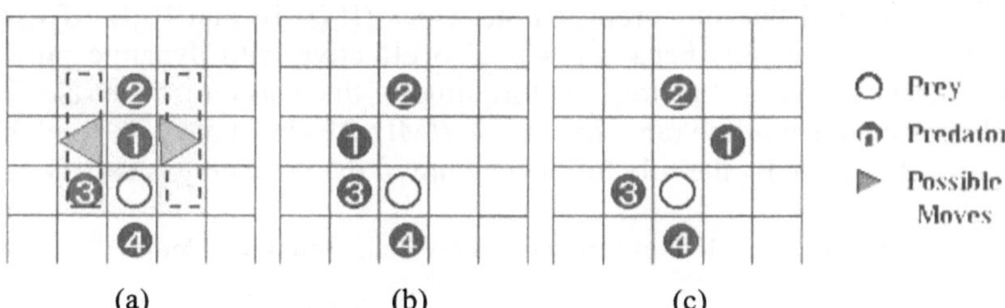

Fig. 10. Improved Evasion Algorithm

4. Experiment

4.1 Capture of Prey at stillness state

The initial game consists of a prey in the center of 15 x 15 Toroidal World, and four predators placed in random without overlapping positions. We change the priority of each predator at random so that they will not overlap for every cycle. The prey is caused to stay in the same position for all of the time. We have examined three strategies: the basic strategy we improved (IMD), and two of our proposed strategies; IMD with random evasion action (IMD-RE), and IMD with improved evasion action (IMD-IE). The result of these strategies on 50,000 studies is shown in Table 1.

We have checked that our proposed strategies (IMD-IE and IMD-RE) obtain 100% of capture regardless of the rule of evasion (See in 3.3.). We have also checked that the average capture time of the evasion by the rule (IMD-IE) is shorter than that of the random evasion (IMD-RE) by about 14%, and the effectiveness of evasion by the rule is certified.

Table 1. Result of experiment 1 (Prey 's algorithm: Stay)

Predator's Algorithm	The Rate of Capture (%)	Average Capture Time (cycle)
IMD	16.47	9.87
IMD−RE	100.0	14.57
IMD−IE	100.0	12.56

4.2 Capture of Prey at random movements

4.2.1 Capture experiment in an infinite plane

We have tested the capture experiments where the prey has random movement, but the prey is caused to stay in the same position for 20% of the time. We also tested experiments that had the same conditions as described in 4.1. The result of these capture experiments on 50,000 studies is shown in Table 2.

We have checked that our proposed strategies (IMD-IE and IMD-RE) obtain 100% of capture, and this behavior was also effective with dynamic prey. We have also checked that the average capture time of the evasion by the rule (IMD-IE) is shorter than that of the random evasion (IMD-RE) by about 17%, and the effectiveness of evasion by the rule for the capture of dynamic prey is certified.

Table 2. Result of experiment2 (Prey 's algorithm: Random)

Predator's Algorithm	The Rate of Capture (%)	Average Capture Time (cycle)
IMD	9.98	10.73
IMD−RE	100.0	63.46
IMD−IE	100.0	52.85

4.2.2 Capture experiment in a limited plane

We have prepared a boundary wall in Toroidal World and restricted the agents' moving range within the limit of 15x15. We have tested the capture experiments with the same conditions as described in 4.2.1. However, we have forbidden the prey to move to the four corners of this field. The result of this capture experiments on 50,000 studies is shown in Table 3.

We have checked that our proposed strategies (IMD-IE and IMD-RE) obtain 100% of capture. Therefore, the behavior we have proposed is effective not only on an infinite plane but also on the limited plane we newly set up.

Moreover, we recognized the new capture action that realized capture of an agent in an edge of wall (Figure. 11).

Table 3. Result of experiment3(Prey 's algorithm: Random)

Predator's Algorithm	The Rate of Capture (%)	Average Capture Time (cycle)
IMD	7. 57	9. 25
IMD−RE	100. 0	35. 55
IMD−IE	100. 0	34. 49

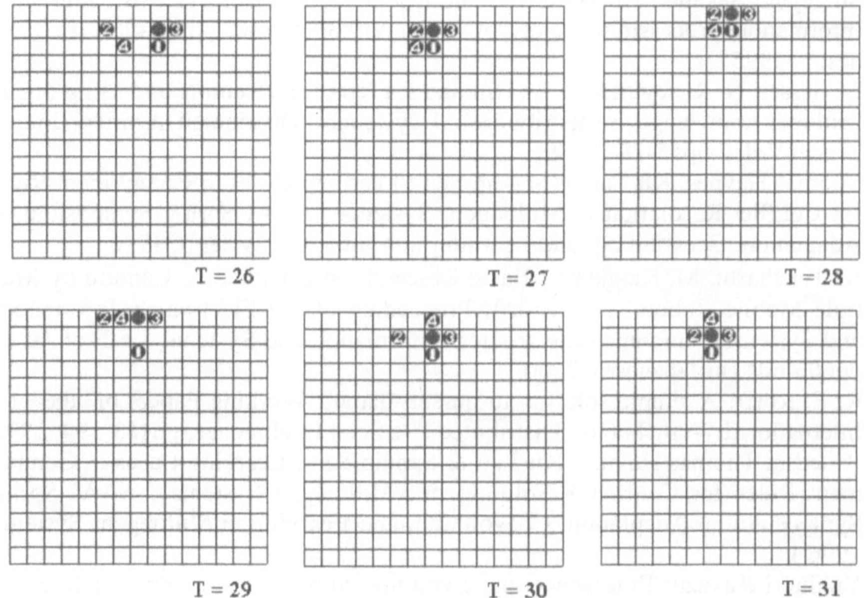

Fig. 11. Capture in an edge of wall

5. Conclusion

As an example of solution to a cooperation problem by multiple mobile robots, we mentioned the problem of capture by multiple mobile robots. We have realized the agents' cooperative action ("Collision Avoidance" and "Dissolution of Compromise") in simulations by using an idea of multithread programming.

We have achieved the dissolution of the agents' deadlock state, and the agents that have no higher functions (such as the learning and the evolutional techniques) but only simple intelligence and function realized the robustness of capture. We have also checked that our proposal behavior is effective not only on an infinite plane but also on a limited plane we newly set up, and recognized a new capture action which resulted in capture of an agent in an edge of wall.

We have change the priority of each predator at random so that they will not overlap for every cycle in these experiments. We are hoping to improve capture time to change their priority.

Reference

1. Shigeyuki Sakane : Decentralized and Cooperative Robot Systems ; Journal of Information Processing Society of Japan Vol.36 No.10, pp953-960 (in Japanese) , 1995.
2. L. Gasser, N. Rouquette, R. W. Hill and J. Lieb : Representing and using organizational knowledge in distributed AI systems ; Distributed Artifical Intelligence, Vol. 2 pp55-78 , 1989.
3. Thomas Haynes, Kit Lau & Sandip Sen : Learning Cases to Compliment Rules for Conflict Resolution in Multiagent Systems ; AAAI Spring Symposium on Adaptation, Coevolution, and Learning in Multiagent Systems, 1996
4. K. Takahashi, M. Kakikura : Basic Research on Cooperative Capture by Multiple Mobile Robots (1st report); Proceeding of the Electronics, Information and Systems Conference Electronics, Information and Systems, I.E.E of Japan, pp603-604 (in Japanese), 2000.
5. R. E. Korf : A simple solution to pursuit games, Working Papers of the 11th International Workshop on Distributed Artificial Intelligence, pp183-194, 1992. (Cited in Thomas Haynes, Kit Lau & Sandip Sen : Learning Cases to Compliment Rules for Conflict Resolution in Multiagent Systems ; AAAI Spring Symposium on Adaptation, Coevolution, and Learning in Multiagent Systems, 1996.)
6. Yukinori Kakazu: Emergence and Evolution in Robotics; Journal of Robotics Society of Japan Vol.15 No.5, pp2-6 (in Japanese), 1997.

Chapter 11
Emergence in Mobility

Interpreting Two Psychological Functions by A Hierarchically Structured Neural Memory Model

Tetsuya Hoya

Laboratory for Advanced Brain Signal Processing BSI-RIKEN, 2-1, Hirosawa, Wakoh-City, Saitama 351-0198 Japan

Abstract. In this paper, a novel neural memory model is proposed. The neural memory model proposed in this paper is based upon generalized regression neural networks (GRNNs) which are the paradigms of radial basis function neural networks (RBF-NNs). With the benefit of their quick learning capability and robustness, the application of GRNNs has been rapidly increased in many disciplines. Then, within the context of a newly proposed hierarchically arranged generalized regression neural network (HA-GRNN), two psychological functions, intuition and attention, are interpreted in terms of the evolution of the HA-GRNN. Within the framework of HA-GRNN, two types of memory, namely both the long and short term memory motivated from biological and cognitive studies, are considered and a dynamic learning system is thus proposed. In the simulation study, the effectiveness of the HA-GRNN in comparison with k-means clustering method is confirmed within the context of pattern classification tasks.

Keywords: Artificial neural networks, memory model, psychological functions, dynamic learning.

1 Introduction

Autonomous robotics can be ultimately defined as such robots that can 'think' and determine the next behavior by themselves without the instruction given by humans. Now, with the recent advancements in both biological and cognitive studies as well as computer technologies, one of which we wish to achieve in near future is develop, what is called, 'brain-style' computers (e.g., [1]), or truly autonomous robotics. It is said that one of the key approaches towards the development of brain-style computing is how to elucidate the mechanism of "intuition" in terms of artificial neural networks. On the other, modeling the notion of "consciousness" has recently been a topic of great interest in robotics [2,3]. Interpreting the notions related to emotional/psychological functions, however, has historically been a controversy among many disciplines from biology to philosophy. In this paper, it is addressed that such psychological functions, "intuition" and "attention", can be interpreted in terms of the evolution of an hierarchically arranged generalized regression neural network (HA-GRNN) model in which each sub-network has memory-based architecture. The evolution process is then justified within the framework of pattern classification tasks. The generalized regression neural networks (GRNNs) [4] fall in the category of radial basis function neural networks (RBF-NNs) [5], while, unlike ordinary RBF-NNs, having a special property that the weight vectors between the RBFs and output neurons are given identical to the target vectors. By exploiting this attractive property, a dynamic neural system can be modeled without any complex mathematical operations.

2 Configuration of a GRNN

A multilayered GRNNs (ML-GRNN) [5] with N_i input neurons, N_h radial basis functions (RBFs), and N_o output neurons is illustrated on the top of Fig. 1. In Fig. 1, each input neuron x_i $(i = 1, 2, \cdots, N_i)$ corresponds to the element in the input vector $\boldsymbol{x} = [x_1, x_2, \cdots, x_{N_i}]^T$ (T: vector transpose), h_j $(j = 1, 2, \cdots, N_h)$ is the j-th RBF (note that N_h is variable), $\|\cdots\|_2^2$ denotes the squared L_2 norm, and the output neuron o_k $(k = 1, 2, \cdots, N_o)$ is given as

$$o_k = \frac{1}{\delta} \sum_{j=1}^{N_h} w_{j,k} h_j, \tag{1}$$

where $\delta = \sum_{k=1}^{N_o} \sum_{j=1}^{N_h} w_{j,k} h_j$, $\boldsymbol{w}_j = [w_{j,1}, w_{j,2}, ..., w_{j,N_o}]^T$, and

$$h_j = f(\boldsymbol{x}, \boldsymbol{c}_j, \sigma_j) = exp(-\frac{\|\boldsymbol{x} - \boldsymbol{c}_j\|_2^2}{2\sigma_j^2}), \tag{2}$$

where $\boldsymbol{c}_j$ is called the centroid vector, σ_j is the radius, and $\boldsymbol{w}_j$ denotes the weight vector between the j-th RBF and the output neurons. As in Fig. 1 on the top, the structure of an ML-GRNN is similar to the well-known multilayered perceptron perceptron neural network (MLP-NN) except RBFs are used in the hidden layer and linear functions in the output layer. In Fig. 1, if the target vector $\boldsymbol{t}(\boldsymbol{x})$ corresponding to the input pattern vector $\boldsymbol{x}$ is given as

$$\begin{aligned} \boldsymbol{t}(\boldsymbol{x}) &= (\delta_1, \delta_2, ..., \delta_{N_o}), \\ \delta_j &= \begin{cases} 1 \text{ if } \boldsymbol{x} \text{ belongs to the class} \\ \quad \text{corresponding to } o_k \\ 0 \text{ otherwise} \end{cases} \end{aligned} \tag{3}$$

and if the centroid h_j is assigned for $\boldsymbol{x}$, $\boldsymbol{w}_j = \boldsymbol{t}(\boldsymbol{x})$, then the entire network becomes topologically equivalent to the one with a decision unit and N_o number of sub-nets as in the bottom of the figure [6]. In summary, the network configuration by means of an ML-GRNN is simply done in the following:

Network Growing: Set $\boldsymbol{c}_j = \boldsymbol{x}$ and fix σ_j, then add the term $w_{jk} h_j$ in (2). The target vector $\boldsymbol{t}(\boldsymbol{x})$ is used as a class 'label' indicating the sub-network number to which the RBF belongs.

Network Shrinking: Delete the term, $w_{jk} h_j$, from (2).

In the above, it is considered that, in hardware implementation, the network growing (learning) can be done very easily and the generalization performance is robust, while conventional neural networks such as multilayered perceptron neural networks (MLP-NNs) with the back-propagation algorithm [7] require iterative training scheme whenever the network configuration is changed and there is always a danger of being stuck in local minima [8].

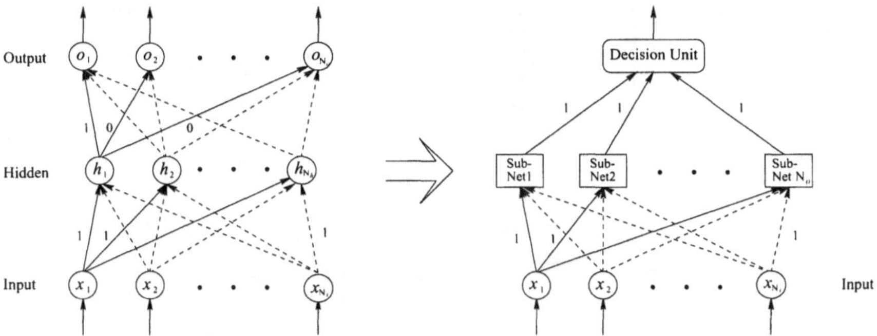

Fig. 1. Illustration of topological equivalence between the ML-GRNN with M hidden and N output units and the assembly of the N distinct sub-networks.

3 An Hierarchically Arranged Generalized Regression Neural Networks

The structure of an hierarchically arranged GRNN (HA-GRNN) is illustrated in Fig. 2. In the figure, a multiple of GRNNs representing long-term memory (LTM) networks (LTM Net (1 to L) in Fig. 2), a modified RBF network representing short-term memory (STM), and a decision unit are used. Moreover, the LTM nets can be subdivided into two parts; one for 'intuitive outputs' (denoted by Region 1 in a circle) and the others (denoted by Region 2). In the second part, each LTM Net (2 to L) has the same structure as in the bottom of Fig. 1, whereas both the STM and LTM Net 1 are given as modified RBF-NNs.

3.1 Structure of the STM Network

The output of the STM network $\boldsymbol{O}_{STM}$ is given in a vector form rather than a scalar value calculated as the sum. of the RBF outputs. The STM network, unlike the LTM nets described later, does not have any sub-nets, namely it is based upon a two-layered structure, with a maximum number of centroids M_{STM}, and functions as a temporal buffer to the LTM nets. The STM has, therefore, a structure similar to a queuing system. The learning of the STM network is summarized as follows:

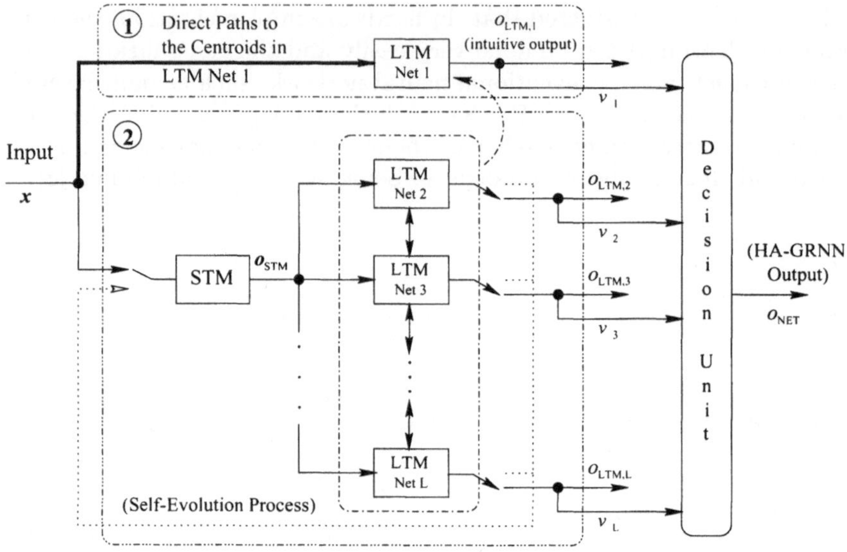

Fig. 2. Schematic representation of an hierarchically arranged GRNN

Step 1: If the number of the centroids is less than M_{STM}, add an RBF with h_i (given in (2)) and $\boldsymbol{c}_i = \boldsymbol{x}$ in the STM. Then, set $\boldsymbol{O}_{STM} = \boldsymbol{x}$.

Step 2: Otherwise,

1) If the activation of the least activated centroid (h_j, say) $h_j < th_{STM}$, replace it with a new one with $\boldsymbol{c}_j = \boldsymbol{x}$ and set $\boldsymbol{O}_{STM} = \boldsymbol{x}$.

2) Otherwise,

$$\boldsymbol{O}_{STM} = \lambda \boldsymbol{c}_k + (1 - \lambda)\boldsymbol{x} \tag{4}$$

where $\boldsymbol{c}_k$ is the centroid vector of the most activated centroid (k-th, say) h_k and λ is a *smoothing* factor ($0 \leq \lambda \leq 1$).

3.2 Structure of the LTM Networks

Similar to the STM, each LTM net in Fig. 2 has a maximum number of the centroids M_{LTM_i} $i = 1, 2, \cdots, L$. The LTM nets, except LTM Net 1, are, in turn, composed of GRNNs rather than RBF-NNs. Therefore, each LTM net is viewed as a collection of the sub-nets plus a decision unit as in the right side of Fig. 1 (except LTM Net 1). In contrast, LTM Net 1 consists of the centroids without a summing operation unit in the output. The output of LTM Net 1, $O_{LTM,1}$ is thus identical to the activation of the most activated centroid (l-th, say) h_l itself chosen by the 'winner-takes-all' strategy.

3.3 Evolution of the HA-GRNN

In the HA-GRNN, the role of the STM is to 'buffer' the incoming input pattern vectors, before storing them to the LTM nets. It is then hypothesized that long-term memory, in itself, has a layered structure representing an hierarchical classification system which is based on the 'significance' or 'attractiveness' of information. In this paper, such a classification system is modeled based on the activation of centroids. In summary, the construction of the HA-GRNN is divided into four phases:

Phase 1: STM (and LTM Net 2) formation ($t = 0$).
Phase 2: Formation of the LTM networks, LTM Net (2 to L).
Phase 3: Reconfiguration of the LTM Net (2 to L) (self-evolution) ($t = t_1$).
Phase 4: Formation of LTM Net 1 ($t = t_2$).

In Phase 1, the STM is formulated in the manner as described in Section 3.1, while LTM Net 2 is also formed by directly assigning the output vectors of the STM to the centroids in LTM Net 2. In the above, t denotes the t-th pattern presentation. The addition of the centroids in Sub-Net i ($i = 1, 2, ..., N_{cl}$, where N_{cl} is the number of classes) of LTM Net 2 is repeated until the total number of centroids in Sub-Net i reaches a maximum $M_{LTM_2,i}$. Otherwise, the least activated centroid in Sub-Net i is moved to LTM Net 3. This process corresponds to Phase 2. Then, Phase 2 is summarized as follows:

Step 1: Provided that $\boldsymbol{O}_{STM}$ belongs to Class i, then, for $j = 1$ to $L - 1$, do the following:
If the num. of the centroids in Sub-Net i of LTM Net j reaches $M_{LTM_{j,i}}$, move the least activated centroid within Sub-Net i of LTM Net j to LTM Net $j + 1$.

Step 2: If the num. of the centroids in Sub-Net i of LTM Net L reaches $M_{LTM_{L,i}}$ (i.e., all the i-th sub-nets within LTM Net from 2 to L are filled), there is no entry to store the newly coming vector $\boldsymbol{O}_{STM}$. Therefore, do the following:

Step 2.1: Discard the least activated centroid in Sub-Net i of LTM Net L.

Step 2.2: Shift all the least activated centroids in Sub-Net i of LTM Net from (L-1) down to 2 into LTM Net from L down to 3, respectively.

Step 2.3: Then, the new STM output vector is stored in Sub-Net i of LTM Net 2.

In Fig. 2, the output of the HA-GRNN O_{NET} is chosen as the largest value among the weighted LTM net outputs $O_{LTM,i}$ ($i = 1, 2, \cdots, L$):

$$O_{NET} = \max(v_1 \cdot O_{LTM,1},\ v_2 \cdot O_{LTM,2},\ \cdots, v_L \cdot O_{LTM,L}), \tag{5}$$

where $v_1 >> v_2 > v_3 > \cdots > v_L$. Note that the weight value v_1 for $O_{LTM,1}$ is given relatively larger than the others. This discrimination indicates the formation of the 'intuitive output' from the HA-GRNN. After the formation of the LTM nets, reconfiguration of the LTM nets is considered in Phase 3 in order to 'shape up' the pattern space spanned by the centroids in the LTM Net (2 to L). This process may be invoked at particular time. During the reconfiguration phase, presentation of any incoming input pattern vector is not allowed to process. In Phase 4, some of the centroids which keep relatively strong activation in a certain period in LTM Net (2 to L) are moved to LTM

Net 1. Each centroid newly assigned in LTM Net 1 eventually forms an RBF-NN and has a direct connection from the input vector $\boldsymbol{x}$.

4 Interpretation of Intuition and Attention

4.1 A Model of Intuition by HA-GRNN

In our daily life, we sometimes encounter such an occasion of which we feel the thing/matter is true but neither can we explain the reason why nor find the evidence or proof of it. This is referred to as the notion of, what is called, "intuition".

> Conjecture 1: In the HA-GRNN context, *intuition* can be interpreted such that, for a particular incoming input pattern vector there exists a certain set of centroids with *abnormally* strong activation within the LTM nets.

The above is drawn from the standpoint that the notion of intuition can be explained in terms of the information processing pertaining to a particular activity of neurons within brain (e.g., see [9]).
The evidence for referring to the output of LTM Net 1 as intuitive output is that LTM Net 1 is formed after a relatively long and iterative exposition of incoming input pattern vectors which results in strong excitation of some centroids in LTM Net (2 to L). In other words, the transition of the centroids from the STM to LTM Net (2 to L) is referred to as *normal* learning process, whereas that from LTM Net (2 to L) to LTM Net 1 gives the chances of generating "intuitive" HA-GRNN outputs.

4.2 Interpreting the Functionality of Attention by HA-GRNN

In the context of HA-GRNN, the model in [11] coincides with the evidence of having a 'hierarchical' structure for representing attention as a function of consciousness. In the HA-GRNN context, the following conjecture can be therefore drawn:

> Conjecture 2: The state of being 'attentive' to something is represented in terms of the centroids within the STM.

Accordingly, the following Phase 5 (at $t = t_3$) is appended to the evolution of an HA-GRNN:

[Phase 5: Formation of Attentive States]

Step 1: Collect m centroids of which number of activation count is the largest within all the LTM nets (2 to L) for particular classes.

Step 2: Add the copies of the m centroids back into the STM, where $M_{STM} - m$ most activated STM centroids are kept untouched. The m centroids so selected remain within the STM for a certain long period, without changing their centroid vectors but the radii.

It is also postulated that the ratio between the m centroids and the rest of the $M_{STM} - m$ in the STM explains the 'level of attention'. Therefore, the following conjecture can also be drawn;

> Conjecture 3: The level of attention can be determined by the ratio between the number of the m most activated centroids selected from the LTM nets and that of the remaining $M_{STM} - m$ in the STM.

Conjecture 3 is also related to the neurophysiological evidence of 'rehearsing' activity [10] in which the information acquired during learning would be gradually stored as a long-term memory after rehearsing. In the HA-GRNN context, an incoming input pattern vector $\boldsymbol{x}$ can be compared to the input information to the brain and are temporally stored within the STM. Then, during the evolution, the information represented by the STM centroids is selectively transferred to the LTM nets in Phases 1-3. In contrast, the centroids within the LTM nets may be transferred back to the STM, because the states being 'attentive' to certain classes is occurred at particular moments. In pattern classification tasks, one may limit the number of the classes to $N < N_{max}$ for representing attention in a way that "The HA-GRNN is particularly attentive to the N classes among a total of N_{max}" in order to compensate for the relatively 'weaker' area of the pattern space.

5 Simulation Study

In the simulation, an HA-GRNN is constructed using the data extracted from SFS database [12]. The data set used consists of a total of 900 utterances of the digits from /ZERO/ to /NINE/ recorded in English by nine different speakers (including even numbers of female and male speakers). The data set was then arbitrarily partitioned into two sets; one for constructing an HA-GRNN (i.e., the incoming pattern set) and the other for testing. The incoming pattern set contains a total of 540 speech samples, where 54 samples were chosen for each digit, while the testing consists of a total of 360 samples (36 samples per each digit). (The evolution within Phase 1 to 4 was therefore eventually stopped at $t = 540$.) Each utterance is sampled at 20kHz and was converted into the input vector of the HA-GRNN with a normalized set of 256 data points obtained by the well-known LPC-mel-cepstral analysis.

5.1 Parameter Setting

In the simulation study, the LTM parameters, $M_{LTM_1} = 5$, and $M_{LTM_2} = M_{LTM,3} = 40$, were used. For the STM, the choices, $M_{STM} = 30$ and $\lambda = 0.6$, were made to sparsely but reasonably cover all the ten classes during the construction. The number of sub-nets in LTM nets was equally fixed to 10 (i.e., for the ten digits). With this setting, the total number of centroids in LTM Net (1 to 3) $M_{LTM,Total}$ yields 85. Then, to give 'intuitive outputs' from LTM Net 1, v_1 was fixed to 2.0, while v_i $(i = 2, 3, \cdots, L)$ were given by a linear decay $v_i = 0.8(1 - 0.05(i - 2))$. For the evolution, the parameters, $t_1 = 200$, $t_2 = 201$, and $t_3 = 300$, were used.

5.2 Simulation Results

To test the classification accuracy of the HA-GRNN, the generalization performance over the testing set was evaluated using only LTM Nets (1 to 3), since the pattern space is formed within LTM Nets (1 to 3). For comparison, a conventional GRNN a total of 85 centroids obtained by the well-known MacQueen's k-means clustering algorithm was also used, which yielded the overall generalization performance of 75.0% as shown in Table 1. During testing, 16 pattern vectors among 360 yielded the generation of the intuitive outputs from LTM Net 1 in which 13 out of 16 patterns were correctly classified. It was then found that the Euclidean distances between the 13 pattern vectors and the respective centroid vectors corresponding to their class IDs (i.e., digit number) are relatively small and close to the minimum. From this observation, it can therefore be said that intuitive outputs are likely to be generated when the incoming pattern vectors are very close to the respective centroid vectors in LTM Net 1. In the simulation, three cases, without any attentive states, with the attentive state of Digit /NINE/ only, and those of Digits /FIVE/ and /NINE/ (by following the procedure in Section 4.2), were considered. For the first case without the attentive states, the generalization performance of the HA-GRNN obtained was 84.4%, which outperforms that of the k-means. This indicates that the HA-GRNN can betterly construct the pattern space in comparison with the GRNN with k-means. For the latter two cases, 10 among the 30 centroids in the STM was fixed and used for representing attention, in order to compensate for the 'weaker' covering by the first setup. For the case of Digit /NINE/ only, the overall generalization performance was improved at 85.3%, while the case of Digits /FIVE/ and /NINE/ was further improved at 86.9% as in Table 2. In the table, it is considered that, since the performance for Digit /NINE/ was not improved more than expected, the pattern space for Digit /NINE/ is much harder to fully cover than other digits.

6 Conclusion and Future Direction

In this paper, the two psychological functions, intuition and attention, have been modeled using a newly proposed HA-GRNN. The HA-GRNN is based upon the GRNN model which is essentially easy to implement in hardware (i.e., only two parameters are required to adjust) and robust for learning (or even for forgetting) and yet can approximate any input-output combinations. The concept of the HA-GRNN and its evolution has been motivated from biological and cognitive studies. It has been justified that the two psychological

Digit	0	1	2	3	4	5	6	7	8	9	Total	Generalization Performance
0	35			1	1						34/36	94.4%
1		17				19					17/36	47.2%
2			28		8						28/36	77.8%
3			3	22	10	1					22/36	61.1%
4					36						36/36	100.0%
5						36					36/36	100.0%
6					1	1	34				34/36	94.4%
7	1		3		3	6		23			23/36	63.9%
8					2	1	1		32		32/36	88.9%
9			1			27				8	8/36	22.2%
Total											270/360	75.0%

Table 1. Confusion matrix obtained by the conventional GRNN using k-means clustering method

Digit	0	1	2	3	4	5	6	7	8	9	Total	Generalization Performance
0	30			1	3		2				30/36	83.3%
1		31			2	2				1	31/36	86.1%
2			31	1	3		1				31/36	86.1%
3				32	4						32/36	88.9%
4					36						36/36	100.0%
5		1			1	33				1	33/36	91.7%
6							32	2	2		32/36	88.9%
7			4					32			32/36	88.9%
8							1	1	34		34/36	94.4%
9		3		1		10				22	22/36	61.1%
Total											313/360	86.9%

Table 2. Confusion matrix obtained by the HA-GRNN after the evolution (with attentive states of digits 5 and 9)

functions, intuition and attention, can be interpreted within the framework of evolution of the HA-GRNN. In the simulation study, the models of both the psychological functions have been introduced to form an HA-GRNN using the data set for digit voice classification tasks. The effectiveness has been investigated and its superiority in comparison with a conventional GRNN using the k-means clustering method has also been confirmed.

Note that the proposed layered-memory concept is not limited to pattern classification oriented tasks but can be widely applicable where the tasks and the goals (or, more generally, 'aims') are appropriately known/given, for instance, various planning tasks for autonomous robotics. In such applications, the incoming input vectors are considered as a set of sequential data points, for example, sensory data to know the position of the robot or the internal states, and to interact with other robots, which can alternatively be represented by the attentive states, while the output sequence from HA-GRNN can directly/indirectly change/set parameters to actually control the movement of the robot or the response to other robots by built-in communication facilities, according to the current situations. With preserving the same attentive states in multiple robots, it is considered that the robots can cooperatively work together for a particular goal which is not able to be achieved by a single robot. Future work is therefore directed to the concrete implementation of the proposed neural architecture towards the development of such autonomous robotics.

References

1. G. Matsumoto, Y. Shigematsu, and M. Ichikawa, "The brain as a computer," in Proc. Int. Conf. Brain Processes, Theories and Models, MIT Press: Cambridge, MA, 1995.
2. I. Aleksander, "Impossible minds: my neurons, my consciousness," Imperial College Press, 1996.
3. T. Kitamura, Y. Otsuka, and Y. Nakao, "Imitation of animal behavior with use of a model of consciousness - behavior relation for a small robot," Proc. 4th IEEE Int. Workshop on Robot and Human Communication, pp. 313-316, Tokyo, 1995.
4. D. F. Specht, "A general regression neural network," IEEE Trans. Neural Networks, Vol. 2, No. 6, pp.568-576, Nov, 1991.
5. P. D. Wasserman, "Advanced methods in neural computing," Van Nostrand Reinhold, New York, 1993.
6. T. Hoya and J. A. Chambers, "Heuristic pattern correction scheme using adaptively trained generalized regression neural networks," IEEE Trans. Neural Networks, Vol. 12, No. 1, pp. 91-100, Jan. 2001.
7. D. E. Rumelhart, G. E. Hinton, and R. J. Williams, "Learning internal representations by error propagation," In Parallel Distributed Processing: Explorations in the Microstructure of Cognition (D.E. Rumelhart and J. L. McClelland eds), Vol. 1, Chapter 8, Cambridge, MA: MIT Press, 1986.
8. S. Haykin, "Neural Networks: A Comprehensive Foundation", Macmillan College Publishing Co. Inc., 1994.
9. M. Minsky, "Emotions and the Society of Mind," in Emotions and Psychopathology, Manfred Clynes and Jack Panksepp, eds., Plenum Press, N.Y., 1988.
10. O. Hikosaka, S. Miyachi, K. Miyashita, and M. K. Rand, "Procedural learning in monkeys - possible roles of the basal ganglia," in "Perception, memory and emotion: frontiers in neuroscience," pp. 403-420, eds. T. Ono, B. L. McNaughton, S. Molotchnikoff, E. T. Rolls, and H. Nishijo, Elsevier, 1996.
11. N. Matsumoto, "The brain and biophysics," Kyo-ritsu Shuppan Press, 1997 (in Japanese).
12. M. Huckvale, "Speech Filing System Vs3.0 – Computer Tools For Speech Research", University College London, Mar. 1996.

Neural Control of Quadruped Robot for Autonomous Walking on Soft Terrain

Shigenobu Shimada[1], Tadashi Egami[2], Kosei Ishimura[1], and Mitsuo Wada[1]

[1] Graduate School of Engineering, Hokkaido University, Sapporo, Hokkaido 001-0008, JAPAN
{shimada,ishimura,wada}@complex.eng.hokudai.ac.jp

[2] Graduate School of Engineering, Kanagawa University, Kanagawaku, Yokohama 221-8686, JAPAN
egami@cc.kanagawa-u.ac.jp

Abstract. In the neurophysiology it has been clarified that some rhythm generators called the Central Pattern Generator (CPG) exists in animal locomotion. The CPG is able to generate various walking patterns without any previous plan and environment model. Using the CPG one can simplify the complicated walking system. In this paper, the CPG model for quadruped static walk is proposed and applied to the quadruped robot, TITAN-VIII. Moreover, we consider the irregular terrain as a walking environment with the sensor to recognize the softness of ground. By using the neural reflex control, it is shown that the robot can autonomously walk the soft terrain.

Keywords: CPG, Ground Recognition Sensor, Reflex, Soft Terrain, Center of Gravity

1 Introduction

In recent years, several researches of walking robot especially on irregular terrain are prosperous. The definition of irregular terrain is "horizontal and hard plane with some differences in height", but it's rare in real environment. The irregular ground condition of which we must take account is "softness". Kaneko et al. [1] discussed the softness and hardness of ground. In their study, assuming that the terrain is accurately known, the trajectory of legs was planned. However, it seems difficult to describe the nonlinear dynamics like walking environment. Kimura et al. [2] proposed a control method of the quadruped robot without planning the trajectory of legs. The control method was based on the adaptive output of central pattern generator (CPG) as torque of joints. However, this idea was only for pitch plane and seemed inadequate for three dimensional adaptation like the soft terrain.

Generally, the walking motion of animals is controlled in the distributed autonomous way. In this paper, we propose a new neural oscillator network of CPG for walking robot on soft terrain. By using the feedback signal from sensors, the robot is able to continue the locomotion. We also show the validity of our control method through the walking experiment of an actual quadruped robot.

2 CPG Model

The CPG model which was proposed by Matsuoka [3] and applied to the bipedal walk by Taga et al. [4] is used in our research. In this model, one neural oscillator is represented by the following differential equations:

$$\tau \dot{u}_i = -u_i - \beta v_i + \sum_{j=1}^{n} w_{ij} y_j + u_0 + F_i \tag{1}$$

$$\tau' \dot{v}_i = -v_i + y_i$$

$$y_i = max(0, u_i) \qquad (i = 1, 8)$$

where u_i is the inner state of the ith neuron; y_i is the output of the ith neuron; v_i is a variable representing the degree of the adaptation or self-inhibition effect of the ith neuron; u_0 is an external input with a constant rate; and F_i is the feedback signal such as a joint angle; β is a coefficient of the adaptation effect; τ and τ' are time constants of u_i and v_i, respectively; w_{ij} is a connecting weight. Each neural oscillator is constituted by a pair of two neurons which have inhibitory actions each other. In each joint, the position instructions in proportion to the inner state of flexor (extensor) neuron drives corresponding flexor (extensor) muscle.

3 Neural Oscillator Network of Walk Gait

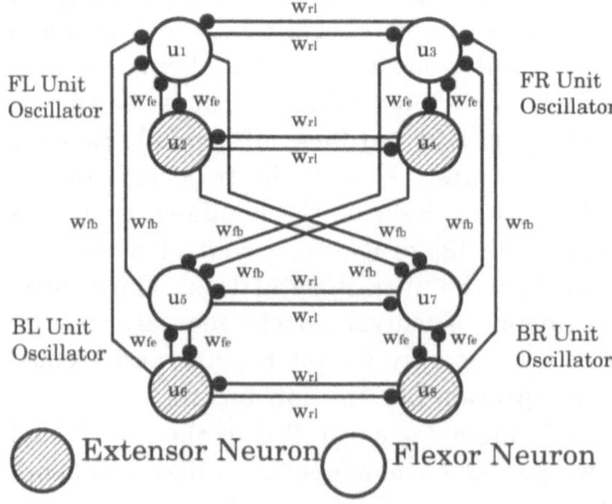

Fig. 1. Constructed CPG

In this study, we consider the irregular terrain as a walking environment. Accordingly, we must have neural oscillator network as a motion of static walk which is most stable own balance in the various gait within a walking period. Following the computer simulations, we determined the parameters of the neural oscillator network of the walking (Fig. 1). The simulation result is shown in Fig. 2. In this figure, FL is the leg of front left; FR is the leg of front right; RR is the leg of rear right; and RL is the leg of rear left. Subscript *command* expresses the position instruction of the CPG to the corresponding leg. It turns out that the swing operation of leg is performed in order of FL, RR, FR, and RL. The output of CPG is nearly 2 seconds period and each

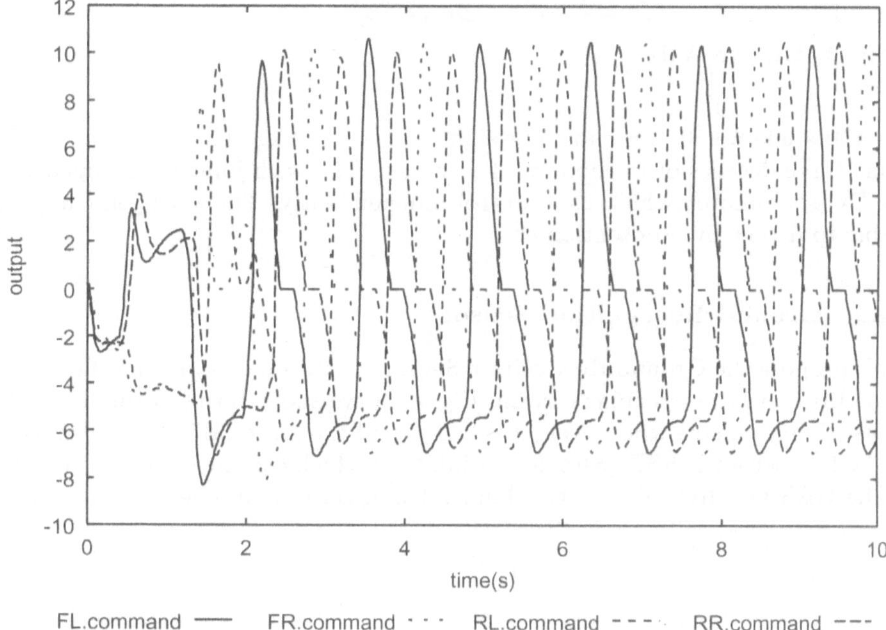

Fig. 2. Simulation result

phase difference is 90 degrees. In summary, the CPG generate walk gait. In the fifth chapter, we apply the simulation result to the actual quadruped robot. In walking experiment, the CPG drives the first joint of quadruped robot.

4 Experiment Apparatus

4.1 Quadruped Robot TITAN–VIII

In this study, we use TITAN–VIII as the quadruped robot(Fig.3). The size

Fig. 3. TAITAN–VIII

is 400 × 600 × 250 mm and the weight is nearly 20 kg. The quadruped robot has three degrees of freedom at the one leg. The each joint is equipped with a DC servomotor which can control independently. The each joint angle is able to detect by a potentiometer.

4.2 Ground Recognition Sensor

We propose the Ground Recognition Sensor (GRS) for recognizing the ground condition on the irregular terrain. Fig. 4 shows the external form. The GRS has 12 film sensors whose weight is 250 g, and the material is resin (duracon). The film sensor is FSR #402 (Interlink Co.) which can measure the pressure. The GRS are divided into two kinds: 4 film sensors at sole of foot is called

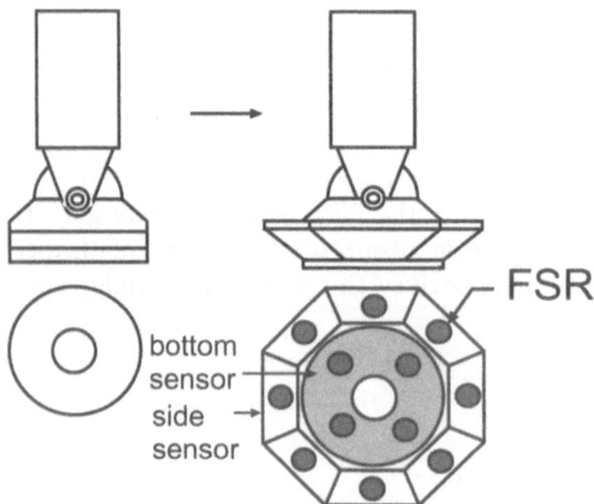

Fig. 4. The shape of ground recognition sensor

bottom sensor, and 8 film sensors at the octagon with inclination is called side sensor. When the GRS touches the hard terrain, only the bottom sensor reacts. When the sensor sinks in the soft terrain, both the bottom sensor and the side sensor react (Fig.5). As a result, the GRS is able to distinguish between hard and soft terrain.

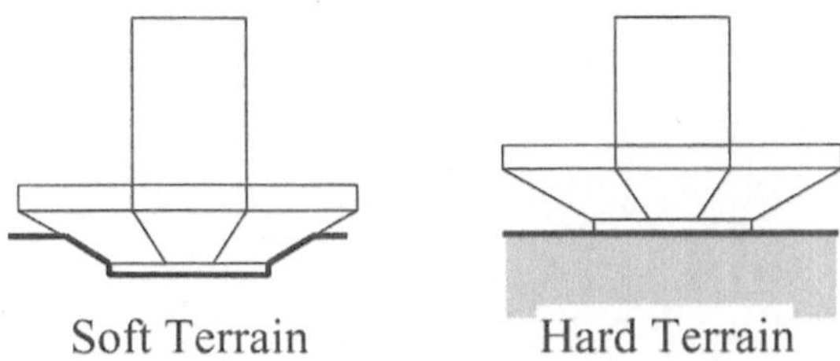

Fig. 5. The function of ground recognition sensor

4.3 Experimental Result of the GRS

We take measurements to make certain that the sensor functions within design specifications. The experiment is carried out by the following procedure:

1. the quadruped robot stands on own 4 legs
2. on the spot, the robot raises and lowers each leg according to given position instruction for 3 seconds
3. the sequence of position instruction is the order of FL, FR, RL, RR
4. we observe only the sensor value of FL

In the experiment, the hard terrain is realized by using horizontal and hard plane, on the other side the soft terrain is realized by using polyurethane form with the hight of 50 mm. The experimental results on the hard terrain is shown in Fig.6. On the hard terrain, only the bottom sensor reacts because legs don't sink in the ground. Therefore, the output of side sensor has been 0 V during the experiment as in Fig.6(b). In Fig. 6(a), when the FL begin the swing operation in around 1.5 s, the output of bottom sensor is 0 V. This result in that the FL touches nothing. In 5 s to 15 s period, the output of bottom sensor is over 3 V because the FL touches the ground. In around of 15 s, the output of bottom sensor is 0 V again. Fig.7 shows the results of experiment on the polyurethane form. When the FL sinks in the ground, the output of side sensor is over 1 V (Fig.7(b)). From this, we conclude that the GRS can distinguish between hard and soft terrain.

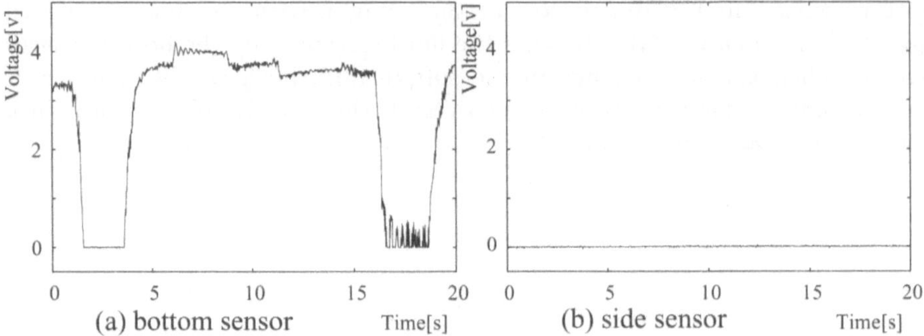

Fig. 6. Hard terrain

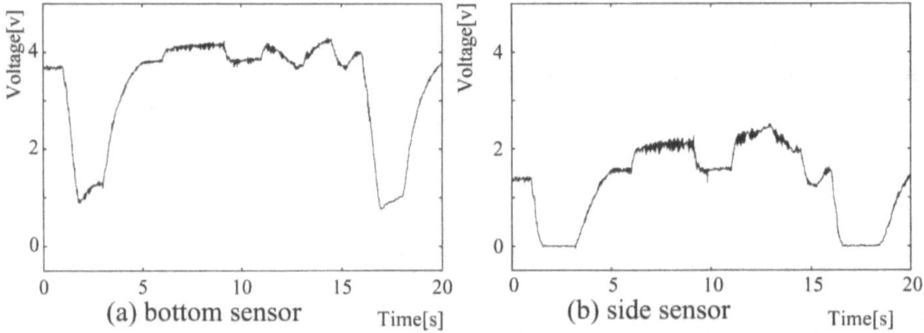

Fig. 7. Soft terrain

5 Walking Experiment with CPG

5.1 Waling Experiment on The Hard Terrain

Let *command* denote the position instruction which is produced by CPG (Eq. 1). The flexor activity period (*command* < 0) and extensor activity period (*command* > 0) of CPG correspond to the leg of support phase and the leg of swing phase, respectively. In walking, we adopt Eq. (2) to the CPG as the feedback signal. The Eq. (2) make the foot position to converge on $y = 0$ of Fig.8.

if *command* < 0

$$\begin{cases} F_i = P_{yi} & (i = 1, 2) \\ F_j = -P_{yj} & (j = 3, 4) \end{cases}$$

if *command* > 0

$$\begin{cases} F_i = -4.0 \times P_{yi} & (i = 1, 2) \\ F_j = 4.0 \times P_{yj} & (j = 3, 4) \end{cases} \tag{2}$$

where P_{yi} is the foot position at the y axis of the body coordinate fixed on leg i and leg j. The leg 1, 2, 3, and 4 expresses the FL, FR, RL, and

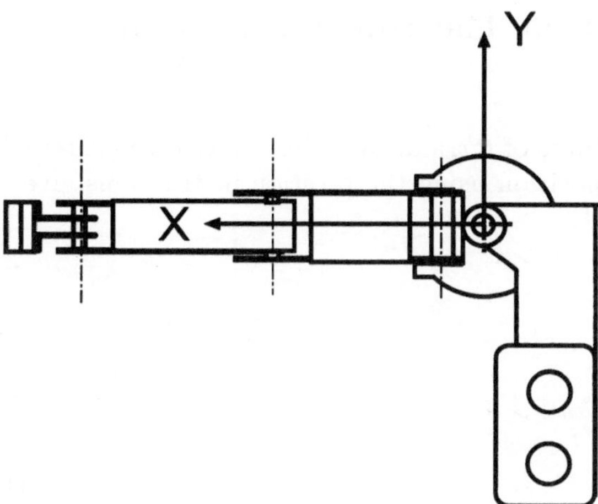

Fig. 8. Axis on walking robot

RR, respectively. The result of walking experiment is shown in Fig.9. Here, *FR.command* is the position instruction of the FR , and *FR.y* is the foot position of the FR on the y axis. It turns out that the leg moves in accordance with the the output of CPG.

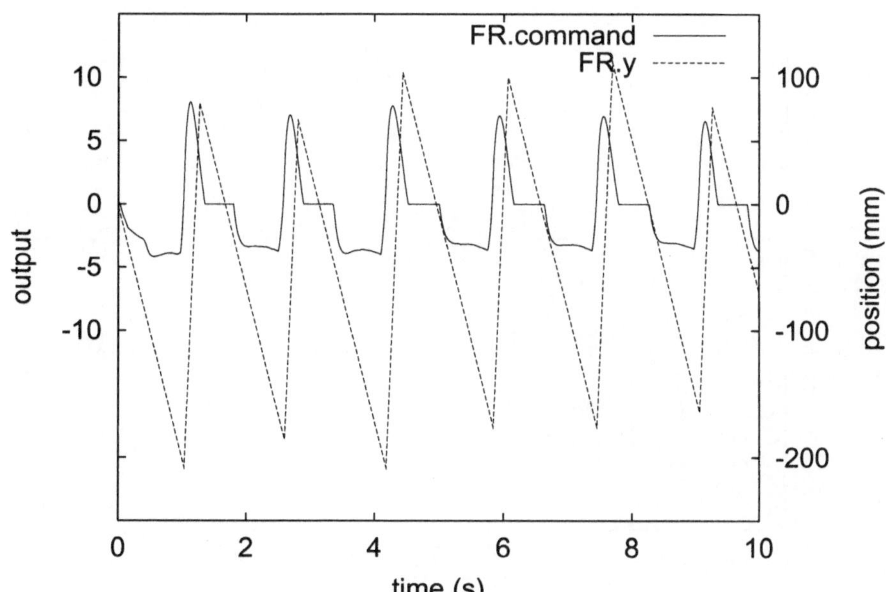

Fig. 9. Results of experiment involving walking on hard terrain

6 Walking Experiment on The Soft Terrain Using Reflex

When the sensor recognizes softness of terrain, the robot move own center of gravity (COG) to more stable position using the position instructions given by a reflex in Eq. (3) and Eq. (4).

if $s\_sens < 0$ and $RL.command > 0$

$$\begin{cases} P_{x\{1,3\}} = P_{x\{1,3\}} + N \\ P_{x\{2,4\}} = P_{x\{2,4\}} - N \end{cases} \tag{3}$$

if $s\_sens < 0$ and $RR.command > 0$

$$\begin{cases} P_{x\{1,3\}} = P_{x\{1,3\}} - N \\ P_{x\{2,4\}} = P_{x\{2,4\}} + N \end{cases} \tag{4}$$

where $P_{x\{1,3\}}$ and $P_{x\{2,4\}}$ are the foot positions of the x axis of the body coordinate fixed on each leg $\{1,3\}$ and leg $\{2,4\}$, respectively. N is the volume of locomotion to the direction of x axis. Subscript s_sens expresses the side sensor value.

Joining the reflex in Eq. (3), (4) with the CPG in Eq. (1), we apply the operation to the quadruped robot, TITAN-VIII. The result of walking experiment is shown in Fig. 10. $FR._{s\_sens}$ is the value of the side sensor at

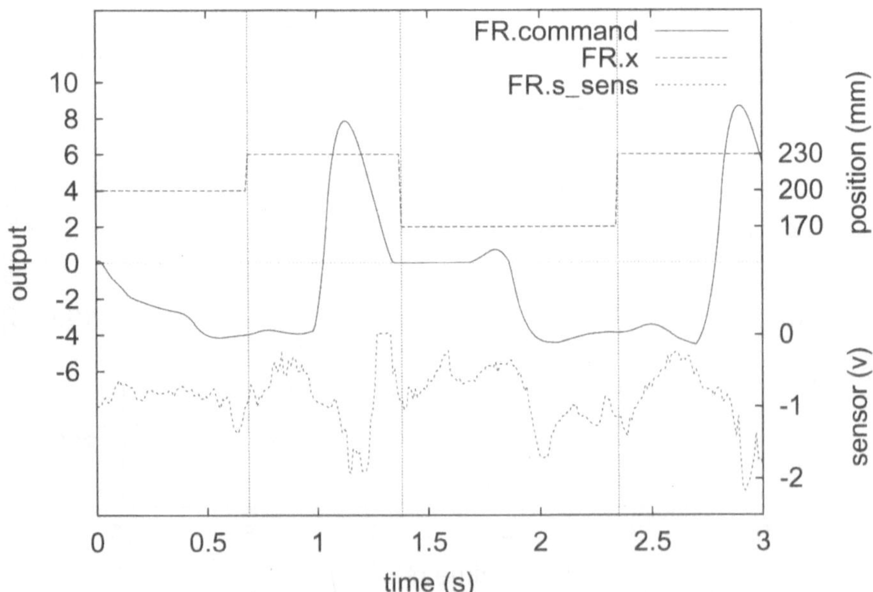

Fig. 10. Experimental result of walking robot on soft terrain

FR. When the sensor touches nothing, the output value is 0 V. When the sensor contacts something, the output is negative value. Then, the Eq. (3) was applied at the period of 0.7 s to 1.4 s. In around of 0.7 s, the sensor value of $FR._{s\_sens}$ increases because the center of gravity (COG) of the robot inclines left-hand. In 1.1 s, $FR._{s\_sens}$ outputs a small value since the RR is in swing operation. In other words, only FR in the right-hand of the robot supports whole body. In around of 1.2 s, the FR is in swing operation. The output value of $FR._{s\_sens}$ is 0 V. It turns out that robot can execute swing operation on the soft terrain. In 1.4 s to 2.4 s is the period of Eq. (4). Then, the Eq. (3) and (4) is applied to the quadruped robot, alternatively. As the result, the walking of the robot is sustained by the adding motion of yaw direction on the soft terrain.

7 Conclusions

In this paper, we showed the robust walking for real environment using the neurally distributed control method. The each joint of the quadruped robot was driven by CPG. The robot has no planning the trajectory of legs. The GRS distinguished between soft and hard on the ground. The side sensor reacted when the ground is soft. On this terrain, COG of the robot moved to a stable position by adding motion to yaw direction as the reflex. CPG attracted the difference of strokes and walking phase which is caused by motion of COG. As these results, the locomotion on a soft terrain by combining CPG with the reflex reacting to the sensor was realized. The parameters of walking robot have infinite varieties when the systems include the ground condition. In spite of these problem, it was shown that the proposed control has adaptability for the irregular terrain such as a soft terrain.

References

1. M. Kaneko, K. Tanie, E. Horiuchi. (1986) A Control Method for Multi-Legged Machine Walking over Soft Terrain – Based on the Consideration of Alternating Tripod Gait – (in Japanese), RSJ, Vol.4, No.3, 231–240.
2. H. Kimura, M. Akiyama, K. Sakurama. (1998) Dynamic Walking on Irregular Terrain and Running on Flat Terrain of the Quadruped Using Neural Oscillator(in Japanese), RSJ, Vol. 18, No.8, 1138–1145.
3. K.Matsuoka. (1987) Mechanisms of Frequency and Pattern Control in the Neural Rhythm Generators, Biological Cybernetics, 56, 345–353.
4. G.Taga, Y.Yamaguchi, H.Shimizu. (1991) Self-organized Control of Bipedal Locomotion by Neural Oscillators, Biological Cybernetics, 65, 147–159.

Kilorobot Search and Rescue Using an Immunologically Inspired Approach

Surya P. N. Singh[1] and Scott M. Thayer[1]

[1] [spns, sthayer]@ri.cmu.edu
Robotics Institute, Carnegie Mellon University, Pittsburgh, PA 15213

Abstract. This paper presents a new concept and simulated results for the distributed coordination of autonomous robot teams via the Immunology-derived Distributed Autonomous Robotics Architecture (IDARA) to perform select search and rescue (SAR) operations autonomously. Primarily designed for the coordination and control of "kilorobot" colonies, this architecture incorporates the immune system's stochastic learning and reaction mechanisms to yield astute and adaptive responses to dynamic environmental conditions. These mechanisms allow the architecture to vary actions from reactive to deliberative to result in a guided, yet stochastic, method that is ideal for dynamic operations such search in unstable terrain. IDARA was evaluated in a variety of SAR problem domains via computer simulations. These tests show that the IDARA framework is robust to noise and does not degrade when coordinating large colonies of up to 1,500 robots. By providing new levels of scalability in noisy environments, IDARA enables the full potential of micro-scale robotic platforms for intelligent exploration, mapping, and SAR operations in a manner not afforded by traditional methods.

Keywords: distributed, multi-robot, search and rescue, artificial immune systems

1 Introduction

Search is an integral aspect of nearly every robotic application ranging from planetary exploration, hazardous environment characterization, urban warfare, to domestic applications. The use of robots for these exploration tasks minimizes human exposure to harmful or tedious operations and may be the only means of performing potentially life-saving operations in constricted and treacherous environments, such as unstable terrain.

In this paper, we consider kilorobotics – large-scale, heterogeneous multi-robot teams having populations in the thousands – for exploration of previously inaccessible and dangerous environments that are complicated by variable, dynamic changes [1]. To fully serve the needs of operators or higher-level layers of autonomy, these robot colonies need a coordination method that distributes tasks such that the environment is fully characterized in a manner that takes into account *a priori* information from available map(s) or heuristic data. Since map-building operations can be handled during the search operation by registration and archiving of sensory data, the deft coordination of very-large distributed robotic colonies will improve performance across a spectrum of SAR tasks including object location and dynamic entry (ingress) and exit (egress) path planning.

Nature is replete with systems where large populations work cooperatively in a cohesive and productive manner to achieve complex goals in a far more efficient manner than may be accomplished individually. The immune system is a remarkable example of a highly scalable distributed control and coordination system that operates in the presence of substantial complexity resulting from environmental uncertainty, noisy inputs, adversarial agents, and external threats [2]. The human immune system is a massively scaled distributed object environment that delivers measured, decisive, and dynamic responses to changing macroscopic and microscopic conditions. As an example, in the time it takes to make a cup of coffee the immune system produces 8 million new lymphocytes and releases nearly a billion antibodies. In other words, the immune system acts like a protective force that monitors the bioenvironment and, depending upon a perceived threat, activates the necessary multi-agent control systems and responses [3, 4].

Just as models of the nervous system can serve as a powerful framework for building deterministic intelligent systems (e.g., neural net classifiers), the immune system serves as a powerful basis for the design of robot architectures that respond and perform learning via stochastic processes [5]. Although its fundamental goal is pathogen/non-pathogen selection and response, the immune system model gives insights to several methods for autonomous multi-robot operations based upon the native exploration methods found within the human immune system [2]. By using this as a basis, kilorobotic systems will be able to more fully exploit the comparative advantages inherent in autonomous multi-robot systems, namely: parallel execution, redundant operations, increased reliability, and robustness to point failures. For SAR operations in particular, the increased number of robots available will give the search method extra degrees of freedom and thus result in a greater potential of finding the object under consideration be it a victim, gas-leak, unexploded mine, or terrain feature.

2 Previous Work & Immune System Overview

Search and rescue using kilorobotic teams is an active area of research. Developments from several areas of robotics research (such as exploration and mapping, artificial immune systems, multi-robot theory/coordination, and microrobotics) can be combined to provide significant guidance on the design characteristics of an immunology-based multi-robotic architecture for this robotic domain [7, 8]. Traditional coordination strategies and behavior-based models of multi-robot control are not well suited for kiloroboitc SAR as they are not designed to support nor take advantage of the vast degrees of freedom inherent with kilorobot populations [8]. Furthermore, traditional methods may not focus at the higher levels of abstraction required for coordinating large populations, which could result in too little importance being placed on group command and control [9].

Multi-Robot Search and Rescue Architecture Developments

Several approaches have been developed to enable multi-robotic search. While these approaches vary in scope and development from basic architectures to complex systems, they are based on the same tenet: that communities of agents working cooperatively towards a common goal will do so more effectively and efficiently than if the same agents worked independently [9, 10].

There are two general approaches to exploration: topological and metric approaches. Topological approaches combine a series of interconnected landmarks that may have been augmented with distance information and/or probability (data confidence) information to yield a final search pattern. By comparison, metric approaches are a simpler representation and essentially view the world as an occupancy grid, which could be modified using additional data (e.g., data confidence) [11]. Several algorithms have been proposed for multi-robot SAR. Refs. [10, 12-13] show that SAR can be considered a special case of the robot exploration problem; however, with the caveat that it must be performed using a directed stochastic search, as the environment is usually quite chaotic. That is, given a general search plan the architecture needs to be stochastic with respect to the exact motions of the robots so that there is significant breadth to the exploration path/area[13].

Previous work suggests that for multi-robotic applications to be effective the architecture's methods and communications need to be compatible with the hardware platforms on which it will likely be executed. As a general rule, the millibots and micro-bots on which this architecture will likely reside will have very limited memory and computational resources (e.g., 8-bit MicroPIC) [6, 14].

Immune System Overview

A brief overview of the immune system is presented as a background on the concepts and methods on which the IDARA metaphor is based. The human immune system works on two levels both with the general goal of pathogen control: a general response mechanism that is not directed at any specific pathogen (i.e., innate immunity) and a specific, anti-body mediated response that encompasses many of the pattern recognition and situational memory aspects that are a core aspect of the human immune system (i.e., acquired immunity). Figure 1 illustrates the specificity ladder connecting early "reactionary" responses and the slower, but highly effective, "deliberative" actions tailored to a specific pathogen or threat.

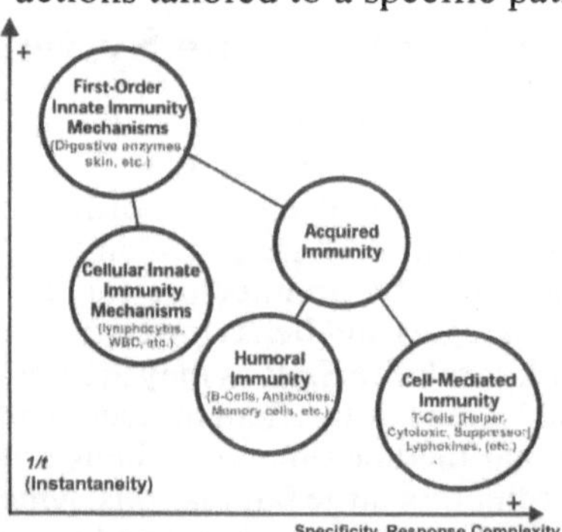

Fig. 1. Cascading Response Model for Immune System Responses
(Response becomes more specific and advanced with time)

Innate immunity is the natural and omnipresent resistance to a variety of pathogens. Its purpose is to act as the first-order, general defense mechanism. These innate mechanisms then couple with principal members of the acquired immune system to form a rapid, yet targeted, response that uses gradient decent as its primary recruitment method. This mechanism operates through self/non-self discrimination and by activating certain general kill mechanisms [3].

In contrast to the innate system, acquired immunity is about higher-level responses to specific and known threats. These responses provide life-long critical immunity (e.g., a person with normal immunity can survive up to 100,000 times the otherwise lethal exposure to a pathogen). There are two types of acquired immunity: humoral (i.e., B-cells and antibody regulation) and cell-mediated (i.e., T-cells proving B-cell assistance and orchestration). Both are initiated by antigens and signaled by antibodies (i.e., the some 10 million Y-shaped molecules that match key proteins to their specifically encoded pathogen) [4]. Together these operations give a recruitment mechanism (clonal expansion) where the recognition of a pathogen sets off a chain-reaction that generates a large population of antibody producing cells specific to the recognized antigen [5]. In addition to scaling the response, this mechanism acts to provide a stochastic form of learning.

The immune system has been the focus of the development of variety of algorithms from negative selection to anomaly detection [2, 5]. As elaborated in [8], the main thrust of this research is not to mimic the immune system's operation, but rather to use its operation as a model for the construction of methods that coordinate large numbers of largely independent agents. Current methods often implement a probabilistic approach based on Jerne's Idiotopic Network Hypothesis these techniques use a simplistic model of the acquired immune system as a control mechanism [15]. For example, this technique has been used to mediating behaviors and to perform various "fuzzy" tasks (e.g., task classification) [16,17].

3 IDARA Architecture Design

IDARA's central tenet is that the principals of immunology offer a promising and scalable analogue for controlling an unprecedented number of robots. While the current IDARA design has focused on the use of these algorithms towards the development of a first-order distributed robotics methodology for use in SAR operations, it is envisioned that the adaptivity and robustness inherent to IDARA can be extended to other robot domains (e.g., to aid in task planning and allocation).

The core of the IDARA architecture is derived from the cascading level of response model for immune system operation, which illustrates a tradeoff between rapid, reactionary actions and slower, deliberate responses (see also Figure 1). IDARA also includes the immune system's stochastic learning mechanism (sometimes referred to as Hebbian learning) and its mechanism for replicating and distributing increasingly specific responses to perceived threats (i.e., the clonal expansion concept). By modeling these three features IDARA provides coordination in a robust, diverse, and non-deterministic manner [5].

The IDARA architecture mimics the immune system's tiered response by having multiple forms of response that vary in their level of instantaneity and complexity. This is a key advance of the IDARA model that enables the consideration of the entire and not just the cell-mediated object recognition mechanisms of traditional immunology-derived approaches (as outlined in [3, 15]). This consideration provides IDARA with a reactionary schema for general responses, which are tuned as it proceeds. In contrast to several mobile robot approaches, this allows IDARA to dynamically vary its response *so that the efficiency of reactive responses can be varied against the specificity of targeted deliberative responses.*

IDARA has multiple means of responding to perceived situations, which are integrated and mixed to yield the final response(s) sent to the robot (see Figure 2). As initial (sensor and/or communication) inputs are received, the more basic and general response actions are used to immediately address the situation. As more information is collected (from multiple sensors or through futher analysis of the sensor data), IDARA's more advanced and specific layers are capable of suggesting specific actions that can be used to either refine the operation of the robot or even to specifically recruit "special" responses (e.g., recruit a robot having specialized sensors, etc.) This enables IDARA to combine deliberative planning with the relative simplicity of reactionary architectures in a unified framework.

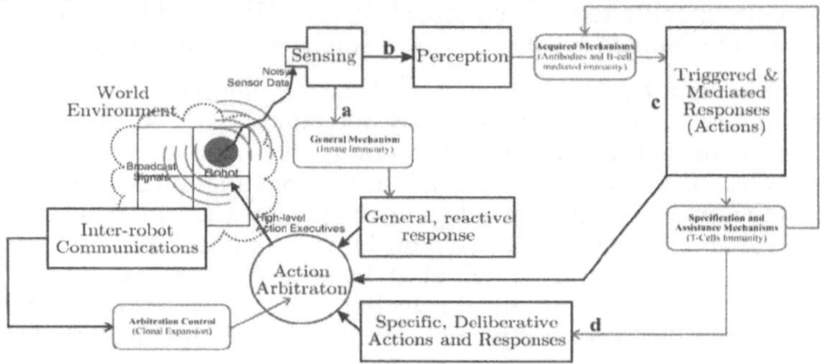

Fig 2. IDARA Software Architecture: (a) Sensor data are used in a rapid general (reactive) mechanism, (b) multiple sensors are combined, (c) this information is then used with known responses to give more complex triggered response, (d) analysis along with learned aspects are lead to specific responses.

IDARA maps different aspects and features of the immune system to various modules and tiers of response ladder architecture and not to entire robots. By placing the analogous operations at a high level of abstraction, the IDARA architecture does not require that the response mechanisms employ low-level mappings of immune theory to specific robotic functions. However, this approach does not preclude this design approach as these mappings can be encapsulated in select response blocks having the onus of control. IDARA uses a heuristically-driven arbitration module to combine the action directives by summing the action and confidence vectors generated by the various control blocks. The result of this procedure is a form of "directed randomness," in which the architecture varies and tunes its response from random exploration to specifically guided actions.

A second feature of the immune system employed by the IDARA architecture is its use of "histamines" as the principal communications and recruitment mechanism. The immune system uses chemical beacons to communicate the presence of a pathogen and to recruit other cells to the site of infection. IDARA's inter-robot communications layer adopted this communications model by broadcasting the robots "histamine" signals in response to threats in the environment. The broadcasting communication model, analogous to citizens' band (CB) radio, is simpler than traditional point-to-point techniques and allows the signals of one robot to be quickly interpreted by neighboring robots. An additional advantage of this model is that third-party robots or operators can "monitor" the ongoing communications without impacting them. The disadvantage of this model is that there is no guarantee that the signal was received correctly and no simple means of responding to a particular robot. However, IDARA's multi-tiered architecture addresses these by making the higher levels robust to errant signals and by using passive communications and actions to aid in local coordination.

By combining the radio signal with the number and direction of signals received, the "histamine" model provides gradient-descent recruitment for IDARA. That is, an available robot will coordinate with this robot (and potentially recruit others) by factoring the strength and the direction of these signals along with other preprogrammed or learned information about the nature of the environment. While not currently addressed by the IDARA architecture, these signals are used extensively in the pattern recognition and learning aspects of the immune system.

While particular details vary with the implementation and arbitration mechanism, a general description of the standard hierarchy used by IDARA is as follows:
Investigative Response – A response for the events when no information is being perceived (e.g., during startup, an environment where antigens are sparsely located, etc.). The objective is to act in a manner that tries to investigate, e.g. through a dispersion mechanism without disrupting the larger system (e.g., randomly selected actions).
General (Routine) Response – In the default case where there is some minimum level of information (i.e., sensor inputs are above some sensitivity threshold), the method applies a general response(s) that may be effective, but may not be optimal or fully exploitive of all the available information.
Triggered Response – When the system can make a deeper inference (e.g., the origin of a gradient communicating a goal point), a pre-planned response is initiated.
Deliberative Reponses – This level of response is the most intensive and complex and includes "machine learning" aspects of the immune system (i.e., the ability to use "memory B-cells" and recall patterns associated with certain cases of a task). In addition, this method would use the internal state and deliberative reasoning to try to develop a more optimal response. This allows IDARA a means of exploring around local-optima that would "trap" the triggered response level.

The IDARA architecture has a variety of features compared to other methods for the coordination of teams of robots. When applied directly at to a population of robots, this architecture will yield a mobile, robust, and adaptive control method, which can combine the functions and critical mass of simple robots to solve complex tasks. One key advantage with large populations is increased robustness and precision. That is, an individual cell in the immune system is not constrained by traditional recovery criteria since the premature death of a cell is a powerful method of controlling an infection. Likewise in IDARA, the failure of an individual (disposable) agent is not detrimental to the entire system and may actually be beneficial to the overall coordinated action. A second feature provided by very-large groups of robots is that such systems will be more economically viable as standard (off-the-shelf), fault intolerant components can be used, as the individual failure modes no longer critically affect the performance.

An additional benefit of this architecture is that many robotics tasks can be encoded in the IDARA framework with minor modification. In general, the process begins by decomposing the task into "antigens" and "cells." The "antigens" would generally be the principal object of the robotic activity that needs to be addressed (e.g., unexplored area, mines, time, costs, etc.). General and specific responses would be encoded as a variety of response "cells" distributed in the network. Sensing modalities would be integrated in the process and would address how "cells" recognize "antigens." Finally, manipulation mechanisms would be encoded as methods for addressing and responding to a "recognized" threat whose specificity would vary depending on the type and nature of this mechanism.

While the IDARA architecture has a number of strengths, especially in the coordination and control of large robot colonies, it is not perfect. One weakness is that agents initially base interaction on Brownian motion until an antigen is found locally and then use local gradient-optimization to follow the signals from initial interactions. This, however, predicates that there is an initial interaction between the two effectors. Thus, this architecture needs an inherent "critical mass" and may not operate well in small populations. For example, when the robots become highly sparse they default to the *Investigative Response* and just move randomly. Further, gradient techniques are only locally optimal, thus to obtain a highly efficient solution, IDARA-based systems need to be somewhat random in their initial motion so that it is fairly well distributed. Finally, the modular aspect of the IDARA architecture only partially aids with the varying computational resources available. While it can handle platform variation, the architecture still needs some intensive processing, especially with deliberative methods initiated after an initial action has taken place that may include planning and optimization sub-routines.

4 Search & Rescue Method Design and Simulation

With the general IDARA architecture in place, we examine its use in the design of an immunology-based SAR method, which we will then characterize via a computer simulation. In particular, we will focus on the SAR operations needed to find an object(s) of interest within a pile of unstable rubble.

IDARA is particularly useful in exploring the highly unstructured environments typical of SAR operations as it uses a stochastic search pattern and can change its coordination strategy as it hones in on the object of the search operation. The control mechanism is the initial placement of antigens in the environment, which serve as goal points to partially direct the search and as indicators of key environmental features. Thus, as the robots (initially moving in an undirected manner) sensed the goal object or other salient "antigens" they are able to broadcast an "antigen found" message, which would spark the architecture's "clonal response" recruitment mechanism. Furthermore, in order to exit the environment the robots would simply reverse the process by associating a second type of "antigen" with the external environment and repeating the search process to find the most appropriate exit, which may not be the point of entry.

Search and Rescue Simulation

The IDARA architecture was applied to this domain (see Figure 3) in a manner that demonstrates the performance of the architecture and its coordination of robot colonies consisting of up to 1,500 robots. Each robot was assumed to independently assess sensor inputs from a differentiating "bump" sensor (i.e., the robot was able to determine if there was an object or robot in an adjacent cell) and beacon signal that it received. With this information, the method calculated future actions

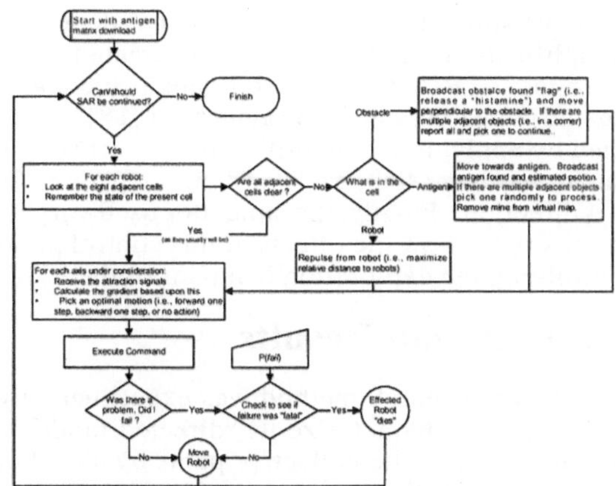

Figure 3: Simulation Flowchart

as a motion vector according to IDARA's three levels of operation (i.e., General, Triggered, and Deliberative). The arbitration mechanism then performs weighted vector summation to obtain the resulting motion. If an object of interest (i.e., an antigen) is identified the method issues a broadcast signal indicating this, which, is used to recruit neighboring robots to search the area of interest.

The simulation was implemented using MATLAB with core libraries as integrated, compiled C++ code [8]. This simulation started with an operator indicating the goal points or areas of interest. These were subsequently tagged as a source of antigens in the environment. Using IDARA the robots then proceeded in a manner to control the antigen source (i.e., the goal). Rescue operations were modeled as a reverse search where the robot would essentially look for the simplest way to approximately return to its initial location. This type of search would help rescue workers and other robots find the most accessible means of exit after the object of the search has been located. The simulator included a Gaussian noise generator that modeled "practical" problems such as senor noise, erroneous transmissions, robot failures, signal decay, and shifting terrain.

In particular, the simulation looked at the facet of SAR operations associated with traversing through a rubble-pile to locate a single object of interest that could be sensed (e.g., warm body, gas leak, etc.). This type of terrain is generally inaccessible to larger robots or SAR crews due to its many narrow, unstable crevices. For simplicity, interaction effects (i.e., reactions in the terrain resulting from robot motions in the terrain) were not included in the simulation.

To begin the simulation, the rubble terrain was modeled in the simulator as a series of vertical cross-sections having random rubble distributions of a set density (see Figure 4). The robots were initially deployed uniformly over the surface of the terrain. The system was instructed (by the operator) to go towards a central goal, which represents the operator's thought as to the likely location of the object.

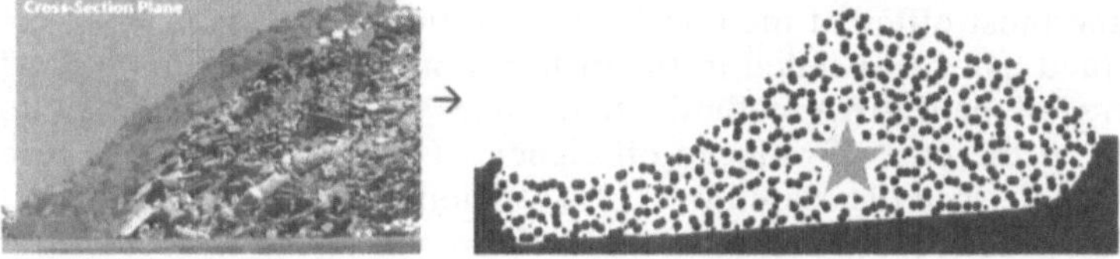

Fig. 4. Rubble-pile cross-section and model that was subsequently studied (star shows goal area).

In general, the simulator was implemented as detailed above with a few computational simplifications. First, the robots estimated the distance to broadcasting robots using the ATRIA (Advanced Triangle Inequality Algorithm) approximate distance and k-nearest neighbor algorithm [18]. Second, 8-way navigation was performed by applying the algorithm twice – once to determine the unit-step action (i.e., forward one step, backward one step, or no action) along the x-direction and the y-direction. Finally, the simulator used varying initial distributions of the antigen density to show the effects of this control parameter on tweaking, but not fully controlling, the exploration behavior of the IDARA-based SAR method.

5 Simulation Results

The IDARA-based search method was experimentally evaluated to validate this architecture and to characterize its "directed randomness" and related performance. Figure 5 shows the collective paths by all robots during the simulation as they proceed towards the goal area (marked by superimposed star). Figure 6 is an iteration-lapse sequence that shows the progression of search as the robots proceeded from their initial position to the goal in the center.

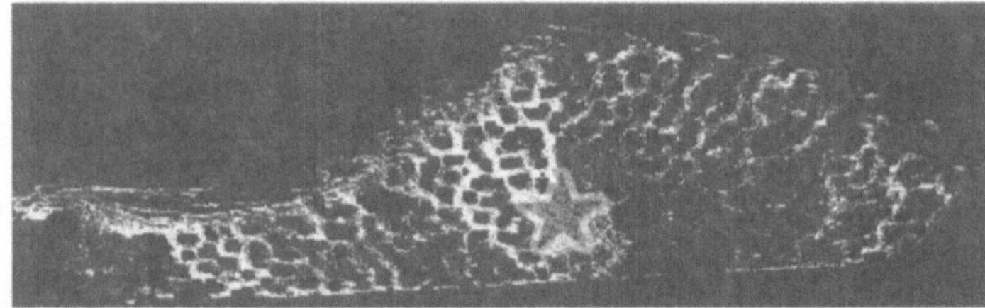

Fig. 5. Visitation Map (areas visited more often are shown in white)

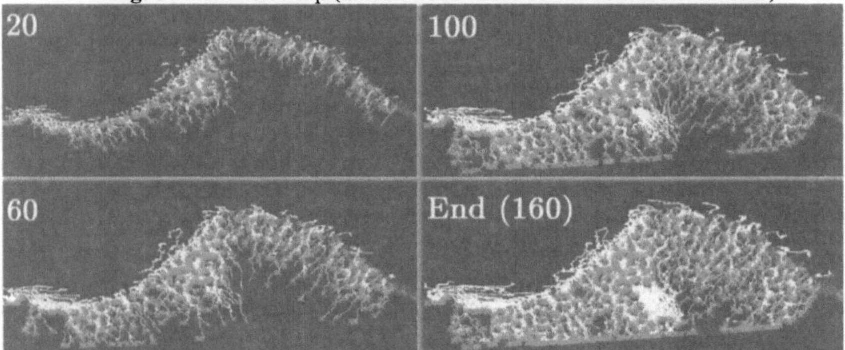

Fig. 6. IDARA progression as seen at the 20, 60, 100, and 160 iteration points

For completely unknown environments a random or uniform exploration strategy provides the most efficient method for exploration. However, when priors are available (and can be encoded in the distribution of the antigens) the "directed randomness" of the IDARA method satisfies user's goals while maintaining global exploration at the cost of reduced efficiency. For example, in the case of the mountain rubble scan the search was 37% efficient; that is, over a third of the motions made by the robots resulted in new information being collected.

6 Conclusions

The results showed that the IDARA framework is a promising technique for coordinating large populations of heterogeneous robots in highly unstable environments. Using the immune system analogues of a specificity response ladder and clonal expansion as a guide, the IDARA coordination architecture was developed for kilorobotic search and rescue activities. The results of the simulation were as hypothesized and show that the IDARA methods were able to efficiently coordinate 1,500 robots in a manner that balanced new and repeated exploration actions such that 37% of actions (on average) were investigating uncharted terrain. IDARA's ability to perform searches in noisy, non-uniform environments was demonstrated by its successful operations (i.e., finding the goal, such as a victim in distress) in purposely noisy conditions. Conventional wisdom gives that a random motion strategy is best exploration environments with large populations. IDARA demonstrates that when priors are available (and can be encoded in the distribution of the antigens) "directed randomness" will provide am intelligent exploration strategy that can efficiently take advantage of this information without becoming dependent on its quality or source.

The IDARA system builds upon immunology models and other related concepts and results in a directed, but flexible, system that mimics that nature of the immune system's control structure [8]. In conclusion, the IDARA method will allow Kilorobotics to be able to more fully exploit the comparative advantages inherent in autonomous multi-robot systems, namely: parallel execution, redundant operations, increased reliability, and robustness to noise.

7 References

1. D. Gage. (2001) Private communications on the term "kilorobot."
2. S. Forrest and S. Hofmeyr. (1999) "Immunity by Design: An Artificial Immune System." *Proc. of the Genetic and Evolutionary Computation Conference (GECCO)*, 1289-1296.
3. N. K. Jerne (1973) "The Immune System," *Scientific American*, **259:52-60**.
4. W. Guyton, J. Hall. (1996) *Textbook of Medical Physiology*. Ninth edition. W.B. Saunders Co., New York.
5. D. Dasgupta. (1999) "An Overview of Artificial Immune Systems and Their Applications," *Artificial Immune Systems and Their Applications*, D. Dasgupta, Ed., 3-21.
6. G. Whitsides, J. C. Love. (1997) "The Art of Building Small." *Scientific American*, **270:39-47**.
7. R.C. Arkin, T. Balch. (1997) "AuRA: Principles And Practice In Review." *J. of Exp. & Theo. AI*, **9:175-188**.
8. S. Singh, S. Thayer. (2001) "Immunology Directed Methods for Distributed Robotics: A Novel, Immunity-Based Architecture for Robust Control & Coordination," *SPIE: Mobile Robots XVI*, **v.** 4573.
9. M. J. Mataric. (1995), "Issues and Approaches in the Design of Collective Autonomous Agents," *Robotics and Autonomous Systems*, **16:321-331**.
10. R. Murphy, J. Casper, M. Micire, J. Hyams. (2000) "Mixed-initiative Control of Multiple Heterogeneous Robots for USAR." *IEEE Trans. on Robotics and Automation*.
11. G. Dedoglu, G. Sukhatme, "Landmark-based Matching Algorithm for Cooperative Mapping by Autonomous Robots," *Distributed Autonomous Robotic System 4*, Eds: L. Parker, G. Beckey, and J.Barhen, 251-260.
12. Kobayshi, K. Nakamura. (1983) "Rescue Robot for Fire Hazards," *Proc. Int. Conf. on Advanced Robotics*,91-8.
13. J.G. Blitch. (1996) "Artificial Intelligence Technologies for Robot Assisted Urban Search And Rescue," *Expert Systems with Application*, **11:2**, 109-124.
14. G. Caprari, P. Balmer, R. Piguet, R. Siegwart. (1998) "The autonomous micro robot 'Alice': A platform for scientific and commercial applications," *Proc. 1998 Int. Symp. on Micromechatronics & Human Sci.*, 231-5.
15. D. Dasgupta, N. Attoh-Okine. (1997) "Immunity-based Systems: A Survey," *Proc. IEEE Int. Conf. on Systems, Man, and Cybernetics*, 1:369-374.
16. Ishiguro, T. Kondo, Y. Shirai, Y. Uchikawa. (1996) "Immunoid: An Architecture for Behavior Arbitration Based on the Immune Networks." *Proc. 1996 IEEE/RSJ Inter. Conf. on Intel. Robots and Systems*, 3:1730-8.
17. Lee, D., Jun, H., et al. (1999) "Artificial Immune System for Realization of Cooperative Strategies and Group Behavior in Collective Autonomous Mobile Robots." *Proc. Fourth Int. Symp. on Art. Life and Robotics (AROB)*, 120-6.
18. C. Merkwirth, U. Parlitz, W. Lauterborn (2001), *TSTOOL User Manual*, 13-18.

Universal Distributed Brain for Mobile Multi-Robot Systems

Peter Sapaty [1,2], Masanori Sugisaka [1]

[1] Department of Electrical and Electronic Engineering, Oita University, Dannoharu 700, Oita 870-1192, Japan, +81-97-554-7831, +81-97-554-7841 (fax)
[2] Institute of Mathematical Machines and Systems, National Academy of Sciences, Glushkova Ave. 42, 03187 Kiev, Ukraine
{msugi, p-sapaty}@cc.oita-u.ac.jp

Abstract. The paper describes a new concept for the creation of a universal distributed brain for mobile multi-robot systems. This brain spatially interprets a special high-level language in which mission scenarios can be efficiently formulated and implemented, with cooperative work of robots being a derivative of the parallel interpretation process. Due to universal nature of the scenario language proposed, which navigates in a unity of physical and virtual worlds and operates with both information and physical matter, the approach may form a new basis for the development and massive production of advanced distributed multi-robot systems.

Keywords: multi-robot systems, distributed brain, scenario language, spatial navigation, parallel interpretation.

1 Introduction

Theoretically, many jobs in dynamic environments can be performed by teams of robots better than by individual robots [1]. And technologically, any number of sophisticated mobile robots can be produced by industry today. But we are still far away from any suitable team solutions that could allow us to use this hardware efficiently, as the teamwork is a complex and insufficiently studied phenomenon, with no universal results proposed so far. A great variety of works in this area [2,3] pursue quite different models, many of them stemming from biology. We do believe that training robots like animals or humans, and especially their teams, to behave reasonably in general situations is extremely difficult, and may not lead to trustworthy practical solutions in the near future.

A considerable increase in the robot's individual and group intelligence at the current stage may be achieved by lifting the level of language in which robots are programmed and tasked. This can be used subsequently as a qualitatively new platform for the creation of advanced robotic systems, which may integrate different existing control models within the same approach or may be based on radically new, higher level, organizational models. We already have such precedents in the history. Only with the invention and introduction of high-level programming languages, like FORTRAN or LISP, real and massive use, as well as production, of computers began.

We pursue a similar approach for multi-robot systems, considering them not as a collection of intelligent individuals with a predetermined functionality, but rather as a parallel machine capable of executing any mission tasks written in a special high-level scenario language [4]. This language, operating with both information and physical matter, and navigating in a unity of physical and virtual worlds, can express semantics of spatial problems directly. Its automatic implementation, which may be environment-dependent and emergent, is delegated to the distributed brain, as shown in Fig. 1.

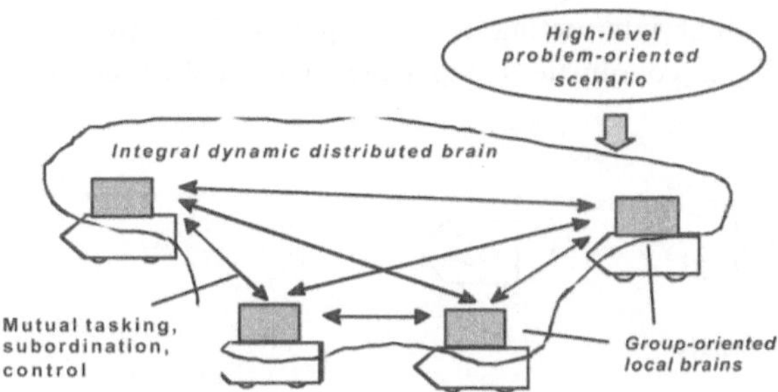

Fig. 1. Group of robots as a parallel distributed machine

The paper describes the latest version of such a language, WAVE-WP (or World Processing), being a further evolution of WAVE [5], with elements of programming in it. It also provides basics of the language spatial interpretation by the distributed artificial brain, and the main organization of the brain. Examples of high-level mission scenarios written in WAVE-WP are given. Application areas outlined.

2 WAVE-WP Key Features

WAVE-WP sets up multiple cooperative activities, or *moves*, in physical, virtual, or combined space. A move can perform certain actions in the place it started, on data at that place, and can also change the current position in space.

General structure. Moves may be applied in a *sequence*, separated from each other by a semicolon, as follows:

```
move1; move2; move3
```

where each new move starts from the place reached by the previous move. Moves may also develop *independently* and *in parallel*, if separated by a comma:

```
move1, move2, move3
```

Composition of moves may be *combined*, such as:

```
move1, move2; move3, move4
```

where moves 3 and 4 will start in parallel from all places in space reached independently by moves 1 and 2. Using parentheses allows for any parallel-sequential composition:

```
(move1; move2), (move3, move4; move5); move6
```

Types of moves. Moves may be of different types, representing various activities in distributed spaces. They may be *hops in physical space*, between points set up by coordinates, as in the following example (# being a hop act):

```
# x1 y1; # x2 y2; # x3 y3
```

This sequence may cause the robot interpreting it move physically between the established points, as shown in Fig. 2 (coordinates being depicted inside the robot box).

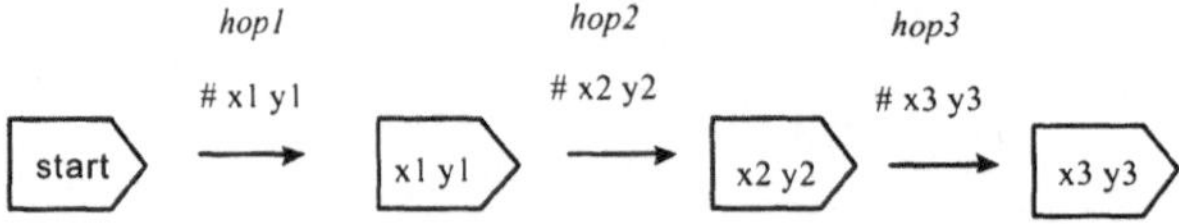

Fig. 2. Robot's movement in physical space

Moves may be *hops in virtual space*, where reaching new points, or information nodes, may either be by direct hops to these nodes, or by traversing existing semantic links between the nodes, as in the following example:

```
direct # a; p # b; q # c
```

with link and node names separated by the hop act. During execution of this program, mobile program agents carrying operations to further nodes will be automatically created and passed (with accompanying data) between robots, if subsequent nodes are located in other robots, as shown in Fig. 3 (skipping the worked parts).

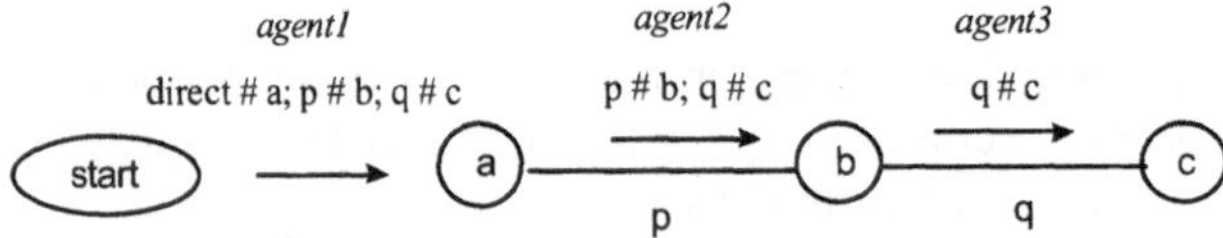

Fig. 3. Movement of mobile agents in virtual space

If adjacent nodes in the sequence to be passed are located in the same robots, the transition will be made within computer memory only, without materializing mobile agents. Moves may also be *data processing & assignment*, such as:

```
N1 = N2 + N3
```

Operating with remote data. Computation & assignment is performed in one place, if all data operands needed for it are associated with this place. If they are located elsewhere, first hops in space may be needed to get the data, as well as to store the result, so the previous move may look like follows:

```
(# x1 y1; N1) = (p #; N2) + (q #; N3)
```

where to get contents of `N2` and `N3`, first hops through links `p` and `q` should be made in the virtual world from the current node. Then the obtained sum should be assigned to variable `N1` set up in a (remote) place of physical space with coordinates `x1y1`. To form the latter node, the current robot must move physically in space or delegate this hop to another robot, passing to it the operation and obtained result first.

Control rules. Programs in WAVE-WP (or *waves*) may be covered by *rules* establishing proper constraints and contexts extending distributed functionality of waves, also allowing the language to be used as a conventional one. The following program (using one of *forward* rules) activates moves 1, 2, and 3 in parallel (which may be arbitrary waves themselves, due to recursion), waiting for termination of all of them.

```
andparallel(move1, move2, move3); move4
```

If all the three moves terminate with success, `move4` will start from all places in space (which may be remote), reached finally by the previous moves.

The next program (using one of *echo* rules) finds and returns minimum of all data items reached or produced by the three moves, which may be local or remote. This may be accomplished in distributed and parallel manner.

```
N = min(move1; move2; move3)
```

A forward rule `create` in the following example:

```
create(direct # a; p # b; (q # c; r ## a), s # d)
```

establishes a context within which the embraced wave, serving as a template consisting of sequential-parallel composition of five hops in virtual space, creates the network topology shown in Fig. 4 (## reflecting a hop to the already existing node, with only a new link created).

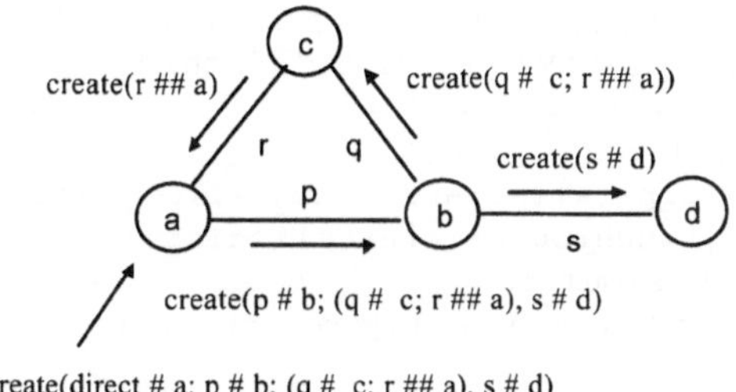

Fig. 4. Creating a distributed virtual world within rule `create`

Stepwise movement of the code inheriting `create` rule is shown at the new links formed within this process. Depending on system resources and additional directives, nodes of this network may be placed in the same robot or may be distributed between robots, and links may happen to connect nodes located in different robots.

Spatial variables. *Variables* (called *spatial*, as they may be scattered throughout the world in WAVE-WP) may be of different types. They may be associated with places and shared by any waves entering these places (nodes), or may be (remotely) heritable and shared by offspring processes only. Others may propagate in space with the program activities as their sole property. Special variables may also access both external and internal environment. Contents of variables may be both information and physical matter (or physical objects).

In the following program, variable F (mobile, or *frontal*) propagates with wave through the virtual network, supporting the lifting and summing of local, or *nodal*, variables N, N1, and N2, respectively, in nodes a, b, and c. The received sum is issued outside the system using *environmental* variable USER in the finally reached node c.

```
direct # a; F = N; p # b; F += N1; q # c; USER = F + N2
```

3 WAVE-WP Language Description

WAVE-WP has simple syntax (suitable for direct interpretation), with the latest version shown below, where coordinated propagation in physical-virtual space is integrated with a collection, return and processing of data obtained in another space propagation.

wave → { *advance* ; }

advance → { *move* , }

move → *constant* | *variable* | {*move act* } | [*rule*] (*wave*)

variable → *nodal* | *heritable* | *frontal* | *environmental*

act → *control-act* | *fusion-act* [=]

rule → *forward-rule* | *echo-rule*

constant → ' {*character*} ' | **{** {*character*} **}** | " {*character*} " | *number* | *address* | *place* | *time* | *speed* | *node-kind* | *doer* | *state* | *special* | {*constant* _ }

nodal → N [*alphameric*]

heritable → H [*alphameric*]

frontal → F [*alphameric*]

environmental → CONTENT | ADDRESS | WHERE | KIND | BACK | LINK | TIME | SPEED | DOER | JOINT | USER | IDENTITY | RESOURCES

control-act → # | ## | ~ | ! ~ | == | ! = | < | <= | > | >= | = | ^ | !

fusion-act → + | - | * | / | ** | : | : : | **|** | % | & | _ | ?

forward-rule → sequence | or | orparallel | and | andparallel | random | repeat | wait | synchronize | create | indivisible | release | quit | simulate

echo-rule → state | rake | min | max | sort | sum | product | count | none

place → { *dimension number* }

dimension → x | y | z | dx | dy | dz | r

time → t *number* | dt *number*

speed → s *number* | ds *number*

node-kind → p | v | vp | pe | ve | vpe | e | ep

address → { *integer* . }

doer → { *alphameric* . }

state → abort | thru | done | fail

special → direct | [*sign*] any | infinite | nil

Words in italics represent syntactic categories, square brackets identify optional constructs, braces show zero or more repetitions of a construct with a delimiter at the right, and the vertical bar separates alternatives (the braces and the bar are shown in bold when used in the language). Others being the language symbols: semicolon allowing for sequential, while comma for parallel (or arbitrary order) invocation of parts, and parentheses are used for structuring of programs. Successive program parts, or *advances*, develop from all nodes of the set of nodes reached (SNR) by a previous advance, and parallel or independent parts, or *moves*, constituting advances, develop from the same starting nodes, adding their SNRs to the resultant SNR of the advance.

Moves can be of three types. First, they can point at a resulting content directly (as a *constant* or *variable*). Second, they can form space navigating & data processing *expressions* consisting of arbitrary moves separated by elementary operations (*acts*) to which the moves may return (local or remote) results. Acts can also assign final results to (local or remote) variables. Third, they can themselves be arbitrary *waves* (parenthesized), optionally prefixed by control *rules*. This simple recursive definition allows for a powerful and compact expression of arbitrary complex space navigation, data processing, and control operations, which can be carried out in fully distributed and parallel mode. There are no type definitions in the language, and any data is *self-identifiable*, being generally represented as a string. Strings in single quotes stand for information, whereas for physical matter or objects double quotes are used. Strings to be potentially used as programs may also be put in braces, which will automatically trigger their parsing and optimization before assignment.

For certain types of information objets (like numbers, node addresses, physical coordinates, time, speed, states, etc.) single quotes may be omitted. *Alphameric* means a sequence of letters or digits. Absolute coordinates are prefixed by `x`, `y`, and `z` (`dx`, `dy`, and `dz` are used for coordinate shifts), and `r` identifies radius of the region with a proper center. Prefixes `t` (`dt`) and `s` (`ds`) define time and speed. Node types (kinds) may reflect physical position (`p`), virtual node (`v`), or execution device (`e`), here robot; the types may be combined: `vp`, `pe`, `ve`, `vpe`, or `ep`. Many elementary acts and rules are inherited from the previous WAVE versions, and more details can be found in [5]. Same concerns the distributed interpretation of the language based on dynamically created spatial tracks, which generalize and process states, channel data flow, support spatial variables, and forward further wave code, which is discussed in detail in [5].

The previous language version was successfully tested via the Internet for distributed databases, network management, distributed simulation of dynamic systems (like battlefields), and collective behavior of mobile agents [5].

4 The Distributed Artificial Brain

Communicating copies of the WAVE-WP interpreter should be installed in mobile components of the distributed robotic system. During the language execution by the network of interpreters, the latter can partition and modify wave programs at runtime, exchanging data, mobile program agents, or just control signals; the interpreters can also force robots hosting them move, perform physical jobs, and pass physical matter (objects) to each other.

The communicating interpreters can be integrated with other, traditional, control and data processing functional units [6,7] in each mobile robot, as shown in Fig. 5 (with M standing for motor control, V for vision, and N for neurocomputers), forming altogether *a distributed artificial brain* of the mobile multi-robot system.

This brain has a dynamic topology, as robots can move at runtime, changing both electronic and mechanical accessibility to each other. Together with the robot's hardware, this forms a spatial machine that can process both information and physical matter, can move in physical space accessing and changing physical environment, and also convert matter to information and vice versa.

This dynamic machine is self-organized and self-controlled by the use of a spatial tracking system accompanying and supporting the distributed interpretation of WAVE-WP. This tracking system allows us to assess distributed situations with making autonomous decisions, collect remote data for processing at proper physical or virtual points, determine the lifetime for multiple spatial variables, and transfer further wave programs to positions in space where success of the previous actions has been achieved. More on the runtime tracking mechanisms may be found in [5].

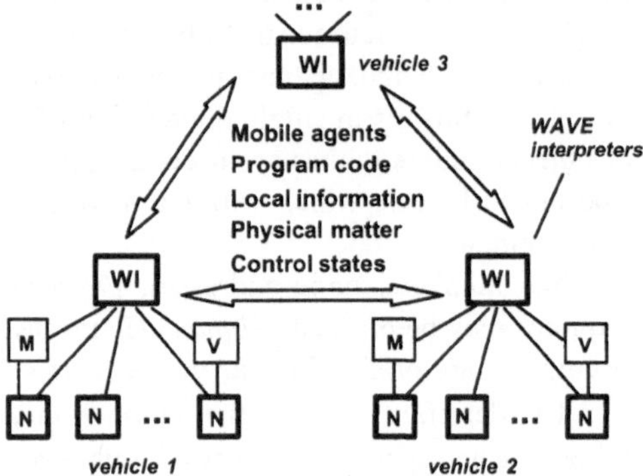

Fig. 5. Distributed brain based on the WAVE-WP interpreter

5 Examples of Mission Scenarios in WAVE-WP

Let us assume that the task is a delivery of some volume of a substance from a starting point to a target, with processing it at the latter. The substance may, for example, be fire-extinguishing chemicals, and the destination may reflect the peak of intensity of a forest fire. Different levels of a solution of this task can be possible.

General solution. For the most general case, the program will be as follows:

```
direct # x-2.3 y0.7; Fload = "60 kg chemicals";
direct # x 0.9 y0.5; Fload ? apply
```

In this program, a hop in space will be performed with the given coordinates –2.3 and 0.7 for the start, and 0.9 and 0.5 for the target. The 60 kg substance is lifted at the start and assigned to mobile variable `Fload`, which will be subsequently processed by the procedure `apply` at the destination (`?` is the procedure invocation act in the language).

Splitting into sub-tasks. Imagine that the delivery of the physical matter may not be possible by a usual transportation on the whole way from the start to the target. Say, the area around the target (as the center of fire) cannot be passed by a vehicle (ground or aerial), and for the last part the matter should use some extra means, say, a shell to be fired at some distance from the target. As there may be limitations on the shell's weight, the whole matter should be split into portions, and put into a number of shells. Also, for improving an impact on the target, the shells should be fired from different positions, being delivered to the latter independently from the starting point, as shown in Fig. 6, and the firing of shells should be synchronized in time.

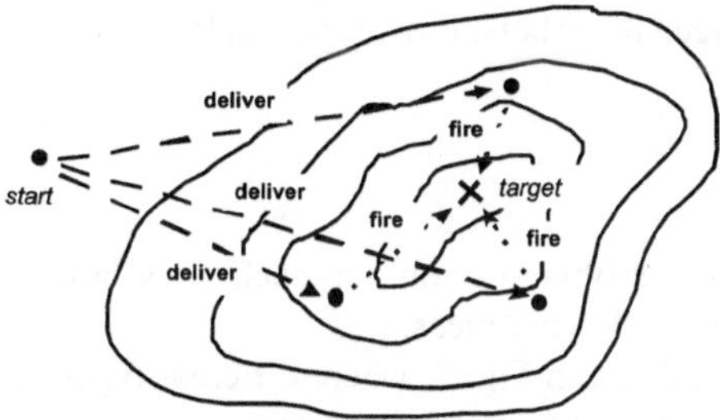

Fig. 6. Splitting into parallel sub-tasks, with independent firing

The program for this scenario, with the chemicals put into 20 kg shells, assigned to the replicating mobile variable `Fload` for a delivery into the three firing positions with certain coordinates, will be as follows:

```
direct#x-2.3y0.7; Fload="20 kg shell";
synchronize(direct#(x1.2y1.2,x-0.1y-0.3,x1.4y-0.3));
(Fload, x0.9 y0.5) ? fire
```

The shells will be fired synchronously (using rule `synchronize`) from these points by procedure `fire` in each of them, with the same target coordinates as a parameter.

Setting specific routes. To add further details to this scenario, let us consider the reaching of each firing point via a sequence of some intermediate way-points before the synchronized firing. The program creating three separate repeating space-propagation branches (using rule `repeat`), with the route coordinates kept in variable `Fpoints` (individually for each branch), will be as follows:

```
direct#x-2.3y0.7; Fload="20 kg shell";
synchronize(
  Fpoints=(x-1.2y1.0,x0.1y1.2,x1.2y1.1),
  Fpoints=(x-1.0y0.1,x-0.1y-0.3),
  Fpoints=(x-1.5y-0.5,x1.4y-0.3);
  repeat(Fpoints!=nil; direct#Fpoints:1; Fpoints:1=nil));
(Fload,x0.9y0.5)?fire
```

where `:` is the indexing act, and used elements are being removed from their sequence. The firing is synchronized upon reaching the last way-points of each route.

Runtime assignment of robots to the scenarios. The scenarios above are written as integral programs reflecting semantics of the tasks to be solved, rather than directives to particular robots. This allows us to abstract from implementation details and concentrate on the mission goals, strategy, and tactics. Assignment of robots to these scenarios will be done automatically. The scenarios can be executed by different numbers of robots, and the failed robots can be substituted at runtime by other robots automatically.

However if needed, the mission scenarios can also be written on the level of cooperating robots, as usual. Such a scenario in WAVE-WP describing the detailed movement of a synchronized column of three robots to the firing positions, with the control infrastructure preliminary set up between the robots, is discussed in [8].

6 Multi-Robot Applications

Multi-robot systems, based on the proposed distributed brain approach, may have a variety of important applications in both civil and defense areas.

A hypothetical robotized hospital [4] is depicted in Fig. 7, where different types of mobile robots, like cleaning (C), life support (L), body state checking (S), may serve patients simultaneously and cooperatively. The robots can autonomously move between patients, avoiding obstacles and each other, and accessing a store of drugs and a litter box. Both scheduled, regular, and emergency scenarios are possible. State checking robots may regularly visit patients and measure body temperature, blood pressure or heart beat. Cleaning robots may periodically search the territory, collecting litter and discharging it into the litter box. Life support robots may deliver prescribed medicine picked up from the store to patients on a schedule. Emergency scenarios may originate from both patients asking for unscheduled assistance, and from robots themselves (say, the state checking ones) discovering non-standard situations.

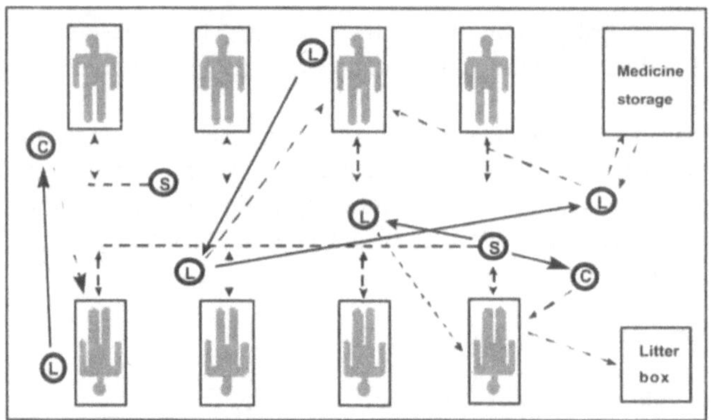

Fig. 7. Multi-robot hospital scenario

Scheduled and emergency scenarios, engaging different robots simultaneously, may include their cooperative behavior, with some patterns reflected by arrows in Fig. 7. Life support or state checking robots may, for example, call cleaning robots to collect accidental waste from patients. Life support robots may be called by both state checking robots and patients, as well as by other life support robots. More than one robot may serve the same patient at the same moment of time. Such scenarios are being successfully programmed in WAVE-WP.

Other applications of the distributed brain approach described may include sentry duties, street cleaning, harvesting, de-mining territories, investigation of other planets or seabed, robotic fishery, future combat systems, international crisis reaction forces, etc., with the use of teams or even armies of cheap mobile devices primarily designed for collective work.

7 Conclusions

We have proposed a new concept for the design of a universal distributed brain for mobile autonomous multi-robot systems, with the key features being as follows.

- The brains of individual robots form altogether a highly integral distributed brain with a unified communication, coordination, and command and control.
- This distributed brain interprets autonomously and in parallel a special high-level mission scenario language WAVE-WP, in which the system plans, goals, strategies, and tactics are effectively formulated and implemented.
- The scenario language has a universal nature, working with both information and physical matter, and covering within the same space-navigating formalism different layers of control of single and multiple robots, making any distributed system solutions integral and seamless.

High-level mission scenarios for multi-robot systems written in WAVE-WP are up to a hundred times shorter than in C or Java, relieving the user from a multitude of usual coordination, synchronization, and data exchange routines. These are shifted now to a more efficient automatic implementation layer, where individual robots are engaged only when and where they are really needed or available. Starting the design of multi-robots from a high-level scenario language expressing overall mission goals, planning and control, and the distributed brain directly executing this language, may allow us to create efficient distributed mobile systems exhibiting highest possible integrity and self-recoverability. This may be hard to achieve by linking robots designed for single use.

References

1. D. W. Gage. (1993) How to Communicate with Zillions of Robots. Proc. SPIE Mobile Robots VIII, Boston, Sept. 9-10.
2. Y. U. Cao, A. S. Fukunaga, A. B. Kahng. (1997) Cooperative Mobile Robotics: Antecedents and Directions. Autonomous Robots 4 (1), 7-27.
3. L. E. Parker. (2000) Current State of the Art in Distributed Robot Systems. Distributed Autonomous Robotic Systems 4, L. E. Parker, G. Bekey, J. Barhen (eds.). Springer, 3-12.
4. P. Sapaty, M. Sugisaka. (2001) Distributed Artificial Brain for Collectively Behaving Mobile Robots. Proc. Symposium & Exhibition Unmanned Systems 2001, Jul. 31-Aug. 2, Baltimore, MD, 18 p.
5. P. S. Sapaty. (1999) Mobile Processing in Distributed and Open Environments. John Wiley & Sons, ISBN: 0471195723, New York, 416p.
6. M. Sugisaka. (1999) Design of an Artificial Brain for Robots. Artificial Life and Robotics 3, 7-14.
7. M. Sugisaka, N. Tonoya, T. Furuta. (1998) Neural Networks for Control in an Artificial Brain of a Recognition and Tracking System. Artificial Life and Robotics 2,119-122.
8. P. Sapaty, M. Sugisaka. (2001) Towards the Distributed Brain for Collectively Behaving Robots. Proc. International Conference on Control, Automation and Systems, ICCAS 2001, Oct. 17-21, Cheju National University, Jeju Island, Korea, 571-574.

Real-time Control of Walking of Insect; Self-organization of the Constraints and Walking Patterns

Masafumi YANO, Shinpei HIBIYA, Makoto TOKIWA, and Yoshinari MAKINO

Research Institute of Electrical Communication, Tohoku University 2-1-1 Katahira Aoba-Ku, Sendai, 980-8577, Japan
email: masafumi@riec.tohoku.ac.jp

Abstract. It is one of the goals of robotics to realize the autonomous robot that can cope with the unpredictably and dynamically changing environment. In order to attain the purpose under the unpredictably changing environment, a robot is usually required to solve the inverse problem. Since the imposed purpose on the system takes first priority, the system inevitably adapts to the unpredictably changing environment to attain the purpose. When the robot attains its purpose, several functions such as keeping its walking velocity, reaching its destination, keeping its proper posture etc., should be well coordinated. To coordinate several functions, the proper constraints should be self-organized from the purposes and the current states of the robot. In addition to the constraints, it is necessary some rules to fulfill these constraints. Here we propose a new real-time control mechanism to solve the inverse problem under the unpredictably changing environment.

Kewords: ill-posed problem, constraints, emergence, gait pattern, polymorphic circuit

1 Introduction

The motion control systems of animals seem to autonomously create appropriate information depending on the purposes self-organized in the system under the unpredictably changing environment. The motor systems of the animals are generally controlled through three subregions in a hierarchical way, the brain, the central pattern generator and the effector organs. The flexibility of the movements is generated by the neural network as a control system, indicating that they can organize the dynamical patterns quickly in response to the changes of the environment. To coordinate the movements of the muscles in response to the unpredictably changing environments, the control system should be indefinite. Indefinite system means that the properties of the elements of the system and the relationship of them are not specified in advance. If the control system is definite, it is impossible to adapt to the unpredictably changing environment. So in the case of dynamical system, the parameters of the equations are determined by the system itself to attain the purpose under unpredictably changing environment. The typical example of the indefinite control system can be seen in central pattern generators termed the polymorphic circuits or multi-functional circuit [1]. Recently, we have demonstrated that the polymorphic circuits can generate various spatio-temporal patterns using a hard-wired model [2]. But the indefinite control system is only one of necessary conditions. To attain the purpose, the proper constraints should be self-organized and fulfilled by the system itself in response to the changes of the purpose and the current environment. In case of straight walking of insect robot, we have already proposed a new method to solve the ill-posed problem. In this case, the purpose is to walk straightly with a constant velocity. So the constraints to attain the purpose are given as the velocities of the strokes of the legs of the two sides. Under unpredictably changing environment, the system requires some rule to satisfy the constraints, then walking patterns of the animals should be emerged as the results of the coordination of the movements of the leg muscles. The constraints on the robot should be contented by optimally integrating each objective function of the elements through competition and cooperation among them. The objective function is derived from the energetics of muscle contraction, in which muscle has an optimal shortening velocity to provide the highest efficiency of the energy conversion. So we introduce "the least dissatisfaction for the greatest number of the elements" rule to generate the walking patterns. This rule is quite similar to the Pareto optimum in the economics and brings forth the cooperation and/or competition among leg movements, resulting in emerging the most efficient walking pattern [3][4].

However, when more complex purposes, such as reaching a destination with a required walking velocity, are imposed on the robot, the constraints to solve the inverse problem cannot be uniquely derived from the purposes. In this case, it is necessary some rules to determine the constraints to coordinate the current state of the system and the purposes. In reaching problem, the constraints required in the motor system are the stroke velocities of legs of the two sides. The constraints are self-organized in the brain by judging whether the current constraints are suitable or not. In this paper, we extend our method to control the motor systems when more complex purpose is imposed on the system under an indefinite environment.

2 Model

Characteristics of animal walking

The motion control of animals are achieved in a hierarchical way of the system composed of higher center of brain, central pattern generator and effector organs. The walking patterns are quickly changed depending on the walking velocities and load [5]-[10]. In the case of stick insect, at high speed the front leg and the hind move simultaneously and the middle antiphasic to the others, forming a tripod to support their body. On the contrary, when they walk slowly, the three legs of each side move metachronally. A pair of legs of the same segment step alternately. As increasing the walking velocity, the insect changes the patterns critically depending on their velocity, resembled to a phase transition. The walking patterns also vary with the load [5][9][10]. In the case of horse, energy consumption during walking does not depend on the walking distance, but almost on the distance.

Higher center; brain

The higher center processes the information from environment, and sends constraints to coordinate the purpose with the environment. When insect walks, a motion planning such as walking velocity and direction is expressed by a small number of degrees of freedom (constraints), which is first made at higher center such as brain. In this case, the constraints are expressed by the velocities of the leg stroke of the two sides. The output made in brain according to this motion plan is sent to the three thoracic ganglions.

Central pattern generator

The walking of the insect is controlled by the three thoracic ganglions, prothoracic, mesothoracic and methathoracic ganglions [11]. These ganglions send motor outputs to control leg muscles and receive the external afferents as to position, load and force of each muscle. These ganglions are internally connected each other through a pair of thoracic connectives. It has been clarified that the well coordinated motion among the legs is organized not only by the neural system composed of the three ganglion, connected through the inter-segmental connectives, but by the mechanical interaction through the movements of legs. Central pattern generators (CPGs) are networks of neurons to control the motor system generating spatio-temporal pattern of neural activities. In this paper, we construct a coupled nonlinear-oscillator system as the polymorphic network, which can produce various walking pattern by modulating the properties of the composing neurons.

Inter-segmental inter-neurons in a thoracic ganglion of locust have been extensively investigated by Laurent and Burrows [12][13]. We adopt fundamentally their results as schematically shown in Fig. 1. In thoracic ganglion, this signal is transformed into rhythmic wave by rhythmic neuron corresponding to a spiking inter-neuron in the ganglion. The rhythmic neuron makes direct synaptic connection with nonspiking inter-neuron (NS neuron), which is great important to integrate the information on the states of muscles and inter-segmental pathway. NS neuron transforms the output of the rhythmic neurons to send the motor neuron. Inter-segmental connections between rhythmic neurons in neighboring CPGs are inhibitive, which produce asynchronous oscillation between neighboring rhythmic neurons. The frequency of the rhythmic neuron determines the walking velocity and the inhibitory interaction among them enables to appear any phase relationship among the movement of legs. In this sense, the rhythmic neuron is a kind of command neuron that receives the information of walking velocity, that is, purpose of

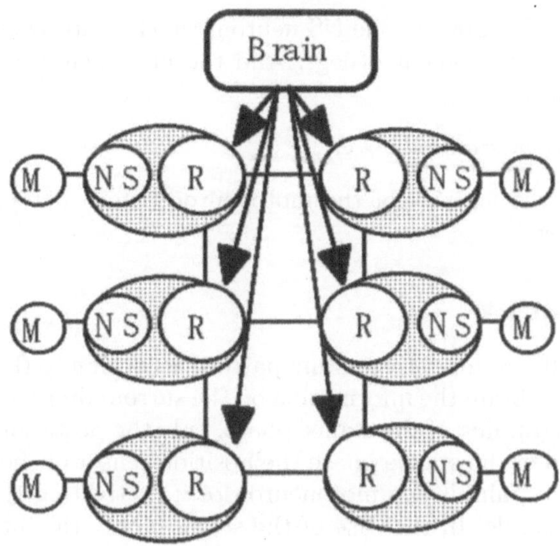

Fig. 1. Inter-segmental connection among CPGs

the animal created in the brain. The spatio-temporal patterns of the movement of legs are emerged by integration of the dynamical information of the effector organs in the NS neurons under the constraint of the purpose.

The equations of rhythmic neuron model are given by

$$\begin{aligned}
\frac{dx_{Ri}}{dt} &= -y_{Ri} - f_R(x_{Ri}) - \sum_j \alpha_{Rij}(x_{Rj} - x_{Ri}) + \beta_{NSi}x_{NSi}, \\
\frac{dy_{Ri}}{dt} &= g_R(x_{Ri}) + D_{Ri}, \\
f_R(x) &= (A_{R1}x^2 + B_{R1}x + C_{R1})x, \\
g_R(x) &= (A_{R2}x^2 + B_{R2}x + C_{R2})x, \qquad (1)
\end{aligned}$$

where x denotes voltage of neuron and D_R is the input to the rhythmic neuron, which determines the frequency of the oscillation.

The NS neurons is given by

$$\begin{aligned}
\frac{dx_{NSi}}{dt} &= -y_{NSi} - f_{NS}(x_{NSi}) + \beta_{Ri}x_{Ri}, \\
\frac{dy_{NSi}}{dt} &= g_{NS}(x_{NSi}) + D_{NSi}, \\
f_{NS}(x) &= (A_{NS1}x^2 + B_{NS1}x + C_{NS1})x, \\
g_{NS}(x) &= (A_{NS2}x^2 + B_{NS2}x + C_{NS2})x, \qquad (2)
\end{aligned}$$

where D_{NS} is the input to the NS neuron, which controls the phase relationship among the movement of legs. And the motoneuron is governed by the following equation

$$x_{mi} = G_{mi} \times sigmoid(x_{NSi}), \tag{3}$$

where x_{mi} is the activity of the motoneuron, which determines the motive force of the leg.

Sensory feedback

In order to self-organize the walking pattern according to the circumstance, it is necessary to obtain the information on the surroundings and the state of the legs. At the beginning of the stance phase, only the posterior muscle shortens, but at the end of the stance phase the position sensor of the posterior muscle should strongly inhibit the motoneuron of it, activating the motoneuron of the anterior muscle. In the case of the swing phase, the interaction between the pair of muscles should be reversed. These interactions can be presented by the direct synaptic connection of the position sensor of each muscle with the motoneurons and by the feedback to the connectives between the nonspiking neuron and the motoneuron as shown in Fig. 1. The hind leg moves antiphasic to the middle, which also moves antiphasic to the front leg, although there is no strong coupling between the hind and the front legs. So, the information required to optimize the efficiency of energy conversion is given as follows; the feedback information from leg to motoneuron is

$$\Delta G_{mi} = k_{\eta}\left(\frac{\partial \eta_i}{\partial f_i} - \frac{\sum_{j \neq i} f_j \frac{\partial \eta_j}{\partial f_j}}{\sum_i f_i}\right). \tag{4}$$

It means that the legs moved synchronously tend to share the load equivalently, where η denote the efficiency curve of the energy conversion of muscle. Each leg requires working more efficiently, so the feedback to NS neuron is

$$\Delta D_{NSi} = k_{D_{NSi}} \frac{\sum_j \frac{\int_0^T f_i \frac{\partial \eta_i}{\partial f_i} dt}{T}}{\sum_i \frac{\int_0^T f_i dt}{T}}. \tag{5}$$

This feedback information determines the degree of the synchronization among the legs.

Turning motion

One of the aims of biological motion is to reach its destination. If more complex purpose such as reaching a destination with a required velocity is imposed on the system, the arrival to the destination takes the priority over all other purposes. The velocity of stance phase of right and left side legs is tuned proportional to the angle between the axis of the body and the direction of the destination, which is feed backed until the second time derivative of the angle becomes zero. The velocities of the two sides are given by

$$V_{leg.req} = V_{body} \pm k_1 \times \theta \pm k_2 \times \ddot{\theta}. \tag{6}$$

If an estimated velocity does not satisfy the required angle, the strength of the feedback should be enhanced.

3 Results

Straight Walking

In case of straight walking, the required velocity is the only purpose of the robot, which is the strong constraint for our model system to attain at any required velocity and any load on the system. Our insect robot can fundamentally generate the two different walking patterns depending on the walking velocities and loads. The walking patterns are characterized by the phase relationship among the six legs, showing the walking pattern of metachronal gait. The phase relationship between the hind and the front drastically changes as the walking velocity increases. As increases the velocity, our robot shows that the front and the hind legs move simultaneously, called tripod gait as reported previously. The gait patterns of the straight walking are shown in Fig. 2.

Turning walk

In this model, the structure of leg is composed of only two muscles, flexor and extensor muscles, so the movements of legs are limited to move parallel to the axis of the body. When the angle between the axis of the body and the direction of the destination is large, the walking velocity should become slower and the gait pattern is metachronal. When is small, the insect can turn at higher velocity with a tripod gait. At intermediate angle, outer side legs and inner side legs take tripod and metachronal gait, respectively, as shown in Fig. 3.

4 Discussions

We have simulated an insect robot as an example that can generate appropriate walking patterns to walk efficiently. Since the walking pattern changes crucially depending on their walking velocities and loads, animals could generate a great number of diversities of walking patterns to adapt the unpredictable changes of their surroundings.

We have also showed that a new control mechanism installed in the insect robot, which can walk attaining more complex purposes of the system as

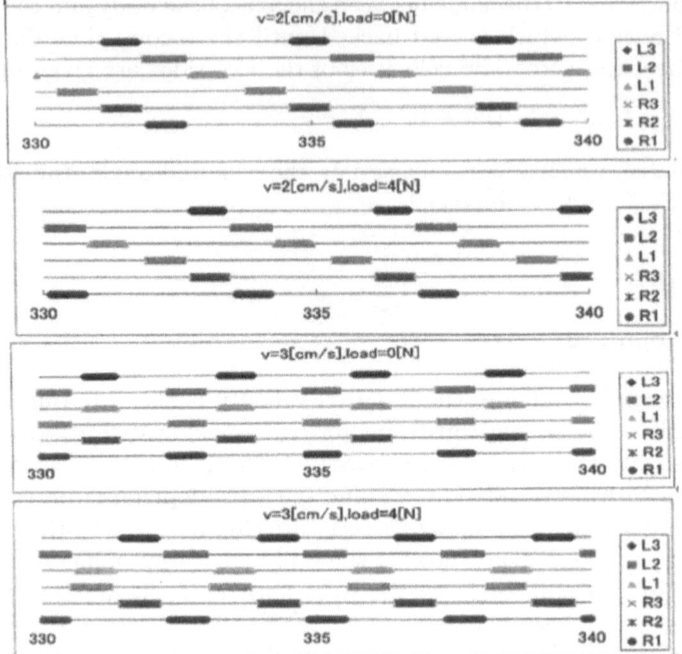

Fig. 2. Gait patterns of straight walk

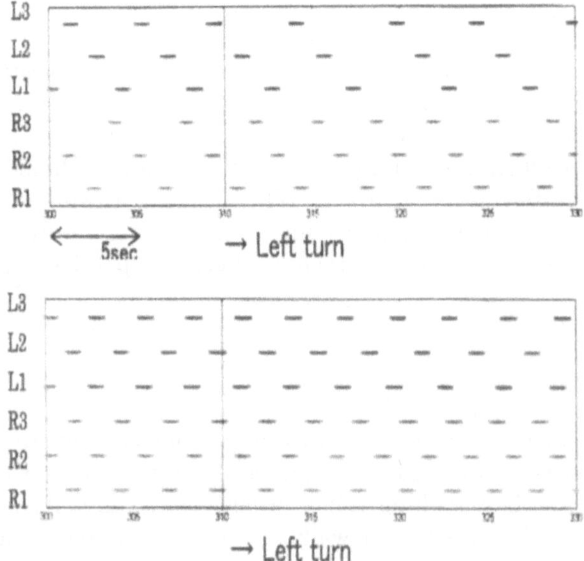

Fig. 3. Gait patterns of turning walk

possible as it can operate at higher efficiency of energy conversion under unpredictable changes of the environment. This control mechanism is derived from a metarule to determine the constraints on the motor system. In case of turning walk, the destination takes the priority over all other purposes. So the constraints are self-organized every moment depending on the current state of the system and the environment to attain the purpose. And the constraints may always fulfilled with more optimal efficiency. As the result the optimal trajectory and the walking patterns emerged.

References

1. Getting PA. and Dekin MS. (1985) Tritonia swimming: a model system for integration within rhythmic motor systems. In: Selverston AI(ed) Model neural networks and behavior. Plenum Press. New York, pp 3-20
2. Makino Y., Akiyama M. and Yano M., (2000). Emergent mechanisms in multiple pattern generations of the lobster pyloric network. Biol. Cybern. **82**, 443-454
3. Kimura S., Yano M., and Shimizu, H., (1993). A self-organizing model of walking patterns of insects. Biol. Cybern. **69** 183-193
4. Kimura S., Yano M., and Shimizu, H., (1994). A self-organizing model of walking patterns of insects II. The loading effect and leg amputation. Biol. Cybern. **70** 505-512
5. Peason, K.G., (1972). Central programming and reflex control of walking on the cockroach. J.Exp.Biol. **56**:173-193
6. Peason, K.G, (1976). The control of walking. Sci. Am. **235**, 72-86
7. Graham, D., (1979a). The effects of circumo-esophageal lesion on he behavior of the stick insect Carausius morosus. I. Cyclic behavior patterns. Biol. Cybern. **32**:139-145
8. Graham, D., (1979b). The effects of circumo-esophageal lesion on the behavior of the stick insect Carausius morosus. II. Change in walking coordination. Biol. Cybern. **32**,147-152
9. Foth E. and Graham D. (1983a) Influence of loading parallel to the body axis on the walking coordination of an insect. I. Ipsilateral effects. Biol. Cybern. **47**:17-23
10. Foth E. and Graham D. (1983b) Influence of loading parallel to the body axis on the walking coordination of an insect. II.Contralateral effects. Biol. Cybern. **48**:149-157
11. Dean, J., (1989). Leg coordination in the stick insect Carausius morosus J. Exp. Biol. **145**, 103-131
12. Laurent, G., and Burrows, M., (1989a). Distribution of intersegmental inputs to nonspiking local interneurons and motor neurons in the locust. J. Neurosci. **8**, 3019-3029
13. Laurent, G., and Burrows, M., (1989b). Distribution of intersegmental inputs to nonspiking local interneurons and motor neurons in the locust. J. Neurosci. **8**, 3030-3039

Chapter 12
Learning in Distributed Robotic Systems

Using Interaction-Based Learning to Construct an Adaptive and Fault-Tolerant Multi-Link Floating Robot

Wenwei Yu[1], Ishioka Takuya[1], Daisuke Iijima[2], Hiroshi Yokoi[1], and Yukinori Kakazu[1]

[1]Hokkaido University, 060-8628 N13 W9 Sapporo, JAPAN
[2]Nango Corporation, Tokyo, JAPAN

Abstract. How to build distributed autonomous systems that can adaptively behave through learning in the real world is still an open problem in the research field. In order to tackle this problem, we constructed a distributed autonomous floating robot that consisted of mechanically linked multiple identical units and proposed a new control framework, adaptive oscillator method, to deal with units' temporal and spatial interaction with their environment. A single model reinforcement learning system was first employed to realize the framework, and a multiple-model reinforcement learning system was proposed further and employed to cope with environmental changes caused by adding obstacles. In order to confirm adaptive behavior acquisition and fault-tolerance in a real environment, we did experiments on target approaching task by using the real floating robot.

1 Introduction

Much attention has recently been focused on distributed autonomous systems (DASs) due to limitation of canonical centralized systems when they are assigned more and more complicated tasks. A DAS does not have a central controller that supervises the whole system. It is composed of structural elements that interact cooperatively or competitively with each other by behaving autonomously but act as one system to enable the objective task to be performed. A DAS therefore is expected to have the following advantages: 1) adaptability to a complex environment based on the self-organizability of the control rules by dynamical interaction of structural elements, and 2) competence to complete the objective tasks by re-configuration of the control rules or morphologies in the cases where there is a partial fault in the system.

Considering these advantages, the usefulness of DASs for robots has been studied by using distributive structured real module robots [1][2][3]. In these works, the possibilities for realizing self-organizability in morpho-generation or re-configuration of the system and movability on various morphologies were discussed; however, there was little discussion of the behavioral design and behavior acquisition in an actual dynamical environment.

In this study, aiming to investigate the behavior acquisition through learning in DASs, we first built a real robot working on the water surface, in which, the influence of friction and heating is weak, so that, we can put our focus on essentials of the control problem. On the other hand, hydrodynamic elements such as viscosity, force of inertia, wave, currents, etc. would make the water environment closer to real environment, and also make the control task much more challenging.

Multi-link robot is one of the simple realizations of module robots, in terms of that 1) each structural element has its own decision making component and identical simple motor function, which means only one elements cannot achieve any effective movement; 2) multiple identical units are mechanically linked in series. For the convenience of description, each structural element will be called a "unit" in the following paragraphs.

One of the most important features expected of a multi-link structured robot is self-organizability of locomotion patterns, which enables the robot to adapt to complex environment by interactions among plural decision-making mechanisms and by various motions generated by multi-part actuators.

Considering the locomotion patterns of the animals living in the water, such as, eel, fish, the rhythmic motion, oscillation patterns would easily come into mind. That is, by wiggling the parts of body rhythmically, those animals can obtain various drag forces, and move freely in the water. In [4], the neurobiological bases of rhythmic motor acts in Vertebrates were analyzed. In [5], the mechanical construction of snake-like robots was described in detail. In [6], a swimming controller for the whole system was evolved for simulated lamprey. Our study attempts to build distributed autonomous multi-link robots that can generate situation-oriented, environment oriented oscillation patterns, by cooperative actions of each unit. Therefore, in this study, as a decision-making framework for each unit, we proposed Adaptive Oscillator Method (AOM), which is based on interaction of one unit with the other units and the outer world. A single model reinforcement learning system was employed to realize the framework. In previous studies, we have confirmed obtainability of target approaching by the floating multi-link robot [7][8]. In this study, the case where there is a partial fault in the system was dealt with.

Also, in the case where there is a sudden change caused by adding obstacles, happened to the external environment, the adaptability of the control framework realized by a 2-model reinforcement learning system was discussed.

2 AOM and its Realization

2.1 Ideal Control Architecture

According to the above analysis, the following requirements for the control of DASs in dynamical environment were to be taken into consideration.

- The control system itself should be distributed, since the number of units is possibly very large, the centralized control will face the combination problem.
- For the control of a model-less system, a kind of adaptation scheme is usually employed to self-organize control rules.
- As an incrementally improved control system, an evaluation is important. In order to avoid optional settings corresponding to situations, the quantities that relate to basic motivations of individuals, such as power generated, energy, should be taken as the evaluation.
- For a large scale (although, we only report 3-4 units experiment results) distributed system, the interaction not only on space, but also along the time axis should be taken into consideration.
- DASs situated in the environment such as the shallow water in a small container, might subject to high order dynamics, so that, high order quantities, for example, speed, acceleration might play certain important roles, therefore should also be taken into account.

2.2 Adaptive Oscillator Method

The basic formulation of AOM is shown in equation (1), for each unit,

$$g(\sum_n k_n \frac{d^n}{dt^n} y_m(t)) + f(I(t)) = \nabla_t e \quad (1)$$

where, y_m is displacement that is mth solution of the equation, when there exist multiple solutions. k_n is a coefficient of term $\frac{d^n}{dt^n} y_m$ that enables the system to reflect the effects of long-term, high-dimensional derivatives. I is an external-effect-input-term to the unit. *g()* and *f()* are transformation function of high-order derivatives and input vector, respectively. The right side of the equation is an item concerning the increase rate of internal energy. Thus, it is assumed that all parameters could be smoothly learned by the effect on energy management.

The formulation is very similar to the equations in Taga's work [12], in which walking gaits in a humanoid-type-two-legged system were studied. Comparing with his work, the features of our work are 1) high-order quantities are described explicitly in the formulation, their effect would be taken into considered if necessary, 2) all parameters will be incrementally identified by a learning method.

2.3 Realizing AOM by a Reinforcement Learning Method

To build up an adaptive control method that can enable each unit of our distributed robot to do behavior acquisition autonomously, an incremental learning scheme should be employed. Supervised learning, such as Back Propagation Neural Network (BPNN), or unsupervised learning, such as Adaptive Resonance Theory (ART) can be considered. However, in this research, Reinforcement Learning (RL) method was chosen, because, RL enables to learn from a scalar value, which meets the requirement of the energy-driven problem solving well. Another feature of RL, that is, prediction based on experience, also fits the multi-step requirement of AOM well.

On the other hand, within the research field of RL, there is no consideration taken for distributed system control and temporal-spatial interaction, therefore, the formulation of AOM will be a good model for RL.

In order to realize the AOM by RL, we transformed the equation (1), so that, the solution can be learned incrementally, according to an expected sum of energy increase. At first, the first term of the left side can be extended and simplified into a weighted sum of the preceding solutions.

$$\begin{aligned}\sum_{n=1}^{\infty} k_n \frac{d^n}{dt^n} y &= k_1 \frac{y_t - y_{t-1}}{dt} + k_2 \frac{\frac{y_t - y_{t-1}}{dt} - \frac{y_{t-1} - y_{t-2}}{dt}}{dt} + \cdots \\ &= l_0 y_t + \sum_{n=1}^{\infty} l_n y_{t-n}\end{aligned} \tag{2}$$

Then, the input vector term of equation (1) can be extended into a vector consisting of sensor information from the neighboring units, and preceding solutions. This range of the neighborhood *p*, and depth of action history, *n*, will be called control structure, in the following explanation.

$$I_t = (S_{self}, S^1_{neighbor}, \ldots S^n_{neighbor}, y_{t-p}, \ldots, y_t) \tag{3}$$

At last, to take the multi-step effect into consideration, we wrote the expected sum of energy increase as the formula (4). Then behavior acquisition based on AOM can be viewed as a RL problem, that is, the y_t can be calculated as an output function *F*, while the output function *F* can be updated according to the derivatives of Ξ with respective to *F*.

$$\Xi_{t+1} = f_{\Xi}(\Xi_t, \nabla_t, e, I_t) \tag{4}$$

3 Experiment Settings

3.1 Hardware Structure of the Floating Robot

In this study, all the units were linked in series, that is, in a two-dimensional chained structure. As shown in Fig. 1-A and 1-B, each round float (diameter: 15 cm) in the robot system is taken as one unit, and a three-unit version was used here [7]. Moving force is generated by paddling of a fin that is fixed under the float. If the motion of each unit generates proper propelling waves, the robot can move, but if not, the robot can hardly move. Additionally, for structural necessity, the robot has a dummy unit at one end of the chain. This dummy unit carries battery cells for driving the robot.

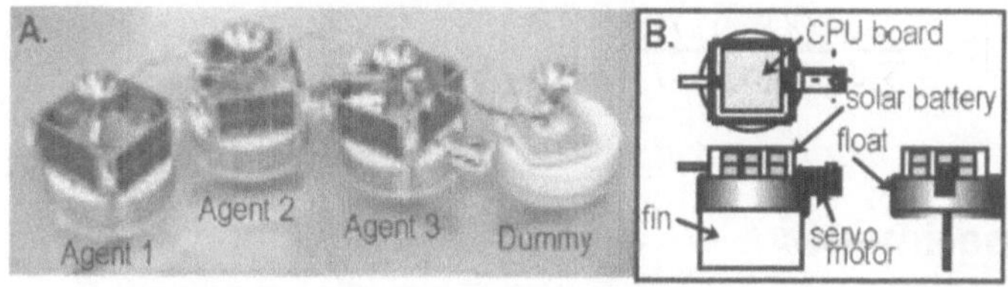

Fig. 1. A, Distributed autonomous floating robot (3 units); B, Structure of a unit

3.2 Q-Learning Implementation

We employed the Q-learning method [9], which is an approximate dynamic programming method in which only value updating is performed based on local information.

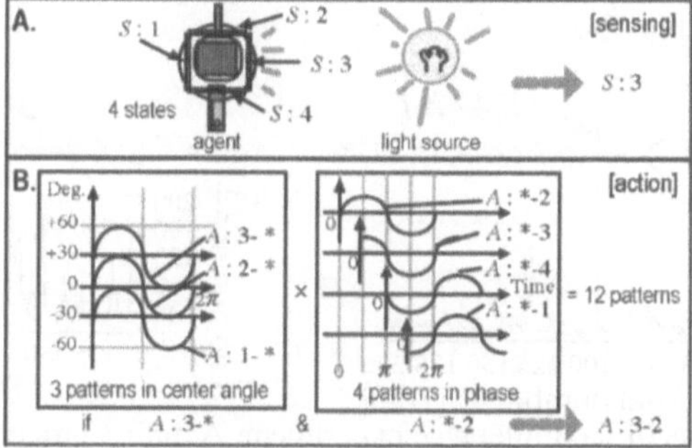

Fig. 2. A, Definition of sensing states; B, Definition of element action patterns

The Q-table is constructed according to the description of section 2.3. The control structure was decided according to the experiment results in [8], that is, n=1, and p=1, expressed by $Q((S_{UnitLeft}, S_{UnitSelf}, S_{UnitRight}), A_{t-1}, A_t)$. The sensing states S are divided into 4 states. Each state denotes one of the 4 directions by which the largest value is sensed (Fig. 2-A). On the other hand, the action patterns A are divided into 12 patterns from the basic sine function that has amplitude of 30 degrees. Twelve variations are achieved by using 3 variations of center angle and 4 variations of phase (Fig. 2-B). The execution term of each oscillation motion per one action is two periods.

To update the Q-value, equation (5) is used. For action selection, the Boltzmann exploration depending on equation (6) is implemented. Here, b is the action for which the current selection probability is maximum, α is the learning rate, γ is the discounting rate, r_t is the reward value, and T is the parameter that decides the randomness of the action. In the experiments, we used α=0.1, γ=0.95, r_t=100, and T=4.0 based on the experiences in previous tests [8].

$$\Delta Q(S_t, A_t) = \alpha(r_t + \gamma \max_b Q(S_{t+1}, b) - Q(S_t, A_t)) \tag{5}$$

$$P(A \mid S) = \frac{\exp(Q(S,A)/T)}{\sum \exp(Q(S,A)/T)} \tag{6}$$

4 Experiments

4.1 Recovery from partial fault

Fault-tolerance tests were performed. In this experiment, one unit was technically "broken", that is, the unit lost its mobile ability and communicative ability. Two cases were studied. In the first experiment, unit 3 (see Fig. 1) was "broken", after 100 trials. In this case, considerably smooth target-approaching behavior was realized in next few trials. The robot reached the goal at about 20 steps after 100 trials learning (see Fig. 3-A).

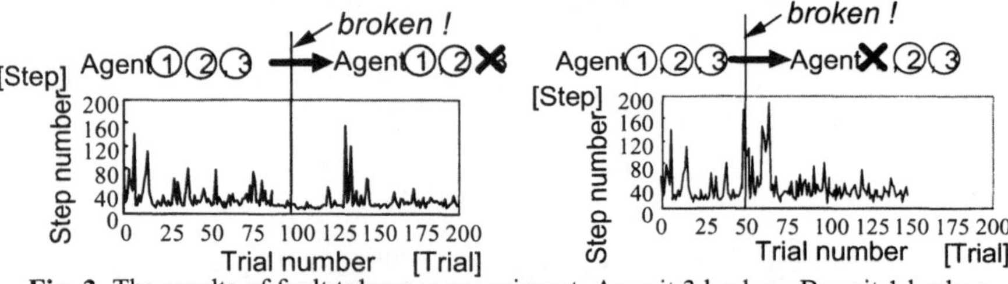

Fig. 3. The results of fault tolerance experiment, A: unit 3 broken; B: unit 1 broken

In the second experiment, the unit 1 (see Fig. 1) was "broken" after 100 trials, however the robot could hardly reach the goal. This shows the fact that, when the environmental oriented behavior is formed, each unit in DARS plays different role. It may need greater real time, which is critical to the real world application to recover from the fault caused by unit 1. So that we decide to perform the test that the fault happens before the task behavior was completely formed, by shifting the broken point to 50 trials. As the result, the robot ultimately reached the goal at 30-40 steps after another 100 trials.

4.2 Obstacle avoidance: Adapt to Changed Environment

4.2.1 Obstacle setting

At first, we set up obstacles and observed the robot's behavior by using the previously acquired action rules [8] in order to discuss what measure should be taken. The obstacle was a fence made of acrylic resin, and a transparent fence was used to avoid light being sheltered away. The width of the entrance was 32 cm, about twice the diameter of each unit.

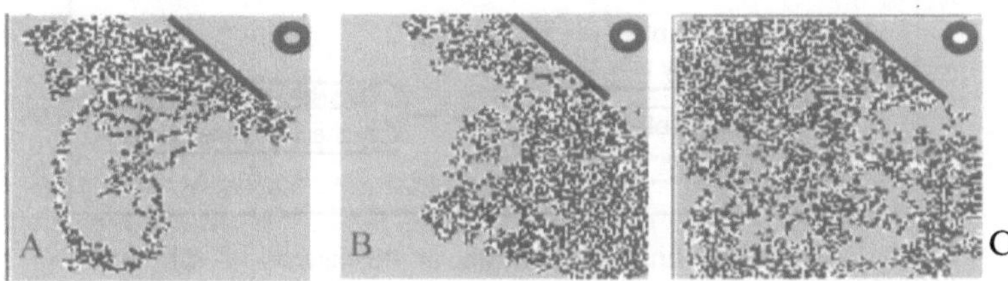

Fig. 4. A, Replay behavior based on past action rules; B, relearning is allowed; C Learning from initial phase

In Fig. 4-A, the robot approached the fence using the locomotion patterns for reaching the target from the left side, but after that, the robot was not able to approach to the target because of the fence. The robot could not reach the target even though 1000 steps were performed.

4.2.2 Proposed Method: "Switching-Q"

The failure is somewhat inevitable, since the environment was changed, while there was no rule for each unit to deal with the change. After this failure, we also did another two attempts. One is, to let the learning continue, after adding the obstacle. And the other is, to let the robot learn from very beginning in the changed environment. However, both attempts failed. Here, by failure, we mean a relative steady state cannot be reached over a preset number of steps.

The reason is that, since the obstacle fence is almost transparent to the robot sensor, so that, in most cases, the robot would not be able to recognize the difference, however, different behavior is required. That is, using only one rule set caused the failures.

The AOM realized till now could not cope with the problem with this type of change, since only one single rule set cannot represent the diversified behaviors required. For this kind of problem, the methods that can switch between multiple rule sets, while effectively reusing previously obtained rules and obtaining new necessary rules, is necessary. In this study, we then proposed a Switching-Q method and presented its usefulness. This can be regarded as an advanced version of AOM realization. For cases involving robot's navigation learning, some hierarchical learning methods have been proposed [10][11]. In these methods, the objective task is decomposed into simpler tasks, and the learning mechanisms are structured hierarchically. Therefore, the more the layer is higher, the more the problem is solved in more abstract level. It has been verified that such a hierarchical learning method works effectively for a centralized controlled systems, but the effectiveness of such a distributed controlled system is not guaranteed. Thus, a new method is needed for a distributed controlled system. In addition, the mechanism of the method should be constructed as simply as possible in order for the method to work effectively in the real world.

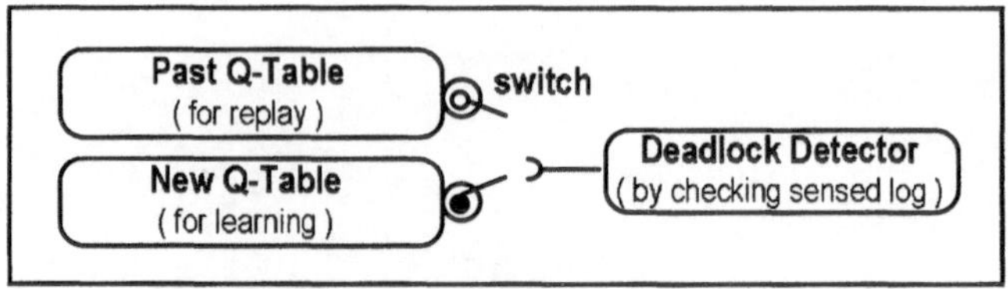

Fig. 5. Switching-Q Learning (in the case of x=2)

Based on the observations described above, it is thought that one general solution for effective learning of motion in more complex environments by reusing previously obtained rules is a method in which the robot first advances by reusing past rules, and if the robot falls into a deadlock situation, it starts to learn by means of acquiring new action rules. For this purpose, a new learning method, which is described by equation (7), was proposed. The method has a deadlock checking mechanism and past and new rules (Fig. 5). The control shifts between the past rules and new rules, depending on the deadlock situation.

$$Switching\_Q = \{Q_1, \ldots, Q_x, Deadlock\_Detector\} \tag{7}$$

The new Q-table for learning is assigned to Q_x, which has the maximum value of x. Here, only the case of x=2 is dealt with, so the actual Q-tables can be represented as follows.

$$Q_1(S_t, A_t): Past\_Qtable \qquad Q_2(S_t, A_t): New\_Qtable$$

Switching-Q works as follows. First, for i steps from the start, the past Q-table is used unconditionally. During the trial, the log of the sensed values in recent j steps, $Sv_{t-j}, \cdots, Sv_t$ is memorized, and in each time step, slope l is calculated by the function $v()$, which calculates the linear approximation of their log data (equation (8)).

$$l = v(Sv_{t-j}, \ldots, Sv_t) \tag{8}$$

The Q-table that will be used in time step t is determined by l. If l is a positive value, the past Q-table is chosen at that time step, and if l is not a positive value, a new Q-table is chosen. That is, Q_t that is chosen at time step t is represented by the following equation. In addition, i and j were set to 9 and 5, respectively, in this experiment.

$$Q_t = \begin{cases} Q_1(S_t, A_t), & 1 \le t \le i \\ Q_1(S_t, A_t), & t > i \cap l > 0 \\ Q_2(S_t, A_t), & t > i \cap l \le 0 \end{cases} \tag{9}$$

If Q_2 is used, the Q-value at that time step is memorized as $Rule_{recent}$, which is the rule that has been used most recently. Here, the Q-value is updated only for Q_2. That is, $Rule_{recent}$ is updated every time when Q_2 is used. Finally, when the task is completed, the reward is given to the rule. The feature of Switching-Q is to active reuse of past experiences. One shortcoming, however, is that an optimal solution is difficult to obtain because the new Q-table tends to learn the parts that are weak points in the past Q-table.

4.2.3 Experiment results

Fig. 6 shows the numbers of steps needed to reach the target in two cases: Obstacle 1 and Obstacle 2. Just after the start of the tests, the robot sometimes needed over 400 steps in Obstacle 1 and over 700 steps in Obstacle 2 to reach the target. However, the number of steps gradually decreased, and after the 130th trial, the robot could complete the task in about 30 steps in both cases.

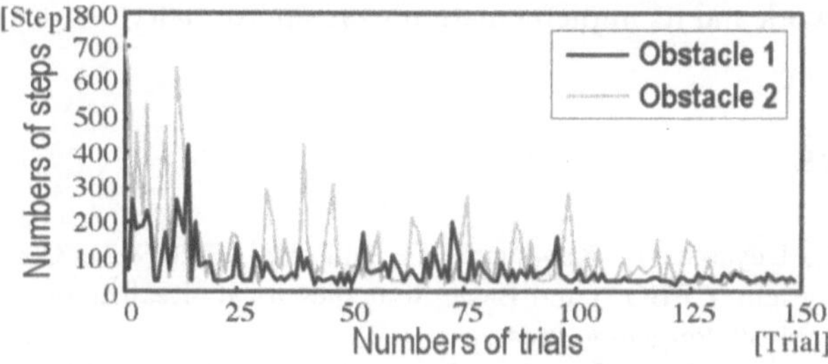

Fig. 6. Steps needed to reach target in two cases: Obstacle 1 and Obstacle 2

Fig. 7 shows the results of replay tests performed by reusing the action rules acquired by Switching-Q. The results show that there are various behavioral patterns.

In Fig. 7-1, the robot goes through the entrance by considerably efficient action sequences, but in Fig. 7-2, the robot reaches the target by tracing a somewhat roundabout way. In any case, these results show that the robot acquired obstacle avoidance behavior that was not observed in the experiment described in 4.2.1.

Under each of the pictures showing the results of behavior, the transition of action rules chosen by each unit is shown. The dark-colored lines represent the past rules, and the light-colored lines represent the new rules. After the first nine steps that was assumed to use the past rules in 4.2.2, it was observed that each unit individually switched from new to past rules, or from past to new rules, based on the deadlock checking.

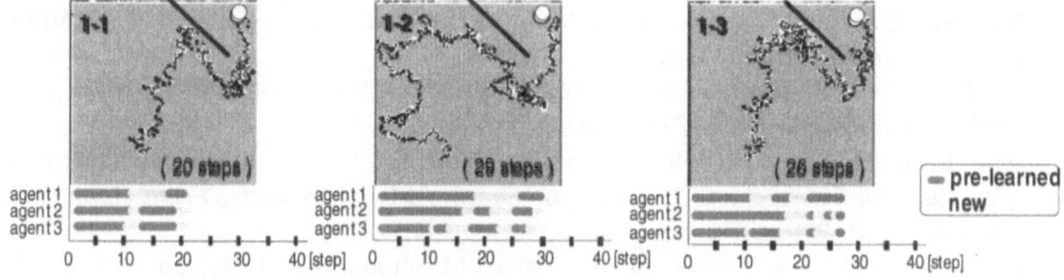

Fig. 7. Replay behavior based on Switching Q, and transitions of chosen action rules in the replay tests

5 Conclusion

In this paper, we proposed a control framework, AOM for DASs. In order to verify the effectiveness of a robot system based on the proposed design method, behavior acquisition tests in target approaching and obstacle avoidance were performed using a distributed autonomous floating robot on the surface of water. For the environment change brought by adding an obstacle, we proposed switching-Q learning method in which pre-learned action rules are effectively used. The results of the experiments showed the dynamics that is necessary for behavioral learning in a DAS was required, and it was verified that the proposed control framework and its implementation were appropriate.

References

1. Fukuda, T., Ueyama, T., Kawauchi, Y., and Arai, F. (1992). Concept of Cellular Robotic System (CEBOT) and Basic Strategies for its Realization, Computers Elect. Eng 18(1):11-39, Pergamon Press.
2. Yim, M. (1993). A Reconfigurable Modular Robot with Many Modes of Locomotion, *Proc. of the JSME Int. Conf. on Advanced Mechatronics,* 283-288.
3. Kokaji, S., Murata, S., and Kurokawa, H. (1994). Self Organization of a Mechanical System, *Distributed Autonomous Robotic Systems*, 237-242.
4. Grillner, S. (1985). Neurobiological Bases of Rhythmic Motor Acts in Vertebrates, *Science* 228:143-149.
5. Hirose, S. (1993). *Biologically Inspired Robots (Snake-Like Locomotors and Manipulators)*, Oxford Univ. Press.
6. Ijspeert, A.J., Hallam, J., and Willshaw, D. (1999). Evolving Swimming Controllers for a Simulated Lamprey with Inspiration from Neurobiology, *Adaptive Behavior* 7(2) 151-172.
7. Iijima, D., Yu, W., Yokoi, H., and Kakazu, Y. (1998). Autonomous Acquisition of Adaptive Behavior for a Distributed Floating Robot Based on the AHC Method, *Intelligent Engineering Systems through Artificial Neural Networks* 8:537-542.
8. Iijima, D., Yu, W., Yokoi, H., and Kakazu, Y. (1999). Autonomous Acquisition of Target Approaching Behavior for Distributed Controlled Swimming Robot (The Case of Presetting Oscillation Action Patterns), *Trans. of the JSME* 65(637):208-215.
9. Watkins, C.J.C.H., Dayan, P. (1993). Technical Note: Q-Learning, Sutton, *Reinforcement Learning*, 55-68.
10. Lin, L.J. (1993). Scaling Up Reinforcement Learning for Robot Control, *Proc. of the 10th Int. Conf. on Machine Learning*, 182-196.
11. Tani, J., and Nolfi, S. (1998). Learning to Perceive the World as Articulated: An Approach for Hierarchical Learning in Sensory-Motor Systems, *From Animals To Animats 5*, 270-279.
12. Taga, G. (1995). A Model of the Neuro-Musculo-Skeletal System for Human Locomotion II. Real-Time Adaptability under Various Constraints, *Biol. Cybern.* 73:113-121.

Neural Networks (NN) Using Genetic Algorithms (GA) and Gradient Back-Propagation (GBP) for an Intelligent Obstacle Avoidance Behavior

A. CHOHRA[1] and O. AZOUAOUI[2]

[1] Ecole Nationale Supérieure d'Ingénieurs (ENSI) de Bourges, Laboratoire de Vision et Robotique (LVR), 10 Boulevard Lahitolle, 18020 Bourges, France, chohra@ensi-bourges.fr or chohra@hotmail.com
[2] CDTA – Centre de Développement des Technologies Avancées, Laboratoire de Robotique et d'Intelligence Artificielle, 128, Chemin Mohamed Gacem, BP 245 El-Madania, 16075, Algiers, Algeria, azouaoui@hotmail.com

Abstract

To ensure more *autonomy* and *intelligence* with *real-time* processing capabilities for the obstacle avoidance behavior of Intelligent Autonomous Vehicles (IAV), the use of Neural Networks (NN) is necessary to bring this behavior near to that of humans in the recognition, learning, adaptation, reasoning and decision-making, and action. In this paper, three (03) supervised learning algorithms namely Gradient Back-Propagation (GBP), Genetic Algorithms (GA) and GA-GBP are suggested to train a NN to learn spatial obstacle avoidance situations. A synthesis of the suggested NN/GBP, NN/GA and NN/GA-GBP is presented where their results and performances are discussed. Finally, a Field-Programmable Gate Array (FPGA) architecture, characterized by its high flexibility and compactness, is suggested for the NN implementation.

1 INTRODUCTION

With increasing demands for high precision autonomous control over wide robotic applications, conventional control approaches are unable to adequately deal with system complexity, nonlinearities, spatial and temporal parameter variations, and uncertainty. Intelligent control which is experiential based rather than model based is designed as a new emerging discipline to overcome these problems. In fact, today researchers have the requisite hardware, software and sensor technologies at their disposal for building intelligent dynamic systems particularly Intelligent Autonomous Vehicles (IAV). They have also in possession of computational tools which are far more effective in the design and development of intelligent systems than the predicate-logic-based methods of traditional artificial intelligence. These tools derive from a collection of methodologies known as *soft computing* implying particularly Neural Networks (NN) which offers function approximation and learning capabilities and and Genetic Algorithms (GA) which provide a methodology for systematic random search and optimization. IAV must be endowed with the obstacle avoidance behavior to have the ability to move and be self-sufficient in partially structured environments. They can then carry out tasks in spatial, underwater and terrestrial environments by themselves like humans exploiting the recent developments in autonomy requirements, intelligent components, multi-robot systems, and massively parallel computers [1–4]. To reach then their targets without collisions with possibly encountered obstacles, IAV must particularly have the capability to achieve the obstacle avoidance behavior with *autonomy*, *intelligence* and *real-time*. Thus, to acquire this behavior while answering these requirements, IAV must be endowed with recognition, learning, adaptation and generalization, reasoning and decision-making, and action capabilities. To achieve this goal, classical approaches have been replaced by current approaches based on *soft computing* implying particularly NN [5–8]. In these approaches, the obstacle avoidance behavior is based on the learning, adaptation and generalization capabilities of NN to avoid collisions. Indeed, NN are recognized to improve the learning, adaptation and generalization capabilities related to variations in environments where information is qualitative, imprecise, or incomplete [2], [9–14].

In this paper, the obstacle avoidance behavior problem of IAV in partially structured environments is stated. To acquire such an intelligent behavior, three (03) supervised learning algorithms namely Gradient Back-Propagation (GBP), GA and GA-GBP are suggested to train a NN to learn spatial obstacle avoidance situations. A synthesis of the suggested NN/GBP, NN/GA and NN/GA-GBP is presented where their results and performances are discussed. Finally, a Field-Programmable Gate Array (FPGA) architecture is suggested for the NN implementation.

2 OBSTACLE AVOIDANCE BEHAVIOR OF INTELLIGENT AUTONOMOUS VEHICLES (IAV)

Recent research on IAV has pointed out a promising direction for future research in mobile robotics where *real-time*, *autonomy* and *intelligence* have received considerably more attention than, for instance, optimality and completeness. Many navigation approaches have dropped the assumption that perfect environment knowledge is available. The representation of the environment knowledge is, in fact, based on acquisition of intelligent behaviors that enable the vehicle to interact effectively with its environment [15]. Consequently, IAV are facing with less predictable and more complex environments, they have to orient themselves, explore their environments autonomously, recover from failures, and perform whole families of tasks in real-time. More, if vehicles lack initial knowledge about themselves and their environments, *learning* and *adaptation* become then inevitable to replace missing or incorrect environment knowledge by experimentation, observation, and generalization. Thus, in order to reach a goal, learning and adaptation of vehicles rely on the interaction with their environment to extract information [16]. Currently, the most obstacle avoidance approaches are inspired from observations of human navigation behavior. Indeed, human navigators do not need to calculate the exact coordinates of their positions while navigating in environments (roads, hallways, etc.). The road-following or the hallway-following behavior exhibited by humans is a reactive behavior that is learned through experience. Given a goal, human navigators can focus attention on particular stimuli in their visual input and extract meaningful information very quickly. In this paper, the partially structured environments such as factories, passenger stations, harbors and airports with static obstacles are considered. In fact, human perceives the spatial situations in such environments as topological situations: rooms, corridors, right turns, left turn, junctions, etc. Consequently, trying to capture the human obstacle avoidance behavior in such environments, several approaches based on a recognition of topological situations have been developed [5], [7], [8], [17]. An intelligent obstacle avoidance behavior can be then acquired using NN based pattern classifiers under *supervised* learning and adaptation paradigms which allow to recognize topological situations from sensor data giving vehicle-obstacle distances.

2.1 Vehicles and Sensors

2.1.1 Vehicles

For the sake of simplicity, the vehicle movements are possible in three (03) directions and consequently three (03) actions A_i (i=1, ..., 3) are defined as action to turn to the Right, action to move Ahead and action to turn to the Left, as shown in Fig. 1. They are expressed by the action vector $\mathbf{A} = [A_R, A_A, A_L]$.

2.1.2 Sensors

To detect possibly encountered obstacles, the perception system is assumed to have three (03) UltraSonic sensors (US) necessary to get distances (vehicle-obstacle) covering the three areas: US_R to get distance d_R on the Right, US_A to get distance d_A in Ahead, and US_L to get distance d_L on the Left, as shown in Fig. 1.

2.2 Partially Structured Environments

The uncertainty of the real world knowledge is the main problem of IAV. This uncertainty is due to the fact that the environment representation is based essentially on the vehicle perception systems. It must be taken into account in the data representation. For the suggested classifiers, the perception system uses US to detect static obstacles giving the vehicle-obstacle distances necessary for the obstacle avoidance behavior. Indeed, the problem of correctly evaluating noisy and incorrect data for the interpretation of US signals is often an encountered one [5]. This problem is taken into account, in this paper, by the use of parallel basis structures of NN with their inherent characteristics of adaptivity and high fault tolerance with respect to defective sensors or noisy sensor data [9], [13], [14].

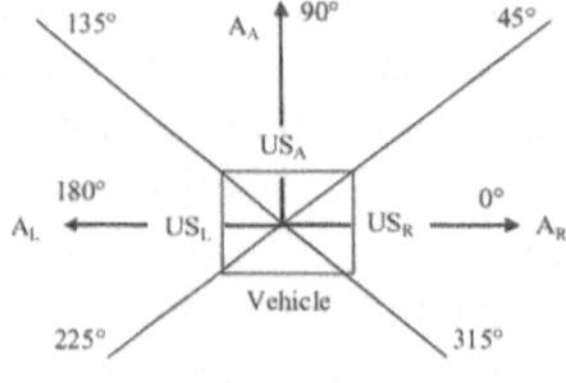

Fig. 1. Vehicle and sensors.

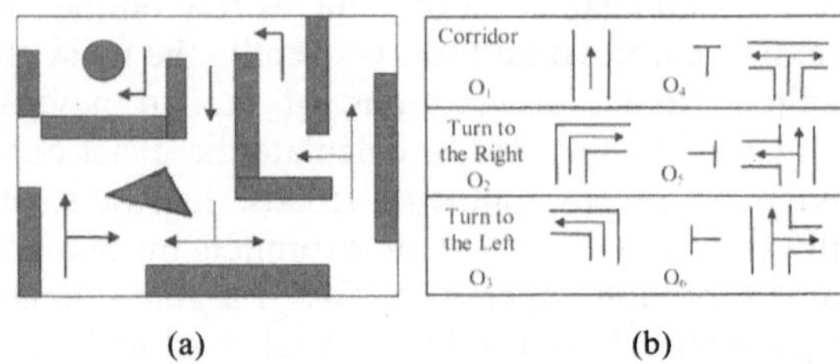

(a) (b)

Fig. 2. (a) A partially structured environment. (b) Obstacle avoidance situations.

2.2.1 Possibly Encountered Static Obstacles

In reality, static obstacles in partially structured environments are of different shapes representing walls, pillars, machines, tables, etc. as shown in Fig. 2(a).

2.2.2 Spatial Situations Structured in Topological Situations

The possible movements of vehicles lead us to structure the spatial situations in six (06) topological situations called obstacle avoidance situations as shown in Fig. 2(b). These situations are defined with six (06) classes O_1, ..., O_j, ..., O_6.

2.3 Training Environment

The training environment must contain all obstacle avoidance situations O_j as shown in Fig. 3. This environment allows vehicle to be in different obstacle avoidance situations (positions). Thus, training examples are defined by randomly selecting fifty four (54) vehicle positions (patterns), where each particular position corresponds to one training example for a particular obstacle avoidance situation. This environment has been used to train the suggested NN/GBP, NN/GA and NN/GA-GBP classifiers.

3 NEURAL NETWORKS UNDER SUPERVISED LEARNING ALGORITHMS

In this Section, the obstacle avoidance problem is solved using pattern classifiers based on NN trained by GBP, GA and GA-GBP to recognize topological situations (obstacle avoidance situations). During the navigation, each IAV must build an implicit internal map (i.e., obstacles and free spaces) from sensor data, update it and use it for intelligently controlling their obstacle avoidance behavior. This behavior is acquired by learning and adaptation of the suggested pattern classifiers from ultrasonic sensor data. For all these classifiers, the input vector is defined by $\mathbf{X} = [X_1, \ldots, X_i, \ldots, X_3]$ while the output vector is defined by $\mathbf{O} = [O_1, \ldots, O_j, \ldots, O_6]$. Indeed, in each step, the vehicle-obstacle distances d_R, d_A and d_L are updated from US_R, US_A and US_L, respectively and used to define the input vector **X**. Thus, for each input vector **X**, the classifier must provide the vehicle with the capability to recognize in which situation O_j it finds itself, see Fig. 2(b), to avoid possibly encountered static obstacles.

3.1 NN/GBP Classifier

NN pattern classifiers exploit NN features such as implicit knowledge representation, learning and generalization from experience (from examples), robustness related to the fault tolerance with respect to defective sensors or noisy sensor data, and massively parallel processing (real-time) [9], [10], [12–14]. The first suggested pattern classifier, NN/GBP, is a multilayer feedforward NN built of three (03) layers as shown in Fig. 4. The distances d_R, d_A and d_L are pre-processed to constitute the input vector **X** of this classifier, where ρ: norm of the input vector **X** and a: input pre-processing factor with a > 1:

$$X_1 = (1/\rho)\exp(-d_R/a),$$
$$X_2 = (1/\rho)\exp(-d_A/a), \qquad (1)$$
$$X_3 = (1/\rho)\exp(-d_L/a),$$

Input Layer: with three (03) input nodes receiving the components of the input vector **X**. This layer transmits these inputs to all nodes of the next layer.

Hidden Layer: with five (05) hidden nodes, where their outputs are obtained using the output sigmoïd function f as follows:

$$net_k = \sum_i X_i W2_{ki}, \qquad (2)$$

$$Y_k = f(net_k), \qquad (3)$$

$$\text{where } f(x) = \frac{1}{1+\exp(-x)}. \qquad (4)$$

Output Layer: with j sigmoïdal output nodes which are obtained by:

$$net_j = \sum_k Y_k W1_{jk}, \qquad (5)$$

$$O_j = f(net_j). \qquad (6)$$

To acquire the obstacle avoidance behavior, this classifier is trained, in the training environment shown in Fig. 3, using the *supervised* GBP paradigm detailed in [10] where the error is computed as follows, where the index ex runs over all examples of the training set:

$$e_{ex} = (1/2)\sum_j (DesiredO_j - O_j)^2. \qquad (7)$$

NN/GBP classifier is trained from fifty four (54) examples of the training set with the input pre-processing factor a = 3. Weights $W1_{jk}$ and $W2_{ki}$ are adjusted from a random weight initialization between [-1, +1] with learning rate $\eta = 0.1$.

3.2 NN/GA Classifier

GA are one of the optimization methods based on the biological evolution process [11], [13], [14]. They are search algorithms that are based on the mechanics of natural selection and natural genetics. They perform a global, random, parallel

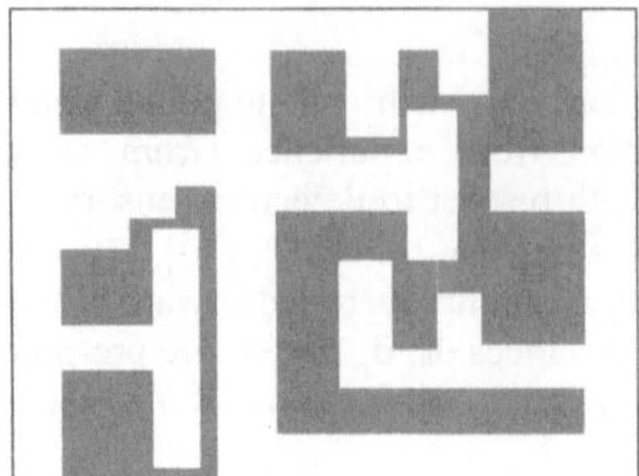

Fig. 3. Training environment.

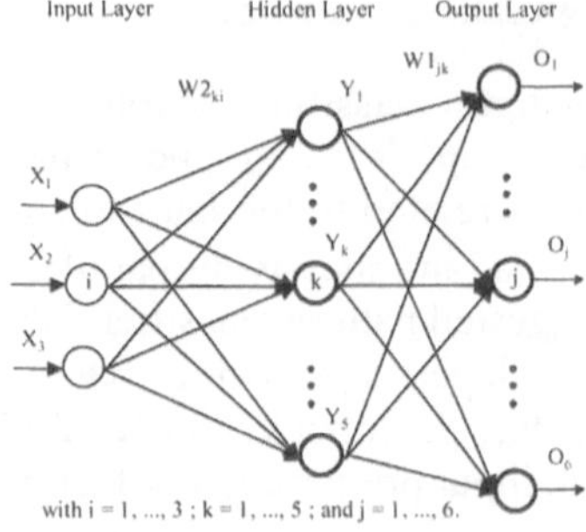

Fig. 4. NN/GBP, NN/GA, and NN/GA-GBP Classifiers.

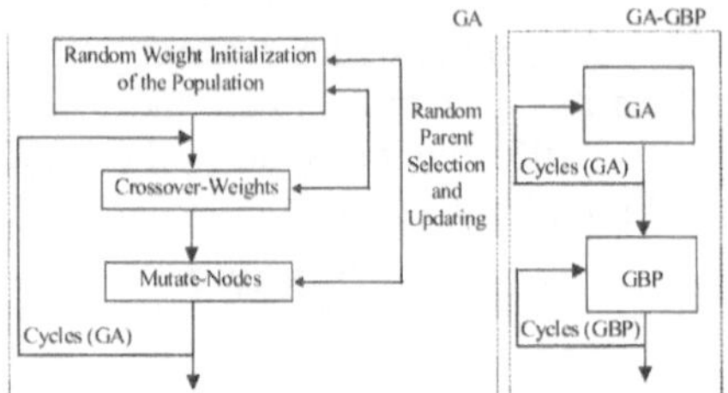

Fig. 5. Synopsis of supervised GA and GA-GBP learning paradigms.

search for an optimal solution using simple computations. Starting with an initial population of genetic structures, genetic inheritance operations based on selection, crossover, and mutation are performed to generate "offspring" that compete for survival ("survival of the fittest") to make up the next generation of population structures [13]. In this Section, GA are used to train a multilayer feedforward NN shown in Fig 4, the second suggested NN/GA classifier, in which the weights are seen to form a parameter space, in order to obtain a best weight configuration [18], [19]. The distances d_R, d_A and d_L are pre-processed as in Eq. (1) to constitute the input vector **X** of this classifier. This classifier is trained, in the training environment shown in Fig. 3, to acquire the obstacle avoidance behavior based on a *supervised* GA learning paradigm as shown in Fig. 5. The population size which is a self-explanatory parameter is set to fifty (50) chromosomes as suggested in [18], [19] and the genetic structures represent weight values of the NN connections, see Fig. 4. Each chromosome (i.e., the weights in the NN) is encoded as a list of real numbers with the error, with the index pop which runs over the chromosome population: pop = 1, ..., 50 and e_{ex} as defined in Eq. (7):

$$E_{pop} = e_1 + \ldots + e_{ex} + \ldots + e_{54}. \tag{8}$$

Crossover: The used operation is the Crossover-Weights [19].

Mutation: The used operation is Mutate-Nodes as detailed in [19]. This operation selects nbr non-input nodes of the network represented by parent chromosome (nbr = 2). For each of ingoing links to these nbr nodes, the operation adds a random value v_{mut} to link's weights from initialization probability distribution. It then encodes this new network on child chromosome.

Fitness function: The Fitness function evaluates each chromosome whether it is desirable or not. It consists of the summation of squared error between desired and current output values of the network defined by:

$$\text{Fitness} = 1/E_{pop}. \tag{9}$$

Selection: Based on the fitness value, each chromosome is selected or not to the next generation. Elite preservation strategy is adopted where the chromosomes having high fitness value remain to the next generation and sampling strategy that the chromosomes are selected randomly remain to the next generation.

NN/GA classifier is trained from fifty four (54) examples of the training set with the input pre-processing factor a = 3 and initial population pop = 50 chromosomes. The weights $W1_{jk}$ and $W2_{ki}$ are adjusted from a random weight initialization between [-5, +5] with the random value $v_{mut} \in [-0.25, +0.25]$.

3.3 NN/GA-GBP Classifier

Since GBP is a gradient descent approach, it has tendency to get stuck in local minima of the error surface and thus not find the global minimum. Moreover, GBP is sensitive to parameters such as learning rate and momentum [20] while the performance of GA does not depend on these parameters. The main advantage of using GA for the training of feedforward NN is that they can find global minima without getting stuck at local minima. However, the GA's problem is that they are inherently slow [11], [14], [20]. In fact, when the search space is large, as is usually the case in training a NN, the GA approach takes a long time to converge. The length of search is due to the optimal generalization of the training process with no a priori knowledge about the parameter space [20]. Another problem of GA is their weakness in performing fine-tuned local search which is widely recognized. Although GA exhibit very fast convergence to a point of approximate solution in a search space, GA themselves do not entail a mechanism for local fine-tuning as seen in GBP.

The third classifier, NN/GA-GBP, is suggested to remedy insufficiencies of slowness of GA and getting stuck in local minima of GBP. In this classifier, NN shown in Fig 4 is trained in two (02) stages. First, GA train weights of nodes to find a location in the weight space which is closed to the optimal solution. Second, GBP starts from that point taking this weight configuration as initial weights and conducts an efficient local search. This combination would be an efficient approach of training NN because it takes advantage of the strengths of GA and GBP (the fast initial convergence of GA and the powerful local search of GBP), and overcomes the weaknesses of the two (02) methods (the weak fine-tuning capability of GA and a flat spot in GBP) [18]. The distances d_R, d_A and d_L are pre-processed as in Eq. (1) to constitute the input vector **X**. This classifier is trained, in the training environment shown in Fig. 3, to acquire the obstacle avoidance behavior using *supervised* GA and GBP learning paradigms, see Fig. 5. NN/GA-GBP classifier is trained by GA from fifty four (54) examples of the training set with the input pre-processing factor a = 3 and initial population pop = 50 chromosomes. The weights $W1_{jk}$ and $W2_{ki}$ are adjusted from a random weight initialization between [-5, +5] with the random value v_{mut} = [-0.25, +0.25]. From the resulting weights and the same examples this classifier is trained again by GBP with learning rate $\eta = 0.1$.

4 SYNTHESIS

In this Section, a synthesis on the suggested NN, trained by the supervised learning algorithms GBP, GA and GA-GBP, is presented where their results and performances are discussed. NN/GBP classifier yields convergence to tolerance $E_T = 0.01$ in well under the cycle number CN (GBP) = 927, see Fig. 6. (a). For NN/GA classifier, the error is equal to 0.244301 after the cycle number CN (GA) = 10000, see Fig. 6. (b). Indeed, this fact confirm that GA are effectively slow and must be trained with more cycles to get the tolerance $E_T = 0.01$. GBP is a steepest gradient descent method which uses several parameters leading to a dependency of the algorithm on the choice of these parameters as well as on the selection of initial weights. Unlike GBP, GA do not involve gradient descent and usually has fewer, less sensitive, parameters suggesting that GA based training may be more robust [20]. However, the problem of using a larger population is that it requires extensive computation for each generation, and, even though the outcome converges into an optimal solution in a few generations, GA using large populations would be outperformed by GBP [18].

The NN/GA-GBP classifier yields convergence to the tolerance $E_T = 0.01$ after a total number of cycles CN (GA-GBP) = 360 with CN (GA) = 70 and CN (GBP) = 290. This combination of GA and GBP gives better result since NN/GA-GBP converges more rapidly than NN/GBP and NN/GA suggesting then that a hybrid learning algorithm (GA-GBP) is more efficient than the use of GBP or GA separately. Learning by GA-GBP can not only remedy the problem of local minimums but also to slowness of GA. In fact, training of NN with GA-GBP shows effectively that it outperforms GA and GBP taken separately., see Fig. 6. (a), (b) and (c). GA converge faster than GBP in its early stages of training. While weakness of GA in local fine-tuning is obvious, although this problem was circumvented in NN/GA-GBP method, the speed of convergence in early stages of training is the crucial factor in evaluating their utility as a method of training NN [18]. As shown in Fig. 6. (c), GA successfully reduce the error at beginning of training while GBP conducts a powerful local search to reach tolerance $E_T = 0.01$.

5 FIELD-PROGRAMMABLE GATE ARRAY (FPGA) ARCHITECTURE

An FPGA architecture, based on Xilinx technology, presented in Fig. 7, is suggested in this Section for the NN implementation, where each neuron is modeled by a Multiply and ACcumulate (MAC) operator which computes the weighted sum. The MAC result points a Look-Up Table (LUT) which implements the sigmoid activation function [21]. This architecture exhibits a high degree of regularity, compactness and can be easily modified (reduce or increase the number of MAC and change the size of the multiplexer) to implement any other type of feedforward NN. Fully integrated digital NN for an intelligent obstacle avoidance behavior has been implemented using only a single XC4062EX Xilinx FPGA.

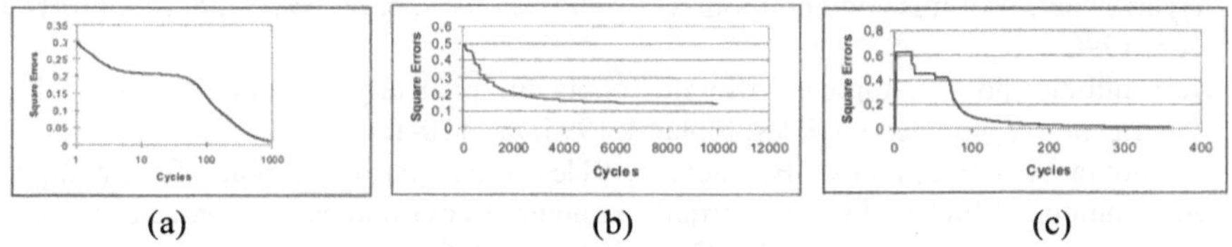

Fig. 6. Error graphs during training: (a) NN/GBP. (b) NN/GA. (c) NN/GA-GBP.

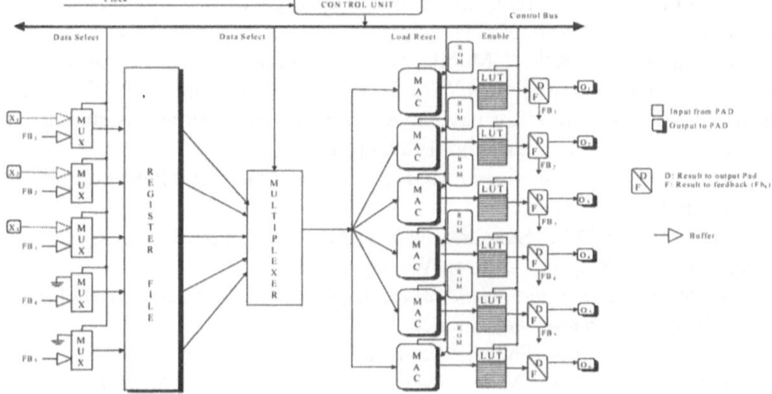

Fig. 7. FPGA Architecture for the NN implementation.

6 CONCLUSION

Results show that suggested NN classifier allows IAV to achieve an intelligent obstacle avoidance behavior in *partially structured environments* using supervised learning algorithms GBP and GA separately or in combination. However, the NN under the hybrid algorithm GA-GBP is more efficient than the others since it combines the complementary strengths of GA and GBP to learn more rapidly the obstacle avoidance situations and therefore converges faster than the others. Also, FPGA architecture suggested for the NN implementation provide IAV with *real-time* processing capabilities making them *more robust* and *reliable*. An interesting alternative for future research is integration of NN, fuzzy systems and GA which will involve further progress in research on intelligent behaviors of IAV.

REFERENCES

[1] O. Azouaoui and A. Chohra, ''Evolution, behavior, and intelligence of *A*utonomous *R*obotic *S*ystems (*ARS*),'' in *Proc. 3*[rd] *Int. IFAC Conf. Intelligent Autonomous Vehicles*, Spain, 1998, pp. 139-145.

[2] T. Fukuda, F. Arai, and K. Shimojima, ''Intelligent robotic system,'' in *Proc. Int.*

IMACS IEEE-SMC Multiconf. Computational Engineering in Systems Applications, Lille, France, 1996, pp. 01-10.

[3] D.B. Marco, A.J. Healey, and R.B. McGhee, ''Autonomous underwater vehicles: Hybrid control of mission and motion,'' *Autonomous Robots*, vol. 3, no. 2/3, pp. 169-186, 1996.

[4] K. Schilling and C. Jungius, ''Mobile robots for planetary exploration,'' *Int. Conf. Intelligent Autonomous Vehicles*, Finland, 1995, pp. 110-120.

[5] A. Chohra, A. Farah, and C. Benmehrez, ''Neural navigation approach for *I*ntelligent *A*utonomous *V*ehicles (*IAV*) in partially structured environments,'' *Int. J. of Applied Intelligence*, vol. 8, no. 3, pp. 219-233, May/June 1998.

[6] H. Herbstreith, L. Gmeiner, and P. Preuβ, ''A target-directed neurally controlled vehicle,'' in *Proc. Int. IFAC Conf. Artificial Intelligence in Real-Time Control*, Delft, The Netherlands, 1992, pp. 67-71.

[7] M. Meng and A.C. Kak, ''Mobile robot navigation using neural networks and nonmetrical environment models,'' *IEEE Control Systems*, pp. 30-39, October 1993.

[8] E. Sorouchyari, ''Mobile robot navigation: A neural network approach,'' in *Proc. Art Coll. Neuro., Eco. Poly.*, Lausanne, 1989, pp. 159-175.

[9] J.A. Anderson, *An Introduction to Neural Networks*, The MIT Press, Cambridge, MA, London, England, 1995.

[10] J.A. Freeman and D.M. Skapura, *Neural Networks: Algorithms, Applications, and Programming Techniques*, Addison-Wesley, 1992.

[11] D.E. Goldberg, *Algorithmes Génétiques: Exploration, Optimisation et Apprentissage Automatique*, Addison-Wesley, France, 1994.

[12] T. Khanna, *Foundations of Neural Networks*, Addison-Wesley, 1990.

[13] D.W. Patterson, *Artificial Neural Networks: Theory and Applications*, Prentice-Hall, Simon & Schuster (Asia) Pte Ltd, Singapore, 1996.

[14] S.T. Welstead, *Neural Network and Fuzzy Logic Applications in C/C++*, Jhon Wiley & Sons Inc., Toronto, 1994.

[15] S. Cherian and W. Troxell, ''Intelligent behavior in machines emerging from a collection of interactive control structures,'' *Computational Intelligence*, vol. 11, no. 4, pp. 565-592, November 1995.

[16] S. Thrun and T.M. Mitchell, ''Lifelong robot learning,'' *Robotics and Autonomous Systems*, 15, pp. 25-46, 1995.

[17] Y.S. Kim, I.H. Hwang, J.G. Lee, and H. Chung, ''Spatial learning of an autonomous mobile robot using model-based approach,'' in *Proc. Int. Conf. Intelligent Autonomous Vehicles*, Finland, 1995, pp. 250-255.

[18] H. Kitano, ''Empirical studies on the speed of convergence of neural network training using genetic algorithms,'' in *Proc. 8th National Conf. in Artificial Intelligence*, Boston MASS, vol. 2, 1990, pp. 789-795.

[19] D.J. Montana and L. Davis, ''Training feedforward neural networks using genetic algorithms,'' in *Proc. 11th Int. Joint Conf. on Artificial Intelligence*, Morgan Kaufman, San Mateo CA., 1989, pp. 762-767.

[20] M. McInerney and A.P. Dhawan, ''Use of genetic algorithms with back propagation in training of feed-forward neural networks,'' in *Proc. Int. IEEE Conf. Neural Networks*, vol. I, California, 1993, pp. 203-208.

[21] O. Azouaoui, ''Neural networks based approach for the manipulators inverse Jacobian problem,'' in *Proc. ICSC Int. Conf. on Neural Computing*, Austria, 1998, pp. 966-972.

Interactive Q-Learning on heterogeneous agents system for autonomous adaptive interface

Yuko Ishiwaka[1], Hiroshi Yokoi[2], and Yukinori Kakazu[2]

[1] Dept. Information Engineering, Hakodate National College of Technology, 14-1,Tokura-cho, Hakodate, 042-8501, Japan, ishiwaka@hakodate-ct.ac.jp
[2] Complex System Engineering Dept., Hokkaido University, North-13, West-8, Sapporo, 060-8628, Japan, {yokoi,kakazu}@complex.eng.hokudai.ac.jp

Abstract. Purpose of this system is to adapt the bedridden people who cannot move their body easily, so the simple reinforcement signals are applied. The application is to control the behaviors of Khepera robot, which is a small mobile robot. For the simple reinforcement signals the on-off signals are employed when the operators as the training agent feels discomfort for the behaviors of the learning agent Khepera robot. We proposed the new reinforcement learning method called Interactive Q-learning and the heterogeneous multi agent system. Our multi agent system has three kinds of heterogeneous single agent: Learning agent, Training agent and Interface Agent. The system is hierarchic. There are also three hierarchies. It is impossible to iterate the many episodes and steps to converge the learning which is adopted in general reinforcement learning in simulation world. We show the results of experiments using the Khepera robot for 3 examinees, and discuss how to give the rewards according to each operator and the significance of heterogeneous multi agent system. We confirmed the effectiveness through the some experiments which are to control the behavior of Khepera robot in real world. The convergences of our learning system are quite quick. Furthermore the importance of the interface agent is indicated. The individual differences for the timing to give the penalties are happened even though all operators are young.

Keywords: Interactive Q-Learning (IQL), POSMDP, heterogeneous multiagent system, Khepera

1 Introduction

The aim of our research is to construct interfaces for bedridden people. Here we treated a task which an autonomous robot takes things instead of an operator. We constructed the heterogeneous multi agent system for interactive learning system to control the robot, and it is satisfied with reduced frustrations of an operator as small as possible. The difference from related interactive system [9] is to incorporate the human in the multi agent system. According to this it is possible for the system to learn quickly with only low-level interaction on-line. We proposed Interactive Q-learning (IQL) to be suitable for this multi agent system. IQL is extended well-known Q-learning, which is one of reinforcement learning [1,2].

In this paper we describe our heterogeneous multi agent system and IQL algorithms in details, then we lead this task to be Partially Observable Semi-Markov Decision Processes (POSMDP) in reinforcement learning. We show the results of this goal-directed task using the Khepera robot for 3 examinees under two kinds of real world. As results the learning convergences have big differences among those examinees. We conclude that the timing of reward is very important factor of this system.

2 System Architecture

The system we aimed is shown in **Fig.1**. **Fig.1** means that an autonomous mobile robot brings the things which bedridden people want instead. When the operator feels discomfort during observation of the robot, the reinforcement signals is transmitted the computer and the penalty is given to the mobile robot. Our objective operator is a person who can hardly move his own body with his mind, especially who cannot speak. For this reason it is impossible for this system to apply gesture [3] using generally. Our system must be satisfied with

1. low-lever interaction like as a on-off switch (EEG, EMG, a button),
2. autonomy when the operator cannot observe the environment,
3. on-line learning,
4. quick convergence of learning.

Nehmzow etc. [8] apply supervised Artificial Neural Networks for interactive robot learning, however we employ reinforcement learning. One big reason is that it is only needed the simple reward to learn, so we can think about electromyogram (EMG)[4], electroencephalogram (EEG) [5] for switches. Second reason is that it is possible to learn autonomously and third reason is to learn without supervising. However we must avoid many episodes usually examined in simulations [10] because the operator must be tired. We propose the Interactive Q-leaning which is extended to satisfy on-line learning and quick convergence. IQL is able to learn quickly in the first episode because of its on-line disposition **Fig.2** shows the internal state expression of each agent.

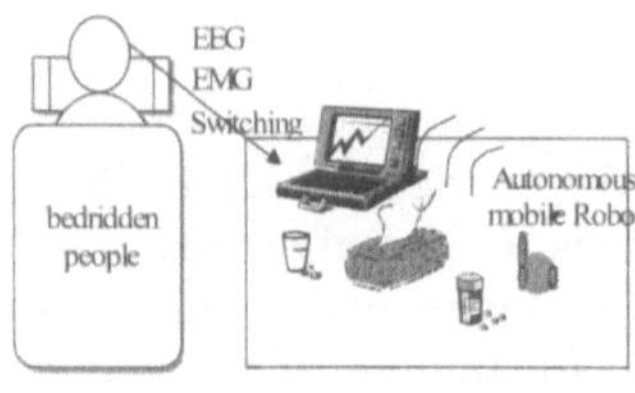

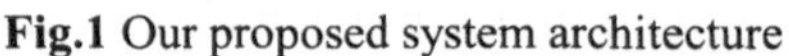
Fig.1 Our proposed system architecture

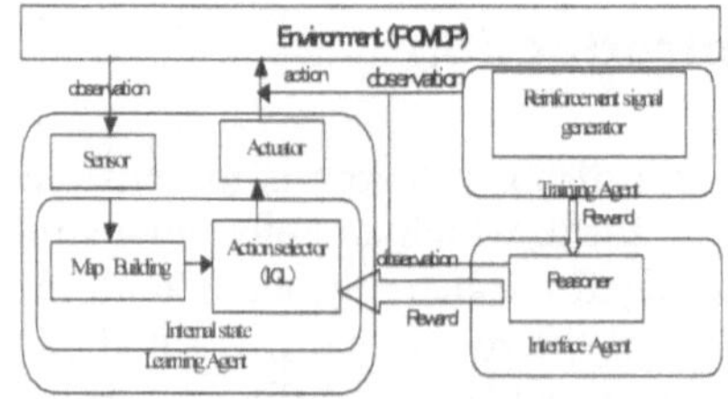

Fig.2 The internal state expression

3 Multi Agent Model with Heterogeneous Single Agents Architectures

Heterogeneous multi agent system is proposed here. This system has 3 kinds of Agent and 3 hierarchies. Learning Agent belongs to the hypostasis hierarchy, Training Agent belongs to the epistasis hierarchy and Interface Agent belongs to the mediation between learning agent and training agent. The features of this model are distributed ness, hierarchy, activeness and selfish. The learning agent in hypostasis hierarchy can do decision making if there are no precise indications from higher hierarchy. Each single agent has the communicator and the decision maker. The architecture of each agent is represented in **Table 1**.

4 Analyzing the Problem in Reinforcement Learning

It is necessary to analyze the general problem in reinforcement learning in order to construct our system. The rewards are caused interactively at the very moment when the operator feels discomfort during his observations of the sequence of robot's behaviors, that is to say the decision processes are chosen randomly. Such model is called Semi-Markov Decision Processes (SMDP). Event-driven SMDP [6] is applied to our system. One more problem is caused from real world. The values observed with infra-red proximity sensors include the errors cannot be ignored. Partially Observable Markov Decision Processes (POMDP) [7] is the mathematical model. Our approach to POMDP is to predict the state transitions of environment by constructing probabilistic the state observations and probabilistic Q-value updating.

The mathematical model treated in our system is called Partially Observable Semi-Markov Decision Processes (POSMDP). The problem description is followings. The system has 6-tuple:

Table 1. Heterogeneous multi agent architecture

	The Communicator:	The Decision Maker
Learning Agent (an autonomous robot)	Sensors (POMDP) Actuators (take)	Map building (automatically mapping with its own sensors) Reasoner (receiving reinforcement signals) Action Selector based on Q-values
Training Agent (an operator)	Sensors (POMDP) Transmitter (sending reinforcement signals to the Interface agent)	Reward Generator (SMDP)
Interface Agent (an interface)	Sensors Receiver (receiving the reinforcement) Transmitter (sending the reinforcement signals to the learning agent)	Reasoner (It judges how the reinforcement signals give to the hypostasis agent.)

$$\langle S, A, P, R, F, O \rangle \tag{1}$$

, where S: the finite set of states, A: the finite set of actions P: the probability of state transitions, R: the finite set of rewards, F: probabilistic function of time transition for each pair of state and action, O: the finite set of observable. Each factor represents as follows

$$s = \{x, y \mid x \in X, y \in Y\}, s \in S \tag{2}$$

$$A = \{a_n^m \mid n, m \in N\} \tag{3}$$

$$P(s' \mid s, a), \quad s \in S, s' \in S, a \in A \tag{4}$$

$$R = \left\{ r_O, r_f \mid r_O = \Pr(O(o \mid s, a)), r_f = r_f(t \mid s, a) \right\} \tag{5}$$

$$F(t \mid s, a), s \in S, a \in A, t \in T \quad \textit{T: terminal state} \tag{6}$$

$$O : S \times A \rightarrow O \tag{7}$$

Transition Probability J from the state s at time t to the next state t' after an action a has been chosen is shown by the following equation,

$$J(t, s' \mid s, a) = P(s' \mid s, a) F(t \mid s, a) \tag{8}$$.

5 Interactive Q-Learning under POSMDP

5.1 Heterogeneous agents architectures

In this paper we adopt the Khepera robot as a learning agent. For the communicator the robot has the eight infrared sensors as Sensors and two wheels with two DC motors as Actuators. In order to build the map automatically in the decision maker the existence probability of the obstacle are calculated by the Gaussian distribution whose input values are obtained from its infrared sensors.

$$P_r(x \mid s) = e^{-\frac{(x-\mu)^2}{2\sigma^2}} \tag{9}$$

, where σ is standard deviation and μ is mean.

Q-learning is employed as action selector. The update rule of Q-values as following;

$$Q(s,a) \leftarrow (1-\alpha) Q(s,a) + \alpha \left[r + \gamma \max_{a'} Q(s_{t+1}, a_{t+1}) \right] \tag{10}$$

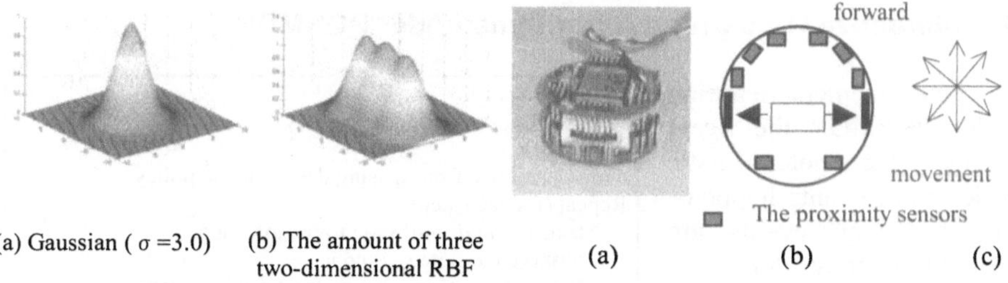

(a) Gaussian (σ =3.0) (b) The amount of three two-dimensional RBF (a) (b) (c)

Fig3 Two-dimensional radial basis functions

, here s: state, a: action, α : step size parameter and γ : discount rate. The relative coordinate from the start point is used as states of learning agent and the movement toward 8 directions is used as actions of it. The updating time t is given from training agent and interface agent, then we obtain our updating equation,

$$Q(s_{t+t_i}, a_{t+t_i}) \leftarrow (1-\alpha)Q(s_{t+t_i}, a_{t+t_i}) + \alpha\left[r + \gamma \max_a Q(s_{t+1+t+t_i}, a_{t+1+t+t_i})\right] \quad (11).$$

Here the range of t_i is $-t_{prev} < t_i < t_{predict}$. The reward parameter is fixed. Interface agent decides the parameters t_{prev} and $t_{predict}$. The negative parameter means the number of steps before the action taken by learning agent at the time t and the positive time parameter is correspondent to number of steps which predicts the action taken at the time t with that the same action will be selected at next step.

For the continuous space we employed radial basis function (RBF) to update Q-values. The equation are shown as

$$f(x) = \sum_{i=1}^{N} c_i \phi(\|x - xc_i\|, \|y - yc_i\|) \quad (12).$$

Here (xc_i, yc_i) means the current coordinate of learning agent, N is the number of data and c_i is the unknown parameter. In this work the two-dimensional Gaussian is employed as the kernel function for RBF. The kernel function is shown in following equation and **Fig3** shows the Gaussian distribution.

$$\phi(x, y) = \exp(-((x - xc)^2 + (y - yc)^2)/\sigma^2) \quad (13).$$

The updating range of Q-value is a circle whose center coordinates is corresponding with the current.

Training agent in this system means that the operator who observes the learning agent. The operators push the switch on line when the behaviors of learning agent make them discomfort. The training agent learns the timing of pushing the switch to bring the learning agent to the goal. The sensors in communicator mean their eyes` and transmitter is to put the switch and the reward generator in decision maker means their feelings. Interface agent decides the range of updating Q-values and it should be learned depends on the timing of pushing switch from training agent.

5.2 Interactive Q-Learning Algorithms under POSMDP

In reinforcement learning under POSMDP the agent follows the deterministic policy until events happens. The following events are adapted to our system,

Event 1: the learning agent detects the obstacle

Event 2: the learning agent receives the rewards from interface.

For the event 1 the reward is calculated with the Gaussian distribution whose max value is 1and for the event 2 the reward is fixed. **Fig.5** shows the interactive Q-Learning algorithms under POSMDP.

Initialize $Q(s,a)$
Repeat (for each episode)
initialize s
choose action a_t from s_t using deterministic policy
Repeat (for each step)
    take action a_t until state transition happens
    observe the state s_{t+1} and r
    If state transition happens from sensors' values
    $Q(s,a) \leftarrow (1-\alpha)Q(s,a) + \alpha[r + \gamma \max_{a'} Q(s_{t+1}, a_{t+1})]$
    If state transition happens from interaction
    choose action a_{t+1} from s_{t+1} using deterministic policy
    $q(s,a) \leftarrow Q(s,a)$
    for all s_{t+i}, a_{t+1} ($-t_{prev} \leqq i \leqq t_{predict}$)
    $Q(s_{t+i}, a_{t+i})$
        $\leftarrow (1-\alpha)q(s,a) + \alpha[r + \gamma \max_{a'} Q(s_{t+1+i}, a_{t+1+i})]$
until s_t is terminal
    $s_{t+i} \leftarrow s_{t+1+i}, a_{t+i} \leftarrow a_{t+1+i}$

Fig.5 Interactive Q-Learning Algorithms under POSMDP

6 Experiments

The application of this system is to control the behaviors of Khepera robot interactively. We let the all tasks to the goal directed problem that is required obstacle avoidance. Training agent knows the location of the goal, but it can partially observe near the goal because the goal is behind an obstacle. We experimented under the some situations using proposed method, IQL. The results are shown and we discuss the difference of learning efficiency depends on the training agents (operators) and depends on how to give the rewards of interface agent.

6.1 Experimental setting and the parameters

Learning agent Khepera robot which is a small mobile robot is shown in **Fig.4 (a)**. The radius of the Khepera is 2.75cm and it has 8 infrared proximity sensors **Fig.4 (b)**. The Khepera has two wheels with two DC motors, and its processor can read the pulses of the incremental encoder. Then it is possible to count the migration length and calculate the current location. For the interaction, the serial line is connected to RS232 port in the host computer. The direction of Khepera's movement is 8 like as **Fig.4(c)**. Khepera robot rotates on the center of it as a shaft to the 8 kinds of degrees and after rotation in the same place, it goes straight 1 step. Those operators are enough just to click the mouse of host computer to give the penalties to the learning agent on line.

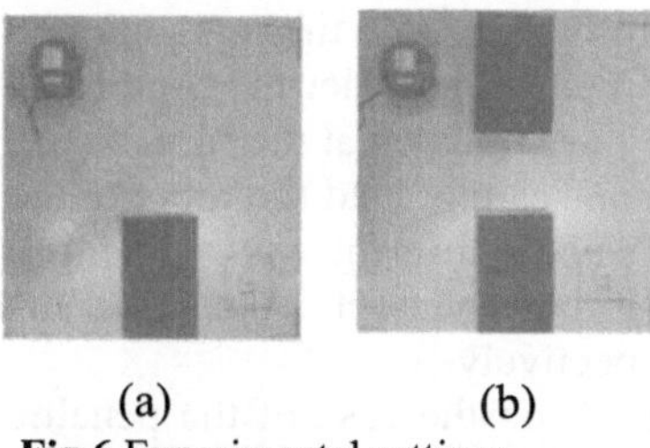

Fig.6 Experimental setting:
(a) Environment1 with one obstacle
(b) Environment 2 with two obstacles

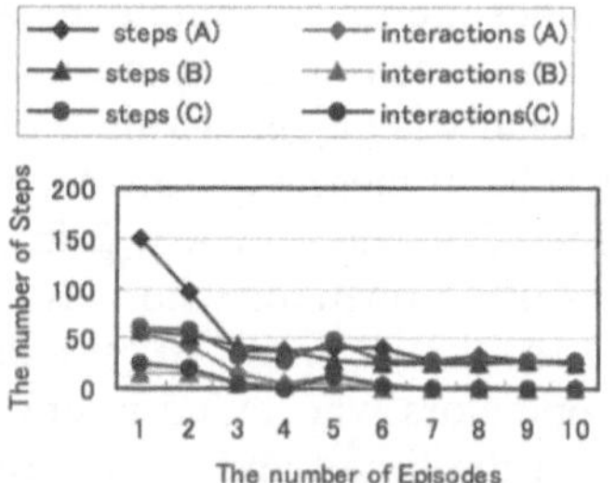

Fig.7 The learning curves for 3

The parameters of Q-leaning in these experiments are α=0.2 and γ=0.8. The existence probability of obstacles is calculated with the Gaussian distribution and the range of rewards are $-1.0 \leqq r_p \leqq 0$. From the operators the rewards are always r=-0.5. The range of Q-value updating is depends on RBF (σ=3.0 and threshold > 0.01). The initial Q-values are 0.0. The state of Q-learning is the x-y position. The robot starts the same position and direction every episode.

The migration length of 1 step is 25% of the size of Khepera and the range of observation internal state Q-values are corresponding with 80% of the robot size in the round shape. One step is treated here from the rotation to moving 25% after decision-making. The setting of each parameter is based on the experience. For the deterministic policy of Q-learning the greedy algorithms are employed. The advance is repeated, until whether it detects the obstacle in front of the Khepera robot or the penalties are given. **Fig.6** shows the experimental setting. Black boxes are obstacles in **Fig.6**.

We show those results of the trajectories and Q-maps in **Fig.10** and **Fig.11**. Each figure of the center shows the trajectories. Each start position is the upper left hand side and goal position is bottom right hand side described squared. Each interaction is shown in small gray circle and 50% over probability of existing obstacles is represented bigger gray circle. The last location of Khepera robot is shown in white big circle. Each state-action Q-values are shown in the same direction of action. White color shows the lower Q-values. Q-values just are updated in the location Khepera has passed.

6.2 The comparison of the learning efficiency between the training agents

In order to compare the learning efficiency between the training agents, for the two kinds of environments the experiments were carried out. The specification of the rewards is fixed in those experiments. The operators as the training agents are three males whose ages are from 18 to20 years old. Q-values of the state action pair are updated by assuming that the same action as the current action will be taken in the next step. The number of episodes is set to10 and the number of steps is set to 150.

The penalties are given simultaneously as following three timings, one previous step, the current step and one next step. In other words if let the time t the time when the penalty is given, Q-values of t-1,t,t+1 are updated at the same time.

As the results of these experiments, two persons cannot lead the robot to the goal after episode 2 in environment 2 in spite of succeeding for environment 1. The trajectories of an operator who was succeed in environment 1 but failed in environment 2 are shown in **Fig.10 (a)** and **(b)** respectively.

For the operators who failed in environment 2, on the case of the penalties are just given at the time t and t+1 the re-examinations were carried out in the environment. **Fig.11** shows the trajectory in each episode and **Fig.7** shows the learning curves of the number of interactions and the steps to achieve the goal. We can find that the learning smoothly advances from **Fig.7** because there is little penalty at the later episodes. Therefore it can be said that the updating of the previous step influenced failed factor. The cause of the differences between operators and environments will be discussed in details later.

7 Discussions

From the experiments in spite of same range and same values updating, the following results are obtained, i.e. 1. Some operators could not lead the learning agent to the goal after episode 2 even though episode 1 was successful, 2. The success or fail is depends on the operators even in same environment. On this cause it is clear that the updating of the previous step is related greatly. We focus on the previous step and discuss the updating problem.

7.1 The differences of environments

Two operators failed in environment 2, however they were succeed in environment1. As this reason, it is mentioned that in the environment 2 the number of selectable path to arrive at the goal is quite smaller than environment 1. It seemed that the updated Q-values on episode 1 disturbed the action selection on episode 2. For example, the Q-values to take action ① in **Fig.8** on episode 1, then on the next episode the action will no be selected because the Q-value is lower than the

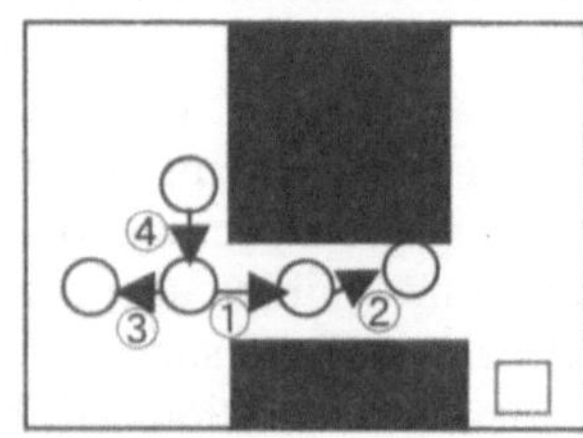

Fig.8 Updating Q-values at Previous steps

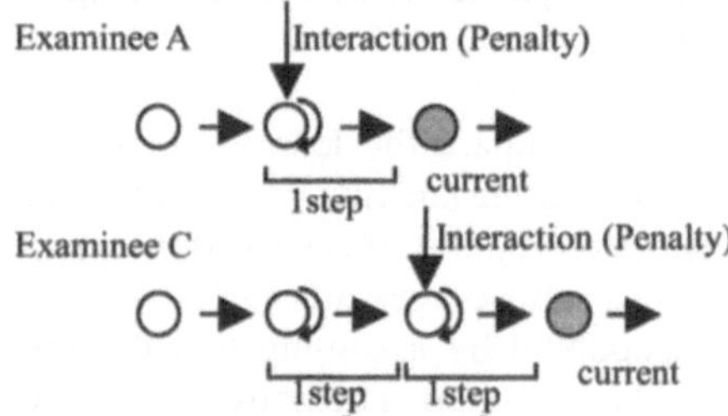

Fig.9 The difference of timing for penalty depends on the examinees

others. Therefore the learning agent tries to take another action, but if the movement of the agent make the training agent discomfort, action③and ④ are also updated. These updating are iterated, then action ④ which is not discomfort for the operators is not selected either. From these reasons it is impossible that the learning agent enters the narrow path. Though the equal phenomenon occurs in environment 1, since there are many spaces to select the paths to the goal, the learning agent chooses others paths autonomously to arrive at the goal. It is confirmed according to compare the trajectories of **Fig.10** to **Fig.11**. In **Fig.11** the same path was selected on later episodes. On the other hand in **Fig.10** the agent could reach the goal, but the every path has changed to the previous episodes.

7.2 The differences of training agents

In spite of same environment and same updating rules, two operators failed and one operator was succeed. We check the timing when the operators switch in order to investigate this cause. The Khepera robot repeats following steps: 1. select the action, 2. turn to the direction, 3. move 1 step. The operators cannot observe the stage 1. Failed operators gave the penalties in stage 2. On the other hand succeeded operator gave the penalties in stage 3. Then the time delay has been generated among operators (**Fig.9**). The time t of succeeded operator was corresponding with the time t+1 of failed operators. If the time t-1 was removed in the case of succeeded operator, the learning efficiency will be quite wrong. It is necessary for the interface agent to decide the range of updating Q-values in order to adapt the individuals.

8 Conclusions

The purpose of this system is to adapt the bedridden people who cannot move their body. We constructed the heterogeneous multi agent system built in the person as one of an agent. Interactive Q-learning are proposed to adopt this system. The strengths of this system are 1. it is just needed low-level interaction like on-off switch, 2. it is possible to learn on line which caused the quick convergence even in one episode, 3. the probabilistic Q-tables are able to deal with some shifts of states (x-y position), 4. when an operator observes a learning agent, it learns interactively and when cannot observe the learning agent, it learns automatically with their own information from environment. The current limitations of this system are 1. when wrong reinforcement penalty is obtained, learning is quite slow, 2. feedback is necessary from the learning agent to an training agent, 3. the interface agent must change how to give the rewards on-line, specially depends on the operators, 4. the start location must be careful to put the real robot. For the future work we explore how to 1. learn interface agent automatically, 2. have the states.

References

1. Watkins C.J.C.H. (1992). Q-Learning, *Machine Learning* 8 p279
2. Sutton, R.S. and A.G.Barto (1998). Reinforcement Learning An Introduction MIT
3. Lee.C and Xu.Y (1996) Online, Interactive Learning of Gestures for Human/Robot Interfaces, In Proceedings, IEEE international Conference on Robotics and Automation, vol.4,pp.2982-2987, MN
4. Yu.W, H.Yokoi and D.Nishikawa (1998). Adaptive Electromyograohic (EMG) Prosthetic and Control Using Reinforcement Learning, *IAS-5,IOS Press*, pp.266-271
5. Ishiwaka,Y. H.Yokoi, and Kakazu,Y.(2000) Adaptive Learning Interface Used Physiological signals, Proceedings *SMC 2000 Conference. Nashville*, USA, pp. 32 – 38
6. Bradtke,S.J and Duff, M.O.(1994) Reinforcement Learning Method for Continuous Time Markov Decision Problems, Advances in Neural Information Processing Systems 7,pp.393-400
7. Parr,R. and Russell,S.(1995) Approximating Optimal Policies for Partially Observable Stochastic Domains, In Proceedings of the International Conference on Artificial Intelligence,pp.1088-1094,Morgan Kaufmann
8. Nehmzow U. and McGonigle B.: "Achieving Rapid Adaptations in Robots by Means of External Tuition", SAB,1994
9. A. Cesta and D. D'Aloisi. :"Building Interfaces as Personal Agents", Sigchi Bulletin, vol.3,1996
10. Wiering M. and Schmidhuber J.: "HQ-Learning", Adaptive Behavior vol6. No.2, 1997

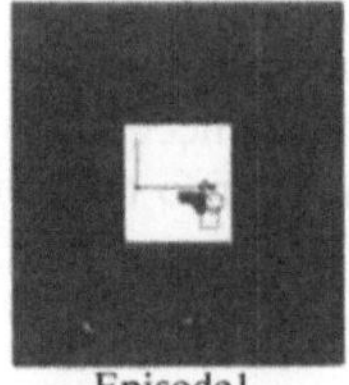
Episode1

Episode5

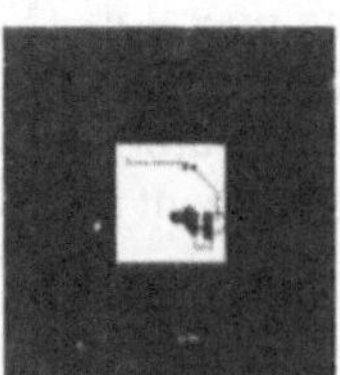
Episode10

Fig.10 (a) At the previous, current and next time simultaneously updating Q-values

Episode1

Episode2

Episode3

Fig.10 (b) At the previous, current and next time simultaneously updating Q-values, after episode 2 the operator failed

Fig.10 The trajectories and Q-maps of same operator who was succeeding in (a) environment 1 but failed in (b) environment 2

Episode1

Episode5

Episode10

Fig.11 At the current and next time only updating Q-values for a failed operator

Author Index